# BIOLOGICAL CONTROL BY AUGMENTATION OF NATURAL ENEMIES

Insect and Mite Control with Parasites and Predators

# Environmental Science Research

Volume 1 – INDICATORS OF ENVIRONMENTAL QUALITY
    Edited by William A. Thomas

Volume 2 – POLLUTION: ENGINEERING AND SCIENTIFIC SOLUTIONS
    Edited by Euval S. Barrekette

Volume 3 – ENVIRONMENTAL POLLUTION BY PESTICIDES
    Edited by C. A. Edwards

Volume 4 – MASS SPECTROMETRY AND NMR SPECTROSCOPY IN
    PESTICIDE CHEMISTRY
    Edited by Rizwanul Haque and Francis J. Biros

Volume 5 – BEHAVIORAL TOXICOLOGY
    Edited by Bernard Weiss and Victor G. Laties

Volume 6 – ENVIRONMENTAL DYNAMICS OF PESTICIDES
    Edited by Rizwanul Haque and V. H. Freed

Volume 7 – ECOLOGICAL TOXICOLOGY RESEARCH: Effects of Heavy
    Metal and Organohalogen Compounds
    Edited by A. D. McIntyre and C. F. Mills

Volume 8 – HARVESTING POLLUTED WATERS: Waste Heat and Nutrient-Loaded
    Effluents in the Aquaculture
    Edited by O. Devik

Volume 9 – PERCEIVING ENVIRONMENTAL QUALITY: Research and Applications
    Edited by Kenneth H. Craik and Ervin H. Zube

Volume 10 – PESTICIDES IN AQUATIC ENVIRONMENTS
    Edited by Mohammed Abdul Quddus Khan

Volume 11 – BIOLOGICAL CONTROL BY AUGMENTATION OF NATURAL ENEMIES
    Insect and Mite Control with Parasites and Predators
    Edited by R. L. Ridgway and S. B. Vinson

A Continuation Order Plan is available for this series. A continuation order will bring
delivery of each new volume immediately upon publication. Volumes are billed only upon
actual shipment. For further information please contact the publisher.

# BIOLOGICAL CONTROL BY AUGMENTATION OF NATURAL ENEMIES

Insect and Mite Control with Parasites and Predators

Edited by

## R. L. Ridgway

*U.S. Department of Agriculture*
*Beltsville, Maryland*

and

## S. B. Vinson

*Texas A & M University*
*College Station, Texas*

PLENUM PRESS · NEW YORK AND LONDON

Library of Congress Cataloging in Publication Data

International Congress of Entomology, 15th, Washington, D. C., 1976.
  Biological control by augmentation of natural enemies.

  (Environmental science research; v. 11)
  "Proceedings of a symposium held at the fifteenth International Congress of Entomology, Washington, D. C., August 19-27, 1976, and other selected papers."
  Includes index.
  1. Insect control–Biological control–Congresses. 2. Mites–Biological control–Congresses. I. Ridgway, Richard L. II. Vinson, S. Bradleigh, 1938-    III. Title.
SB933.3.I57 1976               632'.7                77-14410
ISBN 0-306-36311-9

Proceedings of a Symposium held at the Fifteenth International Congress
of Entomology, Washington, D.C., August 19-27, 1976, and other selected papers

© 1977 Plenum Press, New York
A Division of Plenum Publishing Corporation
227 West 17th Street, New York, N.Y. 10011

Printed in the United States of America

PREFACE

The protection of agricultural crops, forest, and man and his
domestic animals from annoyance and damage by various kinds of
pests remains a chronic problem. As we endeavor to improve production processes and to develop more effective and acceptable
tactics for achieving this protection, we must give high priority
to all potentially useful techniques for the control and management
of insects.

Pest control is recognized as an acceptable and necessary part
of modern agriculture. Methods employed vary greatly and tend to
reflect compromises involving 3 determining factors: technological
capability, economic feasibility, and social acceptability. However, these factors are also subject to change with time since each
involves value judgments that are based on available information,
cost, benefit considerations, the seriousness of the pest problem,
and the political climate. Whatever method is chosen, energy
resources continue to dwindle under the impact of increasing population, and it is inevitable that greater reliance must be placed
upon renewable resources in pest management. One alternative is
the use of a pest management method that uses the energy of the
pest's own biomass to fuel a self-perpetuating control system.

The use of biological control agents for the control of pests
has long been an integral part of the pest management strategy in
crop production and forestry and in the protection of man and
animals. The importance and unique advantages of the method are
well recognized; numerous treatises deal with accomplishments and
methodologies. Also, there have been significant developments in
the past decade. The implementation of new pest management tactics
and concomitant changes in production methods will certainly require
increased reliance on biological control methods. Their appears to
be new opportunities for the use of biological control agents based
on developing research. With increased understanding of the genetics,
the population biology of biocontrol agents, and of the factors that
influence their behavior, the practical application of these methods

v

becomes more complicated but potentially more useful.  In addition,
the use of biological control agents is inextricably interconnected
with social economic and environmental factors that cannot be ignored.

Insects have the potential to increase their numbers dramatically and to adjust their numbers in response to the dynamic environment in which they occur.  Nevertheless, changes in population
numbers often occur slowly because of the continual adjustment caused
by abiotic and biotic factors.  Pest situations arise as a result of
environmental disturbances of an unusual nature or degree.  Although
catastrophic disturbances over which man has little control such as
floods and droughts may induce insect or pest outbreaks, man
generates many of his own pest problems.  The introduction of
potential pests, either intentionally or accidentally, into favorable environments where natural enemies are not present often leads
to serious pest problems.  The growth of susceptible crops or
animals is an essentially monoculture situation where host abundance
allows for a buildup of large pest populations is of common occurrence.  The widespread disruption of the ecosystem that occurs when
crops are planted or harvested destroys not only alternate hosts
for biological control agents but reduces food, shelter, and oviposition material, which seriously limits the effective response
of the biological control agent to pest resurgence.  The use of
pesticides that adversely affect beneficial organisms and that
induce the development of resistance in the pest populations, which
frees the pest from the biological control agent may also occur.

It has been estimated that over 10,000 species of insects and
mites reduce yields of crop plants throughout the world.  In North
America, about 700 species of arthropods are considered serious
pests; of these, about one-third have been accidentally introduced.
Annual losses in agriculture to insects in the United States have
been estimated to amount to 13% of production valued at over 7
billion dollars.  Much of this loss has occurred in spite of the
widespread use of insecticides.  Additionally, there is a real concern for the impact on the environment of the over one billion
pounds of pesticides used in the United States each year.

The biological control of insects is a very broad concept
encompassing a number of strategies and techniques.  Many pest
problems have resulted from the importation of insects from one
region to another where there are no natural enemies.  Many of these
insects are not serious pests in their native habitat, presumably
because they are controlled by natural biotic factors.  One of the
solutions to such a pest situation is the classical approach to
biological control, i.e., the search for and introduction of exotic
biotic agents.  This  classical approach (search, importation,
release, and establishment of the biological control agent of introduced pests) has been particularly successful in the more stable

environments.  It has been less than satisfactory in intensified
agroecosystems.  In fact, it is the existence of annually disrupted
agroecosystems that has emphasized the need for a different approach
to the biological control of pests, whether native or exotic, in
many of these situations.  Such annual disruption often occurs at
a time that is particularly important in the growth of the population
of natural enemies.  Also, in a simplified ecosystem the alternate
hosts or the food needed by the beneficial arthropods may be lacking.

Such situations of the type described have led to the development of different approaches to the biological control of pests in
continually disrupted agroecosystems or in situations where the
beneficial arthropods are unable to permanently establish in or to
rapidly invade expanding pest populations.  In such cases programmed
releases of beneficial insects, sustained reintroductions, and manipulation of the ecosystem through increased diversification providing
supplementary hosts and food sources, attracting and stimulating
natural enemies through the use of kairomones appear to be among
potentially important methods.

Natural enemies of insects and mites not only include beneficial
arthropods but nematodes, microbial agents, snails, and vertebrates.
However, the specific treatment of the augmentation of parasites and
predators (insects, mites, and nematodes) in the current treatise,
was designed to provide a review of the use of these organisms in
augmentations, thus resulting in the first book devoted exclusively
to this important subject.  The book developed from a symposium on
the augmentation of natural enemies held at the 15th International
Congress of Entomology in Washington, D.C. in 1976.  Dr. R. I.
Sailer, University of Florida, and Dr. P. S. Messenger, now deceased,
University of California at Berkeley, provided valuable assistance
to the senior editor in organizing the symposium.

The book is divided into 4 sections.  The first section brings
into focus some of the biological principles that provide a basis
for the augmentation of natural enemies.  The chapter by Huffaker,
Rabb, and Logan (Chapter 1) points out the importance of understanding the principles of population dynamics in any augmentation
program.  Only with thorough and careful analysis of the dynamics
of both the pest and parasite or predator populations can a realistic approach to an augmentation program be undertaken.

Some of the ecological considerations in augmentation such as
reduced insecticide use, selective insecticide use, the importance
of a diversified ecosystem, and the importance of ecotypes of
beneficial insects are pointed out through examples in the Russian
literature by Shumakov (Chapter 2).  The potential for control
through the release of natural enemies is described by Knipling in
Chapter 3.  In Chapter 4, Gordh discusses the importance of bio-

systematics in choosing the proper beneficial arthropod for augmentation purposes. He also points out the importance of biotypes and the difficulty in establishing the identity of these races.

The second section of the book deals with some of the potentially important methodology that may have an important impact on augmentation approaches to control with natural enemies. House (Chapter 5) points out the possibility of improving the vigor, fecundity and longevity of natural enemies through improved nutrition and provides insight into the laboratory rearing of beneficial arthropods. The chapter by Morrison and King (Chapter 6) describes a number of examples of mass rearing of natural enemies for various release programs, and many techniques are described through examples. The importance of maintaining the quality of mass-reared natural enemies is described in Chapter 7 by Boller and Chambers. As these authors note, fecundity, longevity, and normal sex ratio are important, but the ecological adaptability, environmental preference, and behavioral characteristics that influence host or prey selection and specificity must also be considered. The potential of behavioral manipulation of natural enemies through the use of behavioral chemicals is discussed by Vinson (Chapter 8). In Chapter 8, the importance of behavioral and chemical data is stressed and the potential for the use of some kairomones in augmentation as well as the dangers of using these compounds is described.

Section 3 is devoted to describing and discussing the successful augmentation approaches to natural enemy action in various regions of the world. Many of the techniques, procedures, and strategies employed in the development of augmentation programs are described. The successes and methods employed in the U.S.S.R., and reference to much important literature, are provided by Beglyarov and Smetnik (Chapter 9). The situation in the People's Republic of China is reviewed by Huffaker (Chapter 10) who provides insight into the augmentation of natural enemies in a region of the world that has employed some unique systems of mass production.

A review of natural enemy augmentation in Western Europe is provided by Biliotti (Chapter 11), and Scopes and Hussey, in Chapter 12, describes the unique situation in glasshouses where an integrated approach is used that involves both augmentation and pesticides. The success in a closed system such as glasshouses provides clues to the potential for success in the larger, more difficult agroecosystems. Augmentation in the Western Hemisphere is reviewed by Ridgway, King, and Corrillo (Chapter 13). Examples of successes, costs, and potential for improvements are documented. Those interested in augmentation of natural enemies of pests associated with man and animals have a different view, and many examples, particularly with reference to the Diptera, are provided in Chapter 14 by Weidhaas and Morgan. This chapter also includes a discussion of the use of nematodes for mosquito control.

The last section, Chapter 15 by Starler and Ridgway, is concerned with the economic feasibility of mass rearing and release of natural enemies and with problems associated with this kind of investment.  It contains an assessment of the different possible routes that can be taken in establishing natural enemy augmentation programs.  The socio-economic considerations in this chapter are particularly relevant to the development of a viable augmentation approach to control.

The editors would like to acknowledge that some of the material used in this preface came from a report entitled "Biological Agents for Pest Control: Current Status and Future Prospects" developed by the U.S. Department of Agriculture in cooperation with the Land-Grant Universities, State Departments of Agriculture, and the Agricultural Research Institute.  The editors would like to express their appreciation to Ms. Shirley Gray and Ms. Vicki Bienski for providing assistance in developing the book and typing much of the manuscript.  The editors also want to express their thanks to Ms. Patricia Vinson, Ms. Kathryn Edson, Mr. William Worsley, and Jane Wall for their invaluable help and editorial assistance. Chapters 2 and 9 were translated from Russian by Ms. Ruth Busbey and Mr. William Worsley.  The editors also thank their wives for their support and patience.


Beltsville, MD                                              S.B.V.
June 1977                                                   R.L.R.

CONTENTS

Preface . . . . . . . . . . . . . . . . . . . . . . . .    v

I.   Biological Bases for Augmentation

     1.  Some Aspects of Population Dynamics Relative
         to Augmentation of Natural Enemy Action . . . . .    3
         C. B. Huffaker, R. L. Rabb and J. A. Logen

     2.  Ecological Principles Associated with
         Augmentation of Natural Enemies . . . . . . . . .   39
         E. M. Shumakov

     3.  The Theoretical Basis for Augmentation of
         Natural Enemies . . . . . . . . . . . . . . . . .   79
         E. F. Knipling

     4.  Biosystematics of Natural Enemies . . . . . . . .  125
         G. Gordh

II.  Scientific Thrusts Supporting Augmentation

     5.  Nutrition of Natural Enemies  . . . . . . . . . .  151
         H. L. House

     6.  Mass Production of Natural Enemies  . . . . . . .  183
         R. K. Morrison and E. G. King

     7.  Quality Aspects of Mass-Reared Insects  . . . . .  219
         E. F. Boller and D. L. Chambers

     8.  Behavioral Chemicals in the Augmentation
         of Natural Enemies. . . . . . . . . . . . . . . .  237
         S. B. Vinson

III.  Experimental and Practical Applications of Augmentation

      9.  Seasonal Colonization of Entomophages
          in the U.S.S.R.  . . . . . . . . . . . . . . . . . . 283
          G. A. Beglyarov and A. I. Smetnik

     10.  Augmentation of Natural Enemies in the
          People's Republic of China. . . . . . . . . . . . . 329
          C. B. Huffaker

     11.  Augmentation of Natural Enemies in Western
          Europe  . . . . . . . . . . . . . . . . . . . . . . 341
          E. Biliotti

     12.  The Introduction of Natural Enemies for Pest
          Control in Glasshouses:  Ecological
          Consideration . . . . . . . . . . . . . . . . . . . 349
          N. W. Hussey and N. E. A. Scopes

     13.  Augmentation of Natural Enemies for Control
          of Plant Pests in the Western Hemisphere  . . . . 379
          R. L. Ridgway, E. G. King, and J. L. Carrillo

     14.  Augmentation of Natural Enemies for Control of
          Insect Pests of Man and Animals in the United
          States  . . . . . . . . . . . . . . . . . . . . . . 417
          D. E. Weidhaas and P. B. Morgan

IV.  Analysis of Current Uses and Prospects for Expansion

     15.  Economic and Social Considerations for the
          Utilization of Augmentation of Natural
          Enemies . . . . . . . . . . . . . . . . . . . . . . 431
          N. H. Starler and R. L. Ridgway

     Appendix

          Commercial Sources of Natural Enemies in the
          United States and Canada  . . . . . . . . . . . . . 451

     Author Index . . . . . . . . . . . . . . . . . . . . . . 457

     Genus and Species Index  . . . . . . . . . . . . . . . . 469

     Subject Index  . . . . . . . . . . . . . . . . . . . . . 471

# Section I

# BIOLOGICAL BASES FOR AUGMENTATION

CHAPTER 1

SOME ASPECTS OF POPULATION DYNAMICS

RELATIVE TO AUGMENTATION OF NATURAL ENEMY ACTION

C. B. Huffaker, R. L. Rabb, and J. A. Logan

Department of Biological Control, University of California
Berkeley, California, Albany, California 94706   U.S.A.
Department of Entomology, North Carolina State University
Raleigh, North Carolina 27607   U.S.A.

This volume deals with augmentation of natural enemy action –
a pest control strategy having several specific tactical components.
To do so objectively, this strategy must be placed in a realistic
ecological and economic framework broad enough to include situations
where the strategy might be used successfully alone, where it can
be used as one component in an integrated pest management system,
and (very importantly) where it has little promise of successful
application.  The probable direct and indirect effects of its
application must be ascertained and compared with those to be
expected from the application of alternative strategies and tactics.
This comparison must be on a cost-benefit basis from a long-term,
as well as short-term view (the long-term view essentially is that
the practice must not detract from the crop production or profit
potential of the area in which it is used; i.e., it must be both
ecologically and economically sound).

To develop a tightly logical rationale as a justification for
using such a specific strategy as augmentation is not easy, because
this requires an understanding of the basic relationships among
all the important ecosystem components and a knowledge of the
effects on these components separately and interactively of:
(1) "essential" agronomic practices, (2) the proposed augmentation
tactics, and (3) the alternatives to augmentation.

The fundamental importance of understanding the structure,
function, and economics of ecological systems (particularly viewing
populations of pests and parasites as structural and functional
units) in the logical development of pest control methodology
is so widely recognized that to call attention to it again may
seem platitudinous.  However, after paying homage (or lip service)

3

to this premise, it is quite common for those who concentrate
their efforts in developing specific strategies and tactics of pest
control to then forget or ignore ecological principles and consequently proceed to develop unrealistic applications of their methodology.  In retrospect, many attempts to use specific strategies
and tactics (Chant 1964) might never have been made had those
involved given more attention to sound ecological principles.

Chant (1964) noted that no method of direct control that
overlooks the root causes of rise to pest status will be permanent.
He therefore suggested that in order to attain long-term reduction
or regulation of the insect pests (e.g., of peaches) we have three
possibilities:
    "These are:  (a) can we improve our ability to use
    pesticides to solve the problem, (b) are there features
    of the life history of the pest that would permit suc-
    cessful biological control, either as an adjunct to
    chemical control or by itself; and (c) are there [other]
    non-pesticide methods that can be used with or without
    biological control in a truly integrated program, or by
    themselves?"
Clearly, Chant felt that too little attention is commonly given
to establishing the ecological relationships required to answer
(b) and (c), and that the key pest(s) (serious perennial pests)
must be the central consideration.

We shall not attempt a comprehensive review of the basic
ecological principles (Southwood and Way 1970; Rabb et al. 1974;
Huffaker 1974; Price and Waldbauer 1975) but shall suggest certain
perspectives which seem of particular relevance to augmentation.

Huffaker (1974) discussed the major basic biological principles
upon which the tactics and strategies of pest management rest,
directly and indirectly.  He wrote (in abstract):
    "Many [basic principles] are so self evident that
    we often ignore their underlying significance.  They are
    (1) the principle of inherent variation in the genetic
    properties of organisms; (2) the principle that an orga-
    nism must be adapted to its environment and becomes so
    through its evolution; (3) the principle that all orga-
    nisms require adequate nutrition; (4) the principle that
    to perpetuate their kind organisms must reproduce; (5)
    the principle that as organisms are born immature they
    must grow and develop; (6) the principle that life pre-
    sents various compensations tending to correct for ad-
    verse occurrences; (7) the principle that most organisms
    derive their sustenance from other living organisms
    (predation); (8) the principle that organisms commonly
    suffer from depletion of resources by other organisms
    (competition among the same or different kinds of orga-

nisms); (9) the principle that cooperation serves many
species well; (10) the principle that organisms must move
about and do things (possess mobility and other essential
behavior); and (11) the principle of holism and interactions among factors in ecosystems."
It follows that the ecological relationships have not been adequately
considered if the basic information concerning these adaptive and
interactive features remains unknown.


## THE POPULATION SYSTEMS TO BE MANAGED


### General

The targets (pests) as well as the control agents (natural
enemies) of augmentation tactics are both dynamic populations.
Thus, the development and effective use of augmentation depends in
large measure on an understanding of population dynamics.  There
are a number of good papers and texts on this subject (e.g.,
Huffaker and Messenger 1964; Solomon 1969; Southwood 1975; Hassell
1976; May 1976).  A full discussion even on the main features
would be too lengthy to treat here, although in Section C following,
we do give a resume of some main considerations and developments.
Prior to that, however, we begin with a brief general description
of the major categories of factors and their modes of action in
producing change in population size (and quality) and regulation
of mean densities.  We subsequently discuss the application of
the principles and concepts of population dynamics in designing
augmentation procedures.


### Elementary Considerations

Individuals in a species population develop and reproduce only
in habitats to which they are adapted, and they interact with themselves and other factors in those habitats.  The species population,
the habitat and the factors interacting with the population may be
referred to as the species population's "life system" (Clark et al.
1967).

Since suitable habitats and impinging factors do not occur
evenly, population foci are unevenly distributed throughout a
species' geographic range.  Its distribution is determined by the
location of favorable habitats.  Within a habitat, even one uniformly
favorable, individuals may be distributed randomly, or more uniformly
or more clumped to varying degrees, and their precise patterns
may be affected by a diversity of factors.  In any one habitat,
unlimited (exponential) growth is prevented by (1) extrinsic
factors [finite limits of food and space, variability in

favorability of weather and changes in both positive (mutualistic
or commensalistic) and negative (predatory, parasitic and
competitive) interactions with other organisms which do not serve
it as food], and (2) intrinsic attributes (physiological and behavioral adaptations to the extrinsic factors, including adaptation of individuals to varying densities of its own species).

If all environmental factors were continually favorable for
development and reproduction of the population produces generations
which fully overlap, growth would appear as represented by the
familiar logistic equation,

$$\frac{dN}{dt} = rN\frac{K-N}{K}$$

where $\frac{dN}{dt}$ refers to change in population density with change in
time, $r$ is the power of increase and K is a constant for carrying
capacity of the limiting resource.  In this case (unrealistic in
natural habitats and for populations having discrete generations),
density dependent (or negative feedback) mechanisms (intraspecific
competition and associated self-regulatory adaptations) increase
in intensity with population size (N), until the population
reaches its environmental capacity (K), at which time rates of
birth and death become equal and actual (not intrinsic) growth
rate is zero.  Where N = K over extended generations, the age
structure of the population becomes stable (i.e., exhibiting a
constant ratio of all age classes).

The logistic curve is an inappropriate representation for
poikilothermous organisms, whose growth and reproduction are
discontinuous, (the generations not overlapping but discrete) and
closely tied to seasonal environmental changes, and particularly
inappropriate for those poikilotherms whose generation time ($\tau$)
is short relative to the time of habitat suitability (H) (Southwood et al. 1974).  Most insects, being short-lived poikilotherms,
show seasonal patterns of reproduction and death.  This is also
often associated with seasonal patterns of flowering, growth, and
development of their host plants, habitat and food organisms.  Hence,
neither the carrying capacity (K) of their food resources, nor the
age distribution of the population can be said to be stable.  The
food capacity itself may be a population interactive matter
(reciprocal density dependence).

The growth rate of a population and hence the appearance of
its population curve is dependent upon the net reproduction rate
(R) and mean generation time ($\tau$), which can be calculated from
life table data (see Southwood 1966 and Southwood et al. 1974).
Since R and $\tau$ vary widely, in concert with environmental fluctuations, growth rates fluctuate between positive and negative
values, and mean levels also fluctuate.  While this is true,

nevertheless, at a given season of the year an insect population
often tends to reach its lowest level year after year, while at
another season it reaches its highest level year after year.
While these levels may vary rather widely, there is a rather
specific order of magnitude of this variation in relatively
restricted stable environments. Accordingly, the mean level of
abundance (K averaged over many generations or years) is often
referred to as the characteristic abundance in order to emphasize
the fact that each species does have a specific mean level of
abundance and that the regulation of a population between its
characteristic upper and lower levels requires actions of density
dependent regulating factors (usually intraspecific competition
and/or natural enemies) which respond to the density of the
population (MacFadyen 1963, Huffaker and Messenger 1964).

From habitat to habitat throughout the geographic range of
a species its upper and lower limits (and hence the characteristic
density) vary greatly in relationship to the variation in various
extrinsic factors. To understand and explain the actions and
effects of everchanging factors in causing population change on
the one hand and population regulation or stability (as intuitively
suggested by "characteristic" abundance) on the other, is a central
concern of population ecologists.

### Conceptual Basis for Population Theory

As discussed, e.g., by Huffaker and Messenger (1964) and
Southwood (1975), some earlier ecologists [e.g., Uvarov (1931),
Bodenheimer (1928), Andrewartha and Birch (1954)] seemed primarily
interested in explaining changes in population size, rather than
the factors which might explain mean density over long terms
of time. They therefore placed special emphasis on climate,
resources, the ratio of generation time to habitat favorability
time, and effects of factors whose actions were described as
density independent. In contrast, other ecologists (Woodworth
1908, Howard and Fiske 1911, Lotka 1925, Volterra 1926, Smith
1935, Nicholson and Bailey 1935) were more interested in mean
density and the functioning of intraspecific and interspecific
mechanisms that might regulate populations around a theoretical
equilibrium density. The controversies generated by these, at
that time clearly opposing, views fortunately have become less
intense with the development of more holistic concepts of population
dynamics embracing factors contributing to both instability and
change and those contributing to regulation of populations at
characteristic densities (Milne 1957, Huffaker and Messenger 1964,
Clark et al. 1967, Southwood 1975).

Mathematical modeling has been an effective tool in studying
population, community and ecosystem structure and function. Much

of the resultant progress is embodied in, or referred to in the
work of Cody and Diamond (1975), May (1976) and Southwood (1975).
MacArthur and Wilson (1967) made a major heuristic contribution
in proposing the r-K continuum (with species adapted to very
temporary habitats classed as r-strategists and those adapted to
very stable habitats as K-strategists).  Southwood et al. (1974),
Southwood (1975 and 1977), and May (1976) closely considered the
proposition that r and K as well as other more generally recognized
behavioral parameters (e.g., those affecting efficiency of natural
enemies -- Huffaker et al. 1977) are the result of evolutionary
processes.

Southwood (1977) noted with regard to these evolutionary
pressures:
    "...the templet for them, on theoretical grounds
    ...appears to be the durational stability of their
    habitat...Duration stability is defined as $H/\tau$ where
    $\tau$ is the generation time and H is the length of time
    that the habitat location remains suitable for breed-
    ing...the $H/\tau$ spectrum relates to the r-K continuum--
    those species with habitats of low duration stability
    are r-strategists:  they tend to be small, mobile and
    to have a short generation time.  Species with stable
    habitats (high values of $H/\tau$ toward the K end of the
    continuum:  they are larger, more likely to be terri-
    torial than migratory, and have a long generation time)."
The point with respect to size of individuals may be of less relevance for some insect examples than for certain other groups.  We
would also add broad physiological tolerance and high intrinsic
rate of increase (not just short generation time) to high mobility
and short generation time as properties essential to high r strategy
make-up.  We would also say that a good searching capacity within
the habitat, efficient utilization of food, competitive superiority
over contending species in these and other ways, characterize
species of high K strategy make-up.  Such species also have less
need for high habitat to habitat mobility, high intrinsic rate
of increase, or breadth in physiological tolerances.

It seems appropriate here to give a brief summary and explanation of the origins of the conceptual picture of the population
dynamics landscape, as presented by Southwood (1975, 1977), since
this landscape synthesizes certain of the main concepts of the above
authors and the work of a number of others (e.g., Solomon 1949,
Holling 1966, Fransz 1974, Hassell 1975, MacArthur and Wilson 1967)
who have helped to clarify the roles and quantitative effects of
various intrinsic and extrinsic factors that act on a given
population (Figure 1a and 1b, taken largely from Southwood 1977,
but modified for more specific application to Insecta, by transaction
at the stable end of the $H/\tau$ spectrum.)

First, we note that Solomon (1949) described the phases of
events that make up Southwood's landscape, except that he did not
relate them specifically to highly constant and highly variable
environments (as did Huffaker and Messenger 1964) as opposing
faces of the physical conditions (conditioning forces of Huffaker
and Messenger). Moreover, he did not directly relate the issue
of r- and K-selected species or survival strategies (MacArthur
and Wilson 1967) to these two conditions, the widely variable
and the highly constant faces, respectively, as did Southwood.
Solomon did differentiate and name the two basic types of predator
response to prey density as being functional (improved individual
efficiency) and numerical (increase in numbers through reproduction).
He noted that a population's increase may be simply limited,
e.g., by competition for food or other resources as represented
by the K- line in Southwood's landscape (Figure 1a). It might
also be limited by any repressive factor. This term is seldom
used. Beyond being simply restricted or limited in size at a
certain density, the population may be directly, even sharply
reduced at any time by various factors. Thus the landscape
picturing the relationships between population increase and
population size should appropriately include innumerable hairs
and linear pits above and below the surface (i.e., Figure 1a,
but see below and Figures 1b, 2a, b, c). Even density dependent
factors normally are not perfectly density dependent but overshoot
the mark and cause such fluctuation.

Solomon also noted that for various reasons a population may
be freed from control by one or more of its normal (endemic) regulating agencies (e.g., by intercepting climatic effects) and may
escape or be released so that it increases rapidly to a higher,
sometimes outbreak or epidemic density. If the suppressive or
crash forces resulting from a high epedemic status are severe
enough and do not disengage rapidly with decline to low density
(the conservation phase of Solomon) the population may drop to or
below the point (undercrowding or Allee effect) which puts it in
danger of extinction (at least locally).

All of these phases are represented on Southwood's landscape.
It must be noted here that the ridges and ravines of the landscape
(Fig. 1a) do not represent population size, but rather population
growth rates. The front vertical bluff or face represents growth
rate as related to population size under very stable physical conditions; the opposing face (vertical bluff or backside, unshown)
represents the same relationship under highly variable physical
conditions. Figure 1b represents a cross section of the landscape
of Figure 1a in the region of intermediate H/τ. The growth rate
is maximum on the endemic ridge and declines to zero at S under the
action of natural enemies, at which point population size tends to
stabilize. Where natural enemy action is disrupted, the population
size may increase beyond R (release) where the growth rate becomes

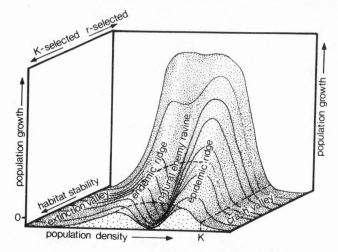

Figure 1a. Three dimensional synoptic model describing population growth rate in relation to population density and various factors of the habitat and survival strategy--the r-K continuum. (Restriction of Southwood's (1977) general model for application principally to insect populations).

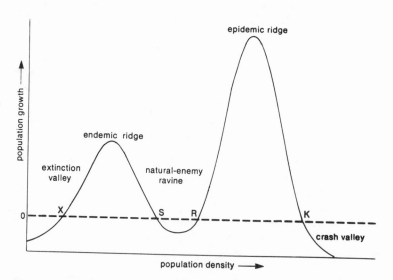

Figure 1b. Mid-section of the synoptic model Figure 1a showing two equilibrium (no growth) points, S, due to natural enemies, and K, due to intraspecific competition--and two unstable points--X, extinction, and R, release to epidemic or outbreak level. (Natural enemies will usually keep populations at the lower, endemic level, but release, R, from such control may occur (see text) and number may increase to epidemic levels.) (After Southwood 1977).

positive and increases to a new peak on the epidemic ridge. The growth rate then decreases to zero at K, which represents the highest plateau in population size for the species in question. Overexploitative features (perhaps a scramble type of competition) involved as the population surpasses the K density causes a drastic decline in population growth rate and the population enters the crash valley of Southwood. Population size, however, in subsequent generations would turn back and it may or may not return to a population size below that of the release position shown at R (Figure 1b). The crash itself may be occasioned by either the overexploitative features noted above, or by factors not represented in the model (Figure 1b), such as the return of conditions favorable to the natural enemies.

It is important to note that the general form of the landscape was derived by Southwood by combining percentage mortality curves in relation to population density (see Fransz 1974) and in relation to interspecific competition (in the prey) (see Hassell 1976), with natality-density curves of the form found in the cabbage aphid (see Way 1968).

In Figure 1a we have modified Southwood's (1977) representation of the landscape as he depicted it for all animals to more nearly represent what we think is typical for action of natural enemies of insects. We note also that Southwood (1977) himself greatly altered his first (1975) approximation of the natural enemy ravine by markedly deepening it in the intermediate range of habitat favorability and extending it considerably in both directions. However, in the most stable and most unstable habitats, Southwood's landscape allows only a single stable point, i.e., on the K-r line, representing intraspecific competition for some resource (K). Only in all ill-defined region intermediate for habitat stability and generation time (H/$\tau$) does he allow that a second point of stability is possible, and that is represented by S in the natural enemy ravine. The worldwide record of biological control shows that stable control at low endemic population levels is more likely to be achieved by natural enemies in very stable environments than in unstable or intermediate ones. Hence, we have extended the natural enemy ravine into the most stable situations for economically important insects. Moreover, we have added to his landscape the alternative bridges shown by dotted lines that pass over the natural enemy ravine in order to retain his representation for those populations that may not be regulated by natural enemies. Such bridges or bypasses of the ravine might cross over the ravine at any position on the landscape and be caused primarily by environmental manipulations antagonistic to natural enemy action. Southwood, himself (1977) suggested that his original landscape is a prototype synoptic conceptualization of the main facets of the major interrelationships. Moreover, it was not intended to represent all the cases that might exist. We note that

the combined predator functional-numerical response of Fransz
(1974) which Southwood used in his landscape formulation includes
only the within-generation aspects of the combined functional-
numerical response [i.e., switching, aggregation (numerical
response of nonreproductive nature), and possibly learning].
It does not include the distinctly delayed reproductive numerical
response of Solomon or Hassell (above) which is commonly the
route to biological control and host or prey regulation at low
endemic densities (Huffaker et al. 1971, 1976).  This delayed,
reproductive numerical response has the potential for deepening
the natural enemy ravine at all positions except perhaps in
the extremely variable habitats (ones of temporary residence
only?), and certainly into the region of very stable habitats.

     We recognize that Southwood's landscape in the region of the
most stable environments applies to many species which have
perfected their own intrinsic regulatory mechanisms (e.g.,
territorial and social behavior restricting reproduction) and which
are not regulated by natural enemies at low endemic levels.  We
believe that the regulation of the densities of most phytophagous
insects of economic significance (and probably many others) in
stable environments is more commonly through the mechanism of
their natural enemies than by their own intrinsic behavioral
mechanisms.  Indeed, the co-evolution with them of highly effective,
highly host specific natural enemies which produce a stable endemic
relationship has probably meant a great reduction of the selective
pressures for the host (prey) species to develop its own highly
sophisticated intrinsic mechanisms of regulation.

     We note that Southwood's, and our own landscape (Figure 1a)
does not provide for the innumerable hairs and linear pits that
would show on its surface as reflections of the various effects
of both density dependent and density-independent factors that
commonly may sharply depress population growth rate or allow for
its increase at various points in time for any particular habitat
stability/generation time gradient.  The linear pits should often
show deeply negative rates of population increase, and the height
of the hairs and depths of the pits would be greater toward the
unstable end of the landscape.  Figures 2a, b, c are presented to
suggest such spines and pits at sections across the landscape
representing stable, intermediate and unstable conditions.

     It is also worthwhile at this point to look at the complexity
of factors and interactions of factors that act on a given species,
and as affected by the species' own population size.  Huffaker
(1963) described the flow of action in such a system (Figure 3).
It is important to note that he categorized five basic intermediate
factors (food, natural enemies, competitors, allies and refuges or
home sites etc.) which may be subject to influence by the pervading
physical environment and which act on the population through

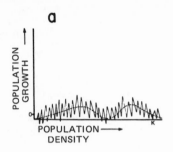

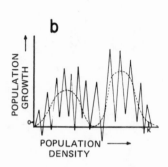

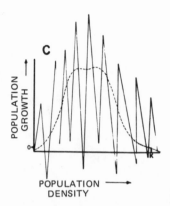

Figure 2.   (a) Foresection (stable habitat), (b) midsection (intermediate habitat), and (c) back section (unstable habitat) of
synoptic model (Figure 1a), showing possible scope of departures
in growth rate from the surface which represents <u>average</u> relationships.   (The departures, however, should be thought of as linear
<u>pits</u> and <u>hairs</u> from the surface, rather than broad hills or ex-
cavations as suggested here.)

effects on its natality, mortality or movements (the paths to change
in population size).   These intermediate factors may also be affected
by the population itself.   Certain of these intermediate factors
also act on others and thus indirectly on the population, either
positively or negatively.   This conceptual picture may be used to
more clearly understand both the direct and indirect effects brought
about by augmentation tactics (e.g. Table 1).

The most basic ideas of the more holistic concepts are, (1)
that no factor acts alone and that factor interactions may be
synergistic or inhibiting with respect to population growth, and
(2) that the relative roles of factors in limitation and regulation
may vary spatially and temporally.   For example:   toward the margins
of a species' geographic range, climatic factors may be largely
responsible for population change; whereas, toward the center

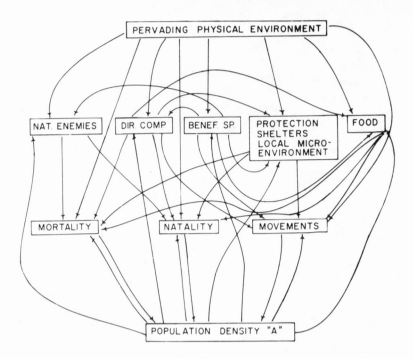

Figure 3.  Actions and interactions of factors influencing a population.

of its distribution where climatic conditions are more uniformly
favorable, biotic factors may be the principal inducers of change
as well as accounting for regulation (Huffaker and Messenger 1964).
The continual maintenance of a population within rather definite
lower and upper limits is possible because of rates of birth, death
and movements change with population density.  Thus, the action
of the total complex of factors impinging on population growth and
density is density dependent.  This action must intensify as
populations increase and relax as populations decrease.  A given
factor, such as a species of natural enemy, may act in a density-
independent way under one set of circumstances and in a density-
dependent manner in another case.  Thus, classifying factors per se,
as density independent or density dependent can be clearly wrong;
classification of the action or role as density dependent or
density independent in the particular situation, however, is
fundamental to understanding population dynamics (Huffaker 1958,
Huffaker et al. 1963, Huffaker and Messenger 1964).  More attention
should therefore be given to the function of factors as the function
varies from situation to situation in relation to population numbers,
quality and behavior.

| Case | Direct Effect on the Density of: The Pest | Direct Effect on the Density of: The Natural Enemy | Possible Indirect Effect on the Density of the Pest (through the Effect on the Natural Enemy) | Combined Effects on the Pest |
|---|---|---|---|---|
| 1 | + | + | - | +, - = ? |
| 2 | + | - | + | +, + = + |
| 3 | + | 0 | 0 | +, 0 = + |
| 4 | - | + | - | -, - = - |
| 5 | - | - | +\| | -, + = ? <br> -, - = - |
| 6 | - | 0 | - | -, - = - |
| 7 | 0 | + | - | 0, - = - |
| 8 | 0 | - | + | 0, + = + |
| 9 | 0 | 0 | 0 | 0, 0 = 0 |

Table 1. Hypothetical direct effects of a control tactic (a) on the population density of a pest and (b) on one of its natural enemies, and (c) the indirect effects on the pest through the effects on the natural enemy.

## Intrinsic Responses to Weather and Resources

Each insect species has evolved responses permitting survival, growth and reproduction only within a specific range of weather and other conditions. Some species have evolved dormancy mechanisms permitting survival under extreme conditions (cold-hot, dry-wet) when such conditions occur regularly or "predictably" in the habitat. Other species have evolved migratory mechanisms which may function in permitting escape from adverse weather or habitat conditions (commonly affected by weather). Weather fluctuations not only affect the rates of birth, death and movement (the three paths to population change, Figure 3) directly, but also can affect the synchrony of an insect's life cycle with those of its hosts and natural enemies, and thus the nature of the resultant interactions.

The effect of resource quality and quantity on insect population behavior varies according to the renewability of each specific resource (e.g., food, water, light, space or abode). A non-renewable resource, such as a suitable nesting place, can be completely utilized without deterioration. When an increasing population completely utilizes such a resource, population growth is limited abruptly. In the intraspecific competition which ensues for nesting sites, for example, losers either make no contribution to future populations, or compete again for sites emptied by previous occupants, or disperse to other habitats, with variable results. Winners in this contest type of competition have their spatial requirements for reproduction met and therefore contribute progeny to the next generation. In habitats where contest competition for non-renewable resources is consistently intense, populations of the species involved tend to be relatively stable. In contrast, a renewable resource such as food may be reduced to low levels of quantity and quality by either the user itself or by other factors, but is not completely used up and will be replenished subsequently. Intraspecific competition for food may be of such a nature, and so intense that no individual gets enough food for normal development and thus extensive starvation and/or reduced reproduction occurs. Populations exhibiting this scramble type of competition (Nicholson 1933) tend to fall sharply after reaching levels when food is removed faster than it is replenished, and striking alterations of abundance and rarity may occur (Nicholson 1954). Certain species have evolved behavioral adaptations serving to change the intraspecific competition from the scramble to the contest type, thus enhancing survival of the population under conditions where individuals compete for renewable resources, such as food, as well as for non-renewable resources. Thus, cannibalism, various forms of territoriality, and certain social or anti-social behavior may function to increase the probability of survival and reproduction by limiting the number of individuals having access to renewable resources. (See Krebs 1972, Price 1975, and Wellington 1977 for a fuller discussion of and references to

intrinsic control mechanisms, including the waxing and waning of
population quality and genetic feedback.)

## Interspecific Interactions

Interspecific population interactions (mutualism, commensalism,
competition, predation in the broad sense) also induce population
increases and decreases.  From the viewpoint of augmentation,
predation (including parasitization) is of greatest interest,
and insights into the mechanisms of predator-prey (host-parasite)
systems have emerged from laboratory and field experiments and
mathematical conceptualizations (modeling).  The most familiar
of the early models were the Lotka-Volterra equations, which upon
solution depict coupled stable oscillations of hypothetical predator and prey populations through reciprocal density-dependent
relationships.  However, such strict stability has not been
demonstrated in simple homogeneous systems, comprised of predators,
prey, and a constant supply of food for the latter.  Huffaker
et al. (1963) were able to prolong the oscillations through
multiple generations by adding various elements of environmental
heterogeneity to the predator-prey system without furnishing
actual refuges or adding immigrants.  Gause (1934) earlier had
prolonged oscillations by adding "immigrant" predators or prey
or by furnishing refuges.  If a user is the principal regulator
of the density of such a renewable resource and is itself entirely
dependent on that resource, the two populations (both organisms)
may be reciprocally regulated by the density-dependent interaction,
as is common relative to certain parasitoids that are host specific
relative to the given habitat organisms.  Total reciprocity is
uncommon, and especially in unstable habitats where predators
of wider prey acceptance may collectively serve regulating roles
by switching from scarce species to more abundant ones (Murdoch
and Oaten 1975).  [They may serve regulating roles also by
readily moving to areas of high prey density from ones of lower
density.]  Yet, reciprocity in interaction is commonly sufficient
in degree to enable a natural enemy to be an effective regulator
of its host's population, and at consistently low, endemic densities.

In a classical paper, Holling (1961) utilized component
analysis procedures in showing that the functional and numerical
responses of Solomon (above) arise basically from the densities of
prey and predator, but yet may be greatly modified by characteristics of the environment, the prey, and the predator--the
former varying spatially and temporally and the latter two
according to the intrinsic attributes of the prey and predator
populations.

One of the primary benefits of natural enemy augmentation
is, in fact, the bypassing or mitigation of the time lag in the

numerical response (primarily the reproductive numerical response)
of the predator (or parasite) to prey density.  For example,
as depicted in Figure 4, the predator population at time t,
would require augmentation by an amount ΔNa (the difference
between predator population at time $t^1$ and $t^2$) to avoid economic
damage.  While Figure 4 represents an idealized interaction seldom
so specifically realized in nature, the principle of using
predator augmentation to offset the time delay in predator response
is none the less valid for populations exhibiting other, more
realistic temporal cycles.

Predator-prey interactions are among the most widely studied
systems in population dynamics.  Classically, the dynamics of two
such populations are represented by models in which the population
densities are reciprocally governed by two first order, coupled,
non-linear differential equations (or more appropriately in the
case of discrete generations, difference equations).  While this
approach has resulted in significant advances in the theory of
population biology, realistic systems representation has by necessity been sacrificed for mathematical elegance (tractability).

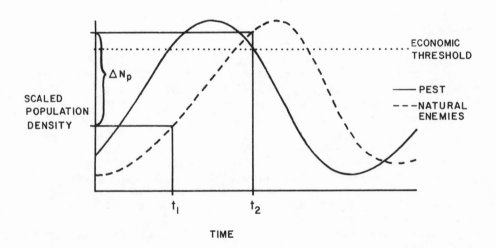

Figure 4.  Hypothetical curves of interacting populations of a
pest and natural enemy.

In fact, the full dynamic behavioral range of even the more simple
mathematical models is only now beginning to be realized (May
1974, 1976).  Computer simulation modeling has provided a means
for representing biological systems in a more realistic manner;
however, the requirements for simulation of spatial heterogeneity
and time-lag effects may tax even the most advanced computer
installations and, more importantly, our biological knowledge.
Mathematical models specifically of the predator (or parasite)
functional response (for example Holling 1966: Hassell and Varley
1969) have provided useful concepts, although very real difficulties
may exist in field evaluation of the necessary parameters and/or
their biological interpretation (Royama 1971: Stinner and Lucas
1976; Huffaker et al. 1977).  Teams of field-oriented researchers
in British Columbia (Frazer and Gilbert 1976; and Gilbert et al.,
1976) and in California have recently grappled with this problem.
A. P. Gutierrez (personal communication) summarizes this work as
follows:

> "Frazer and Gilbert (1976) developed a field
> tested predation model incorporating age dependent
> searching, hunger, and consumption rates.  This
> model describes the interaction of the pea aphid
> (*Acyrthosiphon pisum* (Harris)) and the ladybird
> beetle (*Coccinella trifasciata* L.).  The model
> reduces under special cases to first a Nicholson-
> Bailey model (prey → 0) and to a Lotka-Volterra
> model (predators → 0) (see Gutierrez et al., in
> press).  The model extends many of the methods
> pionerred by Holling to field conditions.  The same
> model has been used in California for examining
> predation in California cotton (R. E. Jones, perso-
> nal communication) and alfalfa (Gossard et al., in
> press; A. P. Gutierrez, personal communication).
> The latter models extend the Frazer and Gilbert
> model to include several prey and several predator
> species.  The complexity of the models has become
> greatly reduced because of the understanding
> obtained from detailed field and laboratory experi-
> ments.  In many cases, it was found that some prey
> and predators did not interact because there were
> size mismatches, the two species were not temporally
> or spatially coincident, or the prey were simply
> nonpreferred.  While parameter estimates are
> difficult to obtain, the effort was not prohibitive,
> and the results appear to be most promising."

Christine Shoemaker (personal communication) has noted
also:

> "Detailed models of crop ecosystems have also
> been developed.  Such models incorporate age structure
> in several populations, the effects of weather on mat-

uration, and the interactions among pest, plant, and
natural enemy populations.  Because of the number of
variables involved in these models and the complexity
of the interactions, they are not amenable to analytical
solution and must be solved numerically by computer.
By incorporating a considerable amount of detail,
predictions of the dynamics of the crop ecosystem
can be made with relatively good accuracy.

"Several simulation models describing the
interactions between at least two populations in an
ecosystem have been developed (Gutierrez et al.
1975 (cotton); Gutierrez et al. 1976 (alfalfa);
Ruesink 1976).  Ruesink's model has since been
used to evaluate the effect of early harvesting
of alfalfa on the alfalfa weevil parasite's
reproductive potential under varying weather
conditions (R. G. Helgesen, personal communication).
By incorporating descriptions of the effects of
management decisions on pest and parasite populations, we can use a model to estimate the final
effect on yield.  For example, in the application
of insecticides, the population dynamics incorporated in the model will describe pest resurgence,
the delay in the buildup of natural enemies,
secondary pest outbreaks, and other aspects of
ecological dynamics which may influence the
effectiveness of an insecticide treatment.

"The use of models to evaluate management
alternatives has frequently involved the use of
optimization methods.  For example, Regev et al. (1976)
determined optimal times of insecticide applications
for controlling a population of alfalfa weevil for
a given weather pattern and a given rate of weevil
migration.  This model did not include natural enemies
because the Egyptian alfalfa weevil in California has
none.  The alfalfa weevil parasite, *Bathyplectes
curculionis* (Thomson), however, was incorporated
into the management model developed by Shoemaker
(1977).  This model calculates optimal timing
of harvests and insecticide treatments for a range
of weather conditions and pest and parasite densities.
Included in the model are the effects of insecticide
treatments and harvests on the parasite population
and the effect of weather on synchrony between the
parasite and host populations.

"Not all management models involve optimization
methods.  A relatively simple model was used to ana-

lyze the use of biological, cultural and chemical
methods for olive pest control comprised of a disease
and two scale insect pests (Shoemaker et al., in
press). In this analysis, it was shown that the
most profitable method of integrated pest management
is the use of cultural control provided by frequent
pruning and the use of biological control provided
by two parasites. Chemical pesticides were still
recommended when necessary."


## Environmental Heterogeneity

Environmental heterogeneity commonly has meant spatial and
temporal variation in the habitat, and should include, but has
not always included, the diversity of organisms and their activities in the ecosystem. We would define the local habitat as
the area within which the localized food gathering movements and
other activities associated with growth, development, reproduction
and survival in adverse periods in that local area occur. Areas
of entirely distinct nature to which a population disperses
over long distances, even though required for continuance, are
envisioned only as the complete habitat or region of overall
residency.

Here we will refer to the local habitat concept. Such local
habitats will tend to be larger for species whose immature and
adult stages utilize different food, or for species which utilize
different food, or for species which utilize different sorts
or places for feeding and dormancy. Environmental heterogeneity
required to satisfy habitat needs of a species also tends to be
greater, of course, for polyphagous species and for multi-voltine
species whose succeeding generations require different conditions
and/or hosts or prey. As Southwood (1975) noted, habitats vary
in their degree of permanence. The complete habitat of permanent
residency of course must include an area sufficient to satisfy
all the needs for year-around living, including food, places
for breeding, refuges from enemies and inclement weather and for
aestivation and/or hibernation. For migrating species it is
easier to visualize different habitats for satisfying these needs
in the separate areas of existence. But since both time intervals
and sets of conditions are involved, the degree of stability or
instability relevant to Figure 1a, must be that resulting from
the variations in both (or all) such required habitats.

If one views the geographical area circumscribed by the
range of a species, one sees an uneven distribution of areas of
permanent residency, with the species population component of
each such area being a deme and each being isolated to a degree
from other demes. The rate of interdeme movement varies greatly,

and though of recognized importance in both intrademe and total
species dynamics, population movement is one of the least understood phenomena in insect ecology.  Dispersal characteristics
and behavior vary widely among species and include innate mechanisms for long-distance migration as well as local inter- and
intra-habitat movement.  Such movement is influenced greatly by
weather and also by density induced behavior, in some instances
at least.

Numbers of a local population change through three routes:
births (B), deaths (D), and movements (M), either into (immigration,
I) or out of (emigration, E) the occupied area.  Many studies
of population dynamics have suffered because of fuzzy definitions
of the target population and overly simplified assumptions
regarding movements (such as, that I=E).  As noted by Price (1975),
"Dispersal is adaptive.  It permits a population to spread as
the population increases, and it enables the colonization of
new sites." Density dependent dispersal would seem to be adaptive
in that it is postponed until the population has reached high
enough numbers that the losses during dispersal to reach the new
sites and resources can be sustained (Kennedy 1972).  While emigrants
represent a short-term loss to a local population they also may
be the source of future residents.  Where there is a local
destruction of a plant and animal community (the habitat) by
natural catastrophic events or by man (as in agriculture),
reinvasion occurs with the sequential establishment of plants,
herbivores and carnivores.  In the agroecosystem, man establishes
the plants (crops) and somewhat inadvertently the associated plants
(weeds or ground cover).  The kinds of herbivorous and carnivorous
invaders, as well as their invasion rates, depend upon the proximity
and size of invader sources and the phenology of the crop and its
attractiveness at the crucial period of dispersal of the prospective
immigrants.  It is a very practical matter to understand these
different rates of invasion, that is, the factors influencing them
as this affects development of pest management systems.  As noted
by Janzen (1968) and Price and Waldbauer (1975), we should view
crops somewhat as islands (though not as strictly isolated),
both on the short-term and the long-term, or evolutionary time
scale.  The theory of island biogeography (MacArthur and Wilson
1967) may have some practical applications in our efforts to
manipulate pest and natural enemy populations (Price and Waldbauer
1975).

## Survival Strategies

Explaining survival is as essential as explaining mortality
if the dynamics of a population is to be understood.  It is the
relative number of survivors that is important.  A change in
mortality from 99% to 98% seems an insignificant difference (ca

1%).  Yet the survivors are twice as many (100% more).  The
equilibrium mortality, that level which just permits replacement
is an exceedingly important statistic in viewing population
dynamics, yet it has been given very little emphasis.

The niche of each species differs from the niches of other
species and so does each species' adaptations, which can be
interpreted as components of its survival strategy.  Filling a
niche (a _role_ in a specific unit of a habitat wherein its superior
adaptation enables it to have the role) under wide and frequent
variations in the favorability of environmental factors requires
a strategy for survival (and role business) quite different from
filling even a very similar role under consistent (stable)
favorability.  The actual niches would thus vary correspondingly
in their total descriptions (niche volumes).  Understanding
survival strategies of pests and natural enemies is important
in the logical development of control strategies and tactics
designed to decrease pest survival.  As advocated by Wellington
(1977), "We must pay more attention to the various attributes
that promote the survival of different species.  That is the
best way to discover the rules of the game of survival which
could help us to counteract specific strategies more effectively
and more consistently."

The deductions from theoretical and empirical studies of
most relevance to biological control are those which deal with
the relative influences of various factors (independent of their
relationship to density) on population size, change and stability in relation to the r-K continuum and the H/$\tau$ spectrum.
Southwood (1977) suggests that "biological control or integrated
control with a significant natural enemy component must be the
dominant strategy" for pests intermediate on the r-K continuum,
but that the probability of successful classical biological
control of r and K pests is low.  We would agree in general for
r pests, but not for K pests.  He also discusses the relevance
of his model in the selection and use of other control strategies.
While Southwood's model and various modifications of it (e.g.,
Fig. la) are useful attempts to synthesize our knowledge concerning the behavior of populations of diverse taxonomic status,
there is a need for further validation of these models and for
development of more specific models for particular groups of
organisms of interest.  The application of generalizations to
specific cases can result in unsound and unfortunate conclusions.

## POPULATIONS IN AGROECOSYSTEMS

The basic principles of population dynamics apply to species
in both unmanaged (natural) and managed (cultivated) areas; however,

man's actions comprise powerful forces which dramatically
alter population behavior in agroecosystems.

The comparative features of natural and agricultural ecosystems were discussed by Southwood and Way (1970) and Rabb et
al. (1976). It will suffice to note:  (1) agroecosystems are
artificially established and maintained by man; (2) they usually
consist of much simpler communities than would normally occupy
the managed area in its natural state; (3) the producer level
commonly consists of a single species (the crop) selected by man
and usually not adapted for permanent residency nor even one-
generation survival without the care of man; (4) the stands of
primary producing species are more uniform in space and age
structure than one finds in most natural communities; (5) ferti-
lization and irrigation influence attractiveness and suitability
of foliage for the herbivores, and influence other facets of the
microenvironment; (6) crop rotation bears no resemblance to the
natural succession that would occur without man's inputs; and (7)
while the cropping practices preclude success of many herbivorous
species that inhabit natural habitats in similar areas, they may
create an artificially super favorable environment for those
herbivores adapted to the crop--that is, through the provision of
more accessible and/or more nutritious food or lush growth which
may itself affect microhabitat favorability, and by the absence
of or suppresseion by man of natural enemies.

Crops tend to attract a rich complex of insectan herbivores
during a season of growth.  Some soil inhabiting herbivores
essentially become permanently established as do some species in
long-lived tree crops.  On the other hand, foliage, stem and
fruit feeders of short-term crops become established at various
times and densities following planting, and in relation to invader
sources (as noted above) and the specific adaptations of each
herbivore to the various phenological aspects of the crop
(some invading and/or being reproductive throughout the period
of plant growth and maturity and others only during a more
limited period, such as flowering).  Naturally, the duration
of their activity is limited to the life term of the crop.

The damage inflicted by a crop's herbivore complex varies
widely among crops and from place to place and time to time in
any one crop.  As noted by Chant (1964), "there are four classes
of phytophagous insects:  (1) serious, perennial pests; (2) in-
termittent but sometimes serious pests; (3) insects that are
rarely if ever pests; and (4) non-residents of the ecosystem
that enter the environment periodically for short times and then
are serious, irregular, or rare pests."  As discussed by Chant,
the class to which a pest belongs will in large measure determine
the practicability of specific control strategies and tactics.

## MODE OF ACTION OF AUGMENTATION

Augmentation, with the purpose of increasing natural enemy effectiveness, is accomplished by repetitive release of natural enemies or their hosts (at strategic times) or use of energy and/or materials in prescribed ways to increase pest mortality by natural enemies. [In contrast, classical biological control involves only the importation and establishment of natural enemies which without further assistance accomplish biological control (or would do so if detrimental chemical control practices were stopped). Such repetitive measures are not required.] There are two general but overlapping categories of augmentation tactics, with several aspects in one of them (see DeBach and Hagen 1964; van den Bosch and Telford 1964; and Rabb et al. 1976):

1. Environmental manipulations
   (a) Provision of alternate hosts for survival and buildup of natural enemies prior to the time the target pest will attain pest status.
   (b) Provision of attractants (e.g. kairomones) rendering the natural enemies more effective.
   (c) Provision of subsidiary foods or other required requisites (food sprays, nesting places, etc.).
   (d) Modification of cropping practices, including change in crop varieties and pest control procedures for other pests, and by planting and cultivating so as to favor natural enemies.
   (e) Addition of the host (pest) insect itself at critical times so as to allow for natural enemy continuity.
2. Periodic releases of a natural enemy for an immediate (by the released individuals) or time-lag (by progeny of released individuals) control effect on the pest population.

However, these categories are useful as a point of departure in understanding the role of such augmentation tactics only if one keeps in mind the critical criterion: that is, that the depressing effect on the pest population must be through natural enemy action induced through repetitive actions by man.

The effects of all crop management (including pest management) actions reverberate throughout the total crop ecosystem (e.g., Figure 3), and it is important to understand the cause and effect pathways followed by the energy and materials (living and non-living) introduced if one is to objectively determine that improved natural enemy action (augmentation) has actually resulted. This is especially true with respect to the augmentation tactics classified under environmental manipulations (above). In applying this criterion, it may be useful first to consider a single situation involving one pest and one natural enemy, and list the possible immediate and direct effects of a tactic (+, -, or 0) on the population density of each, and the possible in-

direct effects on the pest via the effects on the natural enemy,
as indicated in Table 1.

While there are pitfalls in using this table, the perspective
it offers can be useful in recognizing tactics which satisfy the
criteria for augmentation.  Keeping in mind that augmentation
must cause an increase in natural enemy effectiveness (hopefully
sufficient to decrease the pest population), the following
comments are useful.

There is no way of judging from Table 1 itself whether the
positive or negative effects on the pest (direct and indirect
ones) would be greater in the cases where both a negative and
positive effect is entered, and there is no way of judging if
these effects are strong enough to give economic control.  The
table is presented to show how such direct and indirect effects
may coincide to give an effect not suspected from the direct
effects alone.  In Case 1, the negative indirect effect may cancel
the positive direct effect on the pest; if stronger on the natural
enemy the pest could be suppressed; in Case 2, the indirect
effect would add to the positive effect on the pest; in Case 3,
the indirect effect would not add to or detract from the direct
effect; in Case 4, the negative indirect effect would augment
the negative direct effect.  In Case 5, we have the possibility
of a direct negative effect on the natural enemy cancelling the
negative indirect effect on the pest or an additive negative
direct effect, depending upon whether the negative effect on the
natural enemy or the pest is more pronounced; in Case 6, we have
a clear possibility of a zero direct effect on the natural enemy
causing a negative indirect effect on the pest, since this would
improve the ratio of natural enemies to pests, but this would
not result immediately and the immediate effect could be that the
direct effect on the pest alone has suppressed the population
sufficiently below the economic injury level that an indirect
subsequent effect due to the increased ratio of natural enemies
to pests would not be needed (Fig. 5); in Case 7 the negative
indirect effect will dominate over the zero direct effect; in Case
8 the positive indirect effect dominates; and in Case 9 there
are no direct or indirect effects.  Thus, assuming the effects
are strong enough for Cases 2, 3 and 8, the pest will increase;
for Cases 1 and 5 the outcome is uncertain; for Cases 4, 6, 7 and
one possibility for 5, the pest would decrease; for Case 9 there
would be no effect as none was postulated.

In the hypothetical example of an immediate effect of Case
6 (Fig. 5), a control tactic (such as a cultural practice or
selective pesticide) lowered the mean level of abundance of the
pests removed by the natural enemy.  In doing so, however,
the significance of the natural enemy in preventing economic
damage was dramatically and at once increased.  As shown, after

the treatment was applied, the mortality inflicted by the natural enemy then represented the difference between negligible and substantial economic damage (in contrast to pretreatment).  In the strict sense, however, this does not illustrate immediate augmentation.  If the treatment were a cultural practice it would be cultural control; if a selective pesticide, chemical control.  However, it also represents conservation of natural enemies and improvement in relative numbers, and the subsequent effect could readily satisfy the concept of augmentation.

Periodic releases, particularly inundative releases, of biotic agents, both microbial insecticides and predators or parasites, which give temporary, rather immediate reductions in target pest populations, have been referred to loosely as applications of biotic insecticides, implying that the procedures and effects are analagous.  However, this analogy should not obscure the basic differences in the modes of action of the biotic and chemical agents.  Currently used chemical insecticides are most effective

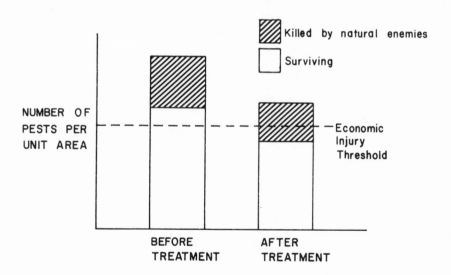

Figure 5.  Diagrammatic representation of one hypothetical possibility of a pest population before and after subjection to a control tactic which has little effect on the percent mortality caused by the natural enemy.

at the time of application, with relative residual effectiveness
declining rapidly.  In contrast, a biotic agent such as *Trichogramma* maintains effectiveness for extended periods due to its
functional or immediate response to host density, or after one
or more generations of the living agent due to its numerical
response.  Timing of the release with respect to synchrony with
the pest stage(s) most susceptible will determine which of these
responses will be of most importance.  Likewise, multiplication
of a microbial agent after its application may intensify and
extend a control effect for some time.

## CHARACTERISTICS OF NATURAL ENEMIES EFFECTIVE

## AS BIOLOGICAL CONTROL AGENTS

Current theory and empirical evidence support the view that
no single factor operates entirely alone in explaining population
changes and stability (Huffaker and Messenger 1964; Southwood
1975).  Rather, a regulating factor commonly acts in relation
to the effects of certain other factors.  Nevertheless, a natural
enemy (or other factor) may impose limits on the amplitude of
population fluctuations and ones having regulatory capacity
may establish specific equilibrium densities characteristic of
and dependent upon, the respective environmental conditions.
Biological control in the economic sense obtains when the suppressive effect of a natural enemy accounts for the difference
between economic and non-economic host populations.  A natural
enemy can accomplish permanent biological control only if it has
evolved morphological, physiological and behavioral adaptations
to its host and other components of its life system by means of
which it can consistently find, attack and suppress its host's
populations and maintain them at sub-economic densities, and
which permit permanent residency with its host.

Attempts have been made to list the characteristics of natural
enemies essential for their effective (economic) control of
their insect hosts or prey (pests).  For example, Huffaker et al.
(1976) summarized the desired characteristics for achieving self-
sustaining (permanent) economic control of a pest as follows:
(1) fitness and adaptability to the various conditions of the host
and the host's environment, (2) searching capacity, (3) power of
increase relative to that of its host or prey, (4) host or prey
specificity, or potential for development of host or prey preference,
(5) synchronization with the host or prey, and the habitat, (6)
density-dependent performance relative to either or both the
host's or prey's density (functionally, reproductively, or aggre-
gatively) and its own density (e.g., mutual interference), (7)
detection of and responsiveness to the condition of the host
(stage and whether previously parasitized or not), and (8) effici-

ent competitiveness with other natural enemies.

While such a list focuses attention on some important and
inevitable aspects of predator-prey (parasite-host) systems, their
utility as criteria in selecting candidate agents for biological
control leaves something to be desired.  Many of these characteristics are represented in any natural enemy that normally attacks
a given host or prey species, but the degree of perfection of the
adaptation may vary greatly, and from one life system to another.
The main problem is that the differences exhibited in these characteristics are often difficult, if not impossible, to explicity
quantify in a practical manner in the field.

It was recognized by Huffaker et al. (1976, 1977) and by
other specialists that fitness and adaptability is a prior
requirement, but a natural enemy may be both fit and adaptable
for survival without being able to maintain economic control of
its host's population.  Various authors have proposed high host
searching capacity as the most important criterion, but high
generation to generation searching capacity also requires high
fitness and adaptability.  Searching capacity also implies certain
other desirable characteristics.  To be an effective searcher,
a parasitoid must not waste time, energy and progeny in superpara-
sitizing hosts; discriminatory or nondiscriminatory behavior
is thus a factor.  Doutt (1964) noted that searching capacity
includes power of locomotion (both for habitat finding and host
finding within the habitat), host perception, power of survival,
aggressiveness and persistence.  These characteristics also are
associated with competitive status.  DeBach et al. (1976) and
Huffaker et al. (1977) suggest that tests to establish the rate
of numerical response of a natural enemy in the field, as related
to increase in host density within a low range of host density,
should test the net effectiveness of the species as against
others, and would represent a test of many of these characteristics collectively, but especially of searching capacity.  The
species exhibiting the better numerical response in such a test
is the better of the natural enemy.

Much emphasis has also been placed on the natural enemy's
power of increase relative to that of the host or prey.  However,
the power of a natural enemy to control its host or prey cannot be
evaluated solely on the simple basis of the comparative fecundities
or intrinsic rates of increase of the host and natural enemy.
Many authors have made this error.  In fact, Messenger (1976)
neglectfully made such a comparison and presented graphs presenting
intrinsic rates of increase plotted against temperature for three
species of aphid parasites and their host, the spotted alfalfa
aphid, *Therioaphis trifolii* (Monell).  He claimed that the
temperature range within which control by these species was possible lay entirely within the respective temperature region where

the intrinsic rate of increase of the parasite exceeded that of
the host, although he (with Huffaker) had earlier noted (Huffaker
and Messenger 1964, p. 107) that to be efficient a parasite need
not possess a power of increase equal to that of its host.  Huffaker and Flaherty (1966) and Laing and Huffaker (1969) emphasized
that other mortality factors may modify this relationship, but
more significantly, that the mortality caused by the parasite
negates that much of the fecundity of the host.  This fact is
automatically built into the interaction in host-parasite
modeling.

     In keeping with the theory of r- and K-selection as related
to the permanence of habitat favorability (Section II, above),
relatively low fecundity or power of increase would suffice in
habitats with stable favorability, where there would be a premium
on high competitiveness and the ability to find hosts at low
densities.  On the other hand, high powers of increase and
mobility would be required for effective natural enemy action in
unstable habitats.  If the habitat has continuity but is subjected
to high or low temperatures that kill most of the natural enemies
during a severe period, a high power of increase may offset this
adversity and is needed if the host is not to escape control by
the enemy.  This is the case in San Joaquin Valley of California
where *Aphytis maculicornis* Masi, an excellent searcher, suffers
extreme mortality during the hot dry summers and its interaction
with olive parlatoria scale, *Parlatoria oleae* (Colvée), is
intercepted.  However, because with the return of favorability
in the fall and spring and because of its very high power of
increase, relative to that of its host, it can regain its status
each season, and, except in certain situations, can maintain a
high degree of economic control of the scale population on a
year-around basis (Huffaker et al. 1962).  [A second species
*Coccophagoides utilis* Doutt, has been introduced in an effort to
improve upon the degree of control, with excellent results
(Huffaker and Kennett 1966).]

     Such parameters as mutual interference, density dependent
aggregation, handling time and searching coefficients can be
measured fairly readily in laboratory studies but this poses
difficult problems in the field, the place for which we most need
such values.  Thus, we must live with the fact that we cannot fully
characterize a universally effective natural enemy for use in
biological control.  Attempts to stereotype the universally effective biological control agent is counterproductive.  There is no
substitute for searching for natural enemies which are uniquely
adapted to the unique life system of the target pest.  In some
but not all cases, such enemies can be found for self-perpetuating
biological control.  In certain other pest situations, augmentation
procedures may make it possible to use certain natural enemies
for pest control in spite of the fact that these same natural

enemies unaided are ineffective.

## SOME KEY POPULATION PERSPECTIVES FOR DEVELOPING AUGMENTATION

A wide-area model of the pest population seems to be the most logical conceptual framework in which to view the potentiality and practicability of augmentation tactics (i.e., manipulation of resident populations of natural enemies indirectly, or the periodic release of natural enemies). Even a crude holistic model (by definition including interacting natural enemies) may provide a basis for partitioning the pest population into ecologically meaningful units as a prelude to selecting specific units as augmentation targets.

Pest species differ widely in the proportion of their total biomass produced on the crop(s) they damage and on other hosts (cultivated and non-cultivated) which they inhabit. In some cases, augmentation tactics may be directed against pest population components on hosts other than the crop(s) to be protected in order to reduce the movement of pests into the crop(s). Where more than one host plant species is involved, a major consideration is that the effectiveness of an enemy against a herbivore is influenced by the physico-chemical characteristics of the plants, some enemies being effective on some plant species but not others. When the natural enemy action must be exerted in the crop itself, cultural practices considered essential for crop production may impose constraints on the use of natural enemies. Thus the model must reflect the cause and effect pathways between these cultural practices and the pest and natural enemy populations. If these constraints are severe and cannot be removed (e.g., the use of a disrupting pesticide to control another pest in the pest complex is required), a tactic dependent for its success on natural enemy action in the crop itself may be impractical.

Augmentation tactics may have a variety of effects on populations of both natural enemies and pests. Thus, it is important to define the type of effects to be expected. The success of certain tactics hinges on the degree to which natural enemy populations can be increased through reproduction in situ (e.g., reproductive increase of a released enemy at least one generation prior to the time of its required attack on the pest, or its reproduction on alternate hosts provided near the target area). In contrast, other tactics may primarily induce shifts of ambient natural enemy populations from one searching arena to another and from one host to another (e.g., strip cutting of alfalfa, or the use of food or chemical attractants, or the provision of nesting sites). Of course, such tactics may also induce increases in the natural enemy populations.

The most desirable effect on the pest population would be to lower its general level to the point that the number of pests invading the susceptible crop would not constitute a problem. However, this can be accomplished only through the application of the tactic (with uniform results) to all population foci throughout a wide area.  Such an area would need either to be isolated from other pest-infested areas or circumscribed by an arbitrarily defined border zone in which control is not expected.  The size and shape of such zones should be determined by study of the dispersal characteristics of the pest species involved under the prevailing weather systems.  A reduction in total generation survival is essential to the success of wide-area pest suppression.

Where a wide-area pest suppression is not feasible, augmentation in some situations can be used successfully against localized pest foci.  In such situations, a natural enemy may be used to change the age distribution of pest mortality without necessarily significantly influencing total generation mortality.  For example, shifting mortality to the egg and early larval stages by releases of appropriate enemies might be sufficient to avoid economic damage, if such is primarily caused by late-instar larvae.

Some natural enemies which are intrinsically incapable of self-perpetuating biological control can be used successfully in repetitive inoculative or inundative releases, if they can be reared satisfactorily (i.e., if sufficient numbers of a given quality can be produced cheaply on a schedule permitting their timely application in target areas).  Consideration in selecting and rearing enemies are discussed in subsequent chapters (also see Rabb et al. 1976 and Stinner 1977).  A major problem is the development of guidelines for deciding the numbers of individual enemies to be used, how they should be distributed, and when they should be released.  Researchers in the People's Republic of China seem to have developed the necessary information for their release of *Trichogramma* against rice and sugarcane pests (NAS 1977 and Huffaker, Chapter 10, this tristise).

As noted previously, mathematical modeling may provide useful insights to the ecological constraints of augmentation programs. Care must be taken, however, in applying models to specific ecological situations.  Some models are of such naive ecological construction (Knipling and McGuire 1968; Knipling 1971) that the resulting inferences are of questionable value.  In fact, a prior judgment of the effectiveness of the biological control agent in a specific interaction based on mathematical reasoning is at best tenuous.  The true power of modeling is best realized only after the <u>potential</u> effectiveness of a biological control agent has been established in the field.  In this case, a posteriori situation, models may indeed provide insights to both the reasons for failure in control and, as well, the possible results of specific augmen-

tation tactics that may be used.  For example, Tanner (1975)
and May (1971, 1976) have shown that a wide class of predator-
prey models may exhibit higher stable point or stable limit cycle
behavior, depending on the relationship between the predator and
prey population growth rates, while Murdoch and Oaten (1975) have
emphasized effects of prey switching.  In other words, if environmental vagaries differentially affect predator and prey dynamics,
an interaction which provides adequate "natural" control in one
situation (stable point or stable limit cycling entirely below the
economic threshold) may not do so in another (unstable or stable
limit cycling with prey densities fluctuating above the economic
threshold); augmentation may be required (Logan 1977).

    Models also may be useful for timing application of augmen-
tation tactics and evaluation of field trials.  For example, in
the system depicted in Figure 6, which illustrates a stable limit
cycle, the prey isocline (Rosenzweig 1969 and 1971) of a predator-
prey interaction provides a measure of the augmentation required
to avoid damaging pest densitites.  The two population densities
will cycle in a counter-clockwise direction and will tend to

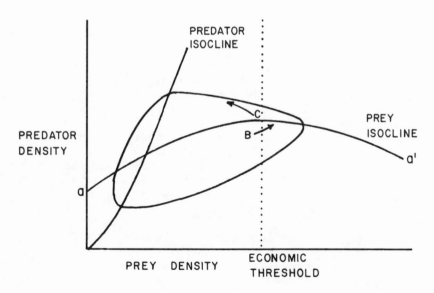

Figure 6.  Trajectory of a stable limit cycle for a predator-
prey interaction (see text).

return to the same cycle after perturbations to the system.
To avoid economic damage, augmentation (addition of predators)
is required at prey density A, i.e., to a density above the
prey isocline (a, a'). Increasing the predator density from
a lower level only to point B would result in both populations
cycling to the right (arrow into the zone of damaging prey
densities); with increase of the predators to density C, the
resultant trajectory would turn to the left, away from the
economic threshold for the prey.

REFERENCES CITED

Andrewartha, H. G., and L. C. Birch. 1954. The Distribution
    and Abundance of Animals. Univ. Chicago Press. 782 pp.
Bodenheimer, F. S. 1928. Welche factoren Regulieren die Indivi-
    duenzahl einer Insektenart in die Natur? Biol. Zentrabl.
    48:714-39.
Chant, D. A. 1964. Strategy and tactics of insect control.
    Can. Entomol. 96:182-201.
Clark, L. R., P. W. Geier, R. D. Hughes, and R. F. Morris. 1967.
    The Ecology of Insect Populations in Theory and Practice.
    Methuen & Co. Ltd., London. 232 pp.
Cody, M. L., and J. M. Diamond (Eds.). 1975. Ecology and Evolu-
    tion of Communities. The Belknap Press of Harvard Univ.
    Press, Cambridge, Mass. 545 pp.
DeBach, P., and K. S. Hagen. 1964. Manipulation of entomophagous
    species. *In* Biological Control of Insect Pests and Weeds.
    Ed.: DeBach, P. pp. 429-58. Reinhold, N.Y.
DeBach, P., C. B. Huffaker, and A. W. MacPhee. 1976. Evaluation
    of the impact of natural enemies. *In* Theory and Practice
    of Biological Control. Eds.: Huffaker, C. B., and P. S. Mes-
    senger. pp. 255-85. Academic Press, N. Y.
Doutt, R. L. 1964. Biological characteristics of entomophagous
    adults. *In* Biological Control of Insect Pests and Weeds.
    Ed.: DeBach, P. pp. 145-67. Reinhold, N. Y.
Fransz, H. G. 1974. The Functional Response to Prey Density in
    an Acarine System. Pudoc, Wageningen. 143 pp.
Frazer, B. D., and N. Gilbert. 1976. Coccinellids and aphids:
    A quantitative study of the impact of adult ladybirds (Coleop-
    tera: Coccinellidae) preying on field populations of pea
    aphids (Homoptera: Aphididae). J. Entomol. Soc. Brit. Co-
    lumbia. 73:33-56.
Gause, G. F. 1934. The Struggle for Existence. Williams and Wil-
    kins, Baltimore, Md.
Gilbert, N., A. P. Gutierrez, B. D. Frazer, and R. E. Jones. 1976.
    Ecological Relationships. W. H. Freeman and Co., Reading and
    San Francisco, 156 pp.
Gossard, T. W., C. G. Summers, and A. P. Gutierrez. In Press.
    Some factors affecting food intake by *Hippodamia convergens*

Guerin-Meneville (Coleoptera: Coccinellidae).

Gutierrez, A. P., J. B. Christensen, C. M. Merritt, W. B. Loew, C. G. Summers, and W. R. Cothran. 1976. Alfalfa and the Egyptian alfalfa weevil (Coleoptera: Curculionidae). Can. Entomol. 108:634-48.

Gutierrez, A. P., L. A. Falcon, W. Loew, P. A. Leipzig, and R. van den Bosch. 1975. An analysis of cotton production in California: A model for Acala cotton and the effects of defoliators on its yields. Environ. Entomol. 4:125-36.

Gutierrez, A. P., Y. Wang, D. W. DeMichele, R. Skeith, and L. G. Brown. In Press. The systems approach to research and decision making for cotton pest control. *In* New Technology of Pest Control. Ed.: Huffaker, C. B.

Hassell, M. P. 1975. Density-dependence in single-species populations. J. Anim. Ecol. 44: 283-95.

Hassell, M. P. 1976. The Dynamics of Competition and Predation. Instit. Biol. Studies in Biology No. 72. Edward Arnold. 68 pp.

Hassell, M. P., and G. C. Varley. 1969. New inductive model for insect parasites and its bearing on biological control. Nature (London) 223:1133-6.

Holling, C. S. 1961. Principles of insect predation. Ann. Rev. Entomol. 6:163-82.

Holling, C. S. 1966. The functional response of invertebrate predators to prey density. Mem. Entomol. Soc. Can. 48:3-86.

Howard, L. O., and W. F. Fiske. 1911. The importation into the Unites States of the parasite of the gypsy moth and the brown tail moth. U. S. Dept. Agric. Bur. Entomol. Bull. 91. 312 pp.

Huffaker, C. B. 1958. The concept of balance in nature. Proc. 10th International Congress of Entomology. (Vol. 2, pp. 625-36).

Huffaker, C. B. 1963. Some contributions of biological control to concepts of population dynamics. Proc. 1st Intern. Conf. Wildlife Diseases, High View, N.Y., 1962. (Micro-card).

Huffaker, C. B. 1974. Some ecological roots of pest control. Entomophaga 19:371-89.

Huffaker, C. B., and D. L. Flaherty. 1966. Potential of biological control of two spotted spider mites on strawberries in California. J. Econ. Entomol. 59:786-92.

Huffaker, C. B., and C. E. Kennett. 1966. Studies of two parasites of olive scale, *Parlatoria oleae* (Colvée). IV. Biological control of *Parlatoria oleae* (Colvée) through the compensatory action of two introduced parasites. Hilgardia 37:283-335.

Huffaker, C. B., and P. S. Messenger. 1964. The concept and significance of natural control. *In* Biological Control of Insect Pests and Weeds. Ed.: DeBach, P. pp. 74-117. Reinhold Publ. Corp., N. Y.

Huffaker, C. B., C. E. Kennett, and G. L. Finney. 1962. Biological control of olive scale, *Parlatoria oleae* (Colvée), in California by imported *Aphytis maculicornis* (Masi) (Hymenoptera: Aphelinidae). Hilgardia 32:541-636.

Huffaker, C. B., R. F. Luck, and P. S. Messenger. 1977. The ecological basis of biological control. Proc. XV Intern. Congr. Entomol., Washington, D.C., Aug. 19-27, 1976. pp. 560-86.

Huffaker, C. B., P. S. Messenger, and P. DeBach. 1971. The theory, ecological basis and assessment of biologiccal control. *In* Biological Control. Ed.: Huffaker, C. B. pp. 16-67. Plenum Press, N.Y.

Huffaker, C. B., K. P. Shea, and S. G. Herman. 1963. Experimental studies on predation: (III) Complex dispersion and levels of food in an acarine predator-prey interaction. Hilgardia 34:305-30.

Huffaker, C. B., F. J. Simmonds, and J. E. Laing. 1976. The theoretical and empirical basis of biological control. *In* Theory and Practice of Biological Control. Eds.: Huffaker, C. B., and P. S. Messenger. pp. 41-78. Academic Press, N.Y.

Janzen, D. H. 1968. Host plants as islands in evolutionary and contemporary time. Amer. Nat. 102:592-5.

Kennedy, J. S. 1972. The emergence of behavior. J. Aust. Entomol. Soc. 11:168-76.

Knipling, E. F. 1971. Use of population models to appraise the role of larval parasites in suppressing *Heliothis* populations. USDA Tech. Bull. 1434:1-36.

Knipling, E. F., and J. U. McGuire, Jr. 1968. Population models to appraise the limitations and potentialities of *Trichogramma* in managing host insect populations. USDA Tech. Bull. 1387: 1-44.

Krebs, C. J. 1972. Ecology: The Experimental Analysis of Distribution and Abundance. Harper and Row, N.Y. 694 pp.

Laing, J. E., and C. B. Huffaker. 1969. Comparative studies of predation by *Phytoseiulus persimilis* Athias-Henriot and *Metaseilus occidentalis* (Nesbitt) (Acarina, Phytoseiidae) on populations of *Tetranychus urticae* Koch (Acarina: Tetranychidae). Res. Popul. Ecol. 11:105-26.

Lotka, A. J. 1925. Elements of Physical Biology. Williams and Wilkins, Baltimore, Md. 462 pp.

Logan, J. A. 1977. Population model of the association of *Tetranychus medanieli* (Acarina: Tetranychidae) with *Metaseiulus occidentalis* (Acarina: Phytoseiidae) in the apple ecosystem. Ph.D. Dissertation, Wash. St. Univ. 137 pp.

MacArthur, R. H., and E. O. Wilson. 1967. The Theory of Island Biogeography. Monographs in Population Biology. Princeton Univ. Press. Princeton, N.J. 203 pp.

MacFadyen, A. 1963. Animal Ecology, Aims and Methods. 2nd ed. Isaac Pitman and Sons, London. 344 pp.

May, R. M. 1971. Stability in multi-species community models. Math. Biosci. 12:59-79.

May, R. M. 1974. Stability and Complexity in Model Ecosystems. 2nd ed. Princeton Univ. Press, Princeton, N.J. 265 pp.

May, R. M. (Ed.). 1976. Theoretical Ecology: Principles and Ap-
    plications. W. B. Saunders Co., Philadelphia. 317 pp.
Messenger, P. S. 1976. Theory underlying introduction of exotic
    parasitoids. *In* Perspectives in Forest Entomology. Eds.:
    Anderson, J. F., and H. K. Kaya. pp. 191-214. Academic Press.
    N.Y.
Milne, A. 1957. Theories of natural control of insect populations.
    Cold Spring Harbor Symp. Quant. Biol. 22:253-71.
Murdoch, W. W. and A. Oaten. 1975. Predation and population sta-
    bility. *In* Advances in Ecological Research. Ed.: MacFadyen, A.
    9:1-131.
National Academy of Sciences. 1977. Insect Control in tne People's
    Republic of China. CSCPRC Rept. No. 2. National Academy of
    Sciences, Washington, D.C. 218 pp.
Nicholson, A. J. 1933. The balance of animal populations. J.
    Anim. Ecol. 2:132-78.
Nicholson, A. J. 1954. An outline of the dynamics of animal popu-
    lations. Aust. J. Zool. 2:9-65.
Nicholson, A. J., and V. A. Bailey. 1935. The balance of animal
    populations. Proc. Zool. Soc. Lond. Pt. 1, pp. 551-98.
Price, P. W. 1975. Insect Ecology. John Wiley and Sons, N.Y.
    514 pp.
Price, P. W., and G. P. Waldbauer. 1975. Ecological aspects of
    pest management. *In* Introduction to Pest Management. Eds.:
    Metcalf, R. L., and W. Luckman. pp. 37-73. John Wiley and
    Sons, N. Y.
Rabb, R. L., R. E. Stinner, and G. A. Carlson. 1974. Ecological
    principles as a basis for pest management in the agroecosys-
    tem. *In* Proceedings of Summer Institute on Biological Con-
    trol of Plant Insects and Diseases. Eds.: Maxwell, F. G.,
    and F. A. Harris. pp. 19-45. University Press of Mississip-
    pi, Jackson.
Rabb, R. L., R. E. Stinner, R. van den Bosch. 1976. Conservation
    and augmentation of natural enemies. *In* Theory and Practice
    of Biological Control. Eds.: Huffaker, C. B. and P. S. Mes-
    senger. pp. 233-54. Academic Press, Inc., N.Y.
Regev, U., A. P. Gutierrez, and G. Feder. 1976. Pests as a com-
    mon property resource: a class study of alfalfa weevil con-
    trol. Am. J. Agr. Econ. 58:188-96.
Rosenzweig, M. L. 1969. Why the prey curve has a hump. Amer.
    Natur. 103:81-7.
Rosenzweig, M. L. 1971. Paradox of enrichment destabilization
    of exploitation ecosystems in ecological time. Science.
    171:385-7.
Royama, T. 1971. A comparative study of models for predation
    and parasitism. Res. Popul. Ecol. Suppl. 1:1-91.
Ruesink, W. G. 1976. Modeling of pest populations in the alfalfa
    ecosystem with special reference to the alfalfa weevil. *In*
    Modeling for Pest Management. Eds.: Tummala, R. L., D. L.
    Haynes, and B. A. Croft. Michigan State Univ. Press, East

Lansing, Michigan.

Shoemaker, C. A. 1977. Crop ecosystems models for pest manage-
    ment. *In* Ecosystem Modelling in Theory and Practice. Eds.:
    Hall, C. A., and J. Day. Wiley-Interscience, N.Y.

Shoemaker, C. A., C. B. Huffaker, and C. E. Kennett. In press.
    A systems analysis of olive pest management.

Smith, H. S. 1935. The role of biotic factors in the determination
    of population densities. J. Econ. Entomol. 28:873-98.

Solomon, M. E. 1949. The natural control of animal populations.
    J. Animal Ecol. 18:1-35.

Solomon, M. E. 1969. Population Dynamics. Inst. Biol. Studies
    in Biology. No. 18. Edward Arnold. 60 pp.

Southwood, T. R. E. 1966. Ecological Methods with Particular
    Reference to the Study of Insect Populations. Methuen and
    Co., Ltd. London, 391 pp.

Southwood, T. R. E. 1975. The dynamics of insect populations.
    *In* Insects, Science and Society. Ed.: David Pimentel.
    pp. 151-91. Academic Press, Inc., N.Y.

Southwood, T. R. E. 1977. Entomology and mankind. Amer. Scien-
    tist. 65:30-9.

Southwood, T. R. E., R. M. May, M. P. Hassell, and G. R. Conway.
    1974. Ecological strategies and population parameters. Amer.
    Natur. 108:791-804.

Southwood, T. R. E., and M. J. Way. 1970. Ecological background
    to pest management. *In* Concepts of Pest Management. Eds.:
    Rabb, R. L., and F. E. Guthrie. pp. 6-28. N. C. State Univ-
    ersity, Raleigh, N.C.

Stinner, R. E. 1977. Efficacy of inundative releases. Ann. Rev.
    Entomol. 22:515-31.

Stinner, R. E., and H. L. Lucas, Jr. 1976. Effects of contagious
    distributions of parasitoid eggs per host and of sampling vaga-
    ries on Nicholson's area of discovery. Res. Popul. Ecol.
    18:74-88.

Tanner, J. T. 1975. The stability and the intrinsic growth rates
    of prey and predator populations. Ecology. 56:855-67.

Uvarov, B. P. 1931. Insects and climate. Trans. Entomol. Soc.
    London. 79:1-247.

Van den Bosch, R., and A. D. Telford. 1964. Environmental modi-
    fication and biological control. *In* Biological Control of In-
    sect Pests and Weeds. Ed.: DeBach, P. pp. 459-88. Reinhold,
    N.Y.

Volterra, V. 1926. Variazoni e fluttuazioni del numero d'individui
    in specie animali conviventi. Atti accad. naz. Lincei Memorie.
    Cl. di. sci. fis. mat. nat. 2:31-112.

Way, M. J. 1968. Intra-specific mechanisms with special reference
    to aphid populations. *In* Insect Abundance. Ed.: Southwood, T.
    R. E. 4:18-36. Roy. Entomol. Soc. Lond. Symp.

Wellington, W. G. 1977. Returning the insect to insect ecology:
    Some consequences of pest management. Environ. Entomol. 6:1-86.

Woodworth, C. W. 1908. The theory of the parasite control of in-
    sect pests. Science (N.S.) 28:227-30.

CHAPTER 2

ECOLOGICAL PRINCIPLES ASSOCIATED WITH AUGMENTATION OF NATURAL ENEMIES

E. M. Shumakov

All-Union Scientific Research Institute of Plant
Protection, Pushkin 6, 188620, Leningrad, U.S.S.R.

The concept of controlling pests of agricultural crops is a
complex ecological and biocenological problem which acquired basic
importance in the 1940's, beginning with the arsenal of chemical
control agents reinforced by synthetic compounds with a broad range
of effects.

The consequences of massive use of insecticides, which affected
not just individual species but whole societies of living organisms,
led unavoidably to the establishment of a new direction in plant
protection, i.e. integrated control, which marked a biocenotic
approach to the problem that is different in principle.

As Newsom (1967) emphasized: "The negative consequences of the
wide use of universal synthetic compounds revealed the exceptionally
important effect of natural enemies on the numbers of harmful arthropods. Massive outbreaks of pests accompanied by the destruction of
predators and parasites forced applied entomologists to recognize
the important role of natural enemies to a greater degree than previously considered.

However, the present state of research on this problem does
not yet permit the discussion of the existence of a single theory
of the population dynamics of insects. The efforts of a number of
ecologists and biocenologists working on the problem of biocontrol
frequently produce data that are difficult to compare and develop
views that are not compatible. Undoubtedly, success of the work in
the field of biocontrol and continuity of investigations are possible
when there are 1) a broad theoretical grasp of the phenomena of the
population dynamics of animals, 2) scrupulous work based on

discovery, and 3) detailing of the cause and effect connections in the complex structures of specific biocenoses.

Among the numerous factors that influence the population dynamics of pests, the role of biotic factors (natural enemies, entomophages) has been subjected to the most careful investigations and has been discussed repeatedly (Southwood 1968, Viktorov 1955, 1960a, 1960b, 1963, 1964a, 1965, 1967, 1969a, 1969b, 1970, 1971, 1975, 1976, Huffaker et al. 1971).

The role of predators and parasites in the fluctuation of the entomophage population has been recognized by some authors as being of major importance, while their role has been denied by others, or has not been taken into account. One of the main reasons for the contradictory points of view has been an insufficient study of the complexes of natural enemies and the specificity of their relationships with different species of harmful insects. Not infrequently the conclusions arrived at in a study of any one or a few similar organisms are generalized and extended to a wide circle of organisms, and universal significance is imparted to these conclusions.

As a result of such a situation the necessity of developing a classification for the factors of population dynamics became obvious. Thus there arose the concept of the occurrence of factors both independent of and dependent on population density (Smith 1935). Acceptance of this alternative implies at the same time acceptance of the concept of automatic regulation of the density of organisms in nature. The appearance and development of these views is associated with the names of Woodworth (1908), Howard and Fiske (1911), Nicholson (1933), Smith (1935), Varley (1947, 1953), Solomon (1958), and later, Huffaker and Messenger (1964), and Huffaker and DeBach (1971); in the U.S.S.R. work in this direction has been done by G. A. Viktorov, as mentioned previously.

The most effective regulation is accomplished when there are reciprocal, density-dependent relationships of entomophages with their hosts (Huffaker and Messenger, 1964). In other words, in the evolutionary process of a biocenosis, relationships between its members that are in related trophic levels are reciprocal and negative. "This means that the host is regulated by its own enemy, and its enemy in turn is limited by host numbers. This means that a phytophagous pest is limited by a predator or parasite to a greater extent than by food." (Huffaker et al 1971.)

Fundamentally, influences modifying pest or entomophage density are of another character. These are influences that are incidental to the population, not connected with its density by a reciprocal relationship. A modifying influence is exerted on the density by climatic factors that act on the organism both directly and

indirectly through a system of biocenotic relationships.  Under
certain conditions, particularly when the natural self-regulating
relationships are disrupted, entomophages can act as a modifying
factor (Viktorov 1965, 1967).

At the present time investigators are being attracted to
investigate the mechanisms of population regulation both within
species and within biocenoses.  Especially interesting is the
existence of thresholds and zones of activity of the basic mechanisms
of regulation of insect populations.  Correlating factual information, G. A. Viktorov (1967, 1969) created a stepwise scheme of regulation (which corresponds to the principle of ultrastability in
cybernetics).

At present data have been obtained that indicate that any group
of populations of a phytophagous species experiences a system of
regulating mechanisms, and that in this process "each mechanism of
regulation can be characterized by definite thresholds and zones of
activity" (Viktorov 1969).  The lowest threshold of activity and
the narrowest zone of activity with respect to the population density of the host are characteristic of polyphagous entomophages.
They are not capable of numerical reaction (Solomon 1949) and do
not respond by an increase in their own numbers to an increase in
the density of the host.

Specialized parasites and predators also are characterized by
a low threshold of activity, and they are capable of regulating
their hosts for a long time at low densities.  This group is distinguished not only by a functional response, i.e. an increase in
the quantity of attacked hosts (Solomon 1969), but also by a
numerical response.  It is precisely this characteristic that guarantees the appearance of their regulatory role over a wider interval
of population density of the host and makes possible, for example,
the suppression of outbreaks of the latter.

At a high level of population densities the regulatory action
of pathogenic organisms may appear.  When the population density
reaches values associated with the destruction of food resources
leading to starvation and mutual depression, the action of a terminal
regulating mechanism, intraspecies competition, sets in which averts
extinction of the population.

However, in a number of species of hosts there is an intraspecies mechanism of self-regulation that can avert the appearance
of intraspecies competition.  It is based on the warning action of
a growing population density and goes into action before the onset
of a food shortage.  For example, this mechanism stimulates the
production of winged individuals and migratory flight in aphids,
deposition of unfertilized eggs that produce males in Hymenoptera

and the formation of gregarious and migratory phases in some species
of locusts and Lepidoptera.

G. A. Viktorov points out that in a complex of mechanisms the
imperfection of each mechanism is compensated for and thus "the
stable existence of a population is guaranteed even in a constantly
changing environment."

However, such a universal classification of factors has proved
to be insufficiently practical when attempts are made to use it in
the analysis of the population dynamics of a specific species of
pests.  Some gap has arisen between the general theoretical position and the comparatively scanty factual information on the ecology
and biocenology of most harmful species and their entomophages.

A study of the population dynamics of such subjects as locusts,
the harmful stinkbug *Eurygaster integriceps* Puton, the genus
*Heliothis*, the codling moth, *Laspeyresia pomonella* (L.), and others
reveals highly specific characteristics of variation of their numbers and a completely different role of individual factors depending
on concrete ecologico-geographic and agrocenotic conditions under
which the populations of the given species of pest exist.  The controlling activity of entomophagous insects as a factor in the population dynamics of phytophagous insects also is subject to the
influence of various conditions, and thus it has a different level
of importance in the population dynamics of different pest species.
An analysis of the ecological and biocenotic situations in which a
number of serious pests are found in the territory of the U.S.S.R.
leads to this conclusion.

In the years after the Second World War the harmful stinkbug
(*Eurygaster integriceps* Puton) became a pest of primary importance
in the south of the European part of the U.S.S.R.  Periods of
massive multiplication of this species and high numbers of it alternated with short-lived declines, which seldom reached depressed
levels.  As noted by G. A. Viktorov (1967), "in the postwar years
the population dynamics of the harmful stinkbug underwent substantial
changes in some parts of our country.  From a pest with irregularly
periodic waves of increase and sharp decline, it became a chronic
pest, the numbers of which vary on a significantly higher level,
surpassing the economic threshold."

In the nation's center of entomological research, the study of
the biology, physiology and ecology of this pest has been expanded;
considerable attention also has been given to its entomophages.
Work since 1940 in the field of biocontrol is characterized not only
by a variety of directions, many of them ecological, but also by
clearly expressed tendencies on the one hand toward the recognition
of fine, sometimes very complex characteristics of parasite-host

relationships, and on the other hand toward the enrichment of knowledge. This knowledge has been used for 1) understanding the process of parasite development, 2) defining concretely the theory of population dynamics of the species, and 3) determining the place and role of entomophagous insects in this phenomenon.

Study of the entomophagous insects of the harmful stinkbug required first of all the establishment of the species affiliation of the harmful stinkbug parasites in the family Scelionidae (Hymenoptera). The status of the taxonomy of this group was far from satisfactory. The needs of the theory and practice of bio-control stimulated the development of taxonomic and diagnostic work on oophagous insects. Of the Soviet investigators, Viktorov (1964, 1967) and Kozlov (1968) made a substantial contribution to the taxonomy of the Scelionidae. With the development of the taxonomy of the Scelionidae, the lists of species of this group that are limited in some degree to development in the body of the pest changed. At the present time the principal species of parasites are already known from the main habitats of the harmful stinkbug and from a number of other Pentatomidae, but the process of finding new species revealed as oophagous insects of the stinkbug continues. In the territory of the U.S.S.R., the harmful stinkbug is attacked by 10 species of oophagous insects and 4 species of parasitic tachinid flies of the subfamily Phasiinae (Shchepetil'nikova 1958, Viktorov 1964, 1967, Kozlov 1968). The dipteran parasites of the harmful stinkbug, *Eurygaster integriceps* Puton include 4 species of tachinid flies of the subfamily Phasiinae, namely the golden phasiid (*Eliozeta* [=*Clytiomyia*] *helluo* F.), gray phasiid (*Alophora* [=*Phasia*] *subcoleoptrata* (L.)), variegated phasiid (*Phasia* [=*Ectophasia*] *crassipennis* F.) and the black phasiid (*Helomyia lateralis* (Meigen)). The golden and gray phasiids are oligophagous; the variegated and black are polyphagous. The tachi-nids, along with the oophagous Telenomus parasites, play a very important role in the population dynamics of the harmful stinkbug. They are parasites of the adults and not only stop the maturation of eggs in the female, but also sharply limit its period of active oviposition. A purposeful investigation of the phasiids was started in the U.S.S.R. only in the middle of our century, when Fedotov (1944, 1947), Rubtsov (1945), and Chernova (1947) began to study phasiids under the conditions of Central Asia. On the basis of a wide and lengthy investigation of the phasiids carried out in the Krasnodar area from 1950-1953, it was established that the 4 species of phasiids mentioned above parasitize the harmful stinkbug. These species differ from each other biologically in all stages of develop-ment (Shumakov 1958). With the aid of determinative tables of endo-parasites of the harmful stinkbug that were drawn up by Rodendorf (1947), it is now possible to easily determine tachinids with respect to their immature stages, which are found upon inspection and dissection of the host. Shumakov (1958) has developed a guide to

the puparia.  Mention must be made of a whole series of works by
Dupuis (1948, 1963) where the author, along with a study of the
taxonomy and biology, introduces new generic names: *Phasia Latreille*
for *Alophora subcoleoptrata* and *Ectophasia* Townsend for *Phasia
crassipennis* (F.).

Investigation of the predators of the harmful stinkbug has
only started.  It has met with considerable methodological difficulties both in the establishment of trophic relationships of the
predators with the stinkbug and in the quantitative evaluations of
the results of the activity of the predators.  Development of knowledge in the biocenotic relationships of the predators and stinkbug
became possible through the use of serological methods of investigation, labeling of insects with radioisotopes, and the use of
different methods of catching insects along with mathematical
analysis.

At present it has been found that many polyphagous predatory
insects and spiders are trophically associated with the harmful
stinkbug.  In 2 regions of the steppe and forest-steppe zones of
the U.S.S.R., 150 species of predators have been discovered; of
them are 90 species of ground beetles (*Carabidae*) (belonging to
35 genera), 45 species of spiders (of 14 families), predatory bugs,
lacewings (*Chrysopidae*), several species of the family Staphylinidae,
ant-like beetles (*Anthicidae*), and a daddy longlegs (*Phalangida*);
among these, feeding on larvae is more widespread, while the adult
insects are subject to attack relatively more rarely (Burov et al.
1974).  Feeding on eggs of the stinkbug also is noted; thus for one
of the regions more than 25 species of predators have been reported
as consumers of eggs (Antonenko 1971).

On wheat fields all the dominant and constant species of ground
beetles and spiders are associated with the harmful stinkbug (Zaeva
1971, Burov et al. 1974).  In the central belt and in the south
European part of the U.S.S.R. on wheat plantings, in the main centers
of massive multiplication of the harmful stinkbug, 31 out of 33
species of beetles of the family Carabidae are regular components
of the biocenosis of wheat fields.  They are characterized by ecological variety; in populations of the dominant species considerable
elevations in numbers are observed that are synchronous with 2
periods; with egg laying of the harmful stinkbug and hatching of its
larvae and with the appearance of larvae of the older instars and
the young insects (Zaeva and Kupershtein 1971).  For the widely
distributed and numerous *Pterostichus crenuliger* Chaudoir it has
been established that in the period of predominance of third-instar
larvae of the harmful stinkbug, when there is a certain absolute
number of predators and prey, the mortality caused by the predator
amounts to 0.4-0.5% of the population of the pest per day.  In this
instance a functional relationship of the level of consumption of

the prey to the level of its numbers has been discovered (Kupershtein 1974).

Thus, the few investigations still devoted to predators
show that there is an important resource in the biological control
of the harmful stinkbug, of which the evaluation of possibilities
has only begun.  It is very probable that the use of this group
of insects in developing an integrated method of lowering the
numbers of the stinkbug can substantially supplement the activity
of parasites.

The population dynamics of the harmful stinkbug has undergone
substantial changes in some parts of the U.S.S.R.  From a pest
with irregular waves of multiplication and sharp decline, it has
become a chronic pest, the numbers of which vary on a level considerably exceeding the threshold of economic importance (Viktorov
1967).

In the nation's entomological research centers, study of the
biology, physiology, and ecology of the harmful stinkbug has been
expanded for the purpose of finding the factors determining the
population dynamics of the pest, the development of long-term
forecasting, and the founding of rational measures for control.

During periods of development of a system of control measures
for the harmful stinkbug in the U.S.S.R. (Starostin and Burov 1976)
it is interesting to trace the general changes in the level of
economic importance of the pest (reflecting basically not only
short-term, but also long-term measures) and the dynamics of chemical and biological control.

The exterminatory measures of the first period (1936-1953),
which were directed mainly against overwintered bugs, were very
incomplete.  At that time work had only begun on evaluating the
possibilities of the biological control.  Directions were laid for
investigations, and ecological approaches to the problem were outlined.

In the following period of control development (1953-1968)
the chemical control method prevailed without limit.  Pesticides
with a broad spectrum of action appeared at the disposal of entomologists.  The reduction of quantitative losses, the substantial
lessening of damage of grain and the preservation of its quality
became a reality.  In the field of biocontrol at that time the
accumulation of factual information on the biology and ecology of
entomophagous insects of the harmful stinkbug was going on.  The
action of contemporary agrotechnology and chemical measures of control against the stinkbug on the local beneficial fauna was being
studied.

Some authors evaluating this period see it in a number of positive aspects and affirm that although the harmful activity of the stinkbug widened, stabilization occurred, and then the density of its populations even decreased, thanks to the wide use of insecticides (Starosti and Burov 1976).

G. A. Viktorov, analyzing the vast domestic information on the harmful stinkbug and its parasites, sees the causes of the conversion of the pest from a periodically important one to a chronic and primary one in the disruption of the biocenotic relationships of the stinkbug with its parasites, which were potentially capable of controlling the numbers of the pest automatically, or at least as well as the pesticides did. G. A. Viktorov (1967) points out the difficulty of the chemical method alone in solving the problem resulting in a steady reduction of the pest population. The widely practiced spring use of toxicants against overwintering insects had consequences the reverse of those expected; the subsequent generation of insects increased in numbers (Viktorov 1960, 1964a; Areshnikov 1967, Fedchenko and Yarynkina 1968, Zaeva 1969). In this case, massive destruction of overwintering populations of parasites and predators considerably decreased the quantity of stinkbug eggs that perished. Thus, according to a number of observations, the infestation by oophagous insects decreased 33-67% (Shchepetil'nikova 1958, Ryakhovskii 1959, Viktorov 1960, Zaeva 1969, Antonenko 1971), and phasiids decreased 50% (Antonenko 1971); in a number of cases the eggs of the harmful stinkbug were not parasitized at all over the course of some time (Zaeva 1969). The number of predators also decreased substantially, 50-83% (Antonenko 1971). In years when there was a depression of the stinkbug population, the coefficient of reproduction of the insects on fields not treated with insecticides varied from 1 to 6 and was 50-92% lower than on treated fields (Antonenko 1971).

In spite of the wide spectrum of predators and parasites of the harmful stinkbug, the principal attention of specialists with respect to biological control has been concentrated for a long time on the study of the oophagous parasites. Their ability to parasitize a considerable portion of the eggs laid by the harmful stinkbug was pointed out long ago and repeatedly. At the end of the past century and the beginning of ours, data were reported for different regions of the country on the parasitization of 85-95% of the eggs of the pest. One of the latest publications reports data on 90% parasitization of eggs collected in the field in the Voronezh district (Zatyamina et al. 1976). However, the parasites are seldom so potent. On vast massive plantings of wheat, distant from forest belts, forests, and orchards, the limiting action of parasites on the stinkbug usually is found to be at a low level or almost without effect. Chemical treatments also are a factor that strongly lowers or nullifies the effect of parasites.

This circumstance stimulated the development of investigations of the causes of variations in the economic importance of oophagous insects of the harmful stinkbug in agrocenoses and the development of the use of native and introduced parasites. The investigations depended on both the effects of abiotic factors on the complex of parasitic species and the biocenotic relationships of this complex not only with the stinkbug but also with a number of other species of Pentatomid bugs. As a result it was established that the harmful stinkbug as a species has a comparatively wide complex of parasitic species associated with it. This complex of parasites affects different populations of the stinkbug in different regions of the country in different ways.

The nature of the effect of abiotic factors on oophagous insects was established both by direct laboratory experiments and by an indirect route by means of an analysis of the geographic distribution, seasonal distribution, and habitat of the species.

The action of abiotic factors on the "phytophagous-parasite complex" is not identical for its members. On the whole the harmful stinkbug is more flexible, or in any case more resistant to extreme conditions of humidity and low temperatures, a fact which is associated with both the physiological properties and over-wintering characteristics of the stinkbug.

A differentiated relationship to environmental factors (Rubtsov 1937, 1938), a partial lack of convergence, and a divergence of optimal conditions for the host and its parasite (Telenga 1953) constitute a phenomenon widely distributed in nature; together they are regarded as a mechanism that averts complete coincidence of the spatial distribution of parasites and host (Shchepetil'nikova 1957, Viktorov 1970) and thus guarantees a stable existence of the system (Flanders and Badgley 1963).

Within the limits of the range of the harmful stinkbug in the territory of the U.S.S.R., in geographic regions that differ with respect to climatic conditions, the specificity both of the species composition of the oophagous insects and also of the relative abundance of their species appears perfectly clearly (Viktorov 1967). There are comparatively more complete data for the 2 widely distributed and numerous species, *Trissolcus grandis* Thomson, and *Telenomus chloropus* Thomson (=*sokolovi* Mayr).

The more or less stable dominance of either the first or second species is characteristic of different regions of the country. The shift in dominant species that is observed in some cases is associated, as a rule, with sharp yearly changes in climatic factors, especially humidity, and sometimes temperature (Viktorov 1967).

*Trissolcus grandis*, the most widely distributed and numerous parasite of the harmful stinkbug in our country (Shchepetil'nikova 1958), predominates in the comparatively arid regions, while the dominance of *Telenomus chloropus* is confined to territories that are more humid during the growing period (Viktorov 1967). It has been established experimentally that in the period of preimaginal development, which takes place in the eggs of the harmful stinkbug, the two species reveal approximately the same relationship to the temperature factor and a different relationship to humidity (Viktorov 1969). *T. grandis* is a xerophilic species, while *chloropus* is to a considerable extent hygrophilic. The adult females preserve these characteristics; at low humidity the individuals of *T. chloropus* die, while the resistance of *T. grandis* is twice as great (Viktorov 1969).

Thus, the key factor was revealed that determines the distribution of these 2 main species of oophagous insects for the U.S.S.R. not only with respect to geographical regions of the country, but also with respect to habitat.

However, it is not rare for a break to take place in these processes at a time that is dependent on climatic factors unfavorably affecting the viability of the parasites. Sometimes the break between the beginning of oviposition by the stinkbug and the appearance of the first parasitization amounts to 2 weeks (Shchepetil'nikova 1958) or a month (Meier 1949). Consequently, the dates of mass oviposition by the oophagous insects and by the host are considerably separated (in the case of a 2-week break), and the maximum oviposition by the parasite occurs during its second generation. In this case a considerable portion of the eggs laid by the stinkbug in the period of massive oviposition do not undergo parasitization. This period of active oviposition (Shumakov and Vinogradova 1958) or of effective oviposition (Viktorov 1967) is very essential in determining the numbers of the new generation of the pest.

The thermal factor determines the times when the adult individuals of the parasite go into hibernation (Shchepetil'nikova 1958) and when they safely complete the period of hibernation. The degrees of cold-resistance of the oophagous insects and the harmful stinkbug approximately coincide. But their winter hardiness is different; the harmful stinkbug is protected from the effect of low temperatures to a considerably greater degree than the oophagous insects, which undergo the winter in less secure shelters than the stinkbug.

Thus, with respect to abiotic factors of the environment, or at least to humidity and temperature, each species of oophagous insect is more stenobiotic than the host, the harmful stinkbug

(Viktorov 1976). But the set of species of oophagous insects that enter into the complex of parasites of the harmful stinkbug in any certain region has a wider amplitude of ecological properties than each of these species individually. In other words, the complex of parasites associated with the harmful stinkbug represents an ecological spectrum that can guarantee a broader reaction to changes in the environment than an individual species and in this way preserve the effect on the pest. We should view from this aspect a number of ecological and ethological properties of the parasitic complex of the harmful stinkbug which determine both the nature of the relationships of the parasites with the host and the degree of effect of the complex on the population density of the pest.

Among these properties we may note and consider the following:

1) Spatial changes in the species composition of the parasite complex which are expressed in its specific nature for more or less vast regions of the country with unique climatic characteristics. The status of the species composition can be expressed by an index of relative abundance of each species, and hence also by the one that predominates among them and determines the nature of the effect on the host.

2) Temporary changes in the species composition according to years and lengths of growing seasons, which determine the level of action of the parasitic complex in a given locality during specific summer periods.

3) The nature of the distribution of various species of the complex of parasites with respect to plantings of grains and row crops, which affects both the parasitization of the host on specific plantings and the successful development of the last generations of parasites in a season; also, the appearance of a change in species of hosts, which is accompanied in most cases by a change in the habitat. These phenomena determine the assurance of an increase or decrease in the numbers of the overwintering populations of parasites and their density in the spring period.

4) The phenomenon of food specialization of oophagous insects, and their relationship with the complex of species of Pentatomidae.

5) Ethological traits that characterize the host selection process, specificity, and consequently also the durability of the host-parasite relationship.

As a rule the specific nature of the species composition within the limits of the central and southern belts of the European

part of the U.S.S.R. applies to several species that are comparatively rare.  Such species as *T. grandis* and *T. chloropus* are
encountered almost everywhere, and changes affecting them take
place on the level of relative abundance, i.e. changes in the
predominant position.  Factual information that illustrates this
position is rather abundant (Viktorov 1967, Ryakhovskii 1959b,
1974).

The ecological spectrum of the parasitic complex of the harmful stinkbug as a whole reacts to weather in a given locality by
means of population changes and changes in its various parts, a
fact which is expressed in a redistribution of the indices of
relative abundance of the species.  Consequently, the complex
preserves the ability to parasitize the eggs of the harmful stinkbug, but the level of parasitization changes.  Thus, the summer of
1957 in the Krasnodar area was unusually dry and in the following
year the usual predominance of *T. chloropus* was replaced by the
predominance of *T. grandis* (Viktorov 1967).  A shift in predominant
species was reported in the Voronezh district in the course of 3
summers, when 3 types of growth period occurred (Zatyamina et al.
1976).  The summer of 1970, which was usual for these places, was
followed by 2 hot and dry summer seasons.  The relative abundance
of *T. chloropus* during this period fell from 85% to 5% while this
index for *Ooencyrtus telenomicidae*, a species resistance to dryness
(Kochetova 1968), rose from 33% to 61%.  *T. grandis*, in comparison
with other ecologically more flexible species, in these years did
not change substantially in abundance, but some species that usually
are not very numerous practically disappeared.  However, the total
numbers of oophagous insects in the dry hot summer fell, and the
parasitization of eggs of the harmful stinkbug decreased from 90%
to 14%.  Thus, the complex of parasites of the harmful stinkbug,
which is adapted to "work in any weather," still acts with different
intensities under different meteorological conditions.

Changes in the species composition of the parasites and
oophagous insects sometimes take place in the course of the growing
season, which occurs with interspecies competition that is especially apparent toward the end of the harmful stinkbug's period of
oviposition, and particularly in those regions where a wider
variety of oophagous species is represented.  Thus, according to
the data of G. A. Viktorov (1967), in the Saratov district *T. grandis*
predominated in the first generation, but a considerable quantity of
the eggs of the harmful stinkbug during this period was parasitized
by other species: *Trissolcus volgensis* (Viktorov), *Trissolcus
scutellaris* Thomson, and *T. pseudoturesis* Rjachovsky.  In subsequent
generations *T. grandis* again predominated, but then the importance
of *O. telenomicida* and *T. chloropus* intensified.  Parasitization of
eggs of the stinkbug in this period rose from 25 to 100%.

Under specific field conditions the effectiveness of the action of the complex of parasites on the stinkbug is significantly associated with the nature of the distribution of species with respect to plantings. The oophagous species are unevenly distributed on plantings of grain in the period when the harmful stinkbug is present. Many investigators report comparatively higher parasitization of the eggs of the harmful stinkbug by oophagous insects on plantings that are in the vicinity of massive tracts or belts (natural or artificial) of trees and bushes, fruit orchards, or simply unworked areas with wild herbaceous vegetation (Kamenkova 1958, Shapiro 1959, Ryakhovskii 1959a, Areshnikov et al. 1971, Zatyamina et al. 1976). Stinkbugs and supplementary hosts live in all these habitats, as do the adult individuals of scelionids in hibernation. Parasitization of the eggs of the pest decreases significantly in places located far from forest plantations and orchards. In conformance with this, there is observed an increase in the numbers of young insects in comparison with the numbers of individuals of the overwintered generation (Shapiro 1959, Ryakhovskii 1959a). The ability of the overwintered individuals of the oophagous insects to populate fields is associated with their habitational adaptability, and particularly with the nature of the place of hibernation. The 2 species of oophagous insects of the most practical importance differ with respect to this characteristic.

In *T. grandis* the overwintered individuals scatter through the fields so that the zone of parasitization of eggs of the stinkbug does not exceed 500 meters from tracts or forest, forest plantings or orchards. Later on, the summer generations of this species may scatter more uniformly through the field (Shchepetil'nikova 1958). However, for a number of regions cases have been reported where even in the summer individuals of this species do not parasitize eggs of the stinkbug on the central belts of plantings (Ryakhovskii 1974). It should be noted that such data relate to regions where *T. grandis* has not appeared in the years of investigations as the predominant species, but has belonged to the subdominant species.

*T. chloropus* is distributed through the fields more uniformly (Shchepetil'nikova 1958), a fact which is connected with its ability to hibernate not only on tracts with woody vegetation, but also in plant residues in the fields.

Frequently, on the border areas of a wheat planting next to a forest, forest belt, or orchard, eggs of the stinkbug can be found that are parasitized by several species of oophagous insects (for example, *T. chloropus, T. grandis,* and *Trissalcus simoni* (Mayr), or *T. chloropus, T. grandis,* and *O. telenomicida*), whereas the central parts of these fields, sometimes not even very wide ones,

are reached only by *T. chloropus* or *T. chloropus* and *T. simoni*.
These observations relate to regions with *T. chloropus* as the
predominant species, where no significant difference is noted in
the parasitization of eggs of the stinkbug throughout the border
and central parts of the fields.  In regions where *T. grandis* and
*T. simoni* predominate, the parasitization of eggs of the pest has
been reported to be higher in the border parts of the field
(Ryakhovskii 1974).

In connection with the characteristics of hibernation of
*T. chloropus*, an increase in its numbers on some fields of grain
crops may be the result of the arrangement of preceding crops,
such as perennial leguminous grasses, alfalfa, sainfoin and row
crops, corn, millet, sunflower, and also tobacco.  The oophagous
species have spread into these fields in previous years in search
of supplementary hosts; here a third and sometimes a fourth generation develops, and here the adult parasites hibernate in plant
residues.  Thus, some plantings of grains growing up on tracts that
are even distant from forests or forest belts result in an increased
number of oophagous insects; the parasitization of eggs of the
harmful stinkbug here may reach 85-95%.  On fields without annual
rotation of grains and some row crops this index was decreased
almost in half (Fedotova 1956, Shapiro 1959, Ryakhovskii 1959a).

Consequently, a high number of parasites at the end of the
growing period results in part from the presence of natural
habitats of different species of Pentatomidae.  When there is a
variety of crops raised by farms there frequently arises a favorable
situation for successful development and population growth of
oophagous insects (Shapiro 1959).

A study of the seasonal dynamics of distribution of oophagous
insect habitats, which was carried out by Zatyamina et al. (1976),
showed that under the conditions of the central belt of the
European part of the U.S.S.R. (Voronezh district) in the spring-
summer period the egg deposits of supplementary hosts are first
parasitized on plantings of perennial leguminous grasses (sainfoin,
alfalfa) and on wild plants in the forest; on winter wheat the
parasitization of eggs was observed mainly on the edges of the
plantings.  In this period the activity of only one species of
oophage *T. chloropus* is reported.  Later, an increase in parasitization of the egg deposits was observed on wheat fields, and it
reached a maximum during the period of oviposition of the harmful
stinkbug.  In this period *T. grandis, O. telenomicida*, and *T.
chloropus* predominated; *T. pseudoturesis* was represented in smaller
numbers, and there was a solitary *T. volgensis*.

In the postharvest period the numbers of oophagous insects
increased on late gathered crops: sunflower, corn, and also on

some wild plants on forest edges, and on unused fields where
insects of the genera *Carpocoris, Palomena,* and *Graphosoma* were
noted.

Stinkbugs, which reproduce on row crops, may be parasitized
to a considerable degree by oophagous species. The eggs of
*Dolycoris baccarum* on sunflower may be 56-76% parasitized by 2
species: *T. grandis* and *T. chloropus.* The parasitization of the
eggs of this insect on corn may reach 80-100%, and here in
addition to the 2 species that have been mentioned are *O. telenomicida* and *T. pseudoturesis,* which are not found on sunflower
(Ryakhovskii 1959a).

The contacts of the oophagous parasites with the eggs of the
harmful stinkbug are comparatively brief. The oviposition of the
harmful stinkbug ceases approximately when the development of the
second generation of parasites is completed. The subsequent
generations, the third and fourth, operate on other species of
stinkbugs. The period of hunting for new hosts is complicated,
associated with migrations and with noncoincidence of the end of
the stinkbug's oviposition and the beginning of oviposition of
other Pentatomidae.

In this case either a delay occurs in the beginning of oviposition by the parasites or they die, sometimes in large numbers
(Shchepetil'nikova 1958).

In connection with this, the numbers of parasites capable of
parasitizing the stinkbug after its appearance on crop plantings
are determined by 2 main points in their life: first, by the
numbers of available supplementary hosts (Kamenkova 1955, 1958,
Shchepetil'nikova 1958), which in a number of cases is associated
with the number of generations of some species of stinkbugs in
which the second generation is facultative and is characterized by
a significant change in numbers in different years (Viktorov 1964a).
These circumstances determine an increase or curtailment of the
numbers of adult individuals of the last generation of oophagous
insects entering into hibernation. The conditions of passing the
winter in a state of hibernation are the second point determining
the numbers of oophagous insects in the spring.

The period of association with forest belts is characterized
by feeding of the adult parasites on late flowering plants and the
utilization of honey dew (Elfimov and Ryakhovskii 1974). Inasmuch
as an increase in numbers of parasites toward the end of the growing period is associated with a number of species of Pentatomidea,
their species composition, numbers, and physiological state, the
nature and status of the forests and field-protecting forest belts
adjacent to plantings are of substantial importance. Forests

planted without underbrush, devoid of borders with bushes and
herbaceous plants, are unfavorable for the development or supplementary hosts of the oophagous species, and thus here the possibility of an increase of the parasites is small (Shapiro 1959).

Most favorable for stinkbugs are fruit-bearing trees that are
well lighted and warmed, and the presence of herbaceous vegetation,
either planted or wild (Elfimov and Ryakhovskii 1974).

According to data of Elfimov and Ryakhovskii (1974), up to
98% of the eggs of a green forest insect, *Palomena prasina* (L.),
which is rather numerous in the central belt and whose egg
deposits are located on elm, linden, and pine, are destroyed by
*T. grandis, T. chloropus* and *O. telenomicida*. The eggs of the sloe
bug, *Dolycoris baccarus*, which is associated with tree species and
mixed grasses also is sometimes parasitized completely. The sorrel
bug, an inhabitant of bushes, and representatives of the genera
*Elasmostethus, Elasmucha,* and *Pentatoma* undergo destruction;
oophagous species also parasitize predatory bugs of the genera
*Troilus* and *Rhaphigaster*.

The range of food specialization of oophagous insects that
parasitize the harmful stinkbug is rather wide and includes various
representatives of the superfamily Pentatomidae. On this basis this
group of parasites is considered either broadly as one of polyphages
within the limits of the stinkbugs (Masner 1958) or as one of oligo-
phages with a wide circle of hosts (Shchepetil'nikova 1958). Experimental data have shown that most of the species of this group have
a definite specialization with specific preference for certain hosts
(Viktorov 1964a). Thus, for example, *Trissolcus simoni* Mayr reveals
close relationships with *Eurydema ventralis* (=*ornatum*), willingly
parasitizes the eggs of *Graphosoma italicus* and *Dolycoris baccarum,*
rarely parasitizes other species of crucifer bugs, and very rarely,
with high mortality, develops in the eggs of the harmful stinkbug
and the maura bug, *Eurygaster marus* L. (Shapiro 1951).

In such species as *T. scutellaris, Teleas reticulatus* Kieffer,
and *T. volgensis* there is a clearly expressed preference for the
eggs of the harmful stinkbug, the sloe bug, and the sharp-shouldered
bug, whereas the eggs of cruciferous insects are seldom parasitized
by these species, and the mortality of the preimaginal stages in the
last case is high, reaching almost 100% (Viktorov 1964a, 1967).
The preferability of the eggs of the harmful stinkbug within the
limits of the group under consideration is manifested by a gradation
from zero to high indices.

G. A. Viktorov (1967) prepared a list of hosts of 11 species of
oophagous insects of the genera *Trissolcus, Telenomus, and Ooencyrtus*
which included 17 species of stinkbugs. In a guide to representatives

of the genus *Trissolcus* inhabiting the U.S.S.R., M. A. Kozlov (1968) presents 15 species that are associated in their development with 17 species of stinkbugs. *T. grandis* is characterized by the widest circle of hosts; it includes 12 species of insects of 8 genera. This species, which is a broad oligophage, shows a clear preference for *Eurygaster intergiseps* Puton, *E. maura,* and *E. austriaca.*

*Telenomus chloropus* and *Ooencyrtus telenomicida* have a large number of hosts; in the first case 7 species of insects, and in the second 6 species.

Greatly limited in number are the hosts of *T. scutellaris* and *T. volgensis,* which each parasitize 2–3 species of insects; *T. pseudoturesis* apparently develops only on the harmful stinkbug. *T. grandis, T. simoni, T. scutellaris,* and *T. chloropus* are the most closely associated with the harmful stinkbug.

Selectivity of a specific circle of hosts is clearly manifested in the oophagous insects, and the development of the preimaginal stages in the eggs of rejected hosts is associated with high mortality; this indicates physiological deficiency of the rejected host for the parasite.

The complexity of the biocenotic relationships of the harmful stinkbug with the complex of oophages and the degree of confinement of this complex to development on the pest are indicated also by the ethological characteristics of the parasitic species associated with searching and host selection. In this process there are sequential behavioral steps: finding the habitat of the host, locating the host (Doymm 1968), and evaluating its quality (Viktorov 1972).

For 2 species of oophages, *T. grandis* and *T. chloropus*, the existence of chemical stimulators for host searching has been established (Viktorov et al. 1975). The females of a whole series of entomophages are attracted to the hosts by odoriferous substances, kairomones, that are secreted by the hosts. The 2 species of oophages mentioned are attracted by the secretions only of active, ovipositing females of the harmful stinkbug. Such secretions are characteristic also of other Pentatomidae: for example, the cabbage bug; its kairomones stimulate searching by the oophagous *Trissolcus victorovi* Kozlov (Buleza 1971).

The determination of the chemical composition of the kairomone of the harmful stinkbug and its artificial production and use under field conditions may be the principal new tactical means in the biological method of controlling the harmful stinkbug (see chapter 8).

Evaluation of the quality of the host is characteristic of species of the genus *Trissolcus*; their females prefer to parasitize eggs in which the embryo is in an early stage of development (Viktorov 1967). *Ooencyrtus telenomicida* selects freshly laid eggs of the host only under conditions where there is a daily possibility of finding the host. Under conditions where there is a rare possibility of depositing an egg (a host is not available every day), the selectivity is lost (Kochetova 1968).

Telenomus develops the ability to distinguish parasitized hosts from healthy ones; this ability is associated with the "labeling" of parasitized eggs by excretions from the ovipositor onto the surface of the egg. In spite of this, within the limits of the group of oophagous insects under consideration some species are characterized by multiparasitism or overparasitism, the ability to parasitize eggs containing a parasite not of their own species.

There is reason to suppose that in the life of many species of parasitic Hymenoptera capable of sharply changing their numbers an important role in preserving populations must be played by sex attractants. This aspect of the vital activity of oophagous insects of the harmful stinkbug still awaits investigation.

Thus, the results of numerous comprehensive, lengthy, and laborious investigations indicate the great potential of the complex of parasites of the harmful stinkbug to limit the numbers of the pest, and in a number of cases to reduce it to a level below economic significance.

The entire sum of accumulated knowledge about the ecology of the harmful stinkbug and the complex of its parasites has allowed investigators to speak variously of the fact that the incapability of oophagous insects to realize their potential of sharply reducing the numbers of the pest has been caused by features of large-scale grain farming, in which no place has been left as a habitat for Pentatomidae, or for hibernation of these insects and many other oophagous species. Sharp disruptions of the biocenotic relationships have led to serious consequences that embrace vitally important spheres of the economic activity of man. The entirety of accumulated information on the problem of the harmful stinkbug is most fully correlated and combined by the theoretical ideas of G. A. Viktorov in a whole series of publications, the most thorough of which is a compendium published in 1967, "Problems of the population dynamics of insects in the example of the harmful stinkbug."

This information, along with a critical analysis of the data and opinions of a number of Soviet and foreign investigators, became the basis for the development by G. A. Viktorev (1971) of a synthetic theory of the population dynamics of insects.

G. A. Viktorov came to the conclusion that on large tracts where grain crops are planted extremely unfavorable conditions are created for parasites of the harmful stinkbug, conditions that do not affect the pest, which comes in contact with the plantings for a comparatively short time, passing the winter outside them. On the contrary, the parasites are subjected to the destructive action of agrotechnical measures and are practically deprived of places to hibernate; the monoculture prevents the necessary contacts with other species of Pentatomoidea that are supplementary hosts of the parasites that inhabit both other crops of the crop rotation and the hibernation habitats of the parasites. All these circumstances can have an effect directed toward decreasing the numbers of the representatives of beneficial fauna.

Territories that preserve favorable conditions for the life activity of the parasites of the harmful stinkbug at one time were sharply curtailed. They have been preserved in a few places, including the south of the Krasnodar area (Kamenkova 1958) and Armenia (Viktorov 1967). Here the parasites are able to realize their most valuable property, their regulating effect on the density of the phytophage. Present day agrocenoses disrupt these relationships and reduce them to the simplest level when the entomophages do not enter in as agents regulating the population density of the harmful stinkbug, but as a factor resembling the pest population, i.e. they affect the density of the pest only incidentally, and are equal in influence to abiotic factors.

Under present day conditions, an escape from the serious situation that has been created has been found in a transition to an integrated method of decreasing the numbers of the pest with every kind of cooperation in realizing the possibilities of natural complexes of parasitic species. In connection with this, the wide development of forest protective belts, habitats for estivation and hibernation not only of the stinkbug but also of its parasites and their supplementary hosts, is very important. Regulation of the biocenotic relationships in these new ecological situations is a matter for future investigations into improving the integrated control of the harmful stinkbug. Careful consideration is now necessary before resorting to the use of insecticides on these belts.

Thus, the transition to the third step of the system of controlling the harmful stinkbug has been accomplished, a step that is marked by a tendency on the one hand toward curtailment of the span of chemical treatments and their limitation to definite stages in the development of the harmful stinkbug, and on the other hand toward the fullest possible preservation of the beneficial fauna (parasites and predators) and the promotion of conditions that favor the fullest possible realization of the potential of natural entomophages.

Thus, the system of the parasitic complex of the harmful stinkbug is very mobile, very sensitive to changes in the abiotic and biotic factors, and also initially secure in its relationships with the harmful stinkbug, having the ability not only to modify, but also to regulate the population density of the pest. Solution of the problem of the harmful stinkbug in our time is tied to the scientific basis for rational use of a large group of potentially effective parasites.

There is a somewhat different situation with respect to the possibilities of protecting fruit crops with biological control agents from such primary pests as the codling moth (*Laspeyresia pomonella*) and a complex of species of mining moths.

The codling moth is a difficult pest; specifically, in controlling it present-day orchards are saturated with insecticides. Massive use of contemporary chemical preparations in orchards has quickly removed from the complex of orchard leafeating pests the species that are easily vulnerable to unfavorable external influences. But literally before our eyes a unique sequence of events has occurred; many old pests have disappeared, and in their place new ones have appeared whose numbers were regulated (before the use of insecticides with a broad range of action) within the limits of economically unimportant levels by certain groups of parasites and predators. The least vulnerable species have remained; the codling moth continues to occupy the first place among orchard pests. Massive destruction of beneficial insects has yielded its fruits in the form of new problems in plant protection.

But the number of cases of massive multiplication of "secondary pests" has not been very great, and in the U.S.S.R. it is limited only to the multiplication of spider mites in orchards and on cotton (Kozlova 1960, 1963, Livshits and Petrushova 1969, Smirnova, 1969, 1975), and also massive damage of the foliage of fruit crops by mining moths (Vinogradov 1971, Kholchenkov, 1974, 1975, Vereshchagina 1972, 1974, 1975).

The relatively low numbers of such cases immediately indicates which species of pests are regulated in nature primarily by natural enemies; it also has become evident that far from all the harmful species have effective complexes of entomophagous insects in a given geographic zone.

Analysis of the possibilities of the biological control the codling moth shows that this species belongs to the group of insects for which a large number (more than 120 species) of entomophagous insects are recorded (Franz 1960, Shteinberg 1962). But in various geographic regions the role of entomophagous insects in limiting the numbers of the codling moth is very different. Under the conditions of the European continent entomophagous insects scarcely act at all on the pest, whereas for the southeastern Kazakhstan, and especially for Uzbekistan, the data are entirely different; although they are not very numerous, they permit us to believe that under these conditions the possibility exists for practically perceptible limitation of the damaging activity of the codling moth.

On the whole, publications on entomophagous insects of the codling moth contain information on the species composition of the comparatively more active or more often encountered species, and also on the parasitization of various stages of the pest, usually expressed in percentage indices. Knowledge of the groups still is only at the point of revealing some biological characteristics of the species that are encountered comparatively more frequently. Serious ecological investigations have not yet been started.

In the European part of the U.S.S.R., namely the Ukraine, 7 species of parasites of the codling moth have been reported: 3 species of Ichneumonidae, 2 species of Braconidae, 1 species of Pteromalidae, and 1 species of Tachinidae.

In one of the orchards not treated with insecticides the parasitization of the codling moth by the tachinid *Neoplectops pomonella* (Schnabl and Mokrzecki) reached 26%, and on small plots 10.5%. Most attention has been attracted by the following parasites: *Pristomerus vulnerator* (Panzer), *Coccygomimus* (=*Pimpla*), *turionellae* (L.), *Trichomma enecator* (Rossi), and *Braunsia* (=*Microdes*) *rufipes* (Nees); the maximum parasitization of the moth (by these species) that was observed was respectively: 3.6%, 3.1%, 3.2% (sic).

This group is very sensitive to treatment with insecticides, and most individuals die under the usual treatments of apple trees (Tkachev 1974).

In Moldavia the species composition of the parasites is wider; it includes 12 species, of which 5 are in common with the Ukraine species. Besides this, snakeflies and earwigs have been reported as predators. The most widely distributed and frequent is *Braunsia* (=*Microdes*) *rufipes*, which may infest 4.3% of codling moth populations. Infestation by *Ascogaster quadridentata* Wesmael reached 4-5% in orchards with chemical treatment, and *Pristomerus vulnerator* (Panzer) reached 3.5-4% (Goncharenko 1971).

In Georgia, Gaprindashvili and Novitskaya (1967) discovered 21 species of entomophagous insects of the codling moth belonging to the Ichneumonidae (3 species), Branconidae (7 species), Tachinidae (2 species), Ostomidae (1 species), Carabidae (3 species), and Trichogrammatidae (1 species).

The most important in lowering the numbers of the codling moth
are the following 7 species: *Ephialtes extensor* Taschenberg,
*Epunctulatus carbonarius* (Christ) (Ichneumonidae), *Macrocentrus
nidulator* (Nees), *Macrocentrus marginator* (Nees), *Ascogaster rufipes*
(Latreille), *Casinaria ichnogaster* Thomson (Braconidae), and *Leskia
aurea* (Fallen) (Tachinidae).

It is reported that these entomophages can show their effectiveness on the codling moth only in those plantings in which insecticidal treatments have not been carried out; in this case there were
on the average 5-6 adults and pupae of the predators per tree and
not more than 11 healthy caterpillars and pupae, i.e. about 80%
less than on trees treated with insecticides.

The number of pests by the end of the season on untreated
plots had decreased in comparison with treated ones about 83%.
The numbers of parasites when chemical toxicants were used on previously untreated plots during the year decreased 88%, and in the
following year only solitary examples of parasites could be found
there.

For the Alma-Altinskii fruit growing zone a characteristic
parasite of the codling moth is *Braunsia rufipes* (Nees) (Braconidae).
Its effectiveness varies from 14 to 60% depending on the nature of
the orchard and the characteristics of the year (Zlatanova 1970).
Of the other parasites, *Ascogaster quadridentata* and *Ephialtes
extensor*, which can decrease the number of caterpillars of the moth
by 45% on the average, are notable (Zlatanova and Lukin 1971). *A
quadidentata* parasitizes the plum moth (up to 88% in some years),
while the index for the codling moth varies from 0.41 to 8%
(Zlatanova 1968).

In Uzbekistan, according to the data of Abdullaeva (1974),
in the Ferganskii Valley 17 species of parasitic Hymenoptera are
associated with the codling moth: 10 species of Ichneumonidae,
2 species of Braconidae, 1 species of Trichogrammatidae, and 3
species of Chalcidoidea.

The most widely distributed in the Ferganskii Valley is a
*Mastrus* sp. Parasitization of caterpillars by it in some orchards
reaches 60%, while in others there is only an individual parasitization.

In the Andizhanskii district *Braunsia rufipes* predominates,
but the picture also is mixed; parasitization from almost zero to
30% has been reported. In the Namanganskii district there is a
predominance of *A. quadridentata,* which destroys up to 40% of the
caterpillars of the pest.

For the Tashkent district an ectoparasite of diapausing caterpillars, *Liotryphon punctulatus* (Ratzeburg), has been reported,
which in the spring and autumn may destroy up to 30% of the pest.
*A. quadridentata* shows a very unstable effect: from 25 to 0%. For
*Braunsia rufipes* very low effectiveness has been reported in the
Tashkent and Ferganskii districts: 0.01 to 3%. Such parasites as
*Pristomerus vulnerator* and species of the genus *Pimpla* in Uzbekistan
do not have practical importance. Abdullaev draws the conclusion
that a complex of parasites "to a certain degree regulates the
numbers of the codling moth." However, attention must be turned to
the extreme sporadicity and inconstancy of the indices of effectiveness. Undoubtedly, there still has been almost nothing done with
respect to studying the ecology of parasites of the codling moth
under the conditions of central Asia.

In orchards of central Europe 5 species of parasitic Hymenoptera
have been reported as parasites of the codling moth: 2 of the family
Ichneumonidae, the most common of them being *Pristomerus vulnerator*
Pans which parasitizes the caterpillars, 1 species of the family
Braconidae, 1 of the family Chalcididae, and 1 species of Tachinidae
have also been reported (Zech 1959). These 5 species are characterized by wide geographical distribution. Either these are transarctic
species (*Trichomma enecator* Rossi, *Perilampus tristis* Mayr) or they
embrace the whole Eurasian continent to China and Japan, as well as
northern Africa, North America, and Australia (*A. quadridentata*
Wesm.). The area of the tachinid *Arrhionomyia tragica* Mg. is not
so wide, being limited to central Europe.

The circle of hosts of the parasitic hymenopteran species that
are under consideration is different. *P. vulnerator, T. enecator*,
and *A. quadridentata* may be assigned to the polyphagous forms.
The first of these parasitizes 2 species of Diptera and 2 species
of Hymenoptera as well as 26 species of Lepidoptera. Two of the
others parasitize only Lepidoptera, but their circle of hosts is
large: 21 species and 31 species out of 8 and 6 families, respectively. More specialized are *Perilampus tristis*, which is
associated with 6 species of the family Tortricidae, and *Arrhinomyia
tragica* Mg., which parasitizes 4 species of the families Tortricidae
and Tineidae. *P. vulnerator* and *A. quadridentata* are encountered
comparatively more often in the orchards of Europe. But apparently
the wide circle of hosts of these parasites and their univoltine
nature are not favorable for effective action on the numbers of the
moth. According to observations by Zech (1959), *A. quadridentata*
parasitizes 4% of the caterpillars of the codling moth, *P. vulnerator* 8%, *T. enecator* 3%, *P. tristis* and *A. tragica* 1% each. As a
whole the greatest parasitization in the orchards that has been
observed did not exceed 16%. For France there are indications of
the same order (Rosenberg 1934).

Somewhat different results were obtained in East Germany when special observations were carried out on the parasitization of caterpillars of the codling moth in 2 orchards considerably separated from each other. One orchard was treated with insecticides, and the second was untreated (Lehman 1968). Here a still lower parasitization of the caterpillars was reported, the species composition of which was less rich than in the previous observations. In an untreated orchard, *Trichomma enecator* during the course of 2 years parasitized 8.5% and 6.3% of the catapillars, whereas in orchards that had undergone treatment with insecticides during the first year of the investigations no parasitized caterpillars were found, and the following year only about 2% of the caterpillars were parasitized by *T. enecator*. Thus, the parasites of the codling moth, as a result of their biological and ecological characteristics, cannot fulfill the role of a factor limiting the numbers of the pest on the territory of the European continent. This circumstance in a certain measure has determined the direction of work on the investigation of biological agents for pest control. There was no basis, as in the case of the stinkbug, for studying in the greatest detail the complex of natural enemies. One species of the genus *Trichogramma* which is encountered in nature, is flexible, and has a number of promising properties as a parasite, was selected. Investigations are being expanded at the present time on the study of its potential for controlling the codling moth.

As concerns the possibilities of parasites of the codling moth under the conditons of central Asia, the widely available information on the high parasitization of the pest gives reason for serious investigations of an ecological and biocenotic order.

In contrast to the codling moth, the mining moths of apple have been dependably regulated for a long time by a complex of parasites. Only pesticides with a wide range of action have freed the fields from the parasite controlling their numbers.

It is characteristic that the appearance of damaging activity of this group of pests has embraced a large territory of the European continent, from the southern regions (Italy) to the central and eastern parts (Zangheri and Ravelli 1975, Ciampolini 1958, 1959, Briolini 1960, Bagglioni 1960, 1961, Celli 1960, Ivanov and Slavov 1975).

In the U.S.S.R. this group of pests started to show great damage in 1961 in the vast fruit farms of Azerbaidzhan (Vinogradov 1971). Serious damage was caused in orchards of the Crimea (Kholchenkov 1970), Moldavia (Bichina 1969), the Kuban (Mormyleva 1971, Georgia (Gokhelashvili 1973), and Kazakhstan (Petrova 1974).

At the beginning the outbreaks of massive pest increase were of a focal nature, but even then they embraced vast tracts. Even in July and August the damaged foliage fell off, and the yield was

decreased quantitatively and qualitatively.  Losses of 34 to 65%
of the crop were reported (Vinogradov 1974, Kholchenkov 1976).
The content of Vitamin C, sugars, and dry substances decreased.
Furthermore, the establishment of fruit buds was partially or
entirely precluded (Mormyleva 1971).

Investigations showed the breadth of the species composition
of the mining moths.  In Azerbaidzhan 7 species of 4 families were
found (Vinogradov 1971); in the Crimea orchards 24 species of 8
families were found, of which 7 species belonged to the group of
potential pests (Kholchenkov 1970).  In Moldavia, the Kuban,
Azerbaijan, and Georgia the most damaging and widely distributed
is *Stigmella malella* (Stainton), whereas in the Crimea along with
this species *Phyllonorycter* (=*Lithocolletis*) *pyrifoliella* Gram
has acquired great importance; in Moldavia *Phyllonorycter*
*corylifoliella* (Hubner) and *Phyllonorycter pyrifoliella* Gram have
shown great damaging activity; in orchards of the Alma-Atinskii
fruit zone *Stigmella malella, Leucoptera* (=*ceniostroma) scitella*
(Zeller), and *Bucculatrix crataegi* Zeller have been reported.

It is very characteristic that the mining moths have attained
their greatest economic importance in the oldest regions of large
commercial cultivated orchards, and specifically on those farms
where intensive treatments of the orchards with insecticides have
been carried out on large tracts for many years.

Food specialization of the group of pests that is under consideration is mainly monophagous, although some species have been
evaluated as narrow or broad polyphages.

The parasites of the mining moths make up a rather large
group of Hymenoptera of the families Braconidae and Eulophidae.

According to data of Kholchenkov (1976), for *Stigmella malella*
25 species of entomophages have been recorded, 8 species of them
in the Crimea, 3 of which are most active; for *Phyllonorycter*
*pyrifoliella* there are 14 species recorded, of which 3 species are
more important.  In untreated orchards of the Crimea parasites
destroy up to 82 to 95% of the caterpillars and pupae of the mining
moths.  In commercial orchards treated for the codling moth 7 to 8
times with Sevin (carbaryl), entomophages are capable of parasitizing
only 3 to 4% of the individuals of the pest.  Under these conditions
the use of integrated control directed at maintaining the activity
of natural enemies of the mining moths is promising.

In Moldavia 2 main species of miners, *Phyllonorycter*
*corylifoliella* and *Phyllonocrycter pyrifoliella*, are parasitized by
a large number of primary and secondary parasites.  Talitskii (1961)
reported 32 species for the first and 19 species for the second.

Seven species are of practical importance; they parasitize the pests differently. The total percentage of parasitism of *Phyllonorycter corylifoliella* amounted to 71%, and in this case the most important species were of the families *Eulophidae*. *P. pyrifoliella* was 84% parasitized, and for this species of the families *Eulophidae* and *Braconidae* had about the same importance.

According to observations by V. V. Vereshchagina and B. V. Vereshchagin in 1963 and 1964, *Apanteles lautellus* Marshall (Braconidae) was predominant in parasitizing the 2 species; in 1965 the chalcids predominated, and the importance of the braconid fell greatly.

*Phyllonorycter pyrifoliella* in 1965 was as a whole 68% parasitized. Under the conditions of Kazakhstan, in the Alma-Atinskii fruit zone there was a rather effective complex of parasites, which included 5 species, most important among them being a representative of the family Eulophidae capable of decreasing the numbers of *S. malella* Gram by 38%. Here also a high mortality of parasites was noted when there was chemical treatment of the orchard (Petrova 1974).

Both Soviet and foreign investigators (Zangheri and Ravelli 1957, Ciampolini 1958, 1959, Bagglioni 1960, 1961, Ivanov 1975) are unanimous in their recommendations for limiting the use of insecticides in orchards for purposes of preserving the numerous and highly effective entomophages of the mining moths.

Thus, the wide circle of mining pests also has a large number of parasites capable of reliably regulating the level of numbers of the pests to where the miners acquire the status of indifferent importance. Under the conditions of an apple orchard the carrying out of integrated control and the reestablishment of the regulating influence of the parasites of mining moths can be accomplished more easily than in the case of the harmful stinkbug, a fact which is associated with the specific characteristics of grain and fruit growing.

The examples considered show that biotic factors represented by parasites of harmful insects may have a completely different significance in the damaging activity of species that are of primary economic importance. If for the harmful stinkbug and the mining moths of apple the natural entomophages represent complexes well suited to activity against the harmful species and are found able to regulate the population density, parasites of the codling moth under European conditions are still very ineffective, although they are very numerous with respect to species composition.

Under the conditions of Uzbekistan the still scanty observations indicate the possibility of highly effective parasites of the codling moth, but this is reported for small tracts and in different years. Here serious ecological investigations of the complex of parasites of the pest are necessary.

On the basis of the concrete examples of the importance of entomophages in the population dynamics of such primary pests as the harmful stinkbug, the codling moth, and the mining moths, it is possible to find approaches to the problem of classification of the biocenotic and agrotechnical phenomena in question. As the initial variant of classification of the importance of entomophages as a factor determining the size of a population, the following may be proposed:

The unusual complexity of the relationships of different species of entomophages and the insect pests that are their hosts creates significant difficulties in evaluating the role of each species of parasites and predators in the population dynamics of the pests. For the purpose of a general evaluation of this role, the following classification of types of biocenotic relationships of pests and entomophages is proposed.

1) The first may be called the type of *Eurygaster integriceps* Puton. This pest has numerous parasites of eggs and adults, which are very common and which effectively lower the numbers of the pest. At the end of the season of oviposition by the harmful stinkbug a moment occurs when 100% of the eggs of this pest are parasitized by Telenominae parasites. In a similar manner in the summer it can be observed that from 50 to 65% of the adult bugs have in their bodies the larvae of parasitic tachinids of the subfamily Phasiinae. Predatory insects destroy a significant number of larvae of the harmful stinkbug. Thus, the pest in all stages of its development is subject to strong suppression by entomophages, and the role of the latter in regulating the numbers of the pest is very important.

2) The second type of biocenotic relationships may be called the type of *Laspeyresia pomonella* (L.). In the U.S.S.R., in the zone with 2 generations of the pest, parasites and predators of the codling moth are rarely observed. Not more than 10% of the caterpillars are parasitized; because the caterpillars are often hidden, their destruction by predatory insects is almost never observed. Chemical treatments of orchards have an insigificant effect on the population levels of the natural enemies of the moth. The numbers of the pest are always high and require constant use of exterminatory measures.

3) The third may be called the type of *Stigmella malella* (Stainton). Normally in orchards where insecticidal treatments are not used, the entomophages depress the numbers of mining moths.

Parasitization of the caterpillars usually is observed at the
level of 80-90%.  However, under the conditions where insecticides
are used the parasitization of the caterpillars sharply decreases
to 3-4% and the numbers of adult mining moths sharply increase.
Thus, the numbers of the pest population are subject to sharp
variations.

LITERATURE CITED

Abdullaev, E. 1974. Parazity yablonnoi plodozhorki nekotorykh rainov Uzbekistana [Parasites of the codling moth of some regions of Uzbekistan]. Pages 10-15 In Ecologiya i biologiya entomofagov vreditelei. Sel'skokhozyaistvennogo kul'tur Uzbekistana [Ecology and biology of entomophages of pests of agricultural crops of Uzbekistan]. Tashkent.

Annual Review of Entomology. 1956-1976: v. 1-20.

Antonenko, O. P. 1971a. Izuchenie khishchnikov vrednoi cherepashki s pomoshchyu radioizotopa $C^{14}$ [Study of predators of the harmful stinkbug with the aid of radioisotope $C^{14}$]. Pages 5-15 In Nauchno-tekhnicheskoi informatsiya instituta sel'skogo khozyaistva Yugo-vostoka [Scientific-technical information of the Institute of Agriculture of the Southeast]. Vol. 5.

Antonenko, O. P. 1971b. Biologicheskoe obosnovanie integriovannoi bor'by s vrednoi cherepashkoi v Saratovskoi oblasti [Biological basis of integrated control of the harmful stinkbug in the Saratov district]. Pages 17-20 In Kratkie tezisy dokladov k soveshchaniem po priemam biologicheskoi bor'by s vrednoi cherepashkoi v integrirovannoi sisteme zashchity zernovykh kul'tur [Short summaries of reports to a meeting on methods of biological control of the harmful stinkbug in an integrated system of protecting grain crops]. Voronezh. Leningrad.

Areshnikov, B. A. 1967. Sostoyanie i perspektivy khimicheskogo metoda bor'by s vrednoi cherepashkoi na Ukraine [Status and prospects of the chemical method of controlling the harmful stinkbug in the Ukraine]. Zashch. Rast. 6: 16-28.

Areshnikov, B. A., I. V. Kovtun, L. B. Rogachaya, and D. M. Feshchin. 1971. Puti snizheniya otritsatel'nogo vliyaniya khimicheskikh obrabotok na poleznuyu entomofaunu pshenichnogo polya v usloviyakh stepnoi zony Ukrainy [Ways of decreasing the negative effect of chemical treatments on the beneficial fauna of a wheat field under the conditions of the steppe zone of the Ukraine]. Pages 20-23 In Kratkie tezisy dokladov k soveshchaniem po priemam biologicheskoi bor'by s vrednoi cherepashkoi v integrirovannoi sisteme zashchita zernovykh kul'tur. Voronezh. Leningrad. [Short summaries of reports to a meeting on methods of biological control of the harmful stinkbug in an integrated system of protecting grain crops].

Bagglioni, M. 1960. Observations sur la biologie de deux mineuses du genre Lithocolletis: L. corylifoliella et L. blancardella (Lep. Gracilariidae) nuisibles aux arbres fruitiers en Suisse romande. Mitt. Schweiz. Entomol. Ges. 32 (4): 385-97.

Bagglioni, M. and G. Neury. 1961. Essai de lutte contre la mineuse sinueuse du feuillage des arbres fruitiers (Lyonetia clerkella L.) Rev. Romande Agric. 17 (5): 43-7.

Bichina, T. A.  1969.  Opyt bor'by s mol'yu-malyutkoi v sadakh
    kolkhoza "Krasnyi sadovod" Tiraspol'skogo raiona [Experiment
    on controlling pygmy moth in orchards of the collective farm
    "Red Horticulturist" of the Tiraspol' region].  Informatsionnyi
    listok Instituta ekonomicheskogo issledovaniya i nauchno-
    teknicheskoi informatsii pri Gosplane Moldavskoi SSR
    [Information leaflet of the Institute of Economic Research
    and Scientific-Technical Information of the State Planning
    Commission of the Moldavian SSR].  No. 135, 8 pp.
Briolini, G.  1960.  Ricerche su quattro specie di Microlepidotteri
    minatori delle foglie di Melo. *Nepticula malella* Stain. e *N.*
    *pomella* Vaugh. (Nepticulidae): *Leucoptera scitella* Zell.
    (Bucculatricidae): *Lithocolletis blancardella* F. (Gracilariidae).
    Boll. Ist. Dei. Entomol. Univ. Studi Bologna. 24: 239-69.
Buleza, V. V.  1971.  Izbiratel'nost' v povedenii samok nekotorykh
    yaitseedov krestotsvetnykh klopov pri zarazhenii khozaev
    (Hymenoptera, Scelionidae) [Selectivity in the behavior of
    some oophages of cruciferous bugs during parasitization of
    hosts (Hymenoptera: Scelionidae)].  Zool. Zh. 50 (12):
    1885-8.
Burov, V. N., I. P. Zaeva, and E. V. Titova.  1974.  Vyyavlenie
    troficheskikh svazei khishchnykh chlenistonogikh s vrednoi
    cherepashkoi s pomoshchyu serologicheskogo i radioizotopnogo
    metodov [Revealing trophic relationships of predatory arthro-
    pods with the harmful stinkbug with the aid of serological
    and radioisotope methods].  Pages 69-81 In Doklady na 26-m
    ezhegodnom chtenii pamyati N. A. Kholodsovskogo [Reports
    at the 26th annual reading in memory of N. A. Kholodsovskii].
    Leningrad.
Celli, G. 1960.  Ricerche sui parassiti di tre Microlepidotteri
    minatori delle foglie di Melo (*Nepticula malella* Staint.,
    *Leucoptera scitella* Zell. e *Lithocolletis blancardella* F.).
    Boll. Ist. Entomol. Bologna, 24: 271-79.
Chernova, O. A.  1947.  Nekotorye dannye o morfologii i
    plodovitosti parasitnykh mukh vrednoi cherepashki [Some data
    on the morphology and fertility of parasitic flies of the
    harmful stinkbug].  Pages 67-74 In Vrednaya cherepaskha
    [Harmful stinkbug].  Vol. 2 Moscow.
Ciampolini, M.  1958.  Osservazion etologiche sulla *Stigmella*
    *malella* (Stainton) (Lepidoptera, Nepticulidae).  Redia 43:
    111-21.
Ciampolini, M.  1959.  I trattamenti contro la *Stigmella malella*
    (Stainton), la *Leucoptera scitella* Zell. e la *Lithocolletis*
    *blancardella* F. in rapport al ciclo evolutuvo dei tre
    insetti.  Redia, 44: 55-75.

Doutt, R. D. 1968. Pages 117–34 In Biologicheskie osobennosti vzroslykh entomofagov [Biological characteristics of adult entomophages]. Biologicheskaya bor'ba s vrednymi nasekomymi i sornyakami [Biological control of harmful insects and weeds]. Moscow.

Dupuis, C. 1948. Nouvelles donnees biologiques et morphologiques sur les Dipteres Phasiinae parasites d'Hemipteres Heteropteres. Memoire presente a la faculte des Sciences de l'Universite de Paris pour l'obtention du diplome d'etudes superieure. Paris, 79 pp.

Dupuis, C. 1963. Essai monographique sur les Phasiinae (Dipteres. Tachinaires parasites d'Heteropteres). Mem. Mus. Nat. Hist. Nat. Nouv. Ser. Ser. A. Zool. 26, 461 pp.

Elfimov, V. I. and V. V. Ryakhovskii. 1974. Dopolnitel'nye khozaeva yaitseedov cherepashki na drevesno-kustarnikovykh statisiyakh [Supplementary hosts of stinkbug oophages in woody and shrubby habitats]. Pages 90–5 In Fiziologicheskye i biologicheskye osnovy zashchity rastenii [Physiological and biological bases of plant protection]. Voronezh.

Fedotov, D. M. 1944. Nablyudeniya nad vnutrennim sostoyaniem imago vrednoi cherepashki Eurygaster integriceps [Observations on the internal condition of adults of the harmful stinkbug Eurygaster integriceps]. Dokl. Akad-Nauk SSR 4 (9): 423–26.

Fedotov, D. M. 1947. Nablyudeniya nad vzaimootnosheniyami mezhdu vrednoi cherepashkoi i ee parazitami mukhami-faziyami i soobrazheniya ob ispol'zovanii fazii v bor'be s cherepachkoi [Observations on the relationships between the harmful stink-bug and its phasiid fly parasites and ideas on use of phasiids in controlling the stinkbug]. In Vrednaya cherepashka [Harmful stinkbug]. Moscow. Vol. 1. 49–66.

Fedotova, K. M. 1956. Effektivnost' telenomusov v snizhenii vrednonosti klopa vrednoi cherepashki na posevakh ozimoi pshenitsy v zavisimosti ot predshect-vennikov [Effectiveness of Telenomus in lowering the damage by the harmful stinkbug on plantings of winter wheat in relation to the precursors]. Byull. Nauchno-Tekh. Inf. Ukr. Nauchno-Issled. Inst. Zashch. Rast. 1: 29–31.

Fedchenko, M. A., and T. V. Yaryshkina. 1968. V bor'be s vrednoi cherepashkoi [Controlling the harmful stinkbug]. Zashch. Rast. 10: 22.

Flanders, S. E., and M. E. Badgley. 1963. Prey-predator inter-action in self-balanced laboratory populations. Hilgardia 35: 145–83.

Gaprindashvili, N. K. and G. N. Novitskaya. 1967. Estestvennye vragi yablonnoi plodozhorki *Laspeyresia pomonella* (Lep., Tortricidae) i vliyanie khimicheskikh obrabotok na ikh poleznuyu deyatel'nost' [Natural enemies of the codling moth *Laspeyresia pomonella* (Lep.: Tortricidae) and the effect of chemical treatments on their beneficial activity]. Entomol. Obozr. 46 (1): 70-4.

Gokhelashvili, R. D. 1973. Estestvennye vragi yablonnoi nizhnestoronnei miniruyushchei moli (*Stigmella malella*) i rezul'taty ispytaniya parazita apantelesa protiv nee v usloviyakh Kartli (Vostochnaya Gruziya) [Natural enemies of the apple under-side mining moth (*Stigmella malella*) and results of testing the parasite *Apanteles* against it under the conditions of Kartlya (Eastern Georgia). Tr. Inst. Sadovod. Vinograd. Vinodel. Gruz. SSR 22: 196-202.

Goncharenko, E. G. 1971. Entomofagi yablonnoi plodozhorki [Entomophages of the codling moth]. Zashch. Rast. 5: 21-3.

Howard, L. O. and W. F. Fiske. 1911. The importation into [the] United States of the parasites of the gypsy moth and the brown-tail moth. U.S. Dep. Agric. Bur. Entomol. Bull. 91, 132 pp.

Huffaker, C. B., and P. S. Messenger. 1964. The concept and significance of natural control. Chap. 4. In Biological Control of Insect Pests and Weeds (DeBach, P., editor) Reinhold Publ. Co., N.Y. 844 pp.

Huffaker, C. B., P. S. Messenger, and P. DeBach. 1971. The natural enemy component in natural control and the theory of biological control. Pages 16-67 In Huffaker, C. B. [ed.] Biological Control. N.Y.-London.

Ivanov, S., and N. Slavov. 1975. Rolyata na poleznana fauna za chislenata dinamika no listominirashchite moltsi [Role of beneficial fauna in population dynamics of leaf-mining moths]. Rast. Zashch. 23 (10): 33-5.

Kamenkova, K. V. 1955. Parazity cherepashki i ikh dopolnitel'nye khozyaeva v predgornykh raionakh Krasnodarskogo kraya [Parasites of stinkbug and their supplementary hosts in foot-hill regions of the Krasnodar area]. Authors abstract of dissertation. Leningrad: 19 pp.

Kamenkova, K. V. 1958. Prichiny vysokoi effektivnosti yaitseedov cherepashki v predgornykh raionakh Krasnodarskogo kraya [Reasons for the high effectiveness of stinkbug oophages in the foothill regions of the Krasnodar area]. Tr. Vses. Inst. Zashch. Rast. 9: 285-311.

Kholchenkov, V. A. 1970. K biologii yablonnoi moli-malyutki (*Stigmella malella*) i o merakh bor'by s nei [The biology of the apple pygmy moth (*Stigmella malella*) and measures for controlling it]. Byull. Gos. Nikitsk. Bot. Sada, 3 (14).

Kholchenkov, V. A. 1973. Vidovoi sostav i troficheskie svyazi miniruyushchikh molei, povrezhdayushchikh plodovye kul'tury v Krymu [Species composition and trophic relationships of mining moths damaging fruit crops in the Crimea]. In Prikladnaya botanika i introduktsiya rastenii. Nauka, Moscow.

Kholchenkov, V. A. 1974. Miniruyushchie moli--vrediteli plodovykh kul'tur [Mining moths--pests of fruit crops]. In Vasil'ev, V. P. [ed.] Vrediteli sel'skokhozyaistvennykh kul'tur i lesnykh nasazhdenii [Pests of agricultural crops and forest plantings]. Vol. II, Kiev.

Kholchenkov, V. A. 1976. Miniruyushchie moli--vrediteli plodovykh kul'tur Kryma. (Fauna, biologiya, mery bor'by) [Mining moths--pests of fruit crops of the Crimea (Fauna, biology, control measures)]. (Author's abstract of dissertation): 1-20, Khar'kov.

Kochetova, N. I. 1968. Postembrional'noe razvitie *Ooencyrtus telenomicida* Vass. (Hymenoptera, Encyrtidae) i osobennosti ego parazitisma v yaitsakh poluzhestkokrylykh (Hemiptera) [Postembryonal development of *Ooencyrtus telenomicida* Vass. (Hymenoptera: Encyrtidae) and characteristics of its parasitism of the eggs of true bugs (Hemiptera)]. Pages 79-92 In Voprosy funktsional'noi morfologii i embriologii nasekomykh [Problems of the functional morphology and embryology of insects]. Moscow.

Kozlov, M. A. 1968. Vidy roda *Trissolcus* Achmead (Hymenoptera, Scelionidae, Telenominae)--parazity yaits klopov-shchitnikov [Species of the genus *Trissolcus* (Hymenoptera: Scelionidae, Telenominae)--parasites of stinkbug eggs]. Tr. Vses. Inst. Zashch. Rast. 31: 204-10.

Kozlova, E. N. 1960a. Pautinnyi kleshchik. [Spider mite]. Pages 148-52 In Obzor rasprostraneniya glavneishikh massovykh vreditelei i boleznei sel'skokhozyaistvennykh kul'tur v 1959 g.i prognoz ikh poyavleniya v 1960 g. [Review of the distribution of the main mass pests and diseases of agricultural crops in 1959 and prediction of their appearance in 1960]. Leningrad.

Kozlova, E. N. 1960b. O taktike primeneniya preparatov vnutrirastitel'nogo deistviya [Tactics for the use of preparations with a systemic action in plants]. Sel'sk. Khoz. Tadzh. 5:17-8.

Kozlova, E. N. 1963. Sravnitel'naya otsenka metilmerkaptofosa na novom emul'gatore [Comparative evaluation of methylmercaptophos in a new emulsifier]. Byulleten' Gosvdarstvennoi komissii po khimicheskim sredstvam bor'by s vreditelaymi, boleznyami rastenii i sornyakami pri Ministerstve Sel'skokhozyaistva SSSR [Bulletin of State Commission on chemical agents for controlling pests, plant diseases, and weeds of the Ministry of Agriculture USSR]. 5 (1): 72-5.

Kupershtein, M. L.   1974.   Ispol'zovanie reaktsii pretsipitatsii
    dlya kolichestvennoi otsenki vliyaniya *Pterostichus
    crenuliger* (Coleoptera, Carabidae) na dinamiku populyatsii
    cherepashki *Eurygaster integriceps* (Hemiptera, Scutelleridae)
    [Use of precipitation reactions for quantitative evaluation
    of the effect of *Pterostichus crenuliger* (Coleoptera:
    Carabidae) on the population dynamics of the stinkbug
    *Eurygaster integriceps* (Hemiptera: Scutelleridae)].   Zool.
    Zh. 53 (4): 557-62.
Lehmann, W.   1968.   Beitrag zur Kenntnis der Parasiten von
    *Laspeyresia pomonella* L. Arch. Pflanzenschutz 4 (2): 131-41.
Masner, L.   1958.   Some problems of the taxonomy of the subfamily
    Telenominae (Hym. Scelionidae) Pages 375-82 In Transaction
    of the 1st international conference on insect pathology and
    biological control.   Prague.
Meier, N. F.   1949.   Rezul'taty primeneniya telenomusa
    (*Microphanurus semistriatus* Nees) v bor'be s cherepashkoi v
    Kazakhstane v 1942 g. [Results of use of *Telenomus
    (Microphanurus semistriatus* Nees) in controlling stinkbug
    in Kazakhstan in 1942).   Tr. Vses. Inst. Zashch. Rast.
    2: 111-13.
Mormyleva, V. F.   1971.   Biologicheskie osobennosti miniruyushchei
    moli-malyutki v plavnevoi podzone Kubani [Biological
    characteristics of the pygmy mining moth in the flooded
    subzone of the Kuban].   Pages 5-7 In Tezisy dokladov 6-i
    nauchnoi konferentsii molodykh uchenykh Vsesoyuznyi Nauchno-
    Issledovatel'skii Institut Zashchita Rastenii [Summaries
    of reports of 6th scientific conference of young scientists
    at the All-Union Scientific Research Institute of Plant
    Protection].   December 1971.   Leningrad.
Newsom, L. D.   1967.   Consequences of insecticide use on non-
    target organisms.   Anns. Rev. Entomol. 12: 257-86.
Nicholson, A. J.   1933.   The balance of animal populations.   J.
    Anim. Ecol. (Supp. to Vol. 2): 132-78.
Petrova, V. K.   1974.   K izucheniyu khal'tsid- parasitov
    miniruyushchikh molei yabloni [Study of chalcid parasites of
    mining moths of apple].   Vestn. Skh. Nauki Kaz. 11: 61-65.
Rodendorf, B. B.   1947.   Kratkoe posobie dlya opredeleniya
    dvukrylykh parazitov vrednoi cherepashki i drugikh klopov
    Pentatomidae [Brief guide for identification of dipteran
    parasites of the harmful stinkbug and other bugs of the
    Pentatomidae].   Pages 75-88 In Vrednaya cherepashka [The
    harmful stinkbug], Vol. 2. Moscow.

Rubtsov, I. A. 1937. O teoreticheskom obosnovanii raionirovaniya vrednykh nasekomykh i prognoze ikh massovogo razmnozheniya. I. Ponyatie plastichnosti, ee rol' v dinamike chislennosti nasekomykh i metody ee izucheniya [Theoretical basis of regionalization of harmful insects and prediction of their massive multiplication. I. The idea of flexibility, its role in the population dynamics of insects and methods for its study]. Zashch. Rast. 14: 3-13.

Rubtsov, I. A. 1938. O teoreticheskom obosnovanii raionirovaniya vrednykh nasekomykh i prognoze ikh mossovogo razmnozheniya. II. Integral'nye klimaticheskie indeksy dlya tselei raionirovaniya vrednykh nasekomykh [Theoretical basis of regionalization of harmful insects and prediction of their massive multiplication. II. Integral climatic indices for purposes of regionalization of harmful insects]. Ibid. 16: 3-20.

Rubtsov, I. A. 1945. O dvukh parazitakh vrednoi cherepashki iz sem. Phasiidae (Diptera) [Two parasites of the harmful stinkbug from the family Phasiidae (Diptera)]. Entomol. Obzor 28 (3-4): 85-100.

Rubtsov, I. A. 1948. Biologicheskii metod bor'by s vrednymi nasekomymi [Biological method of controlling harmful insects]. Moscow.

Ryakhovskii, V. V. 1959a. Effektivnost' parazitov yaits klopa-cherepashki v zavisimosti ot razmeshcheniya posevov [Effectiveness of parasites of the eggs of stinkbug in relation to the arrangement of plantings]. Nauchn. Tr. Ukr. Nauchno-Issled. Inst. Zashch. Rast. 8: 97-116.

Ryakhovskii, V. V. 1959b. Yaitseedy klopa vrednoi cherepashki v USSR [Oophages of the harmful stinkbug in the Ukr. SSR]. Ibid. 8: 76-88.

Ryakhovskii, V. V. 1974. Khozyaino-parazitarnye i statsial'nye svyazi yaitseedov klopov-cherepashek i drugikh klopov-shchitnikov v raznykh zonakh RSFSR [Host-parasite and habitational relationships of oophages of stinkbugs and other shieldbugs in various zones of the RSFSR]. Pages 75-89 In Fiziologicheskye i biologicheskye osnovy zashchity rastenii [Physiological and biological bases of plant protection]. Voronezh.

Shapiro, V. A. 1951. Faktory, sposobstvuyushchie i ogranichivayushchie razmnozhenie parazita *Trissolcus simoni* Mayr v prirode [Factors promoting and limiting the multiplication of the parasite *Trissolcus simoni* Mayr in nature]. Author's abstract of candidate's dissertation. Leningrad. 15 p.

Shapiro, V. A. 1959. Vliyanie agrotekhnicheskikh i lesokhoz-
    yaistvennykh meropriyatii na effektivnost' yaitseddov vrednoi
    cherepashki [Effect of of agrotechnical measures on the
    effectivenss of oophages of the harmful stinkbug]. Pages
    182-91 In Biologicheskie metody bor'by s vreditelyami
    rastenii [Biological methods of controlling pests of plants].
    Kiev.
Shchepetil'nikova, V. A. 1957. Zakonomernosti, opredelyayushchie
    effektivnost' entomofagov [Rules determining the effectiveness
    of entomophages]. Zh. Obshch. Biol. 18 (5): 381-94.
Shchepetil'nikova, V. A. 1958. Effektivnost'yaitseedov vrednoi
    cherepashki i faktory ee obuslovlivayushchie [Effectiveness
    of oophages of the harmful stinkbug and factors determining
    it]. Tr. Vses. Inst. Zashch. Rast. 9: 243-84.
Shchepetil'nikova, V. A., N. S. Fedorinchik, E. M. Shumakov, and
    M. A. Bulyginskaya. 1976. Biologicheskii metod bor'by s
    yablonnoi plodozhorkoi v SSSR [Biological method of con-
    trolling the codling moth in the USSR]. Pages 178-91 In
    Resursy biosfery (itogi sovetskikh issledovanii po
    mezhdunarodnoi biologicheskoi programme [Resources of the
    biosphere (Results of Soviet investigations in the inter-
    national program)]. Vol. 2.
Shumakov, E. M. 1958. Fazii kak parazity cherepashki [Phasiids
    as parasites of stinkbug]. Tr. Vses. Inst. Zashch. Rast.
    9: 313-21.
Shumakov, E. M. and N. M. Vinogradova. 1958. Ekologiya vrednoi
    cherepashki [Ecology of the harmful stinkbug]. Ibid. 9:
    19-71.
Smirnova, A. A. 1969. Ob ustoichivosti pautinnogo kleshcha
    na khlopchatnike k fosfororganicheskim akaritsidam
    [Resistance of spider mite on cotton to organophosphorous
    acaricides]. Pages 124-7 In Voprosy ulucheniya organizatsii
    zashchita rastenii khlopchatnika ot boleznei i sornyakov
    vreditelei. [Problems of improving the organization of pro-
    tection of the cotton plant from pests, diseases, and weeds].
    Tashkent.
Smirnova, A. A. 1975. Ispytanie insektitsidov (karbofosa 40%,
    ritsifona 30%, dilora 15%, antio 70%, galekrona 25%,
    despirola 5%, elokrona 50%) metodom UMO [Testing of insecti-
    cides (carbophos 40%, ricifon 30%, dilor 15%, anthio 70%,
    galecron 25%, despirol 5%, elecron 50%) by the UMO method].
    Pages 110-32 In Itogi gosudarstvennogo ispytanii insektitsidov
    i akaritsidov v 1974 g. [Results of state testing of
    insecticides and acaricides in 1974]. Moscow.
Smith, H. S. 1935. The role of biotic factors in the determina-
    tion of population densities. J. Econ. Entomol. 28: 873-98.
Solomon, M. E. 1949. The natural control of animal populations.
    J. Ani. Ecol. 18 (1): 1-35.

Solomon, M. E. 1958. Meaning of density dependence and related terms in population dynamics. Nature 181: 1778-80.

Southwood, T. R. E. (ed.) 1968. Insect abundance. Oxford, Edinburgh. 160 p.

Starosti, S. P., and V. N. Burov. 1976. Sovremennaya sistema zashchity zernovykh kul'tur ot vrednoi cherepashki i puti ee sovershenstvovaniya [The present system of protecting grain crops from the harmful stinkbug and ways of improving it]. Tr. Vses. Inst. Zashch. Rast. 45: 10-19.

Talitsskii, V. I. 1961. Naezdniki i mukhi-takhiny--parazity vreditelei sada v Moldavii [Ichneumon flies and tachinid flies--parasites of orchard pests in Moldavia]. Tr. Mold. Inst. Sadovod. Vinograd. Vinodel. 7: 132-35.

Telenga, N. A. 1953. O roli entomofagov v massovykh razmnozheniyakh nasekomykh [The role of entomophages in mass multiplication of insects]. Zool. Zh. 32 (1): 14-24.

Tkachev, V. M. 1974. Entomofagi yablonnoi plodozhorki [Entomophages of the codling moth]. Zashch. Rast. 8: 26.

Varley, G. C. 1947. The natural control of population balance in the knapweed gallfly (*Urophora jaceana*). J. Ani. Ecol. 16 (2): 139-87.

Varley, G. C. 1953. Ecological aspects of population regulation. Pages 210-14 In Transactions IXth International Congress of Entomology. Vol. 2. Amsterdam.

Vereshchagina, V. V., E. V. Vereshchagin, and D. I. Savov. 1968. Miniruyushchie moli *Lithocolletis corylifoliella* Haw. i *L. pyrifoliella* Gram. (Lep., Gracilariidae), vredyashchie plodovym kul'turam v Moldavii [Mining moths *Lithocolletis corylifoliella* Haw. and *L. pyrifoliella* Gram. (Lep.: Gracilariidae) damaging fruit crops in Moldavia]. Zool. Zh. 47 (3): 387-94.

Viktorov, G. A. 1955. K voprosu o prichinakh massovykh razmnozhenii nasekomykh [The problem of the causes of massive multiplication of insects]. Zool. Zh. 34 (2): 259-66.

Viktorov, G. A. 1960a. Faktory dinamiki chislennosti vrednoi cherepashki (*Eurygaster integriceps* Put.) na Kubani v 1956-1958 gg. [Factors in the population dynamics of the harmful stinkbug (*Eurygaster integriceps* Put.) in the Kuban in 1956-1958]. Pages 222-36 In Vrednaya cherepashki [Harmful stinkbug]. Vol. 4.

Viktorov, G. A. 1960b. Biotsenoz i voprosy chislennosti nasekomykh [Bicenosis and problems of numbers of insects]. Zh. Obshch. Biol. 21 (6): 401-10.

Viktorov, G. A.  1963.  Kolebaniya chislennosti nasekomykh kak
    reguliruemyi protsess [Variations in numbers of insects as
    a regulated process].  Pages 11-13 In Pyatoe soveshchanie
    Vsesoyuznogo Entomologicheskogo Obshchestva.  Tezisy
    doklady. [Vth Conference of the All-Union Entomological
    Society.  Summaries of reports].  Moscow-Leningrad.
Viktorov, G. A.  1964a.  Faktory dinamiki chislennosti vrednoi
    cherepashki (Eurygaster integriceps Put.) v Saratovskoi
    oblasti v 1961-1962 gg. [Factors in the population dynamics
    of the harmful stinkbug (Eurygaster integriceps Put.)
    in the Saratov district in 1961-1962.]. Zool. Zh. 43 (9):
    1317-34.
Viktorov, G. A.  1964b.  Pishchevaya spetsializatsiya yaitseedov
    vrednoi cherepashki (Eurygaster integriceps Put.) i ee
    znachenie dlya diagnostiki vidov v rode Assolcus Nakagawa
    (=Microphanurus Keiffer) (Hymenoptera, Scelionidae) [Food
    specialization of oophages of the harmful stinkbug
    (Eurygaster integriceps Put.) and its importance for
    diagnostics of species in the genus Assolcus Nakagawa
    (=Microphanurus Kieffer) (Hymenoptera: Scelionidae)].
    Zool. Zh. 43 (7): 1011-25.
Viktorov, G. A.  1965.  Kolebaniya chislennosti nasekomykh kak
    reguliruemyi protsess [Variations in numbers of insects as
    a regulated process].  Zh. Obshch. Biol. 26 (1): 43-55.
Viktorov, G. A.  1967.  Problema dinamiki chislennosti
    nasekomykh na primere vrednoi cherepashki [The problem of
    population dynamics of insects in the example of the harmful
    stinkbug].  Nauka.  Moscow. 271 p.
Viktorov, G. A.  1969.  Vlivanie abioticheskikh faktorov na
    rasprostranenie i chislennost' yaitseedov vrednoi cherepashki
    (Hymenoptera, Scelionidae) [Effect of abiotic factors on the
    distribution and numbers of oophages of the harmful stinkbug].
    Zool. Zh. 48 (6): 841-49.
Viktorov, G. A.  1969a.  Mekhanizm regulyatsii chislennosti
    nasekomykh [Mechanism of regulating the numbers of insects].
    Vestn. Akad. Nauk. SSSR 6: 37-45.
Viktorov, G. A.  1970.  Mezhvidovaya konkurentsiya i sosushchest-
    vovanie ekologicheskikh gomologov u paraziticheskikh
    pereponchatokrylykh [Interspecies competition and coexistence
    of ecological homologs in the parasitic Hymenoptera].  Zh.
    Obshch. Biol. 31 (2): 247-55.
Viktorov, G. A.  1971.  Troficheskaya i sinteticheskaya teoriya
    dinamiki chislennosti nasekomykh [Trophic and synthetic
    theory of population dynamics of insects].  Zool. Zh. 50
    (3): 361-72.

Viktorov, G. A. 1972. Povedenie parazitov-entomofagov i ego znachenie dlya biologicheskoi bor'by s vreditelyami [Behavior of parasitic entomophages and its importance for the biological control of pests]. Usp. Sovrem. Biol. 74, 3 (6): 482-93.

Viktorov, G. A. 1975. Dinamika chislennosti zhivotnykh i upravlenie eyu [Population dynamics of animals and its management]. Zool. Zh. 54 (6): 804-21.

Viktorov, G. A. 1976. Ekologiya parazitov-entomofagov [Ecology of parasitic entomophages]. Nauka. Moscow. 152 pp.

Viktorov, G. A., V. V. Buleza, and E. P. Zinkevich. 1975. Poisk khozyaina u *Trissolcus grandis* i *Telenomus chloropus*-- yaitseedov vrednoi cherepashki [Host-seeking in *Trissolcus grandis* and *Telenomus chloropus*--oophages of the harmful stinkbug]. Zool. Zh. 54 (6): 922-27.

Vinogradov, A. V. 1966. Opasnyi vreditel' yablonevykh nasazhdenii [A dangerous pest of apple plantings]. Sel'skaya zhizn': 5.

Vinogradov, A. V. 1974. Miniruyushchie moli yablonevykh sadov Azerbaidzhana i mery bor'by s nimi [Mining moths of apple orchards in Azerbaidzhan and measures for controlling them]. Author's abstract of dissertation. Leningrad. 22 p.

Woodworth, C. W. 1908. The theory of the parasitic control of insect pests. Science, N. S. 28: 227-30.

Zaeva, I. P. 1969. Sravnitel'naya rol' vesennykh khimicheskikh obrabotok i kompleksa khishchnikov i parazitov v dinamike chislennosti vrednoi cherepashki [The comparative role of spring chemical treatments and of a complex of predators and parasites in the population dynamics of the harmful stinkbug]. Zool. Zh. 48 (11): 1652-60.

Zaeva, I. P. 1971. Puti ispol'zovaniya analiza korrelyatsii dlya izucheniya biotsenoticheskikh svyazei vrednoi cherepashki, khishchnykh zhuzhelits i paukov [Ways of utilizing correlation analysis for studying the biocenotic relationships of the harmful stinkbug, predatory ground beetles, and spiders]. Pages 87-90 In Kratkie tezisy dokladov k soveshchaniem po priemam biologichicheskoi bor'by s vrednoi cherepashkoi v integrirovannoi sisteme zashchity zernovykh kul'tur [Short summaries of reports to conference on means of biological control of the harmful stinkbug in an integrated system of protecting grain crops]. Voronezh. Leningrad.

Zaeva, I. P., and M. L. Kupershtein. 1971. Zhuzhelitsy
    pshenichnykh polei v osnovnykh ochagakh massovogo
    razmnozheniya vrednoi cherepashki [Ground beetles of
    wheat fields in the main foci of massive multiplication of
    the harmful stinkbug]. Pages 90-94 In Kratkie tezisy
    dokladov k soveshchaniem po priemam biologich. bor'by s
    vrednoi cherepashkoi v integrirovannoi sistame zashchity
    zernovykh kul'tur. [Short summaries of reports to a meeting
    on methods of biological control of the harmful stinkbug
    in an integrated system of protecting grain crops].
    Voronezh. Leningrad.
Zangheri, S., and V. Ravelli. 1957. Ricerche sulla morfologia
    e biologia della Leucoptera scitella Zell. (Lepidoptera,
    Lyonetiidae). Redia 42: 167-89.
Zatyamina, V. V., E. R. Klechkovskii, and V. I. Burekova. 1976.
    Ekologiya yaitseedov klopov-shchitnikov Voronezhskoi oblasti
    [Ecology of oophages of stinkbugs of the Voronezh district].
    Zool. Zh. 55 (7): 1001-5.
Zech, E. 1959. Beitrag zur Kenntnis einiger in Mitteldeutschland
    aufgetretener Parasiten des Apfelwicklers (Carpocapsa
    pomonella L.). Z. Angew. Entomol. 44 (1): 203-20.
Zlatanova, A. A. 1968. Ascogaster quadridentatus Wesm.
    (Hymenoptera, Braconidae)--rasprostranennyi parazitoid
    yablonnoi plodozhorki [Ascogaster quadridentatus Wesm.
    (Hymenoptera:Braconidae)--a widely distributed parasitoid
    of the codling moth]. Vestn. Skh. Nauki 8: 98-104.
Zlatanova, A. A. 1970. Biologiya Microdus rufipes Ness (Hymenop-
    tera, Braconidae)--parazitoida yablonnoi plodozhorki v
    Kazakhstane [Biology of Microdus rufipes Ness (Hymenoptera:
    Braconidae)--a parasitoid of the codling moth]. Entomol.
    Obozr. 49 (4): 749-55.
Zlatanova, A. A., and V. A. Lukin. 1971. Sokhranenie parazitov
    yablonnoi plodozhorki pri integrirovannoi zashchity sada.
    [Preservation of parasites of the codling moth in integrated
    orchard protection]. Zashch. Rast. 12.

EDITORS' NOTE

The reference citations for this chapter are reproduced as
received from the author. However, some references cited in the
text are not listed above and some listed references are incomplete.

CHAPTER 3

THE THEORETICAL BASIS FOR AUGMENTATION OF NATURAL ENEMIES

E. F. Knipling

Agricultural Research Service
U.S. Department of Agriculture
Beltsville, Maryland 20705  U.S.A.

Entomophagous parasites and predators developing on their own play a major role in regulating the abundance of insect pests.  The total complex of parasites and predators in a pest ecosystem may involve many species, each varying in importance at different times and places.  However, even a single well adapted species is capable of reducing the steady density of a pest population.  This is clearly indicated by the success that has been achieved by the introduction and establishment of a wide range of parasites and predators for the control of alien pests.  DeBach (1971), Clausen (1956), Sailer (1972), and other authorities list several hundred parasites and predators that have given partial to excellent control of insect pests after their introduction and establishment.

While self-perpetuating populations of parasites and predators achieve adequate control of many potential pests, there are others that are not adequately controlled.  In the United States, these include such important species as the boll weevil, *Anthonomus grandis* Boheman, codling moth, *Laspeyresia pomonella* (L.); pink bollworm, *Pectinophora gossypiella* (Saunders); tropical fruit flies, Tephritidae, *Heliothis* spp.; European corn borer, *Ostrinia nubilalis* Hubner); cabbage looper, *Trichoplusia ni* (Hubner); and a wide range of other pests affecting man and animals.  Inadequately controlled pests are usually found in environments altered by man's activities.  However, even in natural relatively undisturbed environments, many pests annually or periodically reach numbers that are damaging to plants and animals and threaten the health and comfort of man.  Many interacting factors operate in a natural environment to maintain a reasonable balance between competing organisms.  However, a satisfactory balance between beneficial and destructive insects from an

79

ecological standpoint may not meet society's needs for adequate
protection of food crops or its demands for maximum freedom from
attack by pests that affect man's health and comfort.

Insect population ecologists have made extensive studies of
parasite-host and predator-prey relationships in efforts to determine the type of interactions involved in the coexistence of
various organisms in a natural environment. This is a highly
complex subject. There is little likelihood that all of the interacting forces can be identified and it is even less likely that the
important factors can be accurately quantified. Consequently,
population ecologists have relied primarily on rationalization and
theory to describe and appraise the importance of the major interacting forces involved in the maintenance of a reasonable balance
between destructive and useful insects. Howard and Fiske (1911),
Nicholson and Bailey (1935), Smith (1935), Varley and Gradwell
(1970), Holling (1966), Burnett (1959), Huffaker (1971), Solomon
(1957), and other noted insect ecologists have dealt with the
subject.

It would require far more space than can be allotted to this
paper to discuss parasite-host and predator-prey interactions in
detail and to review the opinions of various population ecologists
who have contributed to this subject. It is important, however,
that scientists have sufficient understanding of the major factors
that govern the numerical relationships between a pest species and
its natural enemies so that they can fully appreciate the limitations and merits of natural biological control agents.

In trying to assess the role of certain biological organisms
as natural control agents--particularly their potential when
augmented--I have attempted to identify and quantify the major
influences affecting the relationship between parasites and their
hosts. The subject of parasite-host relationships is discussed in
considerable depth in a forthcoming publication (Knipling 1977);
but it will be dealt with briefly in this publication.

In applying augmentation techniques, I believe the use of
agents that are highly host-specific will produce more effective
and reliable results than will the use of parasites or predators
that attack wide ranges of hosts. The use of host-specific biological agents would also avoid or minimize adverse effects on non-
target organisms. Moreover, the element of host-specificity makes
it possible to develop more accurate estimates of certain parameters
that must be quantified in order to duplicate in models the role
that a biological agent plays in the regulation of a pest species
under natural conditions.

FACTORS INFLUENCING THE EFFICIENCY OF PARASITE POPULATIONS

The following are regarded as the major factors involved in parasite-host interactions and must be quantified to develop models simulating the relationship between a parasite and its host:

1. <u>Density of host per unit of area and time</u>. The density of the host is judged to be the overriding factor governing the ultimate efficiency of a parasite. This determines the number of hosts that can be parasitized, which in turn, governs the number of progeny that will be available to perpetuate the parasite population in the next parasite generation.

2. <u>Density of the parasite population per unit of area and time</u>. The number of parasites present in a given habitat determines the proportion of the habitat that can be searched. This, in turn, determines the percentage or proportion of the host population in the habitat that will be parasitized. However, the density of the host population within the area searched determines the number of hosts attacked and therefore the number of parasites that will be present the next generation when allowance is made for natural mortality.

3. <u>Size and characteristics of the host environment</u>. The ratio of the number of parasites to the size of the ecosystem occupied by the host rather than the ratio of the number of parasites to the number of hosts in an ecosystem determines the efficiency of a parasite population. Whether the host habitat remains constant in size or fluctuates in size during a parasite and host cycle is a major factor governing the efficiency and dynamics of a parasite population.

4. <u>The egg-laying capability of the parasite</u>. The average number of eggs (or larvae in the case of oviparous species) produced by parasites is considered of little or no importance as a factor governing the efficiency of parasites when the host density is low because fewer hosts will be found under the best of conditions than the reproductive potential of the parasite population. But the reproductive capacity of parasites is considered an important factor for aphids, scales, and certain other insects that occur in large numbers in a limited area during periods of high host density. Under such conditions parasites present can be expected to find far more hosts than the number of eggs they can produce. The same factor probably limits the proportion of hosts that will be eaten by predators.

5. <u>Natural hazards to survival of parasite populations</u>. Biological agents, like their hosts, face many environmental hazards that limit the number of progeny that survive to reproduce. The

density of the host in which parasites develop and the density of
the parasites per se will influence the rate of predation and hence
the rate of survival of the parasite population.

    6.  Host-guidance mechanisms.  While random search for hosts
takes place in all parasite-host complexes, the presence of kairomones and other guidance mechanisms is regarded as an important
factor influencing the host-finding efficiency of parasites.  Too
little is known about kairomones and other host-guidance mechanisms
at this time to place quantitative values on this factor at different
host and parasite densities.

## HOST-FINDING EFFICIENCY EQUATION

    If a certain number of parasites in a given area and time is
capable of finding and parasitizing a given proportion of a host
population, it should be possible to make a close estimate of the
proportion of the host population that will be parasitized by more
or fewer parasites in the same area and time by assuming that random
searching, either for the host or host signal, is the normal
behavior of a parasite population.

    For example, if a population of parasites averaging 100 per
acre during a parasite or host generation is capable of searching
30% of the host environment in that area, a population of 200 parasites theoretically should be capable of searching two times the
area searched by 100 parasites.  However, unless the parasites have
developed a mechanism to avoid host seeking in previously searched
areas, the 100 additional parasites will re-search 30% of the area
and re-encounter 30% of the hosts previously encountered.  Therefore,
the total area searched by the 200 parasites will be 30% + 0.30 (70)
= 51% of the area.  Theoretically, only 51% of the hosts, rather
than 60% will be parasitized.  This re-searching factor is responsible for diminishing efficiency of a parasite population that
becomes increasingly important as the density of parasites per unit
of area and time increases.  If the assumption is made that 200
parasites per acre parasitize 50% of a pest population, 400 per acre
would parasitize 75%; 600, 87.5%, etc.  This intraspecific competition factor alone explains why parasitization approaching 100%
throughout a pest ecosystem is not likely to occur.

    Knipling and McGuire (1968) proposed an equation for calculating
the rate of host parasitization to expect from different numbers
of parasites based on the number that will achieve 50% host parasitization in a given area and time.

The equation is:  $E = 1 - q_n^{\frac{P}{}}$, where

    E = the efficiency in form of a proportion
    P = number of parasites in the population
    N = number of parasites required to find 50% of the hosts
    O = 0.5 E(100) = percent efficiency

Basically the same equation had been proposed previously by Nicholson and Bailey (1935) for calculating the host-finding efficiency of increasing numbers of a parasite population. Some changes in the host-seeking behavior of parasites may occur at various levels of density of the parasite and host. Other known or unknown factors may also influence the efficiency of different size parasite populations. However, the use of the formula makes it possible to estimate rates of host parasitization as the number of parasites and hosts per unit of area and time increase or decrease during the seasonal or periodic cycle of the codeveloping populations.

Figure 1 shows the nature of the host-finding efficiency curve of a host-specific parasite based on the formula described.

### ESTIMATING THE BASIC HOST-FINDING
### EFFICIENCY OF HOST SPECIFIC PARASITES

In order to develop models designed to simulate the role of a given parasite as a natural control agent and to depict the effect of additional parasites that are released, it is necessary to have a reasonably accurate estimate of the host-finding efficiency of the parasite.

In view of the many variables that can influence the effectiveness of a parasite in a natural ecosystem, it may seem futile to even attempt to establish a host-finding efficiency value for a parasite purely by rationalization and theoretical modeling procedures. To my knowledge, when studies were first undertaken in an attempt to estimate the host-finding efficiency of different parasites by the procedure to be described, the host-finding efficiency in numerical terms had not been established for any parasite-host complex, either by conventional field release procedures or by theoretical modeling procedures. The first types of parasites considered in this effort were the *Trichogramma*, which attack eggs of a wide range of lepidopterous pest species. The results were published by Knipling and McGuire (1968). Similar studies were subsequently undertaken involving braconid parasites of aphids (Knipling and Gilmore 1971), hymenopterous parasites of *Heliothis* larvae (Knipling 1971) and the tachinid parasite,

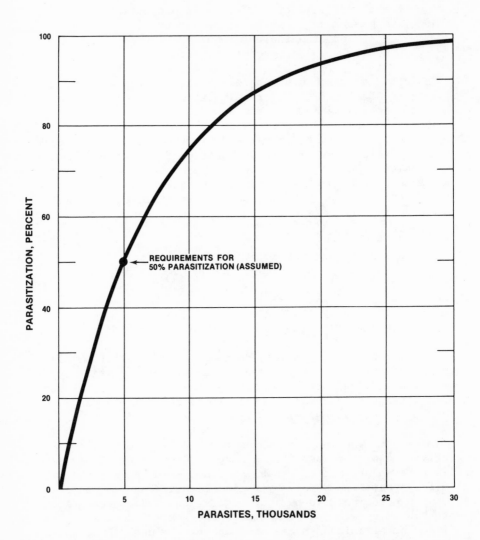

Fig. 1. Host-finding efficiency curve of a host-specific parasite based on the assumption that 5,000 parasites per unit of area and time will be capable of finding and parasitizing 50% of the hosts present. Adapted from Knipling and McGuire, 1968.

*Lixophaga diatraeae* (Townsend), which attacks larvae of the sugar-cane borer, *Diatraea saccharalis* (F.), (Knipling 1972).

To make an estimate of the host-finding efficiency of a parasite, it is necessary to take into account and assign values to all of the previously discussed major factors influencing the efficiency of the parasite when the factors apply to a given parasite and host. The first and most important requirement is a reasonably good estimate of the number of hosts present in a given area each generation during the season or periodic cycle. Estimates must then be made of the number of parasites present during the different host generations. Consideration must be given to the nature and size of the searching area and whether it is variable or constant during the seasonal cycle. Also, estimates are necessary for the rate of survival of developing parasite progeny. Obviously, good information on the life history and habits of the parasite and host is necessary.

In view of the absence of absolute values for host and parasite densities, as well as for other relevant parameters, many trial and error models must be developed by using rough estimates for the different parameters and assuming different host-finding efficiency values before the host-finding efficiency value that seems reasonable will become apparent.

The development of a meaningful host population model, which includes an estimate of the number of hosts present per unit of area (generally 1 acre) each generation, also makes possible reasonable estimates of the number of parasites likely to be present. Estimates of actual numbers of pest insects present per unit of area during periods of scarcity and abundance that were previously made in studies on the feasibility of employing autocidal suppression techniques proved helpful in the development of basic population models for certain pests and their associated parasites.

If a representative host population model is established, information on observed rates of parasitization when related to estimated host-density levels provides a clue for making the first rough estimates of the number of parasites likely to be present per unit of area and time. This, in turn, makes it possible to make preliminary estimates of the host-finding efficiency, expressed as the P-50 value, by assuming values for other relevant parameters, including the rate of parasite survival, size of the searching environment, and egg-laying potential. It may be necessary to consider dozens of models involving different values for the basic parameters before a codeveloping host-parasite model can be selected that seems reasonably accurate.

Despite the lack of field information on actual numbers of both

Table 1. The estimated number of parasites per acre (P-50) value required to parasitize 50% of the host insects present each parasite or host generation.

| Type of parasite | Pest hosts | Search units per acre | Host plants | P-50 value |
|---|---|---|---|---|
| Trichogramma | Heliothis eggs | 1-4 | Cotton | 5,000 |
| | Sugarcane borer eggs | 1-6 | Sugarcane | 5,000 |
| Braconid parasites of aphids | Green peach aphid | 1-5 | Potatoes | 2,000 |
| | Green bug | 1-5 | Sorghum | 2,500 |
| Microplitis | Heliothis larvae | 1 | Variety | 300 |
| Eucelatoria | Heliothis larvae | 1 | Variety | 400 |
| Lixophaga diatraeae | Sugarcane borer larvae | 1-2 | Sugarcane | 600 |
| Brachymeria intermedia | Gypsy moth pupae | 1 | Forest trees | 100 |

hosts and parasites present in pest ecosystems, I am confident
that we know enough about the dynamics of most of the important
pests and some of the associated parasites to assign values sufficiently accurate to eventually arrive at a good estimate of the
host-finding efficiency of a parasite.  If a reasonably good
estimate of the host-finding efficiency of a parasite can be made,
such estimate can be used to appraise the importance of the parasite
as a natural control factor and to calculate the impact the same
parasite can have on the dynamics of a host population when
various numbers are released.

The estimated host-finding efficiency of several types of
parasites have been postulated and reported in prior publications.
These involved *Trichogramma* as parasites of the eggs of *Heliothis*
and the sugarcane borer (Knipling and McGuire 1968); braconid
parasites of the green peach aphid, *Myzus persicae* (Sulzer), on
potatoes and of the green bug, *Schizaphis graminum* (Rondani) on
sorghum (Knipling and Gilmore 1971); hymenopterous parasites of
*Heliothis* larvae (Knipling 1971); and the tachinid parasite
*Lixophaga diatraeae* of the sugarcane borer (Knipling 1972).  In
addition, an estimate has been made of the host-finding efficiency
of *Brachymeria intermedia* Malloch, a hymenopterous parasite of
gypsy moth pupae, *Lymantria dispar* (L.); and of *Eucelatoria* sp.,
a tachinid parasite of *Heliothis* larvae.

The estimated efficiency of the parasites and hosts named is
given in Table 1.  The basic efficiency is expressed as the P-50
value, which is the number of parasites that must be present per
unit of area and time to find and parasitize 50% of the insect
host present.

A detailed discussion of the procedure followed and the
rationale for arriving at the P-50 value for braconid parasites of
aphids; for the tachinid parasite, *L. diatraeae*, which attacks
larvae of the sugarcane borer, and for hymenopterous parasites of
*Heliothis* larvae is given in the publications previously cited.
The codeveloping parasite host models that have been developed for
*Trichogramma* and the sugarcane borer, the tachinid *Eucelatoria* sp.
and *Heliothis* larvae, and the hymenopterous parasite *Brachymeria
intermedia* and gypsy moth pupae will be discussed briefly in this
paper.

## *TRICHOGRAMMA*-SUGARCANE BORER

The number of sugarcane borer eggs shown for the various host
and parasite generations (Table 2) is based on general observations
made by scientists with the U.S. Department of Agriculture laboratory at Houma, Louisiana.  The estimated number of hosts and the

Table 2. Theoretical model of coexisting natural populations of the sugarcane borer and *Trichogramma* parasites on 1 acre in sugarcane fields in Louisiana.

| Host and parasite generations | No. Host eggs | No. Searching units | No. Parasites per unit area | % Parasitization | No. Parasitized eggs | % Parasite survival | No. Parasite progeny for next generation |
|---|---|---|---|---|---|---|---|
| 1 | 15,000 | — | None | — | — | — | — |
| 2: | | | | | | | |
| 1 | 20,000 | 1 | 100 | 1.4 | 280 | 70 | 392[1] |
| 2 | 20,000 | 2 | 196 | 2.7 | 540 | 65 | 702 |
| 3 | 20,000 | 3 | 234 | 3.2 | 640 | 60 | 768 |
| 3: | | | | | | | |
| 4 | 150,000 | 4 | 192 | 2.7 | 4,050 | 55 | 4,455 |
| 5 | 150,000 | 5 | 891 | 11.6 | 17,400 | 48 | 16,704 |
| 6 | 150,000 | 6 | 2,784 | 31.8 | 47,770 | 40 | 38,160 |
| 4: | | | | | | | |
| 7 | 200,000 | 6 | 6,360 | 57.1 | 114,200 | 35 | 79,940 |
| 8 | 200,000 | 6 | 13,323 | 83.3 | 166,600 | 30 | 99,960 |
| 9 | 200,000 | 6 | 16,660 | 89.6 | 179,200 | 25 | 89,600 |

[1] Assumes two parasite progeny for each parasitized egg that survives.

survival values for the parasite differ considerably from the
original model proposed by Knipling and McGuire (1968). However,
the general trend of the parasite population and the rates of parasitization are similar to the estimates made in the original publication.

On the basis of a P-50 value of 5,000 *Trichogramma* per unit of
searching area, the model in Table 2 indicates that a self-perpetuating *Trichogramma* population would reach a parasitization level
near 90% by the fourth or last host generation. The high rate of
natural parasitism is in general agreement with records published
by Jaynes and Bynum (1941). The various rates of parasitization in
the different generations are also in agreement with observations
on natural rates of parasitization made by the Houma, Louisiana
staff. It is important to keep in mind that despite high natural
parasitization rates late in the season, *Trichogramma* do not provide
effective control of the pest because the pest generally reaches
and exceeds the damage level before a high rate of parasitization
occurs. This limitation no doubt applies for many other parasites.
This does not mean that such parasites are of no value in regulating
the abundance of a host. In the absence of a parasite like *Tricho-*
*gramma,* many pests would probably reach higher levels before the end
of a season and would also begin each new season with higher
densities that would result in greater damage to the host crop. It
does mean, however, that such parasite cannot be counted on to keep
a pest host below damage levels.

I believe that the estimated efficiency of *Trichogramma* as a
parasite of the sugarcane borer and the assigned values for other
major factors that influence efficiency will not deviate much from
true values. The parasitization trend developed by the modeling
procedures described was compared with the trend of natural parasitization that authorities working on the sugarcane borer consider
representative. This comparison is shown in Fig. 2.

If the host-egg population, size of the search environment, and
survival values for the parasite progeny in the different generations
are reasonably accurate, the host-finding efficiency value must also
be reasonably accurate. Otherwise, the parasitization trend in a
model will not conform to observed trends in a natural environment.

## *EUCELATORIA* SPP.--*HELIOTHIS* LARVAL POPULATIONS

Several species of *Eucelatoria* are known to parasitize lepidopterous larvae. These tachinid parasites have been investigated
by Jackson et al. (1969). They are not necessarily of major importance as natural control agents for *Heliothis*. It is my view, however, that low parasitization due to a self-perpetuating parasite

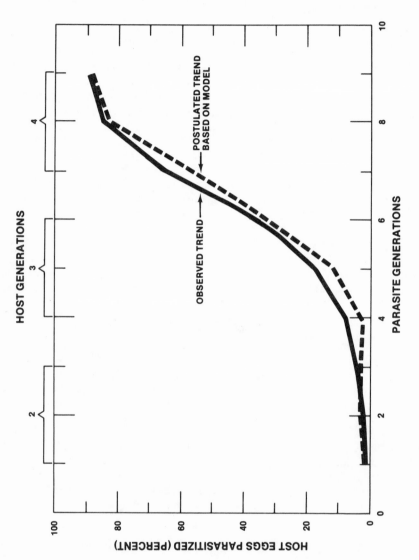

Fig. 2.  Postulated trend of *Trichogramma* parasitism of sugarcane borer eggs based on a simulation model, compared with observed natural trends of parasitism in Louisiana (USA) sugarcane fields.

population should not be regarded as an indication that such parasite will be of no value when populations are augmented substantially at appropriate host-density levels.

Since the host-density is a key parameter that must be established in the development of a parasite-host model, the proposed
host-density levels per acre shown in Table 3 constitute a critical
parameter. Several of my colleagues familiar with the population
ecology and dynamics of *Heliothis zea* (Boddie) and *H. virescens*
(F.) expressed the opinion that the density levels per acre were
probably reasonably representative of average natural density levels
for the larger larvae during a seasonal cycle in areas where these
insects are of economic importance. The density trend shown is not
intended to reflect the numbers present in the most favorable
habitats or the most important host crops, but rather is intended
to be representative of the average for all host plants in a
*Heliothis* ecosystem during years of normal abundance.

The trend of a *Eucelatoria* parasite population per acre and in
the total population and the rates of parasitization, based on a
P-50 efficiency value of 400 parasites per acre per host generation,
are shown in Table 3. It should be noted that the number of acres
of host plants and the survival rates for the parasite have been
given different values in the various host generations. A meaningful
parasite-host model representative of a total ecosystem cannot be
developed unless appropriate allowance is made for the number of
acres of host plants that is in line with the cropping system in
most *Heliothis* ecosystems. The number of acres of host plants
within dispersal range of *Heliothis* in an ecosystem as the season
advances represents the host environment and will determine the
number of hosts and the number of parasites present per unit of area
(acre) each host generation. This, in turn, determines the rate of
parasitization to expect as the crop season advances. There are no
data that can be used to judge the accuracy of the assigned survival
rates for the parasite, but the survival rate of the developing
parasite progeny is assumed to diminish as the season advances.

There is relatively little information on the natural rates of
host parasitization by *Eucelatoria* species. General observations
indicate, however, that the average rates of natural parasitization
are quite low. The model reflects these observations.

It is my conclusion that self-perpetuating populations of
*Eucelatoria* do not parasitize enough *Heliothis* hosts to constitute
a major factor in regulating the abundance of *Heliothis* in most
agroecosystems.

Table 3. *Heliothis* larval populations and a codeveloping natural *Eucelatoria* parasite population (hypothetical).

| Genera-tion | Acres | Hosts per acre | In-crease rate | Para-sites per acre | % Para-siti-zation[1] | Para-sitized hosts per acre | Potential progeny per acre[2] | % Parasite survival to adult | Parasite progeny Per acre | Total |
|---|---|---|---|---|---|---|---|---|---|---|
| 1 | 1,000 | 1,000 | 4.5 | 50 (assumed) | 8.3 | 83 | 249 | 20.0 | 50 | 50,000 |
| 2 | 3,000 | 1,500 | 3.0 | 17 | 2.9 | 44 | 132 | 17.5 | 23 | 69,000 |
| 3 | 4,500 | 3,000 | 1.7 | 15 | 2.6 | 78 | 234 | 15.0 | 35 | 158,000 |
| 4 | 5,000 | 4,500 | 0.9 | 32 | 5.4 | 243 | 729 | 12.5 | 91 | 455,000 |
| 5 | 4,000 | 4,500 | -- | 101 | 15.9 | 716 | 2,148 | Overwintering generation | | |

[1] The parasite is assumed to have a P-50 efficiency of 400 per acre per host generation.

[2] Parasitized hosts average three developing parasite progeny.

### BRACHYMERIA INTERMEDIA--GYPSY MOTH PUPAL POPULATIONS

The introduction of parasites has been one of the major approaches to the control of the gypsy moth. Efforts are continuing in the joint effort by Federal and State agencies to develop control measures for this alien forest pest. Among the several parasites that have become established is *Brachymeria intermedia.*

Information on the life history, habitats and natural rates of parasitization due to *B. intermedia* was obtained by consulting a number of publications including Burgess and Crossman (1929), Dowden (1935), Leonard (1967), Doane (1971). In addition, personal discussions were held with W. W. Metterhouse of the New Jersey Department of Agriculture, who has investigated this parasite in the laboratory and field. A number of authorities on the dynamics of the gypsy moth were consulted in efforts to assign realistic values to the number of host pupae available to the parasite during the different generations in a population cycle. As for all pests, good information is not available on host population densities. However, the densities assigned as shown in Table 4 are believed to be reasonably representative of actual numbers during the periods of minimum and maximum densities in a favorable gypsy moth ecosystem where the pest has been well established for some years. If the values are reasonably valid for the minimum and maximum densities and the number of generations involved in a periodic cycle is also representative, the rates of increase each generation are also likely to be reasonably representative.

I believe that the host population model shown in Table 4 is sufficiently accurate to serve as a basis for making a meaningful estimate of the efficiency of the parasite. If a population of gypsy moth pupae averages about 12,000 per acre when defoliation is expected the next year, we can arrive at a reasonable estimate of the number of egg masses by assigning realistic mortality values to the pupae, adult, and egg states. If only 40% of the pupae at this high density survive to produce adults, the emerging adult population would number 2,400 males and 2,400 females per acre assuming equal numbers of the sexes. If 50% of the females survive to deposit eggs, which seems reasonable based on observations by Beroza et al. (1974) and Cameron (1973), the egg masses would number 1,200 per acre in the late summer. Most authorities consider that 1,000 or even fewer egg masses per acre in certain forested areas will lead to defoliation the following year. Gypsy moth populations generally collapse after the year of defoliation, except in new areas of spread.

Based on the host population trends and with limited information on natural rates of parasitization, I estimated that 100 *Brachymeria* adults per acre during the period of pupal development will be

Table 4. A theoretical model depicting the codevelopment of a gypsy moth pupal population and the parasite *Brachymeria intermedia*, starting with 2 parasites in host generation 2. The parasite is assumed to have a P-50 host-finding value of 100 per acre per host generation.

| Genera-tion | No. Pupae | Expected increase rate | No. Para-sites | % Parasit-ism | No. Para-sitized pupae | No. Parasite survival | No. Parasite progeny | No. Unparasitized host pupae |
|---|---|---|---|---|---|---|---|---|
| 1 | 75 | 6 | 0 | 0 | — | 0 | 0 | 75 |
| 2 | 450 | 4.7 | 2 | 1.4 | 6 | 14 | 1 | 444 |
| 3 | 2,085 | 3.2 | 1 | 0.7 | 15 | 13 | 2 | 2,070 |
| 4 | 6,625 | 1.9 | 2 | 1.4 | 93 | 12 | 11 | 6,532 |
| 5 | 12,411 | 1.0± | 11 | 7.3 | 906 | 11 | 100 | 11,405 |
| 6 | 11,505 | 1 | 100 | 50 | Host population collapse after defoliation due to peaking action of various density-dependent suppression factors | | | |

capable of parasitizing 50% of the hosts.  Different efficiency
values and different survival rates for the parasite were assumed
and tested in preliminary models before arriving at this estimate.
The parasite population is assumed to start with 2 in host generation 2.

There is no information on the number of parasites in natural
populations or on survival rates of parasite progeny from the time
they begin developing in host pupae until adults begin parasitizing
host pupae the next season.  Despite the lack of information on
these parameters, I believe the codeveloping parasite population
shown in Table 4 is realistic.

The rates of natural parasitization shown in the model during
the various host generations would indicate that B. *intermedia* is
a minor factor regulating the abundance of the gypsy moth.  A maximum of 50% natural parasitization (in host generation 6) is in
line with the observations of Doane (1971).  W. W. Metterhouse
(personal communications) has observed that natural rates of parasitization in New Jersey seldom exceed 30 to 40%.  According to
data published by Leonard (1967), natural parasitization by
*Brachymeria* when the host population is low to moderate seldom
exceeds 5%.

On the basis of such general observations on natural parasitization levels, I consider the trends shown in Table 4 to be a satisfactory representation of natural parasitization by B. *intermedia*.
Such parasite developing on its own cannot be regarded as a major
factor in keeping a gypsy moth population below the defoliation
level.


VALIDITY OF THE BASIC HOST-FINDING EFFICIENCY ESTIMATES

The accuracy of the basic host-finding efficiency estimated
for the several parasites is obviously subject to question because
the degree of accuracy of some of the parameters on which the
estimates are based is unknown.  Extensive field release studies
with measurement of rates of parasitization achieved at various
host densities will be necessary to determine if the postulated
values are close to the true efficiency.  It should be emphasized,
however, that efficiency values obtained by the conventional
release of parasites and measurement of rates of parasitization
could also deviate considerably from the true efficiency.  Unknown
and uncontrolled variables may be involved in actual field release
experiments that could lead to errors in the final conclusions that
may be of the same or greater magnitude as the errors that would
result from estimates derived from simulation models.  In my view,
the factor of excessive dispersion of highly mobile parasites from

release areas that may be no larger than a few hundred acres could
cause a greater error in the measurement of the host-finding
efficiency of a parasite than the error that is likely to be made
by using the modeling procedure described.  Some parasites may
readily disperse for several miles and the rate is likely to vary
at different host densities.  Unless the release area is several
miles in diameter, the number that fly out of a small study area
will not be compensated by a comparable number that can enter from
surrounding areas.

It is my opinion that the P-50 values for the various parasites
based on the modeling procedure will fall within the range of 0.5 to
2.0 of the true efficiency value.  In other words, the number of
parasites required to achieve 50% host parasitization will not be
less than half or more than two times the estimated efficiency.
If this is a valid appraisal, I would regard the method employed
to be highly useful in estimating the number of parasites and the
degree of parasitization to expect from a self-perpetuating parasite
population and particularly in evaluating the potential of the
parasite when numbers are augmented by artificial means throughout
a large pest ecosystem.

The rates of parasitization of *Heliothis* eggs resulting from
releases of various numbers of *Trichogramma* reared on eggs of
*Heliothis zea*, indicate that the P-50 value of 5,000 per unit of
area is close to reality (Lewis et al. 1976).  It is estimated by
the modeling procedure that hymenopterous parasites such as
*Microplitis croceipes* (Cresson) and *Cardiochiles nigriceps* Viereck
attacking larvae of *Heliothis* have a P-50 value of about 300.  Lewis
et al. (1972) counted the numbers of adult *Cardiochiles nigriceps*
observed attacking *Heliothis virescens* in a given area on cotton and
then collected larvae to determine the rate of parasitism.  The
rates of parasitism recorded indicate that the efficiency estimate
for these parasites is of the correct magnitude.

Dr. James Smith, with the U.S. Department of Agriculture laboratory at Stoneville, Mississippi, has made some preliminary estimates
of the efficiency of *Lixophaga diatraeae* based on results of field
releases of parasites in Florida for the control of the sugarcane
borer.  The results of his studies have not been published, but in
personal communications he expressed the view that the host-finding
efficiency previously estimated by theoretical models seems close
to reality.

Data obtained by Grimble (1975) on rates of parasitization of
gypsy moth pupae following the release of *B. intermedia* in a forested
area were analyzed.  The releases were made in an area containing
high densities of host pupae.  The purposes of the releases were to
study the dispersal behavior and rates of parasitization produced

by the parasites.  From the number released, the size of the area
where parasites dispersed, and the rate of parasitization recorded,
a P-50 value of 100 for this parasite must be quite accurate.

If the same basic host population models and other values
established for the various parasite-host complexes are used but
the P-50 efficiency is assumed to be much higher or much lower,
the rates of parasitization will deviate greatly from those calculated.

As an example, we will consider the model in Table 3 which
shows a codeveloping population of *Heliothis* larvae and a *Eucelatoria* parasite based on the assumption that the parasite has a P-50
value of 400.  The rate of parasitization according to the hypothetical model would range from a minimum of 2.6 to a maximum of
15.9%.  The total number of parasites in the hypothetical ecosystem,
which is a better measure of the parasite population trend, would
range from 50,000 during the first host generation to 455,000 in
host generation 4.  If the P-50 value of the parasite is assumed
to be 200 in one case and 800 in another, but all other parameters
remain constant, the rates of parasitization will deviate widely
from the rates shown in the model.  Table 5 shows the calculated
parasitization trends for P-50 values of 200 and 800.

The calculations indicate that if the P-50 value for *Eucelatoria*
was of the order of 200, the rate of parasitization would reach
such high level by generation 5 that the overwintering host generation would be virtually eliminated.  On the other hand, if the P-50
value is of the order of 800, the parasite would have difficulty
maintaining a viable population.  It is questionable if such parasite could survive.  The total number of parasites each host generation should be noted in particular in order to fully appreciate the
degree of sensitivity of the efficiency parameter.

If substantially higher or lower values are assigned to the
host density, size of the host-searching environment, and the rates
of survival but the P-50 value of 400 remains constant, the parasitization rates would also differ substantially from those shown
in Table 3.  Therefore, the accuracy of the host-finding efficiency
value is governed by the accuracy of each basic parameter.

I believe, however, that the estimates for the host density,
size of the habitat, and survival rates for the parasites are
sufficiently accurate to conclude that a P-50 value of 400 per acre
per host generation for *Eucelatoria* is a realistic efficiency
estimate.  I have reached the same general conclusion for the
estimated efficiency of other parasites and host complexes that have
been evaluated.

Table 5. The theoretical trend of parasitization of *Heliothis* larvae by *Eucelatoria* parasites assuming that the P-50 efficiency of the parasite is 200 or 300 per acre per host generation. (See Table 3 for the trends based on assumed P-50 value of 400 per acre per host generation)

| Genera-tion | Acres | Hosts per acre | Increase rate | Para-sites per acre | % Para-siti-zation | Para-sitized hosts per acre | Potential progeny per acre | % Parasite survival to adult | Parasite progeny Per acre | Parasite progeny Total |
|---|---|---|---|---|---|---|---|---|---|---|
| **1. P-40 = 200** | | | | | | | | | | |
| 1 | 1,000 | 1,000 | 4.5 | 50[1/] | 14.9 | 159 | 477 | 20 | 94 | 95,000 |
| 2 | 3,000 | 1,500 | 3 | 32 | 10.5 | 153 | 474 | 17.5 | 83 | 249,000 |
| 3 | 4,500 | 3,000 | 1.7 | 55 | 17.2 | 516 | 1,548 | 15 | 232 | 1,044,000 |
| 4 | 5,000 | 4,500 | 0.9 | 209 | 50.1 | 2,255 | 6,765 | 12.5 | 846 | 4,230,000 |
| 5 | 4,500 | 4,500 | -- | 940 | 96.0 | 4,320 | 12,960 | Overwintering generation | | |
| **2. P-40 = 800** | | | | | | | | | | |
| 1 | 1,000 | 1,000 | 4.5 | 50 | 4.2 | 42 | 126 | 20 | 25 | 25,000 |
| 2 | 3,000 | 1,500 | 0 | 8 | 1.0 | 10 | 30 | 17.5 | 5 | 15,000 |
| 3 | 4,500 | 3,000 | 1.7 | 3 | 0.3 | 9 | 27 | 15 | 4 | 13,000 |
| 4 | 5,000 | 3,400 | 0.9 | 4 | 0.3 | 14 | 42 | 12.5 | 5 | 25,000 |
| 5 | 4,500 | 4,500 | -- | 6 | 0.5 | 22 | 66 | Overwintering generation | | |

[1/] Assumed.

While estimates derived through modeling may be subject to considerable error, it may be difficult and costly to make a more accurate efficiency estimate by conventional field releases and measurement of rates of parasitization because of the many variables that are encountered in field studies, especially in small release experiments. Ultimately, however, actual field releases and measurement of rates of parasitization will be the only way to determine how effective and practical the augmentation method will be as a system of pest management. Pending definitive data based on appropriate field studies, it is my opinion that the host-finding efficiency values assigned to the various parasites are sufficiently accurate to make a reasonably good appraisal of the impact on a host population when the natural parasite population is greatly increased by artificial means throughout a pest ecosystem. This will be the subject of analysis and discussion in the next portion of this paper.

### POTENTIAL OF THE PARASITE AUGMENTATION METHOD FOR REGULATING PEST POPULATIONS

As noted in earlier discussions, the augmentation method of insect pest control by the use of parasites has met with little success in the United States in the past. Results obtained by this method of control are discussed by others participating in this symposium (see chapter by Ridgway et al., chapter 13). The lack of success, in my opinion, has been due to inadequate or improper application of the technique rather than inability of well-adapted biological agents to achieve effective insect control by the augmentation procedure. Generally poor results in past efforts may have been caused by one or more of the following factors: (1) The release of inadequate numbers of the parasite; (2) the release of parasites without regard to the need for a favorable host density at the time releases were made; (3) the use of strains poorly adapted to the environment or to the pest host; (4) the release of inferior laboratory-reared insects; and (5) the conduct of field trials on such small scale that excessive dispersal of the parasites from the release area and excessive infiltration of hosts into the release area obscured the results.

The lack of success in past efforts notwithstanding, it is my opinion that when appropriately employed, the augmentation of biological agents can be developed as an effective and practical management system for a number of important insect pests. This method of insect pest control should be particularly appealing because of the sound biological and ecological principles that would be involved.

Three tactics for regulating pest populations by programing releases of parasites may be considered. One system would involve repeated releases of moderate numbers throughout a pest ecosystem during a series of generations beginning when the host population is favorable but still well below economic numbers. The objective would be to maintain a low pest density primarily as a result of the direct effect of the parasites released but with some help from the parasite progeny. A second system would involve the release of substantial numbers of the parasite at a strategic time and would depend primarily on the natural host resources for a rapid and high buildup of the parasite population. The system is not to be confused with the release of small numbers for inoculative purposes only. A third method would involve the release of enough parasites after the pest approaches the economic level to achieve direct and immediate control of the host.

It has been demonstrated repeatedly by many investigators in the field of biological control that the introduction and establishment of a single host-specific parasite will cause a substantial reduction in the steady density of a pest population, even though in many instances the reduction is not enough to provide adequate control. It seems self-evident that if the number of the same biological agent can be greatly increased above normal numbers by artificial means in all portions of the pest environment, this will cause a greater decline in the steady density of the pest than can result from a self-perpetuating population.

We know that the action of self-perpetuating parasite populations is adversely affected by a number of unavoidable natural influences. These include (1) cyclic fluctuations in host numbers that leads to a low parasite population following low periods in a host cycle, (2) the lag time required for a greatly reduced parasite population to reach effective numbers, even after host densities become favorable, (3) high natural hazards to survival of parasite progeny developing in the hosts due to predation (this will limit the number of parasite progeny that can develop, even after host densities reach moderate to high levels), and (4) expansion of the size of the habitat for certain pest hosts and their parasites, which limits the number of parasites available for host seeking per unit of area and time.

If self-perpetuating populations cause substantial reductions in host populations despite such obstacles, there is every reason to rationalize that drastic reductions in host populations will occur if the natural constraints listed are largely overcome by releasing at the proper time much larger numbers of the same parasites that occur naturally. This possibility will be considered later.

## JUSTIFICATION FOR MAJOR RESEARCH EMPHASIS TO DEVELOP
## THE AUGMENTATION METHOD FOR PRACTICAL APPLICATION

Important developments have taken place in the past one or two decades that justify a complete reappraisal of the potential role that parasites and predators can play in pest population management systems.

(1) Despite the advantages of using conventional insecticides to cope with many kinds of pests, there is general realization that continued reliance on insecticides as the principal management technique is an unacceptable long-range solution.

(2) There is increasing appreciation of the vital role that natural biological control agents play in regulating the abundance of many pests, even if they alone do not provide adequate control. Reliance on the augmentation technique would not only provide control by the biological agents released, but would also assure maximum benefit from naturally developing control agents.

(3) Outstanding advances have been made in basic technology for rearing a wide range of insect species. With suitable support for necessary research and development, there should be no technical barriers to practical mass production of hundreds of millions and billions of most parasites or predators.

(4) More information on the host-finding efficiency of certain kinds of parasites based both on theoretical calculations and field observations is being obtained. From such information and estimated potential costs of mass production, it seems economically feasible to release enough of certain key parasites on a programed basis to exceed by many fold the populations that develop naturally.

(5) Enough information is being obtained on the dispersal behavior and dynamics of certain pests to indicate the necessity of a concerted attack on populations in large areas in order to achieve true pest population management.


## THEORETICAL IMPACT OF PARASITE RELEASES
## ON THE DYNAMICS OF PEST POPULATIONS

From the estimated efficiency of certain parasites, it should be possible to make a reasonably good assessment of the impact that programed releases of the parasites can have on the dynamics of the pest hosts. We will select the *Heliothis* species and the previously discussed *Eucelatoria* parasites as one model. This complex is an example of parasites and hosts that have several generations during a season followed by a winter hibernating period and a new seasonal

cycle each year. The gypsy moth and *B. intermedia* will be selected
as another model. The gypsy moth is representative of host species
that have only one generation per year and develop in periodic cycles.

From the theoretical models depicting self-perpetuating populations of *Trichogramma* and the sugarcane borer (Table 2), *Eucelatoria*
and the *Heliothis* species (Table 3), and *B. intermedia* and the gypsy
moth pupae (Table 4), I have drawn the following conclusions:

*It is not possible for self-generating populations of these
parasites to achieve levels of parasitization in time throughout
the ecosystem to prevent the hosts from exceeding the economic
threshold levels.*

The above assessment does not mean that the parasites in question or others facing similar limitations are of no value as natural
biological agents or that this applies to all parasite-host complexes.
I am convinced, however, that even in well balanced ecosystems
certain regulating forces exist that preclude the possibility of
parasites achieving adequate parasitization of many of the major
pests before they reach economic density levels.

On the basis of this hypothesis, we will consider what the
theoretical effect would be if a natural parasite population is
greatly increased by releasing large numbers of the same species at
a suitable time in the host cycle. *Eucelatoria* and *Heliothis* larvae
and *B. intermedia* and gypsy moth pupae will be used as models to
calculate the effects based on various parameters to be outlined.
Other species may be more efficient than those selected as models
but those chosen should serve as suitable models to support the
concepts. The theoretical impact of *Trichogramma* releases for
controlling the sugarcane borer has been considered in a previous
publication (Knipling and McGuire 1963).

## RELEASE OF *EUCELATORIA* FOR THE
## SUPPRESSION OF *HELIOTHIS* POPULATIONS

The basic parasite-host population model shown in Table 3 will
be used to calculate the impact of *Eucelatoria* releases on the
dynamics of a *Heliothis* population. Several assumptions must be
made to establish parameters necessary for the calculations.

(1) The parasite is assumed to have a 50% host-finding efficiency of 400 per acre per host generation. The equation described on
page   will be used to calculate the degree of parasitization
produced by higher or lower parasite populations.

(2) The potential rate of increase of the suppressed *Heliothis*

population each generation, except for generation 1, year 1, will
be 40% higher than for a normal uncontrolled population.  This
adjustment is made to allow for reduced hazards due to other
density-dependent suppression factors, when the host population
exists below the normal steady density range.  A population of
1,000 4th- to 6th-instar larvae per acre on 1,000 acres (a total
of 1 million) is considered a normal or average first-generation
population level in the hypothetical ecosystem.  A population of
4,500 larvae per acre on 4,500 acres (20,250,000 total) is considered a normal last seasonal generation population.  Thus, if
the overwintered survival rate is on the order of 5%, the next
season's population would again be about 1 million.  For the
suppressed population the survival rate of the overwintered pupae
is assumed to be 7%, or 40% higher, than that for an unsuppressed
population.

(3)  Each surviving parasitized host produces 3 parasite progeny.
The natural parasite population in generation 1, year 1, is assumed
to start at 50 per acre.  The surviving rate of the developing
parasite progeny each generation for an uncontrolled population is
indicated in Table 3.  The survival rate of developing parasites in
the population subjected to parasite releases, except for generation
1 of year 1, will be 10% higher each generation than for an uncontrolled population.  This makes allowance for reduced hazards to
the parasite when the host population is suppressed.  The surviving
natural parasite population during the winter is estimated to be
10% of the expected surviving population in generation 5.

(4)  Calculations are based on 1 acre of host plants, but
parasite releases will be made in a closed ecosystem consisting of
the acreage shown in Table 3 for each generation.

Table 6 shows the calculated trend of a *Heliothis* larval population (4th to 6th instar) in three successive years when 500
parasites per acre are released during the first three host generations each year.  The effect of the releases plus the naturally
developing progeny should be noted in comparison with the assumed
normal trend of an uncontrolled population in Table 3.

From the calculated effect, the *Heliothis* population in generation 5 of year 1 would be reduced by about 91%.  When allowance is
made for a 40% higher winter survival than normal, the starting
population in year 2 is estimated to be 87.4% below the normal
steady density for an uncontrolled population.  The host population
in year 2 would remain fairly stable in numbers per acre but because
of increasing acreage of host plants the total host population would
increase slightly but at a greatly reduced rate in comparison with
an uncontrolled population.  The increase by season's end would not
be enough to offset the approximately 93% winter mortality assumed.

Table 6. Calculated impact on *Heliothis* larval population trends (4 to 6 instars) when *Eucelatoria* parasites are released at the rate of 500 per acre per host generation during the first three host generations each year (theoretical).

| Host genera-tion | Host plant acres | Larvae per acre | Potential increase rate | Parasite per acre | | | Percent parasitism | Parasitized hosts/acre | Potential progeny per acre | Percent parasite survival | Total parasite progeny |
|---|---|---|---|---|---|---|---|---|---|---|---|
| | | | | Released | Natural | Total | | | | | |
| **Year 1** | | | | | | | | | | | |
| 1 | 1,000 | 1,000 | 4.5 | 500 | 50 | 550 | 61.6 | 616 | 1,848 | 20 | 369,000 |
| 2 | 3,000 | 576 | 4.2 | 500 | 123 | 623 | 66.0 | 380 | 1,140 | 19.3 | 660,060 |
| 3 | 4,500 | 548 | 2.38 | 500 | 147 | 647 | 67.3 | 367 | 1,101 | 16.5 | 817,493 |
| 4 | 5,000 | 384 | 1.26 | 0 | 163 | 163 | 24.6 | 94 | 282 | 13.7 | 193,171 |
| 5 | 4,500 | 405 | -- | 0 | 43 | 43 | 7.2 | 29 | 87 | Overwintering generation | |
| **Year 2** | | | | | | | | | | | |
| 1 | 1,000 | 126 | 6.3 | 500 | 4 | 504 | 58.2 | 73 | 219 | 22 | 48,180 |
| 2 | 3,000 | 111 | 4.2 | 500 | 16 | 516 | 59.1 | 66 | 198 | 19.3 | 114,642 |
| 3 | 4,500 | 127 | 2.38 | 500 | 25 | 525 | 59.7 | 76 | 228 | 16.5 | 169,291 |
| 4 | 5,000 | 97 | 1.26 | 0 | 34 | 34 | 5.7 | 6 | 18 | 13.7 | 12,330 |
| 5 | 4,500 | 128 | -- | 0 | 3 | 3 | 0.5 | <1 | 2 | Overwintering generation | |
| **Year 3** | | | | | | | | | | | |
| 1 | 1,000 | 40 | 6.3 | 500 | <1 | 500 | 58.0 | 23 | 69 | 22 | 15,180 |
| 2 | 3,000 | 35 | 4.2 | 500 | 5 | 505 | 58.0 | 20 | 60 | 19.3 | 34,740 |
| 3 | 4,500 | 41 | 2.38 | 500 | 8 | 508 | 58.5 | 24 | 72 | 16.5 | 53,460 |
| 4 | 5,000 | 36 | 1.26 | 0 | 11 | 11 | 1.8 | <1 | 2 | 13.7 | 1,370 |
| 5 | 4,500 | 49 | -- | 0 | <1 | 1 | 0 | 0 | 0 | Overwintering generation | |

Consequently, the starting population in year 3 would be even lower than in year 2.

It is pointed out that the decline in the population takes into account a considerably higher than normal increase rate each generation for the suppressed population. If the potential increase rates each generation and the higher winter survival rate assumed for the suppressed population were applied to an uncontrolled population, the maximum and minimum steady density population would be approximately five times as high as for a normal uncontrolled population. This may be a higher adjustment than is justified for pests like *Heliothis*, which are affected by a wide range of general parasites and predators, the abundance of which is not likely to be greatly reduced during the season because of a substantial reduction of only one of their many hosts or prey.

The trend of the hypothetical *Heliothis* population subjected to the parasite releases, in comparison with an uncontrolled population, is shown graphically in Fig. 3.

From the assumed effect of the parasite releases, a *Heliothis* population would continue to decline each year unless replenished by immigrants. Theoretically, the parasite release rate assumed for the model would be higher than necessary for effective management of *Heliothis* populations. Calculations, not shown, indicate that a *Heliothis* population would continue to decline each year if only 400 parasites per acre were released during each of the first three generations. Calculations also show that a population would be held essentially stable at about 90% below the normal steady density range if the release rate was reduced to 300 per acre.

The model is based on an assumed steady density *Heliothis* population. The populations of these pests fluctuate widely in different habitats and in different years. However, if conditions should be unusually favorable for the host, this would also make conditions more favorable for the released parasites. This should result in greater suppressive action than calculated. Theoretically, it would be difficult for the *Heliothis* hosts to make enough adjustment in the rate of survival to overcome the suppressive effect of the high and constant parasite populations maintained in the early generations.

## RELEASE OF *BRACHYMERIA INTERMEDIA* FOR THE SUPPRESSION OF GYPSY MOTH POPULATIONS

As before, the naturally developing parasite-host population in Table 4 will be used to calculate the effect of release of adult *B. intermedia* at the rate of 100 per acre. As previously noted,

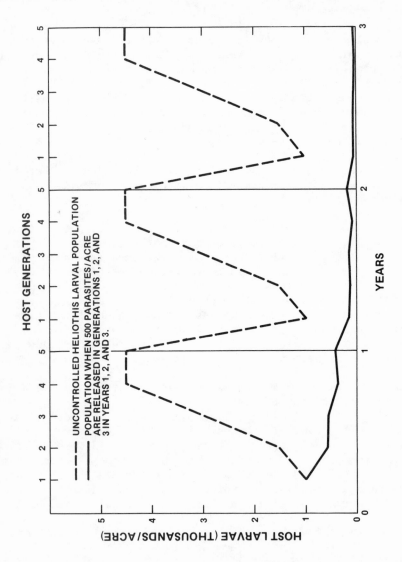

Fig. 3. Theoretical seasonal trends in an uncontrolled *Heliothis* larval population, and a population subjected to the release of 500 adult *Eucelatoria* parasites per acre of host plants in each of the first three host generations.

the P-50 value for this parasite is estimated to be 100 per acre
per host generation.  Releases are programed during only one generation (year).  The natural host population is assumed to provide
resources for subsequent development of parasite progeny.

The impact on the gypsy moth population will be calculated
when releases are made in generations 1, 2, or 3, in order to show
that the host density is a critical factor governing the dynamics
of the parasite population and the role it can play in host
suppression.

The following assumptions are made:

(1)  A representative normal uncontrolled population of the
gypsy moth and the associated natural parasite population will
develop as shown in Table 4.  Calculations are based on the number
of hosts and parasites on 1 acre of forest habitat, but parasite
releases would be made in a large gypsy moth ecosystem so that there
would be equal opportunity for released parasites and their progeny
to move into or out of any given portion of the release area.

(2)  The parasite is assumed to have one generation per year
and the only significant host resource is the gypsy moth.  The parasite survives the winter as adults.

(3)  The potential rate of increase of the suppressed population
will be calculated on the basis of the rates of increase for the
normal pupal populations existing at various densities as shown in
Table 4.  For example, if a pupal population exists at a level of
4,709, this would fall between the level of 2,035 in generation 3
of an uncontrolled population when the increase rate is assumed to
be 3.2-fold and the level of 6,625 in generation 4 when the increase
rate is 1.9-fold.  The difference in the number of pupae at the two
density levels is 4,590, and the difference in the increase rates
at these two density levels is 1.3-fold.  A population existing at
4,709 would be 2,674 above the level of 2,035 when the increase rate
is 3.2-fold.  Therefore, the rate of increase is calculated to be

$$3.2 - \frac{2,674}{4,709} \times 1.3 \text{ or } 3.2 - 0.7 = 2.5$$

Adjustments in the survival of the parasite progeny at various
host density levels are calculated in a similar manner.

(4)  A pupal density less than 6,000 per acre is assumed to be
noneconomic; a host density ranging between 6,000 to 10,000 is
assumed to cause considerable ragging of foliage; and a density above
10,000 is assumed to cause defoliation.

The Impact of Parasite Releases in Host Generation 1
When the Natural Population Averages 75 Pupae Per Acre

Table 7 shows the effect of parasite releases during host generation 1 when the new host cycle is assumed to start at a level of
75 pupae per acre.

The theoretical impact of parasite releases made when the host
density is at such low level is a dramatic portrayal of the influence the host density has on the dynamics of a parasite population.
Even though the 100 parasites per acre that were released would
theoretically parasitize 50%, or 38, of the available hosts, a
survival rate of only 15% of the potential progeny would reduce
the next year's parasite population to 6 per acre. The number of
parasites would parasitize only 4%, or 9, of the available hosts
in generation 2 (year 2). Even though the host density is considerably higher in year 2 than in year 1, this decrease in the rate
of parasitization would result in a further decline in the parasite
population to an average of 1.3 per acre in year 3 and less than 1%
parasitization. The host population in the meantime could increase
at virtually a normal rate after the initial impact of the 100
released parasites. It would be year 5 in the population cycle
before there would be a strong upward trend in the number of parasites produced naturally. But the increase starts from a low parasite population and comes too late for the parasites to have a significant effect on the growth rate of the host population. Defoliation would occur in year 6 because of the presence of 11,654 pupae
by year 5 when only 4% are assumed to be parasitized. Theoretically,
enough parasite progeny would develop in year 5 to parasitize 30.3%
of the pupae in year 6, but this would occur after the larvae defol-
iated the forest.

The model depicted in Table 7 raises a fundamental question.
Is this typical of the usual development of a host-specific parasite
population following the collapse of a cyclic host population? It
is my theory that the type of density relationship between a host
and a host-specific parasite population depicted in the model has
evolved as a natural regulating mechanism for many parasite-host complexes. Such relationship would assure a successfully coexistence of
both the host and any associated parasite that is completely dependent
on the host for its existence. Regardless of the number of parasites that may be produced when the host population reaches the peak,
a collapse in the host population, which might be caused by many
density-dependent factors other than the parasite, would assure that
a comparable reduction in a host-specific parasite population would
follow one or two generations later. If factors independent of
density are favorable and the influence of density-dependent factors
is relaxed, this would give the host every opportunity to again begin
a new and successful growth cycle with minimal control in the early

Table 7. The theoretical impact on the trend of a gypsy moth pupae population following the release of 100 *Brachymeria intermedia* adults per acre during host generation 1.

| Genera-tion | No. Pupae | Expected increase rate | No. Para-sites | % Parasit-ism | No. Parasi-tized pupae | % Parasite survival | No. Parasite progeny | No. Unparasitized host pupae |
|---|---|---|---|---|---|---|---|---|
| 1 | 75 | 6 | 100 | 50 | 38 | 15 | 6 | 37 |
| 2 | 225 | 5.5 | 6 | 4 | 9 | 14.6 | 1.3 | 216 |
| 3 | 1,188 | 4.0 | 1.3 | 0.9 | 11 | 13.6 | 1.5 | 1,177 |
| 4 | 4,709 | 2.5 | 1.5 | 1.0 | 47 | 12.4 | 6 | 4,662 |
| 5 | 11,654 | 1.2+ | 6 | 4.0 | 466 | 11.2 | 52 | 11,138 |
| 6 | 13,425 | | 52 | 30.3 | Probable collapse of host population | | | |

generations by any and all of the biological agents that are strongly
or completely dependent on the host for their existence.

If the model is valid in principle, this would not only provide
a graphic illustration of the mechanism of action of density-dependent suppression forces for cyclic pests, but it would clearly
indicate how and why the augmentation of key natural parasites in
substantial numbers could drastically alter the normal ecological
balance between a host and an associated parasite.

It is my conclusion that the release of B. *intermedia* when the
host density is minimal would be highly inefficient.  The final
result is no higher degree of control than might be expected from
a self-perpetuating population starting with 2 parasites per acre
in year 2.  If a natural population of B. *intermedia* normally
exists at about 100 or more adults per acre the year following a
collapse of the gypsy moth population and the new host cycle
normally starts at a level of about 75 pupae per acre, the parasite
population would decline to an average of about 2 per acre and then
begin to increase only when the host population reaches a moderate
level.  This is why the basic codeveloping parasite host population
shown in Table 4 is assumed to start with 2 parasites in host generation 2.

Parasite Releases When the Host
Density is 450 Per Acre (Year 2)

Table 8 shows the effect of parasite releases in host generation
2, when the density is assumed to be 450 per acre.  This level of
density is not optimal but is much more favorable than a density of
75 pupae per acre.  The number of surviving progeny resulting from
the assumed 50 percent parasitization rate would not equal the
number released, but enough would be produced to lead to high rates
of parasitization before the host population reaches a high level.
The number of host pupae that escape parasitism would be low enough
to assure a subeconomic population.

A point might be made that is relevant to the concept of gypsy
moth management via releases of a parasite such as B. *intermedia* or
any other similar species.  If the assumed normal minimum host
density after a host population collapse is realistic and if other
parameters are also realistic, it would be impossible for a host-
specific parasite to reach a natural population level by host generation 2 that even approaches a released population of 100 per acre.
Hence, we cannot expect a self-perpetuating host-specific parasite
population to become an important host regulating factor during the
early generations of the host cycle.  The only way such parasite
could be present in effective numbers throughout a large host

Table 8. The theoretical impact on the trend of a gypsy moth pupae population following the release of 100 *Brachymeria intermedia* adults per acre during host generation 2.

| Generation | No. Pupae | Expected increase rate | No. Parasites | % Parasitism | No. Parasitized pupae | % Parasite survival | No. Parasite progeny | No. Unparasitized host pupae |
|---|---|---|---|---|---|---|---|---|
| 1 | 75 | 6 | 0 | 0 | 0 | — | — | 75 |
| 2 | 450 | 4.7 | 100 | 50 | 225 | 14 | 32 | 225 |
| 3 | 1,058 | 4.1 | 32 | 19.9 | 210 | 13.6 | 29 | 348 |
| 4 | 3,475 | 2.8 | 29 | 18.2 | 632 | 12.7 | 80 | 2,843 |
| 5 | 7,959 | 1.7 | 80 | 42.6 | 3,390 | 11.8 | 400 | 4,569 |
| 6 | 7,766 | 1.8 | 400 | 93.7 | 7,276 | 11.9 | 866 | 490 |
| 7 | 880 | 4.3 | 866 | 99 | 871 | 13.8 | 120 | 9 |

ecosystem would be to augment numbers by artificial means or for
large numbers to migrate into the area from some distant source.


Parasite Releases When the Host
Density is 2,115 Per Acre (Year 3)

The release of parasites in host generation 3 (Table 9) would
be favorable for an immediate increase in the parasite population.
The number of progeny produced would exceed the number released and
would cause a higher rate of parasitization of the next host generation than the original released population.  The rate of increase
of the host population would be slow because of the fairly high
rates of parasitization, but enough hosts would be present to result
in a continuing increase in the parasite population which would
reach a sufficiently high rate of parasitization to cause a collapse
in the host population before it reaches the defoliation level.
The number of pupae escaping parasitization in the various generations would not even approach the density assumed necessary for
defoliation.

It is concluded that the release of parasites when the host
density is of the order of 2,000 per acre would result in near
optimum effect.  Calculations are not shown, but the release of
parasites in host generation 4 when the host density is assumed to
exist at 6,768 would be too late for optimum effect.  While the
release of parasites at this host density would prevent damage
because of the 50% parasitization caused by the released parasites,
the host population that escapes parasitization would be high enough
to approach the economic damage level before enough parasite progeny
develop to cause a host population collapse.

For an overall perspective of the degree of control resulting
from B. *intermedia* releases at different host density levels, it is
necessary to consider the number of pupae that escape parasitization.
The data in Tables 7, 8, and 9 are shown graphically in Fig. 4.  The
graph shows the number of pupae that escape parasitization each
generation after releases are made at different host density levels.
The number of pupae that escape parasitization in a population
subjected to a normal self-perpetuating patasite population (Table
4) is also shown for comparison.

If the graphs in Fig. 4 reasonably reflect results that can be
obtained by releasing B. *intermedia* at the proper time at the rate
of 100 per acre in a large gypsy moth ecosystem, this would be an
excellent example of the potential of the augmentation system for
insect control.

Data are not presented, but on the basis of the assumed

Table 9.  The theoretical impact on the trends of a gypsy moth pupae population following the release of 100 *Brachymeria intermedia* adults per acre during host generation 3.

| Generation | No. Pupae | Expected increase rate | No. Parasites | % Parasitism | No. Parasitized pupae | % Parasite survival | No. Parasite progeny | No. Unparasitized host pupae |
|---|---|---|---|---|---|---|---|---|
| 1 | 75 | 6 | 0 | 0 | 0 | — | 0 | 75 |
| 2 | 450 | 4.7 | 0 | 0 | 0 | — | — | 450 |
| 3 | 2,115 | 3.2 | 100 | 50 | 1,058 | 13 | 138 | 1,057 |
| 4 | 3,384 | 2.8 | 133 | 61.6 | 2,084 | 12.7 | 265 | 1,300 |
| 5 | 3,638 | 2.8 | 265 | 84.1 | 3,060 | 12.6 | 386 | 578 |
| 6 | 1,620 | 3.6 | 386 | 93.1 | 1,508 | 13.3 | 200 | 112 |
| 7 | 402 | 4.9 | 200 | 75 | 301 | 13.9 | 42 | 101 |

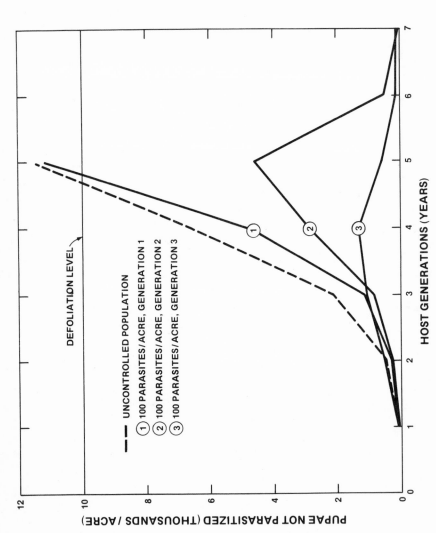

Fig. 4. Theoretical influence of the density of gypsy moth pupae on the dynamics of a *Brachymeria* parasite population and rates of parasitism in subsequent host generations following the release of adult parasites in various generations in the periodic cycle of the host. Data for the graphs are shown in Tables 4, 7, 8, and 9.

efficiency of the parasite, it should be possible to obtain
adequate control by releasing fewer than 100 parasites per acre if
releases are made at a highly favorable host density. Also, it
should be possible to achieve immediate control of a threatening
gypsy moth population by releasing for 1 year a larger number of
parasites per acre. Based on the efficiency equation, if a release
of 100 parasites per acre results in 50% parasitization, a release
of 300 per acre should result in about 87.5% parasitization. This
level of parasitization should assure adequate control of a population that would otherwise lead to defoliation the following year.
Enough parasite progeny should also develop to cause a collapse of
the population the second year following such release.

## GENERAL DISCUSSION

Pilot experiments and field observations will be necessary
with various parasite-host complexes to determine if the results
obtained by the modeling technique are reasonably predictive of the
results of regulating pest populations by augmenting natural parasite populations. The quantitative values assigned for host densities and trends, the number of parasites in the initial population,
survival rates for the parasites, and the nature of the host-searching environment, which are necessary parameters for estimating the
basic efficiency of a parasite, are all subject to considerable
error because definitive information on these key parameters is
lacking for every major pest and the associated parasites that have
been considered. If the error in assigned value for each parameter
is substantial and in the same direction, the error in the estimated
host-finding efficiency would be quite large. As previously dis-
discussed, a precise estimate of the efficiency of a parasite cannot
be assured when accurate information on key parameters does not
exist. But, if the estimated efficiency does not deviate by more
than a factor of about 2 from the true efficiency, the modeling
technique used should still be of great value in estimating the
role that self-perpetuating populations of certain parasites can
play as natural control agents. More importantly, a reasonably
good estimate of the efficiency of a parasite will indicate the
potential such parasite has in an augmentation system.

Codeveloping parasite-host models have been developed that
duplicate the trends and rates of parasitization generally observed
under natural conditions for several parasite-host complexes. I
have most confidence in the accuracy of models depicting populations
of *Trichogramma* and the eggs of the sugarcane borer (Table 2 and
Fig. 2); *Lixophaga* and larvae of the sugarcane borer (Knipling
1972); *Brachymeria intermedia* and gypsy moth pupae (Table 4); and
braconid parasites of aphids (Knipling and Gilmore 1971). If the
potential efficiency for these and other parasites is reasonably

close to reality, this in turn would indicate that the suppression
models will also indicate the effect on the dynamics of the pest
populations that can be expected to result if substantial numbers
of parasites are released throughout the ecosystem during critical
periods in the pest cycle.

It is my conviction, however, that the true efficiency of a
parasite and the influence it will have on the dynamics of a host
population cannot be determined accurately unless parasite releases
are made throughout a pest ecosystem or at least in a large segment
of a pest ecosystem and at various host densities.  For highly
mobile species, the dispersion of either or both the parasites and
host, which may aggregate many miles after several generations, is
likely to preclude accurate measurement of the effect of parasite
releases in areas consisting of several hundred or even several
thousand acres.

The parasite augmentation approach offers suppression characteristics highly important in practical pest management.  If a biological agent is largely or completely pest specific, this would
assure minimal adverse effect on general insect parasites and
predators.  This is a benefit largely lost when broad-spectrum
chemicals are used.

The augmentation method provides a system of control that
would have maximum effect on a pest population when and where it is
most urgently needed.  While simulated suppression models are assumed
to be representative of average numbers of a pest in a given area
and time, it is recognized that pest densities in a natural ecosystem are likely to vary from habitat to habitat within the total
pest ecosystem.  With increasing knowledge of the existence and
nature of kairomones that stimulate host seeking and serve as a
host-guidance mechanism (Lewis et al. 1975), it seems probable
that highly mobile parasites or predators will tend to concentrate
in nitches having the highest host densities.  This characteristic
could be of great practical significance in effective pest management.  Calculations have been made, but not included in this paper,
which indicate that the overall efficiency of a parasite population
is increased substantially if host distribution is variable, but
the number and intensity of host seeking by a released parasite
population varies according to host density.  No other system of
insect control possesses the unique characteristic of producing the
highest suppressive action in localized habitats having the highest
pest density.

The immediate parasitism produced by parasites (or predators)
that are released is one of the important suppression effects resulting from the augmentation method.  However, the host furnishes the
natural resources for the development of parasite progeny that can

continue suppressive action in subsequent generations. Thus, the method provides for lasting suppressive action that will become increasingly effective if the host population remains favorable. This is another unique characteristic not possessed by any other insect-control technique, with the exception of certain insect microbial agents.

The characteristic of maximum effectiveness at the highest host density together with selective action against the target pest makes the augmentation system highly compatible for use with autocidal and sex attractant methods of pest population management. These two relatively new techniques of insect control now under investigation on a wide range of important pests are also target-pest specific. Their suppressive action is also density-dependent, but they have minimal efficiency at high pest densities and maximum efficiency at low pest densities. Since the augmentation system has opposite efficiency characteristics, parasite releases and concurrent or sequential use of autocidal or pheromone control techniques could provide a highly complementary suppressive system. Concurrent suppression due to the released biological agents will potentiate the efficiency of the other two techniques.

Theoretical calculations presented in this publication strongly support the concept of insect control by augmenting natural enemies. However, there are also sound reasons based on rationalization for having confidence that the augmentation of well chosen parasites can be developed as an effective and practical system for the management of a wide range of pests that normally are not adequately controlled by natural enemies. It is well established that a single species of parasite or predator developing on its own can reduce the steady density of its primary host, even though the reduction may be inadequate or come too late to assure that the host will not exceed the economic damage level. There is every reason to assume that if the same biological agent is appropriately distributed in a pest ecosystem in numbers that exceed by manyfold the natural population of the agent, this will completely alter the normal parasite-host density relationship which limits the degree of action of natural self-perpetuating parasite populations.

The logic of such rationalization may be challenged on the grounds that the results of past efforts to control insects pests by augmenting the number of parasites have in general been negative. To my knowledge, no experiments have ever been undertaken, however, to determine the impact on a host population if a host-specific parasite is increased by artificial means throughout a pest ecosystem by as much as 10 to 50 times above normal numbers during the early generation of a pest population cycle. Until such experiments are conducted with suitable parasites and on a suitable scale, it would seem that the potential of the augmentation system cannot be assessed.

Many augmentation efforts have involved the release of what I
consider inadequate numbers of parasites in areas that may not have
involved more than a few hundred acres of crops, even though the
parasite and its host may be capable of dispersing for miles.  Such
experiments are common because of budget limitations.  The degree
of dispersion of *Pediobius foveolatus* (Crawford) observed by
Stevens et al. (1975) following the release of this parasite of the
Mexican bean beetle, *Epilachna varivestis* Mulsant, is just one indication of the degree of mobility that certain kinds of parasites
have in a pest ecosystem.

The release of large numbers of *Trichogramma* and perhaps
several other types of parasites or predators that are not highly
mobile, may be an effective and practical procedure for the control
of certain pests in small areas.  For many pests, however, I
believe that the augmentation techniques will not prove effective
and practical unless one or perhaps several well adapted and
reasonably host-specific parasites are released in substantial
numbers in an organized way in all or a large portion of a target-
pest ecosystem.

I recognize that the results of theoretical studies of the
nature reported in this paper are of no practical value to pest
management unless it can be shown that the economics of augmenting
natural enemies is favorable.  Thus, it seems important to analyze
this aspect of the problem even though good information is lacking
on potential costs for the mass production of various kinds of
parasites.

On the assumption that the predicted effects of parasites are
realistic, we will consider the potential costs and benefits of
releases of *Eucelatoria* parasites that attack the larval stages of
*Heliothis zea* and *H. virescens* and releases of *Brachymeria intermedia*
that parasitize the pupal stage of the gypsy moth.  For an economic
analysis, we will estimate the potential costs and benefits of the
augmentation technique in comparison with the costs and benefits of
control by the use of chemical insecticides.

According to estimates issued by the California Department of
Food and Agriculture, the cost of control for *H. zea* in that state
in 1974 was $22 million.  Despite this high cost, losses due to the
pest were estimated to be $47 million.  The hypothetical model
(Table 6) predicts that effective management of *H. zea* can be
achieved by releasing 500 *Eucelatoria* adults per acre during each
of the first three pest generations.  I have estimated that the
accumulated *Heliothis* host-plant acres in California's agricultural
areas during the first three generations will be of the order of 5

million.  Thus, on this basis the total number of parasites
required would be 2.5 billion.  Some progress has been made in
colonization procedures for *Eucelatoria* (Bryan et al. 1969).  No
concerted effort has yet been made, however, to develop mass
production technology for these and other *Heliothis* parasites on a
large scale.  It is my opinion that if a concentrated program of
research and development were undertaken by qualified scientists
and engineers, the chances would be good that a parasite like
*Eucelatoria* could be mass produced at a cost of $3 per thousand,
or $3 million per billion parasites.  On the basis of this rough
estimate, a management program for this major pest in California
would cost $7.5 million per year.  This would be only about one-
third the estimated cost of control by the use of chemicals.

Of greater economic significance, if the estimated impact on
the *H. zea* population projected in Table 6 could be realized in
practice, the pest population should be maintained at a level of
no economic consequence, except possibly on the most sensitive
crops.  This would mean that most or all of the estimated loss of
$47 million could be avoided.  Moreover, for those who are con-
cerned over the effect of insecticides in our agricultural environments, the development and use of a pest-specific method of control
may be equally or more important than the economic benefits of such
system of pest management.

California was selected for consideration of the economic
aspects in this case.  However, in the southern and southeastern
states, where both *H. zea* and *H. virescens* are pests of major importance on several crops, annual expenditures for chemical control
probably approach $100 million, and losses in spite of control may
amount to several hundred million.  What would it mean in terms of
economics and environmental quality improvement if half the cost of
chemical control were devoted each year to the mass production and
appropriate release of about 15 billion parasites of such species
as *Eucelatoria* spp. and *Microplitis croceipes* for the management of
these pests?  This question, in my view, deserves due consideration
and study.

Admittedly, the estimated efficiency of the parasites in question and the rearing cost estimates may be overly optimistic.  On
the other hand, they could be conservative.  Certainly, a procedure
for managing some of our major pests which could cost substantially
less than the present cost of chemical control and which could
return large economic and environmental benefits deserves more than
passing consideration in the long-range plans for dealing with
insect problems.

The economics of gypsy moth control by the release of a parasite like *B. intermedia*, also seems favorable when related to costs

of control by methods now employed.  Substantial progress has been
made in the development of rearing methods for this parasite, even
though mass production methods on a large scale in well designed
and equipped parasite-rearing facilities have not yet been investigated.  According to personal communications with W. W. Metterhouse
the parasite can be reared readily on larvae of the greater wax-
worm moth, *Galleria mallonella* (L.).  The larvae of this species
are also used for rearing the Cuban fly, *Lixophaga diatraeae*, which
parasitizes the sugarcane borer, *Diatraea saccharalis* (F.).  The
staff at the U.S. Department of Agriculture Bioenvironmental
Insect Control Laboratory at Stoneville, Mississippi, has made
substantial progress in rearing the parasite on the greater waxmoth.
If produced by the hundreds of millions per year in well designed
facilities, I believe that it would be realistic to assume that
mass production and release costs for *B. intermedia* could be as
low as $3 per 1,000 parasites.

If the release of as few as 100 *Brachymeria* parasites per acre
at the appropriate host density would have the impact shown in
Table 9, effective control could be achieved at a cost of 30 cents
per acre.  If this cost could be realized, it would be only a
fraction of the costs of control by the use of insecticides, which
is likely to amount to several dollars per acre.

According to the theoretical suppression model referred to,
the parasite augmentation system of control would probably not have
to be repeated more frequently than every 6 years.

The parasite release procedure would have to be based on good
information on the ecology and dynamics of gypsy moth populations.
Because of the mobility of the parasites, releases would no doubt
have to be made in large areas, where surveys indicate a threat of
economic populations.  Unlike chemical control, which has the advantage of producing immediate protection from populations that have
already reached the economic damage level, the parasite augmentation procedure would have to be implemented in areas where economic
damage is anticipated within 2 or 3 years.  However, the economic
and environmental benefits of the parasite augmentation system would
seem to justify this approach to gypsy moth management.

The two parasite-host complexes analyzed from the standpoint
of potential costs and benefits of the augmentation of natural
enemies as a system of management are only examples of the potential
of this approach to insect control.  Much research, both basic and
applied, is needed to develop effective and practical augmentation
systems.  Formidable problems may have to be resolved in perfecting
the techniques for each program that is undertaken.  Operational
programs would have to be well organized and executed with precision.
Yet, when analyzed in comparison with the costs and benefits of

current methods of insect control, the estimated costs of the augmentation method could be in error by a factor of five and still
offer a better solution to some of our important pest problems than
the uncoordinated chemical approach, which is not only costly and
ecologically undesirable but often only partially effective.

## LITERATURE CITED

Beroza, M., C. S. Hood, D. Trefrey, D. E. Leonard, E. F. Knipling,
    W. W. Klassen and L. J. Stevens. 1974. Large field trial
    microencapsulated sex pheromones to prevent mating of the
    gypsy moth. J. Econ. Entomol. 67: 659–664.
Bryan, D. E., C. G. Jackson, and A. Stoner. 1969. Rearing cotton
    insect parasites in the laboratory. U.S. Dept. Agric. Prod.
    Res. Rpt. 109. 13 p.
Burgess, A. F., and S. S. Crossman. 1929. Imported insect enemies
    of the gypsy moth and the brown-tail moth. U.S. Dept. Agric.
    Tech. Bull. 86: 116–113.
Burnett, T. 1959. Experimental host-parasite populations. Annu.
    Rev. Entomol. 4: 235–250.
Cameron, E. A. 1973. Disparlure: a potential tool for gypsy moth
    population manipulation. Bull. Entomol. Soc. Am. 19: 15–17.
Clausen, C. P. 1956. Biological control of insect pests in the
    continental United States. U.S. Dept. Agric. Tech. Bull. 1139.
    151 p.
DeBach, P. (ed.). 1964. Biological control of insect pest and
    weeds. Reinhold Publishing Corp., New York. 844 p.
DeBach, P. 1971. The use of imported natural enemies in insect
    pest management ecology. p. 211–233. In Proc. Tall Timbers
    Conf. Ecol. Animal Control by Habitat Management. No. 33.
Doane, C. C. 1971. A high rate of parasitization by *Brachymeria
    intermedia* (Hymenoptera: Chalcidae) on the gypsy moth. Annu.
    Entomol. Soc. Amer. 64: 753–754.
Dowden, P. B. 1935. *Brachymeria intermedia* (Nees) a primary para-
    site and *B. compsilurae* (Cwfd.) a secondary parasite of the
    gypsy moth. J. Agric. Res. 50: 495–523.
Grimble, D. G. 1975. Dispersal of released *Brachymeria intermedia*.
    State Univ. of N.Y. Applied Forestry Research Institute
    Research Note No. 16. 4 p.
Holling, C. S. 1966. The functional response of invertebrate pred-
    ators to prey density. Mem. Entomol. Soc. Can. (48): 3–86.
Howard, L. O., and W. F. Fiske. 1911. The importation into the
    United States of the parasite of the gypsy moth and the brown-
    tail moth. U.S. Dept. Agric. Entomol. Bull. 91, 344 p.
Huffaker, C. B. (ed.). 1971. Biological control (Proceedings of
    an AAAS Symposium on Biological Control, held at Boston, Mass.,
    Dec. 30–31, 1969. Plenum Press, New York and London, 511 p.
Jackson, C. G., D. E. Bryan, and R. Patana. 1969. Laboratory
    studies of *Eucelatoria armigera*, a tachinid parasite of
    *Heliothis* spp. J. Econ. Entomol. 62: 907–910.
Jaynes, H. A., and E. K. Bynum. 1941. Experiments with *Trichogramma
    minutum* Riley as a control of the sugarcane borer in Louisiana.
    U.S. Dept. Agric. Tech. Bull. 743. 43 p.
Knipling, E. F. 1971. Use of population models to appraise the
    role of larval parasites in suppressing *Heliothis* populations.

U.S. Dept. Agr. ARS Tech. Bull. 1434, 36 p.

Knipling, E. F. 1972. Simulated population models to appraise the potential for suppressing sugarcane borer populations by strategic releases of the parasite *Lixophaga diatraeae*. Environ. Entomol. 1: 1-6.

Knipling, E. F. 1977. The basic principles of insect population suppression and management. U.S. Government Printing Office, Washington, D.C. (in press).

Knipling, E. F., and J. E. Gilmore. 1971. Population density relationships between hymenopterous parasites and their aphid hosts--a theoretical study. U.S. Dept. Agric. ARS Tech. Bull. 1423, 34 p.

Knipling, E. F., and J. U. McGuire. 1968. Population models to appraise the limitations and potentialities of *Trichogramma* in managing host insect populations. U.S. Dept. Agr. Tech. Bull. 1387, 44 p.

Leonard, D. E. 1967. Parasitism of gypsy moth in Connecticut by *Brachymeria intermedia*. J. Econ. Entomol. 60: 600-601.

Lewis, W. J., A. N. Sparks, R. L. Jones, and D. J. Barras. 1972. Efficiency of *Cardiochiles nigriceps* as a parasite of *Heliothis virescens* on cotton. Environ. Entomol. 1: 468-471.

Lewis, W. J., R. L. Jones, D. A. Nordlund, and A. N. Sparks. 1975. Kairomones and their use for management of entomophagous insects: I. Evaluation for increasing rates of parasitization by *Trichogramma* spp. in the field. J. Chem. Ecol. 1: 343-347.

Lewis, W. J., D. A. Nordlund, H. R. Gross, Jr., W. D. Perkins, E. F. Knipling and J. Voegle. 1976. Production and performance of *Trichogramma* reared on eggs of *Heliothis zea* and other hosts. Environmental Entomol. 5: 449-452.

Nicholson, A. J., and V. A. Bailey. 1935. The balance of animal populations. Proc. Zool. Soc. London, Part I: 551-598.

Sailer, R. I. 1972. A look at USDA's biological control of insect pests: 1888 to present. Agric. Sci. Rev. First Quarter 1972, p. 15-27.

Smith, H. S. 1935. The role of biotic factors in the determination of population densities. J. Econ. Entomol. 28: 873-898.

Solomon, M. E. 1957. Dynamics of insect populations. Annu. Rev. Entomol. 2: 121-142.

Stevens, L. M., Steinhauer, A. L., and Coulson, J. R. 1975. Suppression of Mexican bean beetle on soybeans with annual inoculative releases of *Pediobius foveolatus*. Environ. Entomol. 4: 947-957.

Varley, G. C., and Gradwell, G. R. 1970. Recent advances in insect population dynamics. Rev. Entomol. 15: 1-24.

CHAPTER 4

BIOSYSTEMATICS OF NATURAL ENEMIES

Gordon Gordh

Systematic Entomology Laboratory, IIBIII
Agricultural Research Service
U.S. Department of Agriculture
Washington, D.C. 20560  U.S.A.

INTRODUCTION

Insects are man's primary competitor.  In an effort to compete
successfully with insects for food and fiber, man has employed numerous techniques for the suppression of pest insect populations.
Biological control is one approach that has been used with some
success against many insect pests.  Augmentation is the phase of
biological control concerned with the manipulation of natural
enemies to make them more efficient in applied biological control
programs.  DeBach (1964) observed that the possibilities of augmentation have not been explored thoroughly.  The field of biosystematics is similarly poorly explored but shows great promise as
a research tool for biological control workers.  In many instances
biosystematics may be the only course available to biological control
workers.

Biosystematic studies are not restricted to biological control
programs (cf. Usinger et al. 1966).  However, these comments are
directed at the biological control of insects.  The approach taken
to the biological control of weeds is somewhat different (Huffaker
1964), but the principles of biosystematics are equally applicable.
The greatest abundance of information about biosystematics concerns
entomophagous insects, so that aspect will be emphasized here.

This paper is intended to supplement Schlinger and Doutt
(1964) by providing pertinent references developed since 1964.  For
internal continuity in this paper I will distinguish between
biosystematics and conventional taxonomic studies, outline several
taxonomic problems associated with the identification of natural

enemies used in biological control, and demonstrate the relevance
of biosystematics to biological control, including ways in which
biosystematics can be used to supplement augmentation programs.

## BIOSYSTEMATICS AND CONVENTIONAL TAXONOMY

During the past 10-15 years use of the term biosystematics
has become widespread.  The term and the meaning it conveys has
not always been appreciated despite the fact that the philosophy
and techniques have been covered extensively by Schlinger and Doutt
(1964) and DeBach (1974).  Moreover, the utility and effectiveness
of a biosystematic approach to taxonomic problems with entomophagous
insects, and the distinction between biosystematics and conventional
taxonomy has not been improved by authors who incorporate the word
into their titles and then produce conventionl taxonomic studies.

The importance of taxonomy to biological control has been
cited on numerous occasions (Clausen 1942; Sabrosky 1955; Schlinger
and Doutt 1964; Delucchi 1967; DeBach 1969; Rosen and DeBach 1973).
Conventional taxonomic studies are conducted at all levels of the
taxonomic hierarchy; biosystematic studies focus on the specific
and infraspecific levels only.  Biosystematic studies approach the
generic and suprageneric levels only in that they may be used to
collate information about several species that form natural groups.
Biosystematic studies attempt to collect characters from many independent sources because the independence of characters strengthens
the confidence of the biosystematist in determining taxon limits.
Specifically, the objectives of the biosystematic studies should be:
(1) identify specific and infraspecific taxa or biological entities,
and (2) determine the relative effectiveness of the natural enemies
involved in biological control.

Biological control workers frequently refer to races, strains,
biotypes, ecotypes, semispecies, cryptic species and sibling species.
These are the working units of biological control and their correct
identity is fundamental because no matter how similar representatives
of two populations are morphologically they can be profoundly different biologically.  The importance of these differences as they
relate to biological control will be demonstrated throughout this
chapter.

Sibling species are defined by Mayr (1969) as morphologically
identical ("or nearly so") species that are reproductively isolated.
Isolating mechanisms may be sexual incompatibility, ethological
incompatibility, or genetic incompatibility.  Messenger and van den
Bosch (1971) consider cryptic species to be those that appear to be
sibling species but which on closer examination have minor, but
consistent structural differences.  Semispecies, as related to

biological control, are populations that have achieved partial
reproductive isolation (DeBach 1969).  In this sense the definition
of semispecies is somewhat more restrictive than outlined by Mayr
(1963).  Often, the exchange of genetic material can be in one
direction only (Hall et al. 1962; Mackauer 1969; Rao and DeBach 1969
a,b,c).

The remaining terms have been used interchangeably and have not
been defined adequately in biological control.  This problem occurs
because (1) each term is borrowed from another field, (2) of the
failure of early workers to apply the terms uniformly and unambigu-
ously, (3) of a lack of knowledge about the animals concerned, and
(4) of the biological diversity exhibited by natural enemies.
Clearly there is a need for terms that have no standing in nomen-
clature, but the problem is beyond the scope of this text.

The term strain has sometimes been used interchangeably with
what today may be regarded as sibling species (Hafez and Doutt
1954; DeBach and Fisher 1956).  For the present discussion strains
are considered laboratory-selected cultures of species that have
biological features peculiar to themselves.  Mayr (1954, 1963, 1969)
considered the term race.  In taxonomic studies races have sometimes
been regarded as equivalent to subspecies.  Races here are considered as naturally occurring genotypically or phenotypically
distinct populations that are often geographically isolated but not
reproductively isolated.  Ecotypes are considered synonymous with
biotypes that are populations adapted to different ecological
habitats or environments.

Different biological attributes exhibited by infraspecific
entities influence the level of success achieved by biological
control programs.  Mackauer (1972) observed that biological control
today is little more than intellectual blindman's buff.  Biosystematists can change this situation by providing accurate identifica-
tions of these entities and demonstrating the relative effectiveness
of an entity in applied biological control programs.  When diagnostic
morphological characters are not available, biosystematic techniques must be employed to isolate, identify, and monitor specific
and infraspecific taxa of natural enemies.


                            IDENTIFICATION

Identification is an activity of taxonomy that involves the
referral of specimens to a described taxon, determination of specimens to an undescribed taxon, or the detection of an earlier misidentification.  Accurate identification of specimens is the most
critical aspect of any biological control program (DeBach 1959,
1960; Schlinger and Doutt 1964) and can mean the difference between

establishment or failure to establish (Hall et al. 1962; DeBach
1974).

Classical biological control theory dictates that many pest
problems originate from accidental introduction of a species into
an area where it did not occur and where it evades its natural
enemy complex (Franz 1961; DeBach 1964).  The pest is not a problem
in the area of endemicity because its populations are regulated by
an effective natural enemy complex.  The biological control worker
must determine the origin of the pest insect.  This necessitates
accurate identification of the pest and its natural enemies and
requires information relative to their distribution and the distribution of their relatives.  Incorrect identification of either the
pest or its enemies can result in (1) searching for natural enemies
in the wrong area, (2) failure to select the appropriate natural
enemies, (3) failure of the natural enemy to establish after importation, (4) failure to achieve a high level of success or (5)
failure to monitor correctly the natural enemies after release.  A
few examples will demonstrate the importance of accurate identification.

*Diatraea saccharalis* (F.) has been a pest of sugar cane on
Barbados for more than 50 years.  For over 20 years a species
identified as *Trichogramma minutum* Riley was cultured for inundative
release with up to 300 million individuals released annually.  In
reality the parasite involved was *T. fasciatum* (Perkins).  This
parasite was ineffective in controlling the sugar cane borer, and
its propagation was terminated.  The misidentification of this
parasite was unfortunate.  It has not been determined whether release of *T. minutum* would have been more effective because inundative release of egg parasites has been generally abandoned (Alam
et al. 1971; Bennett 1971; Simmonds 1972; Metcalf and Breniere
1969).

The periodic, accurate identification of a parasite can also
be important.  In a biological control program for *Quadraspidiotus
perniciosus* (Comstock) near Heidelberg, Germany a parasite thought
to be *Prospaltella perniciosi* Tower was released during 1956-1958.
In reality, another species, *P. fasciata* Malenotti, had contaminated the laboratory culture of *P. perniciosi* and subsequently dis-
placed that species.  *Prospaltella fasciata* is an ineffective
parasite against *Q. perniciosi* in Europe (Rosen and DeBach 1973).

Presently one of the most serious pests of citrus in California is the California red scale, *Aonidiella aurantii* (Maskell).
The most effective natural enemies of this pest belong to the
genus *Aphytis* Howard but repeated misidentifications of several
species of *Aphytis* for *A. mytilaspidis* (LeBaron) and *A. chrysomphali* (Mercet) precluded the establishment of the most promising

natural enemies for over 50 years (Compere 1961).

Misidentification is a constant problem in biological control
because the few taxonomists working on nautral enemies are spread
too thin, identification loads are unreasonably large, and the
taxonomic foundation of many groups is so poor that misidentifications are inevitable. Biosystematic studies can aid the biological
control worker by revealing obvious biological, behavioral, and
ecological differences among natural enemies that are similar
morphologically. Such information can often stimulate a taxonomist
to search for reliable diagnostic criteria or undertake revisionary
studies.

Clausen (1940) and Askew (1971) discussed the taxa of entomophagous insects. Although entomophagous habits are widespread in
the insects and representatives of several orders have been used
in biological control with varying success, most insects effectively
used in biological control are parasitic Hymenoptera (DeBach 1974).
Several problems associated with the identification of these insects
must be understood to comprehend why biosystematic methods are
necessary for the solution of taxonomic problems associated with
these natural enemies.

Accuracy of identification relates directly to the soundness
of the taxonomy of the group. If fundamental taxonomy is not
sound, then the probability of accurate identification is reduced.
The present state of taxonomic knowledge for many groups of parasitic Hymenoptera is not sound. This problem prompted Compere
(1969) to observe that on the basis of original descriptions, most
species of chalcidoids are unrecognizable. I would extend this
observation to all parasitic Hymenoptera.

Additional reasons for difficulty with the early taxonomy of
many groups of parasitic Hymenoptera are: (1) Parasitic Hymenoptera are among the smallest insects, and without good optical
equipment the characters used for definition and identification
often cannot be seen. Optical equipment in the last century was
not adequate, and many species were described before 1900. (2)
Early workers placed emphasis on color characters in the descriptions of new species, but color is not always reliable in defining
species. (3) Type-specimens are not always available for comparison. When original descriptions are not adequate for subsequent
species recognition, then type-specimens must be examined before
an identification can be made. For these reasons revisionary
studies are necessary before names can be applied correctly. For
a discussion of these and related problems see Bouček (1958),
Kerrich (1962), Delucchi (1967), DeBach (1969), Compere (1969),
and Oman (1974).

There are also difficulties associated with the natural
enemies themselves.  The number of morphologically distinct, undescribed species of parasitic Hymenoptera can only be speculated.
Presently about 15,000 species of Ichneumonidae are known.  Townes
(1969) estimated that 60,000 species of Ichenumonidae exist.
Presently about 10,000 species of Chalcidoidea are described, but
I (Gordh 1975, in prep.) estimate that there are more species of
chalcidoids than ichneumonids.  Other groups of parasitic Hymenoptera
are similarly poorly known.

Sibling species and cryptic species further complicate the
problem.  The importance of sibling species to medical entomology
has been noted by Coluzzi (1970), and their importance to biological control has been noted by DeBach (1969).  Presumably sibling
species and cryptic species represent recent or rapid speciation.
It is apparent that the parasitic Hymenoptera are rapidly speciating or that they are speciating more rapidly than many other
taxa (Askew 1968; Gordh 1975, in prep.).  Basically this means
that many species have achieved reproductive isolation but that
they have not become differentiated morphologically.

Sibling species, cryptic species, and semispecies are the
bane of museum taxonomists because morphological characters are
difficult to detect (cryptic species) or do not exist (sibling
species and semispecies).  This is not a criticism of museum
taxonomists, but points to the problem associated with the identification of many natural enemies.  Sibling, cryptic, and semispecies
are especially troublesome in the areas of foreign exploration,
importation, and evaluation of the effectiveness of these natural
enemies at the release site.  Supplemental characters must be used
to discriminate among taxa and these characters sometimes are
found using biosystematic techniques.  These techniques include:
(1) hybridization studies, (2) biological studies, (3) behavioral
studies, (4) biochemical studies, and (5) cytological studies.

## TOOLS AND TECHNIQUES USED IN BIOSYSTEMATICS

Biosystematic studies often use the same tools as conventional
taxonomic studies.  These include good optical equipment, a good
synoptic collection, access to taxonomic literature, and a background in taxonomic procedure.  More recently the scanning electron
microscope has provided an extension of a taxonomist's vision, and
this tool has been used extensively.

The preservation and distribution of voucher specimens is also
important for biosystematic studies because such specimens can
provide clues to the identity of species involved in earlier
studies.  Voucher specimens also provide material in the search for

morphological characters.

In addition to literature, collection, and optical equipment, the laboratory culture of natural enemies is essential for biosystematic studies. Some natural enemies are easily cultured, some are difficult to culture, and some have defied culture. For a discussion of the problems associated with the culture of entomophagous insects see Finney and Fisher (1964).

*Hybridization Studies.* Biological control workers probably adhere more strictly to the biological species concept than do other entomologists. They must because subtle differences in the genetic constitution of populations can alter the biological attributes of a population and thus change its effectiveness in biological control. Experimental hybridization tests are fundamental to any biosystematic study because they provide a quantitative assessment of the genetic relationships among populations.

*Trichogramma* Westwood is one of the most popular genera of parasitic Hymenoptera because its species are propagated throughout the world for inundative control of numerous lepidopterous pests. *Trichogramma* typifies the taxonomic predicament found in many groups of parasitic Hymenoptera and demonstrates the absolute dependence on biosystematic studies for correct identification. The taxonomy of the genus is exceptionally difficult because of the minute body size, the great variability of some characters used for identification of taxa, and the absolute dependence on excellent, slide-mounted material for identification. Unquestionably, many published reports concerning *Trichogramma* spp. are based on misidentifications.

The taxonomic situation of *Trichogramma* was summarized by Flanders and Quednau (1960) and Quednau (1961). Recently, the taxonomy has been substantially improved through hybridization studies of Nagarkatti and Nagaraja (1968, 1977), Oatman et al. (1968, 1970), Nagarkatti and Fazaluddin (1973), and Oatman and Platner (1973). These studies have demonstrated that several undescribed species exist, and that semispecies, sibling and cryptic species are prevalent. Moreover, these studies have substantially aided the search for reliable morphological characters.

Hybridization studies enabled Nagarkatti and Nagaraja (1971) to discover reliable male genitalic characters. This situation is ironic because genitalic characters are generally useless in the parasitic Hymenoptera, and taxonomists would not ordinarily examine them. A conventional taxonomic analysis probably would not recognize these characters because the study would be based on preserved specimens collected from different localities and different hosts. Combined with cryptic species, this situation would generate sufficient variation as to make detection of these characters difficult.

However, examination of genetically authentic, cloned specimens for characters shows that genital characters are reliable.

Hybridization studies combined with careful morphological analysis has resulted in the publication of keys to the New World species of *Trichogramma* (Nagaraja and Nagarkatti 1973). Although not all the problems with *Trichogramma* are solved, it is now possible to evaluate its species as biological control agents more objectively.

The reproductive relationship between two entities can range from complete reproductive isolation to complete sexual and genetical compatibility. Rao and DeBach (1969 a,b,c) have shown that within the *Aphytis lingnanensis* Compere complex the entire spectrum of reproductive relationships can be found. Many of the taxa are important in the biological control of various armored scale insect pests throughout the world, and knowledge of their genetical relationships is essential.

Hybridization tests are also important in understanding the genetical relationships among populations of one morphological species. Khasimuddin and DeBach (1976 a) showed that three morphologically indistinguishable populations of the arrhenotokous *A. maculicornis* (Masi) from Pakistan, Persia, and southern California are reproductively isolated and should be considered sibling species.

Thus hybridization tests are useful to biosystematic studies in detecting sibling species (Khasimuddin and DeBach 1976 c). For instance, Hall et al. (1962) showed that two sibling species, *Trioxys pallidus* (Haliday) and *T. utilis* Muesebeck, were very different with respect to host preference. Reproductive isolation between the species was only partial because female progeny were produced when *T. pallidus* females were crossed with *T. utilis* males, but reciprocal crosses yielded no female progeny.

Semispecies can only be identified using biosystematic techniques. Khasimuddin and DeBach (1976 b) studied the relationship between naturally occurring, sympatric populations of *Aphytis mytilaspidis* (LeBaron). Again, by supplementing hybridization tests with host preference and surviorship curves at different temperatures, they concluded that semispecies were involved. One population showed promise in biological control; the other did not. The effectiveness of one population over another can be determined only by the release of these natural enemies. Failure to release the appropriate population reduced the likelihood of effective biological control.

In some instances biosystematic revision may be the only approach available to taxonomic problems involving natural enemies. *Muscidifurax* Girault and Saunders has been used in some biological

control problems involving synanthropic Diptera, but its effectiveness has been uneven.  It remained a monotypic genus for 60 years
until Legner (1969) and Kogan and Legner (1970) found at least
five cryptic or sibling species through hybridization tests and
other biosystematic techniques.  Accurate application of *Muscidifurax* spp. to appropriate pest situations will enhance the effectiveness of these natural enemies.

   *Life Histories, Host Relationships, and Biological Observations.*
Although hybridization studies frequently are necessary for the
identification of many natural enemies, they do not provide sufficient information to evaluate the potential usefulness of these
natural enemies.  Biological information must be collected from
field and laboratory studies on all natural enemies.  Important
kinds of information include:  (1) host preference, (2) life-
table data, (3) life history, (4) description of immature stages,
and (5) survivorship curves.

   The usefulness of host preference as a taxonomic tool is not
a new idea.  Ratzeburg (1852) and Foerster (1856) were aware of its
potential value.  Flanders (1953) has used host and instar preference and integrated them with other biological and morphological
characters to develop a key to some genera of Aphelinidae.  While
the key is not entirely accurate, it does point to the usefulness
of such an approach to taxonomic problems associated with biological control.  Ferriere (1962) observed that the host record of a
chalcidoid (or any other parasitic Hymenoptera) is more important
than the description of a new species.

   The value of host preference as a taxonomic tool can only be
established by providing accurate data on host association.  Gordh
(1975) estimated that hosts are unknown for about 25% of the described Chalcidoidea genera.  Other groups of parasitic Hymenoptera
are slightly better known, but their host records are still in-
credibly poor.

   To weigh host preference as a taxonomic tool, it is necessary
to identify the factors involved in determining host preference.
Baker (1976) provides extensive data on some scale-insect parasites,
and similar studies should be made for all natural enemies used in
biological control.

   Extensive and detailed comparative life history studies are
also important to biosystematics because they provide supplementary
information regarding biological characteristics of natural enemies.
Quednau (1955, 1956 a,b) recognized this problem with *Trichogramma*
and attempted to distinguish among species on the basis of development time, host preference, and related biological parameters.
These kinds of information can be used in a comparative fashion to

evaluate potential usefulness of a natural enemy (Nesser 1973);
Gerling and Limon 1976).

Detailed study of the immature stages and life history can
also be useful when made on a single species because such studies
can be used by other workers to compare their own data to determine
the relative potential of a natural enemy (Gordh 1976; Gordh and
Lacey 1976; Gordh and Evans 1976; Cushman and Gordh 1976).

Infraspecific groups of natural enemies can have different
biological characteristics that would go undetected if biosystematic
studies were not undertaken.  For instance, the Chinese and Japanese
races of *Comperiella bifasciata* Howard are morphologically indistinguishable.  They freely hybridize in the laboratory, and host
preference tests have shown that the Chinese race will develop on
*Aonidiella aurantii* and *A. citrina* (Coquillett), but the Japanese
race will develop only on *A. citrina*.  The evolutionary implications
of such a system are not clear, but in the field the two races appear to maintain their genetic integrity (DeBach 1974; pers. commun.).

The pragmatic importance of accurately identifying biological
races cannot be over emphasized.  Hafez and Doutt (1954) found
several "races" of *Aphytis maculicornis* that parasitize *Parlatoria
oleae* (Colvee).  Through careful biosystematic study it was determined that only one, the Persian "race," was effective in biological control of this pest in California.  Later, Khasimuddin and
DeBach (1976 c) employed additional biosystematic tests on three
morphologically indistinguishable populations of *A. maculicornis*.
Supplementing hybridization tests with host preference and survivorship data, they concluded that three sibling species were involved.

Similar problems are found with morphologically indistinguishable populations of *Prospaltella perniciosi* Tower.  Biological
studies have shown that uniparental and biparental forms of this
parasite exist.  Flanders (1950) noted that two "races" of this
species exist, one that attacks California red scale and one that
attacks San Jose scale, *Q. perniciosus*.  Schlinger and Doutt
(1964) consider these to be sibling species.

*Behavior*.  Behavior is a complex discipline that has been extensively explored with non-entomophagous insects.  Its usefulness
to biosystematics is not fully appreciated, but its potential as a
biosystematic tool is substantial.  Iwata (1972 - translated into
English 1976) has reviewed the behavior of Hymenoptera, and this
work should be consulted by biosystematists.

Some natural-enemy taxonomists are reluctant to accept behavioral characters because they suspect behavioral characters are
more variable than morphological characters.  Both behavioral characters and morphological characters have a genetic basis so both
sources of characters are useful in biosystematics.  Moreover,
laboratory propagation enables the biosystematist to replicate
studies, generate large amounts of data for quantitative analysis,
and examine the factors influencing variability.

Courtship behavior may have promise as a biosystematic tool
in sexually reproducing natural enemies at the specific and infraspecific levels.  A generalized account of sexual behavior in
parasitic Hymenoptera was given by Matthews (1975).  Van den Assem
(1975) observed that many parasitic Hymenoptera have elaborate,
species-specific displays that involve leg, wing, and antennal
movements.  When properly characterized, such species-specific
characters are important in biosystematic studies because they
permit the biological control worker to identify closely related
taxa without elaborate taxonomic analysis. For instance, Evans and
Matthews (1976) found an undetected cryptic species of *Melittobia*
*chalybii* Ashmead on the basis of male courtship behavior.  This
demonstration of the pertinence of sound behavioral observation as
a biosystematic tool is important because the species are sympatric
and can simultaneously develop in the same wasp cell.  If this
problem had been treated by conventional taxonomic analysis, the
conclusion would have been that there was variation in male antennal characters.  Moreover, *M. chalybii* has been reported from
several hosts throughout the United States, but in all likelihood
several sibling species and cryptic species are involved (Dahms,
*in litt.*).  The implications of this problem for biological control
are self evident.

Gordh and DeBach (in prep.) found that several species of the
*A. lingnanensis* complex can be distinguished on the basis of qualitative and quantitative aspects of male courtship behavior and female response to that behavior.  All species in this complex are
morphologically similar, and several sibling species are involved.
Careful and detailed description of this behavioral inventory provides a reliable suite of characters that can be used by field
workers to distinguish species without elaborate preparation of
preserved specimens.

Mackauer (1969) found differences in courtship and copulation
time for three species of *Aphidius* Nees parasitic on the pea aphid.
Hybridization studies showed only partial reproductive isolation
between two of the "species" and thus were relegated to the status
of semispecies.

Goodpasture (1975) also found that courtship behavior was di-
agnostically useful in *Monodontomerus* Westwood, and that it may be
important in inferring phylogenetic relationships.

Van den Assem and Povel (1973) and van den Assem (1974, 1975)
also believe that study of sexual behavior in many related species
may uncover common behavioral features and thus infer phylogeny.

*Biochemical Techniques.* Electrophoresis is one of several
electrokinetic phenomena utilized by biochemists.  Its relevance
as a powerful analytical tool has been recognized by geneticists
and evolutionary biologists (Bullini and Coluzzi 1973; Burns 1975).
The diagnostic capability of this tool in the biosystematics of
natural enemies has not been generally recognized although its
potential is great, especially when supplemented with other characters.  Many groups of natural enemies are exceptionally suitable
for electrophoretic analysis because generation time is short, they
can be propagated in large numbers, and hybrids can be analyzed.
Electrophoretic analysis is well suited to many groups of natural
enemies because analysis requires small amounts of tissue for
sampling.

Proteins are most easily analyzed by electrophoresis because
of their electrical charges.  Proteins are also an important source
of characters because of their fundamental role as building blocks
of organic life.  Of the several kinds of proteins (histones, nu-
cleoproteins, enzymes), enzyme systems are probably best studied
because of the ease with which they can be analyzed and because of
substrate specificity.  Electrophoretic analysis of isozymes (forms
of an enzyme that have different isolectric points) can be used to
distinguish among populations, identify species, and cluster them.

The genus *Colias* Fabricius holds about 60 species, most of
which are found in North America.  The taxonomy of the genus is
difficult, and there have been questions about the status of some
species (*C. eurytheme* Boisduval and *C. philodice* Godart).  Electrophoretic studies have shown on the basis of an isozyme of dimeric
esterase that it is possible to distinguish among species of *Colias*
(Burns and Johnson 1967; Johnson and Burns 1966; Burns 1975).

Isozymes have also been used to detect sibling species in the
medically important *Aedes mariae* (Sergent and Sergent) complex.
Coluzzi and Bullini (1971) have shown through laboratory and
field studies that *A. zammitti* Theobold and *A. mariae* are nearly
identical in habitat preference and morphology, but that they can
be distinguished on the basis of an isozyme at the phosphogluco-
mutase locus.

There are other important implications of electrophoretic
techniques.  Heretofore it has been difficult or impossible to

evaluate the effectiveness of different ecotypes when released at various localities.  Electrophoresis provides a reliable technique for analyzing changes in gene frequency for given loci and thus provides a monitoring technique of unparalleled value.

Chromatography is another biochemical tool that shows some promise in the biosystematic analysis of natural enemies.  It has been used in mosquito taxonomy (Micks 1954; Micks et al. 1966; Micks et al. 1967) so it may prove useful in studies of population differences, subspecies differences, and species differences.

*Cytogenetics*.  Cytological information has not been used extensively in biosystematic studies of natural enemies partly because techniques were not refined for many groups.  Recently, Crozier (1975) summarized existing information concerning the cytogenetics of the Hymenoptera and that paper should be consulted by biosystematists.  Types of cytological information useful for biosystematic studies include karyotypes and chromosomal polymorphism.

Karyotypes can be useful in determining systematic relationships and when analyzing many closely related species they may be informative in pointing to the direction of evolution.  Unfortunately, karyotypical information is meager on parasitic Hymenoptera.  Crozier (1975) enumerates the species for which karyotypes have been determined. Goodpasture and Grissell (1975) estimated that less than 0.001% of the described species of chalcidoids have been karyotyped.  All other groups of parasitic Hymenoptera are equally poorly known.  Clearly, more study of this problem is necessary before the utility of karyotypes in biosystematic studies can be appreciated.

Cytological studies are sometimes helpful in making taxonomic decisions involving natural enemies.  A significant problem in many taxa of natural enemies is the existence of arrhenotokous and thelytokous populations that are morphologically indistinguishable. The proximate and ultimate evolutionary significance of such trends is not clear, but these different types of parthenogenesis have been correlated with different potential in some biological control programs (Hafez and Doutt 1954; Khasimuddin and DeBach 1976c; Rossler and DeBach 1972b).

Arrhenotokous and thelytokous populations that are morphologically indistinguishable cause problems for some taxonomists because if one adheres to the biological species concept, and if such populations are shown to be reproductively isolated, then they must be considered different species.  It is also sometimes difficult to determine the parthenogenetic mode because the sex ratios are often

strongly skewed.  The functional status of rare males must be
demonstrated.

Most workers have generally assumed that no gene exchange occurs between arrhenotokous and thelytokous populations, but this
must be demonstrated in each case.  In a comprehensive cytological
analysis, Rossler and DeBach (1972a, 1973) examined the reproductive
relationship between an arrhenotokous and a thelytokous population
of *A. mytilaspidis*.  Their studies showed that only partial reproductive isolation exists between these populations and that thelytokous females are capable of utilizing sperm from arrhenotokous
males.  Such information could only have been obtained with
cytological studies.

## BIOSYSTEMATICS AND AUGMENTATION

The responsibilities of the taxonomist in a biological control
program have not been clearly defined.  Most workers believe that
the taxonomist's involvement ends with the identification of imported or field-recovered natural enemies.  However, if biosystematic techniques must be applied to solve identification problems,
then the biosystematist may be in the best position to evaluate
some aspects of an augmentation program because the biosystematist
has accumulated many of the tools necessary for the identification
of closely related, morphologically similar natural enemies.

DeBach and Hagen (1964) discussed various approaches to the
manipulation of natural enemies.  As an extension of taxonomic involvement, the biosystematist can facilitate augmentation by:  (1)
hybridization studies to improve genetic fitness; (2) analysis of
genotypes of natural enemies; and (3) determining the feasibility
of artificial selection of natural enemies for improvement of the
resistance to various environmental conditions and improvement of
biological attributes.

*Hybridization*.  Intraspecific hybridization has been suggested
as one way of enhancing the usefulness of natural enemies through
the development of coadapted gene complexes.  This concept stems
from the successful hybridization of many horticultural varieties.
Hybridization has been tested using non-entomophagous insects
(such as *Drosophila* spp.) with varying degrees of success, but
the utility of hybridization to biological control remains to be
established.  From laboratory studies it is evident that hybridization can produce positive results with some natural enemies and
negative results with others (Legner 1972).  Hoy (1975 a,b) studied
several geographically isolated populations of the gypsy moth parasite *Apanteles melanoscelus* (Ratzeburg) and found different per-
centages of diapause among populations.  Under laboratory conditions

diapause was almost completely eliminated.  Through hybridization
of three populations, Hoy noted an "improvement" in the percentage
of *A. melanoscelus* entering diapause, and suggested that hybridization may be an important consideration in improving biological
characteristics for inundative control.  In another study Khasimuddin and DeBach (1976 a) produced hybrids of *Aphytis maculicornis*
which had greater fecundity and produced more progeny than either
parental population.

Interspecific hybridization of natural enemies is unusual,
but it also may prove potentially useful in augmentation. Nagarkatti (1970) crossed females of *Trichogramma perkinsi* Girault with
males of *Trichogramma* sp. (designated D-67) and produced thelytokous hybrids.  Although several explanations of this phenomenon
are possible, hybridization appears to be the most likely, and
thus studies along this line should be pursued.

*Genotype Analysis*.  Analysis of genotypes is another way that
biosystematic techniques can be used in augmentation.  Genetic considerations are related to biological control programs in three
ways:  (1) importation of the correct genome; (2) laboratory propagation and maintenance of the correct genome; and (3) colonization
of the correct genome.  Controversy exists over the appropriate-
source-population importation strategy of natural enemies, but the
hypotheses have not been tested (Remington 1968; Lucas 1969; Whitten
1970).  The theoretical problems involved in mass propagation of
natural enemies have been considered by Boller (1972) and Mackauer
(1972, 1976), but the suggestions have not been tested.  Genetics
of colonization have been treated by Wilson (1965) and Force (1967).
All of the recommendations involved must be tested with biosystematic
techniques such as hybridization studies and biochemical analysis.

*Artificial Selection*.  Aritficial selection also has been
suggested as a method of improving the effectiveness of natural
enemies.  Some workers believe that it is possible to change biological attributes of a natural enemy and thereby improve its effectiveness (Box 1956; DeBach 1958), whereas others believe that
artificial selection for the improvement of attributes can lead to
the deterioration of others (Simmonds 1963).  Few empirical data
have been developed on natural enemies to suggest the correctness
of either contention.  Laboratory studies have shown that it is
possible to select for insecticide resistance (Wilkes et al. 1952),
sex ratios (Simmonds 1947; Wilkes 1964), searching ability (Urquijo
1951), host preference (Allen 1954), and resistance to temperature
extremes (Wilkes 1942; White et al. 1970).

One of the few field applications of selection for temperature
extremes and hybridization was unsuccessful.  Ashley et al. (1974)

attempted to hybridize and select for temperature extremes to improve the fitness of arrhenotokous *Trichogramma pretiosum* Riley.
Control groups were more successful at parasitizing host eggs than
the experimental groups in the field. However, selection studies
should be continued and tested to determine the efficacy of such
an approach.

In summary Carter (1970), Force (1974), and Beirne (1975) have
pointed to the low percentage of successful establishment of natural
enemies in applied biological control. Many of the reasons for
this poor percentage of establishment have been outlined above.
Biosystematic studies should be an integral part of biological
control and could be responsible for improving the record of
establishment and level of success achieved in biological control
programs.

## REFERENCES CITED

Alam, M. M., F. D. Bennett and K. P. Karl. 1971. Biological control of *Diatraea saccharalis* (F.) in Barbados by *Apanteles flavipes* Cam. and *Lixophaga diatraeae* T. T. Entomophaga 16 (2): 151-158.

Allen, H. W. 1954. Propagation of *Horogenes molestae*, an asiatic parasite of the oriental fruit moth, on the potato tuberworm. J. Econ. Entomol. 47 (2): 278-281.

Ashley, T. R., D. Gonzales and T. F. Leigh. 1974. Selection and hybridization of *Trichogramma*. Environ. Entomol. 3 (1): 43-48.

Askew, R. R. 1968. Considerations on speciation in Chalcidoidea (Hymenoptera). Evolution 22 (3): 642-645.

Askew, R. R. 1971. Parasitic Insects. American Elsevier Publ. Co., New York, 316 pp.

Assem, J. van den 1974. Male courtship patterns and female receptivity signal of Pteromalinae (Hym., Pteromalidae), with a consideration of some evolutionary trends and a comment on the taxonomic position of *Pachycrepoideus vindemiae*. Netherlands J. Zool. 24 (3): 253-278.

Assem, J. van den 1975. Temporal patterning of courtship behavior in some parasitic Hymenoptera, with special reference to *Melittobia acasta*. J. Entomol. 50 (3): 137-146.

Assem, J. van den and G. D. E. Povel. 1973. Courtship behaviour of some *Muscidifurax* species (Hym., Pteromalidae); A possible example of recently evolved ethological isolating mechanism. Netherlands J. Zool. 23 (4): 465-487.

Baker, J. L. 1976. Determinants of host selection for species of *Aphytis* (Hymenoptera: Aphelinidae), parasites of diaspine scales. Hilgardia 44 (1): 1-25.

Beirne, B. P. 1975. Biological control attempts by introductions against pest insects in the field in Canada. Canad. Entomol. 107 (3): 225-236.

Bennett, F. D. 1971. Current status of biological control of the small moth borers of sugar cane *Diatraea* (Lep. Pyralidae). Entomophaga 16 (1): 111-124.

Boller, E. 1972. Behavioral aspects of mass-rearing of insects. Entomophaga 17 (1): 9-25.

Bouček, Z. 1958. Zur Taxonomie der entomophagen Insekten, besonders der Hymenopteren in Europa. Trans. 1st. Int. Conf. Ins. Path. Biol. Cont., Prague, pp. 349-353.

Box, H. E. 1956. The biological control of moth borers (*Diatraea*) in Venezuela. Battle against Venezeula's cane borer. Part 1. Preliminary investigations and the launching of a general campaign. Sugar 51: 25-27, 30, 45.

Bullini, L. and M. Coluzzi. 1973. Electrophoretic studies on gene-enzyme systems in mosquitoes (Diptera: Culicidae). Parassitologia 15 (3): 221-248.

Burns, J. M.   1975.   Isozymes in evolutionary systematics, pp. 49–
    62.   *In*:  Isozymes IV.   Genetics and Evolution.   Academic
    Press, New York.

Burns, J. M. and F. M. Johnson.   1967.   Esterase polymorphism in
    natural populations of a sulfur butterfly, *Colias eurytheme*.
    Science 156 (3771):   93–96.

Carter, W.   1970.   The dynamics of entomology.   Bull. Entomol. Soc.
    Amer. 16 (4):   181–185.

Clausen, C. P.   1940.   Entomophagous Insects.   McGraw-Hill Book
    Co., New York.   688 pp.

Clausen, C. P.   1942.   The relation of taxonomy to biological con-
    trol.   J. Econ. Entomol. 35 (5):   744–748.

Coluzzi, M.   1970.   Sibling species in *Anopheles* and their im-
    portance in Malariology.   Misc. Publ. Entomol. Soc. Amer. 7
    (1):   63–72.

Coluzzi, M. and L. Bullini.   1971.   Enzyme variants in the study
    of precopulatory isolating mechanisms.   Nature 231 (5303):
    455–456.

Compere, H.   1961.   The red scale and its insect enemies.   Hilgardia
    31 (7):   173–278.

Compere, H.   1969.   The role of systematics in biological control:
    a backward look.   Israel J. Entomol. 4:   5–10.

Crozier, R. H.   1975.   Animal Cytogenetics.   Vol. 3:   Insecta 7.
    Gebruder Borntraeger Berlin, Stuttgart.   95 pp.

Cushman, R. A. and G. Gordh.   1976.   Biological investigations of
    *Goniozus columbianus* Ashmead, a parasite of the grape berry
    moth, *Paralobesia viteana* (Clemens) (Hymenoptera: Bethylidae).
    Proc. Entomol. Soc. Wash. 78 (4):   451–457.

DeBach, P.   1958.   Selective breeding to improve adaptations of
    parasitic insects.   Proc. 10th Int. Congr. Entomol., Montreal
    4:   759–768.

DeBach, P.   1959.   New species and strains of *Aphytis* (Hymenoptera,
    Eulophidae) parasitic on the California red scale, *Aonidiella
    aurantii* (Mask.), in the Orient.   Ann. Entomol. Soc. Amer. 52
    (4):   354–362.

DeBach, P.   1960.   The importance of taxonomy to biological control
    as illustrated by the cryptic history of *Aphytis holoxanthus*
    n. sp. (Hymenoptera: Aphelinidae), a parasite of *Chrysomphalus
    aonidium*, and *Aphytis coheni* n. sp., a parasite of *Aonidiella
    aurantii*.   Ann. Entomol. Soc. Amer. 53 (6):   701–705.

DeBach, P.   1964.   The scope of biological control.   pp. 3–20 *in*:
    Biological Control of Insect Pests and Weeds, P. DeBach, ed.
    Chapman Hall Ltd., London.   844 pp.

DeBach, P.   1969.   Uniparental, sibling and semi-species in rela-
    tion to taxonomy and biological control.   Israel J. Entomol.
    4:   11–28.

DeBach, P.   1974.   Biological Control by Natural Enemies.   Cam-
    bridge Univ. Press, London.   323 pp.

DeBach, P. and T. W. Fisher. 1956. Experimental evidence for sibling species in the oleander scale, *Aspidiotus hederae* (Vallot). Ann. Entomol. Soc. Amer. 49 (3): 235-239.

DeBach, P. and K. S. Hagen. 1964. Manipulation of entomophagous species. pp. 429-458 *in:* Biological Control of Insect Pests and Weeds, P. DeBach, ed. Chapman Hall Ltd., London. 844 pp.

Delucchi, V. L. 1967. The significance of biotaxonomy to biological control. Mushi (supl.) 39: 119-125.

Evans, D. A. and R. W. Matthews. 1976. Comparative courtship behavior in two species of the parasitic chalcid wasp *Melittobia* (Hymenoptera: Eulophidae). Anim. Behav. 24 (1): 46-51.

Ferrière, Ch. 1962. La taxonomie des insectes entomophages. Verh. 11th. Int. Kong. Entomol., Wien 3: 290-292.

Finney, G. L. and T. W. Fisher. 1964. Culture of entomophagous insects and their hosts. pp. 328-355 *in:* Biological Control of Insect Pests and Weeds, P. DeBach, ed. Chapman Hall Ltd., London. 844 pp.

Flanders, S. E. 1950. Races of apomictic parasitic Hymenoptera introduced into California. J. Econ. Entomol. 43 (5): 719-720.

Flanders, S. E. 1953. Aphelinid biologies with implications for taxonomy. Ann. Entomol. Soc. Amer. 46 (1): 84-94.

Flanders, S. E. and W. Quednau. 1960. Taxonomy of the genus *Trichogramma* (Hymenoptera, Chalcidoidea, Trichogrammatidae). Entomophaga 5 (4): 285-294.

Foerster, A. 1856. Hymenopterologische Studien. II. Chalcididae und Proctotrupii. Aachen, 152 pp.

Force, D. C. 1967. Genetics in the colonization of natural enemies for biological control. Ann. Entomol. Soc. Amer. 60 (4): 722-729.

Force, D. C. 1974. Ecology of insect host-parasitoid communities. Science 184 (4137): 624-632.

Franz, J. M. 1961. Biologische Schädlingsbekämpfung. Paul Parey, Berlin, 302.

Gerling, D. and S. Limon. 1976. A biological review of the genus *Euplectrus* (Hym.: Eulophidae) with special emphasis on *E. laphygmae* as a parasite of *Spodoptera littoralis* (Lep.: Noctuidae). Entomphaga 21 (2): 179-187.

Goodpasture, C. 1975. Comparative courtship behavior and karyology in *Monodontomerus* (Hymenoptera: Torymidae). Ann. Entomol. Soc. Amer. 68 (3): 391-397.

Goodpasture, C. and E. E. Grissell. 1975. A karyological study of nine species of *Torymus* (Hymenoptera: Torymidae). Canad. J. Genet. Cytol. 17 (3): 413-422.

Gordh, G. 1975. Some evolutionary trends in the Chalcidoidea (Hymenoptera) with particular reference to host preference. J. New York Entomol. Soc. 83 (4): 279-280.

Gordh, G.   1976.   *Goniozus gallicola* Fouts, a parasite of moth
    larvae, with notes on other bethylids (Hymenoptera: Bethylidae;
    Lepidoptera: Gelechiidae).   U.S. Dep. Agric. Tech. Bull. 1524,
    27 pp.

Gordh, G.   in prep.   Chalcidoidea.   *in:* Hymenoptera Catalog of
    North America North of Mexico.   K. V. Krombein et al., eds.
    U.S. Gov. Print. Off.

Gordh, G. and P. DeBach.   in prep.   Courtship in the *Aphytis*
    *lingnanensis* group, its potential usefulness in taxonomy, and
    a review of sexual behavior in the parasitic Hymenoptera
    (Chalcidoidea: Aphelinidae).   Hilgardia.

Gordh, G. and H. E. Evans.   1976.   A new species of *Goniozus* im-
    ported into California from Ethiopia for the biological con-
    trol of pink bollworm and some notes on the taxonomic status
    of *Parasierola* and *Goniozus* (Hymenoptera: Bethylidae).   Proc.
    Entomol. Soc. Wash. 78 (4):  479-489.

Gordh, G. and L. A. Lacey.   1976.   Biological studies of *Plagio-*
    *merus diaspidis* Crawford, a primary internal parasite of di-
    aspidid scale insects (Hymenoptera: Encyrtidae; Homoptera:
    Diaspididae).   Proc. Entomol. Soc. Wash. 78 (2):   132-144.

Hafez, M. and R. L. Doutt.   1954.   Biological evidence of sibling
    species in *Aphytis maculicornis* (Masi).   (Hymenoptera: Apheli-
    nidae).   Canad. Entomol. 86 (2):   90-96.

Hall, J. C., E. I. Schlinger and R. van den Bosch.   1962.   Evidence
    for the separation of the "sibling species" *Trioxys utilis* and
    *Trioxys pallidus* (Hymenoptera: Braconidae, Aphidiidae).   Ann.
    Entomol. Soc. Amer. 55 (5): 566-568.

Hoy, M. A.   1975a.   Hybridization of strains of the gypsy moth
    parasitoid *Apanteles melanoscelus,* and its influence upon
    diapause.   Ann. Entomol. Soc. Amer. 68 (2): 261-264.

Hoy, M. A.   1975b.   Forest and laboratory evaluations of hybridized
    *Apanteles melanoscelus* (Hym.: Braconidae), a parasitoid of
    *Porthetria dispar* (Lep.: Lymantriidae).   Entomophaga 20 (3):
    261-268.

Huffaker, C. B.   1964.   Fundamentals of biological weed control.
    pp. 631-649 *in:*   Biological Control of Insect Pests and Weeds.
    Chapman Hall Ltd., London.   844 pp.

Iwata, K.   1972.   (Evolution of Instinct.   Comparative Ethology of
    Hymenoptera) (Translated from Japanese, 1976) Amerind Publ.
    Co., Ltd., New Delhi.   535 pp.

Johnson, F. M. and J. M. Burns.   1966.   Electrophoretic variation
    in esterases of *Colias eurytheme* (Pieriedae).   J. Lepid. Soc.
    20: 207-211.

Kerrich, G. J.   1962.   An assessment of the significance of colour
    in the systematics of the Hymenoptera parasitica.   Verh. 11th
    Int. Kong. Entomol., Wien 3:   302-303.

Khasimuddin, S. and P. DeBach.   1976a.   Biosystematic studies of
    three allopatric populations of *Aphytis maculicornis* (Hym.:
    Aphelinidae).   Entomophaga 21 (1):   81-92.

Khasimuddin, S. and P. DeBach. 1976b. Biosystematic and evolutionary statuses of two sympatric populations of *Aphytis mytilaspidis* (Hym.: Aphelinidae). Entomophaga 21 (1): 113-122.

Khasimuddin, S. and P. DeBach. 1976c. Hybridization tests: A method of establishing biosystematic statuses of cryptic species of some parasitic Hymenoptera. Ann. Entomol. Soc. Amer. 69 (1): 15-20.

Kogan, M. and E. F. Legner. 1970. A biosystematic revision of the genus *Muscidifurax* (Hymenoptera: Pteromalidae) with the description of four new species. Canad. Entomol. 102 (10): 1268-1290.

Legner, E. F. 1969. Reproductive isolation and size variation in the *Muscidifurax raptor* complex. Ann. Entomol. Soc. Amer. 62 (2): 382-385.

Legner, E. F. 1972. Observations on hybridization and heterosis in parasitoids of synanthropic flies. Ann. Entomol. Soc. Amer. 65 (1): 254-263.

Lucas, A. M. 1969. The effects of population structure on the success of insect introductions. Heredity 24 (1): 151-157.

Mackauer, M. 1969. Sexual behavior of and hybridization between three species of *Aphidius* Nees parasitic on the pea aphid (Hymenoptera: Aphidiidae). Proc. Entomol. Soc. Wash. 71 (3): 339-351.

Mackauer, M. 1972. Genetic aspects of insect production. Entomophaga 17 (1): 27-48.

Mackauer, M. 1976. Genetic problems in the production of biological control agents. Ann. Rev. Entomol. 21: 369-385.

Matthews, R. W. 1975. Courtship in parasitic wasps. pp. 66-86. *in:* Evolutionary Strategies of Parasitic Insects and Mites. P. Price, ed. Plenum Press, New York, 224 pp.

Mayr, E. 1954. Change of genetic environment and evolution. pp. 157-180 *in:* Evolution as a Process. J. S. Huxley, A. C. Hardy, and E. B. Ford, eds. George Allen and Unwin, Ltd., London. 367 pp.

Mayr, E. 1963. Animal Species and Evolution. Belknap Press, Harvard Univ., Cambridge. 797 pp.

Mayr, E. 1969. Principles of Systematic Zoology. McGraw-Hill Book Co., New York. 428 pp.

Messenger, P. S. and R. van den Bosch. 1971. The adaptability of introduced biological control agents. pp. 68-92 *in:* Biological Control, C. B. Huffaker, ed. Plenum Press, New York. 511 pp.

Metcalf, J. R. and J. Breniere. 1969. Egg parasites (*Trichogramma* spp.) for control of sugar cane moth borers. pp. 81-115 *in:* Pests of Sugar Cane, J. R. Williams et al., eds. Elsevier Publ. Co., Amsterdam. 568 pp.

Micks, D. W. 1954. Paper chromatography as a tool for mosquito taxonomy: the *Culex pipiens* complex. Nature 174 (4422): 217-218.

Micks, D. W., A. Rehmet, J. Jennings, G. Mason, and G. Davidson.
    1966. A chromatographic study of the systematic relationship
    within the *Anopheles gambiae* complex. Bull. World Health Org.
    35: 181-187.

Micks, D. W., J. Jennings, A. Rehmet, G. Mason, and G. Davidson.
    1967. Further chromatographic studies of the systematic re-
    lationship within the *Anopheles gambiae* complex. Bull. World
    Health Org. 36: 308-318.

Nagaraja, H. and S. Nagarkatti. 1973. A key to some New World
    species of *Trichogramma* (Hymenoptera: Trichogrammatidae),
    with descriptions of four new species. Proc. Entomol. Soc.
    Wash. 75 (3): 288-297.

Nagarkatti, S. 1970. The production of a thelytokous hybrid in
    an interspecific cross between two species of *Trichogramma*
    (Hym.: Trichogrammatidae). Curr. Sci. 39 (4): 76-78.

Nagarkatti, S. and M. Fazaluddin. 1973. Biosystematic studies on
    *Trichogramma* species (Hymenoptera: Trichogrammatidae). II.
    Experimental hybridization between some *Trichogramma* spp.
    from the New World. Syst. Zool. 22 (2): 103-117.

Nagarkatti, S. and H. Nagaraja. 1968. Biosystematic studies on
    *Trichogramma* species: 1. Experimental hybridization between
    *Trichogramma australicum* Girault, *T. evanescens* Westwood, and
    *T. minutum* Riley. Comnwlth. Inst. Biol. Cont. Tech. Bull.
    10: 81-96.

Nagarkatti, S. and H. Nagaraja. 1971. Redescriptions of some
    known species of *Trichogramma* (Hym., Trichogrammatidae),
    showing the importance of the male genitalia as a diagnostic
    character. Bull. Entomol. Res. 61 (1): 13-31.

Nagarkatti, S. and H. Nagaraja. 1977. Biosystematics of *Tricho-
    gramma* and *Trichogrammatoidea* species. Ann. Rev. Entomol. 22:
    157-176.

Nesser, S. 1973. Biology and behavior of *Euplectrus* species near
    *laphygmae* Ferriere (Hymenoptera: Eulophidae). Entomol. Mem.
    Dep. Agric. Tech. Serv. So. Afr. 32: 1-31.

Oatman, E. R., P. D. Greaney, and G. R. Platner. 1968. A study
    of the reproductive compatibility of several strains of
    *Trichogramma* in southern California. Ann. Entomol. Soc. Amer.
    61 (4): 956-959.

Oatman, E. R., G. R. Platner, and D. Gonzalez. 1970. Reproductive
    differentiation of *Trichogramma pretiosum*, *T. semifumatum*, *T.
    minutum*, and *T. evanescens*, with notes on the geographical
    distribution of *T. pretiosum* in the southwestern United States
    and Mexico (Hymenoptera: Trichogrammatidae). Ann. Entomol.
    Soc. Amer. 63 (3): 633-635.

Oatman, E. R. and G. R. Platner. 1973. Biosystematic studies of
    *Trichogramma* species : 1. Populations from California and
    Missouri. Ann. Entomol. Soc. Amer. 66 (5): 1099-1102.

Oman, P. W. 1974. Identification and classification in pest management control. pp. 77-86 *in:* Proceedings of the Summer Institute on Biological Control of Plant Insects and Diseases. F. G. Maxwell and F. A. Harris, eds.

Quednau, W. 1955. Uber einige *Trichogramma*-wirte und ihre Stellung im Wirt-Parasit-Verhaltnis. Ein Beitrag zur Analyse des Parasitismus bei Schlupwespen. Nachrichten. deutsch. Pflanzensch. 7: 145-148.

Quednau, W. 1956a. Die biologischen Kriterien zur Unterschiendung von *Trichogramma* Arten. Zeitsch. Pflanzen. Krankh. 63: 334-344.

Quednau, W. 1956b. Der Wert des Physiologischen Experimentes zur das Artsystematik von Trichogramma (Hym. Chalcididae). Berlin Hundertjahr. Deut. Entomol. Ges. 30: 87-92.

Quednau, W. 1961. Die problematik der nomenklatur bei den *Trichogramma*-arten. Entomophaga 6 (2): 155-161.

Rao, S. V. and P. DeBach. 1969a. Experimental studies on hybridization and sexual isolation between some *Aphytis* species (Hymenoptera: Aphelinidae). I. Experimental hybridization and an interpretation of evolutionary relationships among the species. Hilgardia 39 (19): 515-553.

Rao, S. V. and P. DeBach. 1969b. Experimental studies on hybridization and sexual isolation between some *Aphytis* species (Hymenoptera: Aphelindiae). II. Experiments on sexual isolation. Hilgardia 39 (19): 555-567.

Rao, S. V. and P. DeBach. 1969c. Experimental studies on hybridization and sexual isolation between some *Aphytis* species (Hymenoptera: Aphelinidae). III. The significance of reproductive isolation between interspecific hybrids and parental species. Evolution 23 (5): 525-533.

Ratzeburg, J. T. C. 1852. Die Ichneumonen der Forstinsecten. Berlin, 238 pp.

Remington, C. L. 1968. The population genetics of insect introduction. Ann. Rev. Entomol. 13: 415-426.

Rosen, D. and P. DeBach. 1973. Systematics, morphology, and biological control. Entomophaga 18 (3): 215-222.

Rössler, Y. and P. DeBach. 1972a. The biosystematic relations between a thelytokous and an arrhenotokous form of *Aphytis mytilaspidis* (LeBaron) (Hymenoptera: Aphelinidae) 1. The reproductive relations. Entomophaga 17 (4): 391-423.

Rössler, Y. and P. DeBach. 1972b. The biosystematic relations between a thelytokous and an arrhenotokous form of *Aphytis mytilaspidis* (LeBaron) (Hymenoptera: Aphelinidae) 2. Comparative biological and morphological studies. Entomophaga 17 (4): 425-435.

Rössler, Y. and P. DeBach. 1973. Genetic variability in a thelytokous form of *Aphytis mytilaspidis* (LeBaron) (Hymenoptera: Aphelinidae). Hilgardia 42 (5): 149-176.

Sabrosky, C. W.   1955.   The interrelations of biological control
     and taxonomy.   J. Econ. Entomol. 48 (6):   710–714.
Schlinger, E. I. and R. L. Doutt.   1964.   Systematics in relation
     to biological control.   pp.   247–280 *in:*   Biological Control
     of Insect Pests and Weeds.   P. DeBach, ed. Chapman Hall Ltd.,
     London.   844 pp.
Simmonds, F. J.   1947.   Improvement of the sex ratio of a parasite
     by selection.   Canad. Entomol. 79 (3):   41–44.
Simmonds, F. J.   1963.   Genetics and biological control.   Canad.
     Entomol. 95 (6):   561–567.
Simmonds, F. J.   1972.   Approaches to biological control problems.
     Entomophaga 17 (3):   251–264.
Townes, H.   1969.   The Genera of Ichneumonidae, Pt. 1.   American
     Entomol. Inst., Ann. Arbor, 300 pp.
Urquijo, L. P.   1951.   Aplication de la Genetica al aumento de la
     eficacia del *Trichogramma minutum* en la lucha biologica.
     Bol. de Patol. veg. y Entomol. Agr. 18:   1–12.
Usinger, R. L., P. Wygodzinsky, and R. E. Ryckman.   1966.   The
     biosystematics of Triatominae.   Ann. Rev. Entomol. 11:   309–
     330.
White, E. B., P. DeBach, and M. J. Garber.   1970.   Artificial
     selection for genetic adaptation to temperature extremes in
     *Aphytis lingnanensis* Compere (Hymenoptera: Aphelinidae).
     Hilgardia 49 (6):   161–192.
Whitten, M. J.   1970.   Genetics of pests in their management.   pp.
     119–135 *in:*   Concepts of Pest Management,   R. L. Rabb and
     F. E. Guthrie, eds. No. Carolina St. Univ. Press.
Wilkes, A.   1942.   The influence of selection on the preferendum
     of a chalcid (*Microplectron fuscipennis* Zett.) and its sig-
     nificance in the biological control of an insect pest.
     R. Soc. London, Proc. (B) 130:   400–415.
Wilkes, A.   1964.   Inherited male-producing factor in an insect
     that produces its males from unfertilized eggs.   Science 144
     (3616):   305–307.
Wilkes, A., P. Pielou, and R. F. Glasser.   1952.   Selection for
     DDT tolerance in a beneficial insect.   *in:*   Conference on
     Insecticide Resistance and Insect Physiology.   Natl. Acad.
     Sci., Natl. Res. Coun. Pub. 219:   78–81.
Wilson, F.   1965.   Biological control and the genetics of colon-
     izing species.   pp.   307–329 *in:*   The Genetics of Colonizing
     Species.   H. G. Baker and G. L. Stebbins, eds.   Academic Press,
     New York.   588 pp.

# Section II

# SCIENTIFIC THRUSTS SUPPORTING AUGMENTATION

CHAPTER 5

NUTRITION OF NATURAL ENEMIES

H. L. House

Agriculture Canada, Smithfield Experimental Farm

Box 340, Trenton, Ontario, K8V 5R5 Canada

Because parasitoids[1] are living animals, nutrition inevitably plays an important role in augmenting these natural enemies of pests. For nutrition is about nourishment; that is, it is the action or processes of transforming substances found in foodstuff into body materials and energy to do all the things attributed to life. Nutritional requirements depend on the synthetic abilities of the organism and the basis is genetical. Therefore, through nutrition we have a direct and essential connexion between an environmental factor, foodstuff, and the vital processes of the insect organism.

In 1929 Uvarov stated that nutrition and metabolism should be regarded as a key to successful control of injurious insects and to the progress of industries dependent upon the products of useful insects. Yet 43 years later it was quite valid for Beck (1972) to state that knowledge that a given insect displays a nutritional requirement for a substance or group of compounds has seldom if ever led to a practical method of insect control.

My contention with respect to the use of nutrition in insect control is that we are too disadvantaged because our understanding

_____

[1] In this discussion the term parasitoid is intended to include insect predators and endo-and ectoparasites regardless of taxonomic position and definite differentiation in feeding habits. In any case, owing to intergrading on all sides, it is difficult to establish criteria to distinguish predators from parasites in Insecta. And there is really no sound reason from a nutritional standpoint to make such distinction to date.

of insect nutrition is under-developed, it has not been applied
properly to the problems of pest management and control, and we do
not regard foodstuff of insects in the proper light.  Not infrequently we even seem naive about these matters.  We seem prone to
look to nutrition as a lethal assault weapon rather than to regard
it as a support weapon whose role in pest control is most reasonably
supplemental, integrative, and cooperative.  Its efficaciousness
may very well be in augmenting parasitoids in pest control.  However,
our understanding of the nutrition of parasitoids is especially
noteworthy because it is so limited.  And moreover we know very
little about the actual foodstuff of parasitoids.


                          PRESENT POSITION


                             Foodstuff

    Foodstuff is beyond the present topic; nevertheless, as food
and nutrition are entwined, a digression of a few pertinent
general remarks may not be amiss.

    Natural foodstuffs possess distinguishing properties including
characteristic form, texture, and often chemical feeding attractants,
repellents, and deterrents.  These usually are not nutritional
factors and play little if any significant role in nutrition *per
se*.  To be food, however, foodstuffs must possess nutritionally
important substances, or nutrients, that include protein or amino
acids, carbohydrates, lipids, vitamins, and minerals that can be
obtained through digestive processes to nourish the consumer.  As
a rule we know insect foodstuffs better by name, site, or other
identification (see: Brues 1946) or by some characterizing specific
chemical factor associated with the feeding activities of particular
insect (see: Dethier 1947, 1970) than we do by nutrition properties.

    Food abundance is commonly recognized as one of the major
biotic factors regulating the numbers and population fluctuations
of insects, but the quality of the foodstuff is seldom if ever
given due consideration in these matters.  Food quality depends on
many factors including digestibility, the kinds and amounts of
nutrients available upon digestion of the foodstuff, and how well
the nutrients qualitatively and quantitatively meet the nutritional
requirements of the animal in question.  Chemical analysis of a
foodstuff is not necessarily the most practical measure of the
usefulness of the food.  Chemical analysis can be misleading
because it may make little or no distinction between nutritive
and non-nutritive substances; and it does not show the kind and
amount of nutritive material that the animal can derive by
digestion of the foodstuff.  There are many ways of measuring the
usefulness of a foodstuff (see: Maynard and Loosli 1962; Morrison

1941; Evans 1939; Waldbauer 1968). Among the criteria for the usefulness of foodstuff determined by feeding tests are (1) the
coefficient of digestibility - i.e. the percentage of the foodstuff
digested, (2) the efficiency of food conversion - i.e. the percentage of the food that is converted into body material (measured as
weight gain), (3) the total digestible nutrients - i.e. the sum of
all the digestible organic nutrients - protein, fiber, nitrogen-
free extract, and fat (the latter being multiplied by 2.25), and
(4) the nutritive ratio - i.e. the ratio of digestible protein to
digestible non-nitrogenous nutrients in the foodstuff. Some of
these measures were used to determine the usefulness of leaves of
various plants as food for several insects (Evans 1939; Waldbauer
1964; Soo Hoo and Frankel 1966). Digestibility and efficiency of
food conversion differed with the foodstuff and with the insect
concerned (House 1972a, 1974). Very few attempts have been made
to measure and compare the quality of hosts for a parasitoid by
these means. Smith (1965) determined the efficiency of food
conversion in third instar *Anatis mali* Auct. reared on the aphid
*Acythosiphon pisum* (Harris) to be 39% and on the aphid
*Rhopalosiphum maidis* (Fitch) 29%; that of *Coleomegilla
maculata lengi* Timberlake on *A. pisum* 40% and on *R. maidis* 30%.
However, as the efficiency varied with the instar, the efficiency
of food conversion averaged overall about 20% in each of these
coccinellids. He noted that these coccinellids in this work
seemed to convert their foodstuffs, aphids, less efficiently than
certain other insects converted their foodstuff. These coccinellids
converted the aphids less efficiently than different phytophagous·
insects usually do the leaves of most plants (House 1974). The
nutritive ratio is a measure of usefulness of foodstuff for
nutrition much used in livestock feeding but much neglected in
entomology. It is a most practical measure of the quality of a
foodstuff for nutrition because it takes into account specifically
the chief classes of nutritionally important organic components of
the foodstuff and the digestibility of each. Thus it is a somewhat analytical measure that provides some understanding of the
nutritionally important properties and proportions of nutrients in
the foodstuff. The nutritive ratio of one foodstuff often differs
from that of another, even one quite similar in kind; and that of
any given foodstuff often differs among different animals (Morrison
1941). With the cow the nutritive ratio of cabbage leaves (1:4.9)
is much wider than (1:0.44) with the larva of *Pieris brassicae*
(Linnaeus) (Morrison 1941; Evans 1939). Therefore, this indicates
by the ratio, or balance, of nutrients obtained by digestion that
cabbage leaves are quite a "different foodstuff" for the cow than
for the European cabbageworm.

The various measures above provide but a base for understanding
the usefulness of a given foodstuff for nutrition of a specific
animal. Nevertheless, these measures would be useful to entomologists
especially concerned with mass-rearing insects. One measure,

expressed as the coefficient of digestibility, determines only
the digestibility of a foodstuff. The others show adumbratively
that a given foodstuff may lack - or digestive processes fail to
obtain - all needed nutrients and whether or not the foodstuff
provides useful nutrients in amounts, especially proportions, can
be readily made, and objectively provide quantitative data about
insect foodstuff. Thus entomologists have an objective means,
expressed as a statistic, for measuring and comparing the usefulness of varieties of foodstuffs for nutrition of an insect; for
example, should it be necessary in laboratory-rearing of high-
quality host insects to substitute the foliage of some other plant
and find the best alternative to the insect's customary kind.
Data obtained by any of the above measures concerning a given foodstuff and mammal will not likely be applicable to an insect.
Similarily, data pertaining to the usefulness of a given foodstuff
for a specific insect may not be applicable to another insect.

In general:-Foods of animal origin are high in protein and
low in carbohydrate; consequently mammalian predators have evolved
accordingly well-developed proteolytic systems and poorly developed
glycolytic mechanisms (Brambell 1972). Fats contribute an important
part of the energy and are unlikely to be deficient in essential
fatty acids. The amino acid levels in food of animal origin are
well-balanced and the predator needs only to break down the protein
into amino acid constituents, absorb them, and reassemble them into
its own body proteins. Vitamin requirements are met from supplies
within the body tissues of the prey. Therefore, animals that eat
food of animal origin need consume only about half the volume of
food needed by herbivores because the amino acids of plant proteins
are in different proportions than those of the animal, and plant
foods contain most of their energy in the form of saccharides and
cellulose (Brambell 1972). It is probably for these reasons that
an endoparasitoid can thrive on the limited food supply provided
by the relatively small body volume of its host.

The foodstuffs of parasitoids seem *prima facie* to be of
animal origin. Some may feed on such foodstuff throughout their
larval and adult feeding stages. *Calosoma sycophanta* (Linnaeus)
is an example of one that probably feeds exclusively on animal
foodstuff (Brues 1946). However, most parasitoids are carnivorous
only during the larval stages. With most hymenopteran parasitoids
the larva eats food of animal origin, but the adult has no predatory
habit except in the special sense of the female feeding at oviposition punctures on the body fluids of the host. Many species of
paraistoids feed on a mixed diet of plant and animal origin. The
larva and adult of many coccinellids that feed on scale insects,
plant lice, and aphids also feed on pollen and fungi. Both the
male and female adult of many parasitoids feed on pollen and sugary
plant juices; though, as noted, the female of some may feed
especially during oogenesis on host fluids. Examples include

*Itoplectis conquisitor* (Say) and *Scambus buolianae* (Hartig).
*Orgilus obscurator* Nees is an example of an adult parasitoid that
does not feed on host materials (Leius 1960, 1961a, 1961b).
Leius (1960) pointed out that adults of some hymenopteran
parasitoids visit flowering plants of many different families to
obtain nectar and pollen.  However, a considerable number of
species with short mouth parts are restricted to certain plants.
He showed that both sexes of the parasitoid *O. obscurator* were
attracted only to umbelliferous flowers, which have exposed
nectaries, and preferred wild parsnip, *Pastinaca sativa* Linnaeus.
Two others, *I. conquisitor* and *S. buolianae*, were attracted to
flowers of Umbeliffera, but were attracted also to the flowers of
other families.  Adult female *S. buolianae* showed a preference for
particular kinds of flowers as well as for body fluids of particular
hosts (Leius 1967b).  Moreover, he showed that parasitism of
*Malacosoma americanum* (Fabricius) and of *Carpocapsa pomonella*
Linnaeus was increased many-fold in an orchard by the presence of
wild flowers as food sources for the adult parasitoids (Leius 1967a).
His findings are consistent with those of other workers and suggest
that when particular food sources are absent or their distribution
is limited, adult parasitoids may not be able to find adequate
food supply for their survival.  Failure to recognize that food
sources for the adult parasitoid were needed in the environment of
pest insects was possibly a reason why many attempts at biological
control have failed, according to Beirne (1962).

We do not know the actual foodstuff of a great many parasitoids,
usually of the adult, or for that matter of a great number of
economically important insects.  We should not overlook the fact
that a great number of parasitoids, usually in the adult stage,
wholly or partly, eat foodstuff of plant origin.  It is important
to know what an insect eats; for what it eats determines it
economic importance.  According to Leuis (1961a), the feeding of
the female parasitoid on host fluids increases the value of some
species as control agents; for this feeding may be more important
than parasitism in many cases as a cause of host mortality.  Moreover, understanding fully what a parasitoid eats suggests what its
nutritional requirements are.


Insect Nutrition

Our present understanding of insect nutrition includes the
following:-The feeding habits of a great many species of insects
differ widely between the larval and the adult stage, and so their
nutritional requirements met by diet may differ accordingly (House
1972a).  It is recognized that insects, like other animals, generally
accumulate stores of nutritive material in their body tissues during
active feeding for mobilization later in the life histroy of the
insect, and certain species harbor symbiotic microorganisms in their

gut or body tissues.  These nutrient reserves and symbiotes are
covert sources of nutritive material that play a role in nutrition
and tend to obscure the need for vitamins and other nutrients (House
1972a).    Because of nutrient reserves the requirement for some
essential nutrients may not be expressed always within one generation
or developmental stage.  In many species the adult is very dependent
upon nutrient reserves.  Normally, as will be discussed later, this
dependence and the extent of gonadal development in the young adult
determines the nutritional requirements that must be met by diet.
Nutrition is not a 'fixed structure', but rather it is effected by
a 'flow' of metabolites in dynamic equilibrium, as elucidated by
Schoenheimer (1942).  Materials of food origin are constantly being
built into body tissues, removed, and reused again elsewhere as may be
needed until depleted.  Therefore, one can not say that a given
nutrient is needed exclusively for pupation or for oogenesis, though
symptomatically it may appear so (House 1963, 1972a).  Symptomatically,
the need for a given nutrient may be expressed differently in
different insects (House 1963).  Moreover, the criteria commonly
used in insect nutrition work to determine the nutritional importance
of at least some food substances may fail to show the nutritional
status of a given substance.  For example, that a deficiency of
vitamin A, or its carotene precursor, results in faulty vision and
structure of the compound eye was generally missed because the
criterion used in insect work usually was growth rate or some other
such readily obvious effect.  As the nutritional requirements of
insects become understood through various criteria, the similarity
of their requirements to those of higher animals, including warm-
blooded vertebrates, becomes increasingly noteworthy.

We may state our present position as follows:-Work has advanced
greatly in some ways since Uvarov's 1929 review.  Much expertise
was demonstrated in developing hundreds of artificial diets for
numerous insect species, including several parasitoids (House
1967a; House *et al.* 1971; Singh 1974; Vanderzant 1973; Thompson
1975).  This pursuit is really dietetics, i.e. the art and science
dealing with the application of the principles of feeding behavior
and of nutrition to feeding individuals on concocted food regimens.
Sucessful development of a suitable artificial diet very often
depends as much or more on the non-nutritive factors of the diet,
as well as on the technique for presenting the food material to
the insect, as it does on its nutrient content *per se*.  There are
now artificial diets for many kinds of insects that are useful hosts
for parasitoid rearing (House 1967a; House *et al.* 1971; Singh 1974).

In the few examples of dietetics with parasitoids the dietary
mixtures of amino acids, vitamins, and minerals for the dipteron
*Agria housei* Shewell [=*A. affinis* = *Pseudosarcophaga affinis*] were
the same as those used later for the hymenopterans *I. conquisitor*
and *Exeristes roborator* (Fabricius), and essentially very similar
to those for other insects (House 1972a; Yazgan 1972; Thompson

1975; Vanderzant 1968). Basically, a suitable artificial diet for
nutrition depends on the nutrient content; for which there seems
to be extant an universal, approximately appropriate composition
(see: House 1966a, re concept of an "universal diet"). All insects
seem to require about the same kind and number of amino-acids,
lipids, vitamins, and so forth in some sort of ordinary proportional
relationship in quantities, on a dry-weight basis, though
nutritionally suitable relationships may depend on dietary
composition and vary with the insect (House 1967a, 1969, 1972a,
1974). The donor of these nutrients may be an artificial diet
or species of plant or animal; for a protein of high quality for
nutrition or a vitamin can be equally well utilized irrespective
of whether obtained from the chemist or from tissues or a plant
or animal. For practical artificial diets to routinely rear an
insect the nutrients, for example, amino acids, need not be supplied
as mixtures made up of nutrients in highly-refined, chemically-
pure, form; provided that whatever food materials used contain
the needed nutrients, and that they can be made available to the
insect by its digestive processes. For example, the parasitoid
*A. housei* was reared on chemically-defined artificial diets (House
1954a, 1964, 1966b, 1966c) and, as pork liver is rich in nutrients,
it and the parasitoids *Sarcophaga aldrichi* (Parker), *Kellymyia
kellyi* (Aldrich), and *Pimpla turionella* (Linnaeus) were reared on
pork liver (House 1966a). However, despite sufficient nutrients
to meet the insect's nutritional requirements, to be successful
an artificial diet must meet the insect's feeding requirements.
Feeding requirements are those chemical and physical factors that
afford taste, texture, and other properties so important to induce
normal feeding activities. For example, the feeding requirements
of an insect that has sucking mouth parts will differ from those
of one that has biting and chewing mouth parts. In some cases
feeding activities may depend on the inclusion in the diet of some
chemical feeding stimulant familiar to the insect or on the exclusion
of something that deters normal feeding. Sometimes no more than the
texture, form, or surface of the diet determines whether or not
normal feeding, and consequently normal nutrition, results. In the
example given above the larva of *A. housei*, *S. aldrichi*, and *K.
kellyi* fed readily on slabs of fresh pork liver, but for *P.
turionella* the liver had to be coagulated by heat and, with a
physiological saline solution, ground into a fine-textured slurry.
Furthermore, for example, diets developed at Belleville, Ontario,
for *Musca domestica* Linnaeus and for *A. housei* differed only in the
concentration of agar used to give the diets suitable consistency
(House 1966a). With 0.75% of agar the larvae of *A. housei*
had no difficulty, but that of *M. domestica* submerged itself, remained so, and drowned. However, it was found that with 1.5% of
agar the diet was rigid enough to permit air to follow along the
borings made by *M. domestica* to prevent suffocation; so the diet
was then successful. Yazgan (1972) reared *I. conquisitor* similarily
on an artificial diet containing only 0.5% of agar in order that the

larva, which has a blind gut, would accumulate a minimum of in-
digestible material.  But to induce feeding it was necessary to
grind the semi-gelled diet to a slurry, thus changing its surface
texture.  Thompson (1975) found that *E. roborator*, unlike *I. conquisitor* and *A. housei*, would not feed on agar-gelled diets.  However, feeding activities resulted and the diet was successful when
drops of the nutrient solution were put on a disc of cotton dental
roll on which albumin or Sephadex$^R$ gel was deposited as a "retaining agent" for the nutrients.  Hopefully, this short excursion into
the vast complex subject of dietetics may show how success in
developing artificial diets can depend on small unusual matters.
It by no means exhausts the problems; for they differ with the
insect and with the characteristics and properties of the food
materials and diet.  Even the processing and techniques for preparing materials for artificial diets, whether highly refined or
not, can determine the success of a diet (Vanderzant 1974).
Consequently, the problem of concocting a successful artificial
diet for a particular insect can be very complex; too great
and really beyond the subject of the nutrition of natural enemies.
But in any case, clear understanding of the insect's feeding
habits, activities, and requirements seem essential, and often
much ingenuity is needed to meet them.  To this end effective
ideas often may be obtained from the work of others, as compiled
by House (1967a), House *et al*. (1971), and Singh (1974).  One of
the latest discussions is the review by Vanderzant (1974) of many
aspects of insect dietetics.

    Of course, cooperatively with much of the work on dietetics
was the determination of the qualitative nutritional requirements.
Consequently, the qualitative nutritional requirements of many
insects representing different taxa and feeding habits are understood in considerable detail, if not completely, in a few species.
However, in most species only the larval requirements are known
with data usually based on experience with only one generation
(House 1972a, 1974).  This is the prevailing work on insect nutrition.

    Most of the data on quantitative nutritional requirements are
not particularly meaningful nutritionally except to concoct diets.
For almost all the data on the so-called quantitative nutritional
requirements of insects pertain to the dietary compositional level
found suitable for rearing a particular insect on a particular
dietary composition and physical properties at a particular
temperature rather than to the amount of the nutrient involved in
nutrition of a particular function or activity, as is the case
with mammals (for examples with mammals see: Altman and Dittmer
1974; Typpo and Briggs 1967).  As Sang (1956a) stated, one series
of values that may be of use when comparing nutritional needs of
insects is the amount of each constituent needed to produce one
gram of viable pupae.  Usually with insects the rate of intake -
say to produce a gram of body weight - was not considered.

Taking this into account Sang (1956a) calculated that to produce
one gram wet weight of pupae of *Drosophila melanogaster* Meigen
the following amounts of vitamins in micrograms were needed:
0.6-1.0 of thiamine, 2.4-4.0 of riboflavin, 3.0-5.0 of nicotinic
acid, 4.5-8.5 of pantothenic acid, 0.07-1.2 of pyridoxine, 0.05-
0.08 of biotin, and 0.6-1.0 of folic acid.  Such data on insects
are scarce.  Some work on insects shows that the amounts of different nutrients required for optimal nutrition are relative to
various factors including the amount of some other nutrient(s)
involved, the balance of nutrients, and temperature (Sang 1956a,
1959, 1962; House 1966b, 1972b, 1976).  It is this relativity to
various factors that with few exceptions have been overlooked.
Because we are so lacking in data on the real quantitative
nutritional requirements of insects we do not know their exact
quantitative requirements for specific activities and metabolic
processes.  This is a most important gap in our understanding of
the nutrition of insects; not only for insight into nutrition
biochemistry, but especially for dietetics.  For without data
on the real quantitative requirements of the insect for specific
actions, including optimal rate of growth or of development
or of oogenesis and fecundity, we are unable to expertly design
scientifically artificial diets, as balanced rations, to achive
optimal nutrition and increased performance, as in work with domestic
animals.

Therefore, we understand insect nutrition mostly on the bases
of dietetics and of qualitative requirements; that is to say on
the most simple and elementary level.  For not withstanding some
new developments and progress (Dadd 1973), entomology has little
more than merely broached three important matters concerning the
nutrition of insects; namely, (1) their actual quantitative
nutritional requirements for particular activities, such as for
growth and development, for oogenesis, for fecundity; (2) the effects
of quantitative nutritional factors on their nutrition, behavior,
activities, and vigor; and (3) the matter of food quality.  These
would appear to be generally of great importance, and they would
have considerable ecological significance (House 1969; Dadd 1973).
In this immature position of insect nutrition we unquestionably
are handicapped to attempt to augment natural enemies by applying
nutritional assistance.


## NUTRITION OF PARASITOIDS AND OTHER INSECTS:  A COMPARISON

Parasitoids and other insects must obtain from environmental
sources protein, carbohydrates, lipids, minerals, vitamins, water,
and perhaps other substances.  The real question is:  which particular constituents of foodstuffs and materials are essentially
important for nutrition; i.e. what are the qualitative nutritional
requirements?  It is among the qualitative requirements that we

might expect to find distinguishing differences that relate to taxa and feeding habits.

The qualitative nutritional requirements of insects usually were determined by feeding tests using more or less chemically well-defined artificial diets with the nutrient in question omitted or added. The criterion with larvae usually was the effect of the omission or addition on growth and development. Sometimes with larvae other criteria were used. For example, with deficiencies of certain fatty acids, wing deformities in the adult were produced and vitamin A deficiencies resulted in optic deformities in the adult. The criteria used with adults usually were the effects on maintenance, longevity or reproduction.

The qualitative nutritional requirements of many insects have been reviewed elsewhere (see: Altman and Dittmer 1968; House 1969, 1972a, 1974; Dadd 1973). A generalized summary of these may be stated as follows: All immature insects need essentially a sterol, almost always cholesterol or the like; one or two exceptions are unsettled cases. The nitrogen source of larvae generally includes 10 essential amino acids; namely, arginine, histidine, isoleucine, leucine, lysine, methionine, phenylalanine, threonine, tryptophan, and valine. Larvae need a number of water-soluble vitamins; usually including biotin, choline chloride, folic acid (pteroylglu-tamic acid), nicotinic acid, pantothenic acid, pyridoxine, ribofla-vin, and thiamine. However, this list of vitamins may be modified by one or two deletions and/or additions that pertain variously among species; for example, a few species need inositol, and some phytophagous species need ascorbic acid (vitamin C). Of course, all larvae need certain inorganic salts and presumably water from some source. Many need carbohydrate, usually essentially a hexose monosaccharide; but no specific one. Some need certain fatty acids; usually one or more of the polyunsaturated ones such as linoleic and linolenic. Diptera and a few Coleoptera need certain components of nucleic acid for normal growth and development rate. There is increasing evidence that a number of insects may need the fat-soluble vitamins A and E ( α- tocopherol).

The qualitative nutritional requirements of only a very few parasitoids are determined. The most penetrating work on parasitoid larvae may be summarized as in Table 1.

Table 1 shows that the larva of the dipteron parasitoid *A. housei*, of the hymenopteran *E. roborator*, and of the neuropteran *Chrysopa carnea* Stephens required the same 10 essential amino acids commonly required by other kinds of insects (House 1954b, 1972a; Thompson 1967a; Vanderzant 1973). Like other larval insects *A. housei* required an essential sterol, such as cholesterol or others having certain like structural configurations, as well as a number of inorganic salts. Presumably this is true for other parasitoids

because, as noted above, a sterol and mineral are needed by all
insects, and were always in successful diets for rearing parasitoids.
Both *A. housei* and the hymenopteran *I. conquisitor*, like many other
insects required the vitamins biotin, choline chloride, nicotinic
acid, pantothenic acid, pyridoxine, riboflavin, and thiamine (House
1972a, Yazgan 1972).  However, in neither was vitamin $B_{12}$ nor
inositol essential; but as in some insects, *I. conquisitor*
required folic acid.  In *A. housei* vitamin A has a beneficial
effect on the rate of development and vitamin E was essential
in the larva to permit subsequently normal reproduction in the
adult female (House 1965a, 1966c).  As mentioned above, both of
these fat-soluble vitamins were required in a few other kinds of
insects either in the larva or in the adult (see:  House 1972a,
1974).  But in *I. conquisitor* neither vitamin A, E, nor C had any
apparent effect, as was the case in many insects.  In *A. housei*,
*I. conquisitor*, and *E. roborator* carbohydrate specifically glucose,
had a beneficial effect.  However, a requirement for carbohydrate
might not be essential except under particular conditions related
to nutrient levels and balance and to temperature (House 1954c,
1976; House and Barlow 1956; Yazgan 1972; Thompson 1976b).  In
*I. conquisitor* the essential fatty acid was linolenic acid, but
in *A. housei* oleic acid, rather than polyunsaturated fatty acids,
had the most pronounced effect along with palmitic acid and
stearic acid on growth and development rate (Yazgan 1972; House
and Barlow 1960).  No fatty acid requirement has been found in *E.
roborator*; triolein had no effect (Thompson 1975, 1976b).  Although the fatty acid requirement in *I. conquisitor* follows the
rule, that in *A. housei* is not unique; for the fatty acid requirements of the sawtoothed grain beetle, *Oryzaephilus surinamensis*
(Linnaeus) were the same as those of *A. housei* (House and Barlow
1960; Davis 1967).  Moreover, the dipteron *A. housei* required
nucleic acids - essentially certain nucleotide components; but
*I. conquisitor* did not (House 1964; Yazgan 1972).  However, though
*D. melanogaster* and *A. housei* as well as *O. surinamensis* needed
nucleic acids, the nucleic acids and their specific components that
each species utilized differed to some extent (Sang 1957; House
1964; Davis 1966).  We may conclude that specific requirements of
this sort vary a bit among species irrespective of taxonomic
position and feeding habit.

In general:  The qualitative nutritional requirements of adult
insects seem simple and are often fewer than those of the larva.
But often we are deceived into concluding that the requirements of
the adult are few and simple.  Reserves of nutritive material,
accumulated in body tissues during the larvae feeding stages and
mobilized later to nourish the adult - and the egg - are a covert
source of dietary supplement (House 1966a, 1972a).  The nutritional
requirements of the adult depend on the extent to which development
of reproductive organs continue into the adult stage and as gonadal
demands may differ with gonadal development and sex, requirements

Table 1.  Qualitative nutritional requirements of parasitoid larvae.

| Nutrient[1] | *Agria housei*[2,3] (Diptera) | *Itoplectis conquisitor*[4] (Hymen.) | *Exeristes roborator*[5] (Hymen.) | *Chrysopa carnea*[6] (Neur.) |
|---|---|---|---|---|
| Amino acids:- | + | !? | + | + |
|    arginine | + | | + | + |
|    histidine | + | | + | + |
|    isoleucine | + | | + | + |
|    leucine | + | | + | + |
|    lysine | + | | + | + |
|    methionine | + | | + | + |
|    phenylalanine | + | | + | + |
|    threonine | + | | + | + |
|    tryptophan | + | | + | + |
|    valine | + | | + | + |
|    others | ±[7] | | - | - |
| Carbohydrates:- | +[8] | +[8] | +[8] | ? |
|    glucose | U | U | U | |
|    fructose | U | | | |
|    sucrose | U | | | |
|    others | U | | | |
| Lipids:- | + | + | !? | !? |
|   sterols:[9] | + | !? | !? | !? |
|    cholesterol | U | ? | ? | ? |
|    others | U | | | |
|   fatty acids: | + | + | - | ? |
|    arachidonic | - | | | |
|    linoleic | - | | | |
|    linolenic | - | + | | |
|    oleic | ± | | - | |
|    palmitic | ± | | | |
|    stearic | ± | | | |
| Vitamins:- | + | + | !? | !? |
|    ascorbic acid | | - | | |
|    biotin | + | + | | |
|    choline chloride | + | + | | |
|    folic acid | - | + | | |
|    nicotinic acid | + | + | | |
|    pantothenic acid | + | + | | |
|    pyridoxine | + | + | | |
|    riboflavin | + | + | | |
|    thiamine | + | + | | |
|    α-tocopherol | +[10] | - | | |
|    others | ± | - | | |

(continued)

Table 1.  Qualitative nutritional requirements of parasitoid
larvae.  (cont'd).

| Nutrient[1] | Agria housei[2,3] (Diptera) | Itoplectis conquisitor[4] (Hymen.) | Exeristes roborator[5] (Hymen.) | Chrysopa carnea[6] (Neur.) |
|---|---|---|---|---|
| Nucleic acid:- | + | - | | |
| component: | | | | |
|   adenylic acid | + | | | |
|   quanylic acid | + | | | |
|   cytidylic acid | + | | | |
|   uridylic acid | + | | | |
|   nucleosides | - | | | |
|   purine bases | - | | | |
|   pyrimidine bases | - | | | |
| Minerals:- | + | !? | !? | !? |
|   potassium | + | | | |
|   phosphorus | + | | | |

[1] + = required; - not required; ± beneficial; ! = presumed to be
required because it is universally required throughout Insecta;
? possible requirement, i.e. it or a source was a component of
the diet; U = utilized.
[2] Formerly named Agria - or Pseudosarcophaga - affinis.
[3] See House (1972a, 1974) for references to particular work.
[4] Yazgan (1972).
[5] Thompson (1975, 1976a).
[6] Vanderzant (1973).
[7] Alanine, glycine, serine, and tyrosine were beneficial, but their
essentiality is doubtful owing to the composition of the diet
at the time (see House 1954b).
[8] Sugar was beneficial and may or may not be required; its effects
vary with the dietary level of amino acids.
[9] Despite a couple of unsettled cases all insects probably require
sterols having in particular usually a molecular configuration
similar to cholesterol:  see House (1974) for differences and
a couple of unique cases that do not conform.
[10] Vitamin $B_{12}$(cyanocobalamine (see: House 1972a) and vitamin A
(House 1965a, 1966c) had a beneficial effect.

may differ accordingly.  The qualitative nutritional requirements
are usually fewer than those of the larva, but may differ widely
among species according to taxa, metamorphosis, and feeding habit
(House 1972a).  For example, the adult of the sheep bot fly,
Oestrus ovis Linnaeus, and including others of Oestridae and
Gastrophilidae, do not feed at all (Brues 1946).  Adult Lepidoptera

need only sugar; for they possess only glycolytic digestive enzymes
(Snodgrass 1961). The adult of many kinds of insects may sustain
itself on sugar, but otherwise requires a variety of different
nutrients including protein, salts, and vitamins, especially for
fecundity in the female (House 1958).

Among parasitoids the adult female of many chalcids, braconids,
pteromalids, and ichneumonids feeds on the nutritively rich body
fluids of its host, but in certain species additional carbohydrate
food is essential as well as being the common food of the male (Leius
1961a). The adult of both sexes of *I. conquisitor* fed on sugar and
plant protein. Both sexes required carbohydrate, but the female fed
on host fluids before and during oviposition period. For maximum
fecundity adult *C. carnea* needed protein (Hagen 1950; Hagen and Tassan
1966). Tauber and Tauber (1974) showed that though five closely related species of chrysopids had similar adult feeding habits, they
varied in their specific dietary requirements for mating and oviposition. Symbiotes harbored in the crop of adult *C. carnea*, which
feeds on honeydew, probably are a source of essential vitamins and
amino acids that may be lacking in honeydew (Hagen and Tassan 1972).
Nutrient reserves are another source of cooperating supplements.
Bracken (1965, 1966) showed that for normal fecundity the adult
female hymenopteran parasitoid, *Exeristes comstockii* (Cresson)
required amino acids, salts, and the vitamins folic acid, pantothenic acid, and thiamine. Fatty acids and a number of other vitamins including vitamin E were not needed. He showed that the adult
female had low reserve of protein, salts, and vitamins. Without
these vitamins in its food the rate of oviposition was not immediately affected. In a few days, however, hatching of eggs decreased
sharply and soon ended; presumably as exhaustion of its reserves
of vitamins in tissues progressed to depletion. Atallah and Killebrew (1967) showed, by feeding $C^{14}$-labelled acetate, that the adult
coccinellid predator, *Coleomegilla maculata* (DeGeer) likely needed
the amino acids isoleucine, phenylalanine, threonine, and valine;
but apparently not lysine, and the essentiality of alanine, arginine, and histidine was doubtful.

It follows, therefore, that when the qualitative nutritional
requirements of insects of different taxa and feeding habits are
compared, there is a noteworthy 'rule of sameness' found among them
(House 1966a). As would be expected there are admittedly some differences in requirements found among various species of insects as
there are among those of birds and of mammals. Perhaps the most
notable deviations are a requirement for nucleic acids in Diptera
and a few Coleoptera, that for carnitine (vitamin $B_T$) in Tenebrion-
idae, and that for vitamin C only in certain species that feed on
green plants (House 1972a, 1974). According to Sang (1959), differences resulting from evolution - the taxonomic differences - may be
more likely apparent at the level of metabolism in the way and purpose for which nutrients are utilized. Nevertheless, as a reason-

able generalization the qualitative nutritional requirements of insects do not differ very importantly with taxonomic position or with
feeding habit.  In any case, <u>on the basis of qualitative nutritional
requirements there has appeared no unique or special requirement
that distinguishes parasitoids from nonparasitoids</u>.

According to Shteinberg (1955), where specialization and adaption to a narrow range of hosts are farthest evolved, development
has been in the direction of the parasite mastering the microen-
vironmental conditions encountered on the host.  Biochemical properties of the body tissues of the host were secondary factors.
Because the endoparasitoid *I. conquisitor* could be reared without
an insect host, Yazgan (1972) concluded that the importance of the
host to this parasitoid is nothing more than a nutrient source and
an environment for pupation.  Parasitoids are not highly specialized
parasites such as are found in certain other forms of life.  And
we should not expect anything unique or significant about insect
tissues as foodstuff. The nutrient composition of prey tissues and
the nutritional requirements of parasitoids appear to be well-
matched.  As noted above, food of animal origin provides a highly
nutritious foodstuff unlikely to be lacking or very deficient in
essential nutrients.  Nutritive contributions from symbiotes, as is
the case in certain insects that normally feed on foodstuff deficient
in some nutrients, are not likely needed in the predaceous stage of
parasitoids.  However, it should be noted that *C. carnea* possesses
symbiotes while in its non-predaceous form (Hagen and Tassan 1972).

## THE QUANTITATIVE NUTRITIONAL FACTOR: NUTRIENT BALANCE

The balance of nutrients is the dominant quantitative nutritional
factor.  As mentioned above, quantitatively nutritional requirements
depend on and vary proportionately with various factors that affect
metabolic activities.  For normal nutrition, nutrients are required
in some suitable balance, or proportions in order to effect efficient metabolism.  An unsuitable balance can increase the quantitative requirements for some nutrients and oversupply others.  This
results in wasteful inefficient food utilization, limits the number
of individuals that a given quantity of foodstuff can support, and
has detrimental effects on the insect organism (Gordon 1959; Sang
1959; House 1965b, 1966b).  The crux of the matter of nutrient
balance is quantitatively the 'fit' of each nutrient to the metabolic demand for each.  Consequently, proportional relations
among nutrients can enhance or impair nutrition and affect
the activities of the insect concerned.  A parasitoid on natural
foodstuff probably is not likely to be confronted with a lack
of an essential nutrient, but it may be confronted with deficiencies.
Nutrient imbalance in a foodstuff is really a nutrient deficiency;
that is too little of a particular nutrient relative to the
quantity of some other(s) needed to meet the nutritional requirements

of the consumer for normal nutrition.  Therefore, a most important
aspect of nutrition is nutrient proportionality because of
its effects on insects and consequent probable ecological significance concerning them (House 1966d, 1969).  The following examples
support this view.

Work on the spurge hawkmoth, *Hyles* [=*Celerio*] *euphorbiae* (Linnaeus), showed that feeding activities and the nutrition of an insect
can depend on the quantitative composition of foodstuff (House 1965b).
When fed a suitable balance of nutrients for normal nutrition, the
amount of food eaten by the larva was inversely proportional to the
nutrient level of the food without affecting the efficiency of food
conversion or rate of body-weight gain of the larva.  However, on an
unsuitable balance food consumption, utilization, and weight-gain
were much decreased.

Work on the parasitoids *A. housei*, *I. conquisitor*, and *E. roborator* showed that the rate of development depended on the proportions
of nutrients (House 1966b; Yazgan 1972; Thompson 1976b).  Several ratios of nutrients were equally suitable for optimal rate of development, but others resulted in suboptimal rates.  Moreover, in *A. housei*
the amount of body protein laid down varied wth the balance of nutrients.  In *I. conquisitor* the balance of nutrients affected the sex
ratio of individuals successfully reared.

Host diet can affect parasitoids.  Smith (1957) found that the
sex ratio, reproduction rates, and size and survival of the female
compared with the male varied in several species of parasitoids with
the food plant of their host, the California red scale, *Aonidiella
aurantii* (Maskell).  In other work the sex ratio and fecundity of
*Bracon brevicornis* Wesmael varied with the food of its host *Corcyra
cephalonica* (Staunton) (Mathai 1972).  The rate of survival and consequent emergence of *Aphaereta pallipes* (Say) varied with the balance of nutrients in the chemically-defined diet of its host, *A. housei* (House and Barlow 1961).  More *Aphaereta* (55%) were obtained from
the host reared on a diet containing 2.0% of amino acids and 4.0%
of glucose than (44%) when reared on a diet containing 3.0% of amino
acids and 0.5% of glucose.

Other work showed that an insect can detect the nutritional value of its foodstuff and chose among variants of it on the basis of its
nutritional value (House 1967b, 1970, 1971a, 1971b).  *A. housei* presented with an array of different artificial diets in close juxaposi-
tion avoided a nonnutritive medium and a diet lacking an essential
amino acid and chose to feed mostly on the diets having a balance of
nutrients that in feeding tests was best for normal growth and development.  Atwal and Sethi (1963) correlated the preference of *Coccinella septempunctata* Linnaeus for certain species of aphids with the
protein content of the prey that was most like that of the predator.

Heat tolerance of *A. housei* varied with the balance of fatty acids (House et al. 1958). Larvae were reared on diets containing different mixtures of saturated and unsaturated fatty acids in varied proportions, then exposed to $45^\circ$C, and the time required to kill 50% of them was measured. Larvae reared on a predominately saturated fatty acid mixture survived 41 minutes longer than those on a predominately unsaturated mixture.

Temperature, probably because of its effects on metabolism, affects nutrition and its requirements. Work in which *A. housei* was reared on two different nutrient balances in $15^\circ$C and $30^\circ$C, respectively, showed that the rate of development on one diet at a high rearing temperature can be relatively superior to that on the other diet, but at low rearing temperature relatively inferior to that on the other diet (House 1972b). In other words the most suitable balance of nutrients depended on the temperature. Other relationships and interactions implicating nutrient balance and nutritional requirements with high and low rearing temperatures were shown in *A. housei*, including a marked beneficial effect of - if not a definite requirement for - a carbohydrate in the presence of a relatively low dietary level of amino acids in low temperature (House 1966d, 1976).

We now have had two views of insect nutrition; namely, that of the qualitative factor, and that of the quantitative factor. Nothing significant readily suggests how either of these factors can lead to a practical direct method of insect control as associated with insecticides. Chances for any selective means are lessened by the similarity of qualitative nutritional requirements among insects irrespective of taxa and of feeding habits. The qualitative factor has an "all or nothing" effect. Certainly, unfulfillment of a qualitative nutritional requirement makes unimpaired nutrition impossible. The qualitative factor is necessary for nutrition and it is manipulative to some extent; but nevertheless it is a static factor of limited use insofar as achieving what nutrition can do to augment natural enemies. On the other hand, quantitative nutritional requirements differ among insects. It is significant that these requirements can vary even within an insect owing to variant physiological and environmental conditions that affect nutrition, and that the effects of varying quantities of nutrients may importantly vary accordingly. From above work on natural enemies of a weed and of insect pests, we saw that the quantitative factor in nutrition is a manipulative, effective, and dynamic factor to reckon with concerning the vital processes and behavioral activites of a parasitoid.

APPLICATION OF NUTRITIONAL PRINCIPLES TO AUGMENT NATURAL ENEMIES

Faced with a scarcity of data to the contrary, we may conclude

that very little use has been made of nutritional means to augment
natural enemies.   The question may well be:-How can we use our present understanding of insect nutrition to augment natural enemies
in biological control programs?

The first step is to start looking on nutrition with a different perspective.  Nutrition is the process of nourishing.  Nutritionally important substances hardly can be included in the same
contextual associations as insecticides.  Should a nutrient or its
antagonist be found that somehow would control insects, its general use likely would create problems as grave as those created by
conventional chemical insecticides.  For nutrition and its requirements in insects, higher animals, and man are much too similar to
avoid probable hazardous consequences over a wide range of animal
life.  Possibly even the noteworthy nutritionally essential sterol
requirement of Insecta can not be exempted.  The fact that nutrition
is the process of nourishing may suggest the feasibility of dis-
arranging it by somehow manipulating the nutrient composition of
foodstuff to effect nutritional impairment of pest species.  This
might be possible in some special cases (Pratt *et al*. 1972).  But
in most cases nutritional impairment of pest species very likely
would be very difficult and often impossible.  Furthermore, that
which was to be protected very likely would be much damaged before
nutritional impairment of its pest could be effective.  Conversely;
the application of nutritional principles to increase the vigor,
performance, and effectiveness of natural enemies surely is quite
feasible.

Obviously, much can be done with simple stratagems implicating
nutrition to assure the establishment and concentration of a natural enemy in the environment of a pest.  According to Leius (1960),
establishment of hymenopteran parasitoids depends on the presence of
suitable kinds of flowers that provide needed nutrients to the adult
parasitoid.  Bracken (1965) suggested that a simple artificial diet,
composed of the essential nutrients associated with fecundity, may
be useful for hymenopteran parasitoids while awaiting optimal conditions for release or for supplementary food until they became established in the field.  Smith (1965) suggested that yeast would be
a useful supplement  for the coccinellid *C. maculata lengi*.
Use of nutritive sprays for adult chrysopids in the field increased
the concentration of population and promoted fecundity of these
parasitoids (Hagen and Tassan 1970; Hagen *et al*. 1971; Hagen and
Hale 1974).  From work on chrysopids, Tauber and Tauber (1974) concluded that even with closely related species the influence of diet
on reproductive behavior should be considered for each species
whenever insects are to be reared in the laboratory, mass-produced,
released, or managed in the field.

Laboratory rearing of parasitoids, of course, is commonly practiced.  Here dietary and nutritional factors can be manipulated most

readily to augment natural enemies. Nevertheless, many workers
found that laboratory rearing changed parasitoids somehow so that
they did not thrive or were less effective than wild populations
(Mackauer 1976). Behavioral changes stem from conditioning to laboratory rearing conditions and from strong selection pressures for
that part of the population most adapted to these conditions (Boller
1972). Consequently, peculiar phenotypic and genotypic strains e-
volve that differ in various ways from parental or wild types. Recent discussions of problems resulting from laboratory rearing did
not mention nutritional matters or else alleged generally that laboratory food sources were suitable (Boller 1972; Mackauer 1976).
Mackauer (1976) stated, the degree to which an insect population responds to the ambient conditions of the laboratory - and indeed
the responses possible - is ultimately under genetic control. A
genetic basis for nutritional requirements was shown by various
workers whereby the nutritional requirements in insects differed
qualitatively and quantitatively with genetic strain (House 1974).
With respect to heterosis, or hybrid vigor, Sang (1956b) showed
that the amount of nicotinic acid needed by a hybrid strain of
D. melanogaster was midway between that of the parental strains,
and as was the case with other vitamins sometimes the hybrid's
requirement was more or less than that in both parental strains.

From examples in the previous section on the quantitative nutritional factors, it would seem likely that the effects of temperature, cage crowding and resultant competition for food, the degree
of selection and evolution of strains, and so forth imposed by the
highly specialized laboratory environment can affect metabolism,
nutrition, and the nutritional requirements of the insect. Conversely, one may expect that the effects of the laboratory environment can be affected by nutritional factors to improve the insect.
Both parasitoid and host are included; for the quality of a laboratory reared host may be changed likewise. This situation calls
for consideration of the diet of the insect. Customarily, parasitoid
larvae are reared on laboratory-reared or field-collected hosts and
infrequently on artificial diets (the latter are not infrequently
used to rear host insects). The parasitoid adults usually are fed
honey or sugar somehow and seldomly little more. With such foodstuff
we take for granted that because its supply is unvarying its quality
for nutrition is constant and good, in as much as large numbers of
the insect can be reared on it. Often the foodstuff is a variety
substituted for the usual natural kind; for example, a strange host
species substituted for parasitism or merely a solution of sucrose
for the adult. A foodstuff, natural or artificial, may be attractive
under laboratory conditions but nutritionally marginal to produce a
high quality insect; adequate perhaps for normal rate of growth and
development but inadequate to provision sufficient nutrient reserves
in the larva to produce a superior adult. We may well suppose this.
Kajita (1973) found that longevity and fecundity decreased from
normal in the parasitoid Apanteles chilonis Munakata when laboratory

reared on a diet-fed rather than field-collected host, *Chilo suppressalis* Walker.  In such cases, however, we can not be certain that the artificial diet of the host was suitable for optimal nutrition.  It was shown elsewhere using chemically-defined diets that the diet of the host can affect the parasitoid reared on the host as one diet was found better than the other for parasitoid yield (House and Barlow 1961).

Wilkes (1947) found that the sex ratio of selected, laboratory-reared *Dalbominus* [= *Microplectron*] *fuscipennis* (Zetterstedt) on a field-collected host shifted toward the production of relatively more males than females.  A possible explanation of this may be strictly genetic.  Another could be the 'fit' of the proportions of nutrients to the requirements of the organism; namely that initially the balance may be unsuitable for the requirements of the organism, or if initially suitable, the requirements of the organism may change with various factors including genetic.  It is not uncommon to find that the sex ratio of insects produced varied with the quality of food, as shown in *I. conquisitor* reared directly on chemically-defined diets with different proportions of nutrients (Yazgan 1972).  One might expect that the sex ratio might change because the quantities of nutrients provided by the usual foodstuff no longer met the proportional quantities of nutrients required for normal nutrition and sex ratio (of surviving larvae) in the geno-and/or phenotypes produced by selection and laboratory rearing.  The author sees no reason for regarding insects in the present context as unique or as having some mystic about them. He happens to believe that insects, like domestic animals, are amenable to our will in problems of this kind.  Similar problems were encountered initially in animal husbandry and were dealt with sensibly by research and eventually overcome.  To assume that any foodstuff fed to parasitoids in the laboratory is suitable for optimal nutrition of the laboratory-reared individuals is presumptuous and lacking discernment.  One consequence of our present understanding of the nutritional requirements and foodstuff of parasitoids is that we can not take for granted the suitability of a foodstuff for optimal nutrition.  With parasitoids, as with domestic animals, there is no reason to believe that their characteristic natural foodstuff provides the optimal nutrition that it is possible to achieve.  Certainly, the nutrition afforded the domestic hen by its pickings of insects, seeds, and other natural foodstuffs is inferior to that now enabled on scientifically concocted rations of highly nutritious grains, meat scraps, and vitamin and mineral supplements.

In the countless attempts to mass-produce insects, seldom has much consideration been given to their nutrition beyond what was minimally necessary and obvious in providing a foodstuff.  But food is not synonymous with nutrition, nor is adequate synonymous with optimal nutrition.  In short, we provide foodstuff and over-

look optimal nutrition.  But we expect optimal results.  Seldom do
we look further into the matter for effective factors for maximum
vigor or fecundity.  For example, researches on *A. housei* led to
the development of what was considered a good artificial diet, containing a mammalian salt mixture, because the larvae grew fast and
large (House 1954a; House and Barlow 1960).  However, further work
showed that the rate of growth and development and of body-protein
content varied with slight variations in the balance of nutrients
and with the salt mixture (House 1966b).  In particular, a new
salt mixture developed expressly for *A. housei* increased the body
weight of the larvae and their protein content and decreased their
water content to near normal (House and Barlow 1965; House 1966b).
But, the inclusion of vitamin E, which required one other criterion,
oogenesis, was necessary in order to score wholesome nutrition
(House 1966c).  This points up the fact that normal wholesome
nutrition can only be determined by a variety of criteria, and
that the suitability of a foodstuff can not be judged by just one
criterion.  But, has it not been the common practice in the laboratory to rear parasitoids on some more or less nutritionally
nondescript foodstuff – with yield in numbers as the criterion,
than to release them and trust to luck?

The second step, if we would change our perspective of insect
nutrition to augment natural enemies, is to regard parasitoids in
much the same light as domestic animals.  Because of convenience
and other advantages, and despite the intrinsic shortcomings to
date, the laboratory rearing of parasitoids is inevitably here to
stay.  We will do this for the very reasons that we have confined
other useful animals for handy management in breeding and rearing
of poultry, cattle, swine, horses, fur-bearing animals, game animals, birds, and fish, silkworms and so forth.  Brought thus under
confined and unnatural conditions these animals, like laboratory-
reared insects, may inbreed, suffer nutritional disease, lose vigor,
decline in performance, and so forth.  Or they can be managed
carefully if not scientifically to avoid such deleterious effects,
and in fact managed to enhance their value.  The similarity here
between housed farm animals and laboratory-reared insects suggests
that for our purposes parasitoids should be regarded in much the
same light as domestic animals.  This suggests then the use of
specially developed dietary regimens for parasitoids to affect
their activities and performance much like the feeds specially
designed to promote egg-production in chickens, milk and butter-
fat production in cattle, and, as Legay (1958) pointed out, the
attempts to develop special food regimens to industralize silkworm raising.  Yokoyama (1973) also described the major work on
silkworm nutrition and efforts to improve silk production and
sericulture through nutrition for industry.

One can hardly envisage how specialized dietary regimens for
parasitoids are possible using insect hosts in the laboratory.  We

need greater understanding of parasitoid nutrition and foodstuff
and greater control over the nutritive intake of a parasitoid than
an insect host permits in order to improve parasitoids nutritionally.  It would appear that the most rewarding course for us to
follow was blazed by work on nutrition and dietetics of domestic
animals.  This means then necessary work to determine, mostly
quantitatively, the nutritional requirements (say) for highly
proteinaceous body tissues and full nutrient reserves in the larva,
or for high levels of adult emergence, or for maximum fecundity,
or whatever; perhaps including tolerance to relatively extreme
ambient temperature, or adaptability to environmental conditions.
It means also determination of the specific needs for the nutrition and performance of the adult itself to assure normal or even
prolonged longevity, or perhaps oogenesis and viable eggs as shown
by Bracken (1965, 1966).  An artificial host, into which the parasitoid would oviposit and in which the young would develop to the
imaginal stage, would be most useful for both experimental and
practical use.  Highly refined artificial diets probably will be
initially essential for research, but with improved understanding
of the specific needs for specific activities equally suitable,
less refined, practical foodstuffs will suggest themselves to meet
optimally the needs of the parasitoid.  Of course, this entails a
very great deal of work that admittedly may not be worth the effort
in some cases other than in major programs of raising parasitoids.

However, the practicability of using an artificial diet, even
a highly chemically-defined synthetic one, for routinely rearing a
host insect or directly a parasitoid without resort to an insect
host should not be ignored.  Admittedly, use of such diets for these
purposes may be frowned on as artificial diets present technical
difficulties and have not been always too successful.  Intrinsically, their use *per se* for these purposes is not where the fault
lies but rather in the fact that in most cases the diet was in some
way suboptimal.  There is no reason why a host reared on an artificial diet should not enable nutrition of a parasitoid equal to
that obtained on a field-collected wild host.  The nutrition of a
parasitoid reared directly on an artificial diet can be equal to
that obtained on a good laboratory-reared or wild host.  Whether
for host or parasitoid there is no reason why artificial diets can
not be developed and improved scientifically by proper special
combination of effective nutrients to provide optimal nutrition even
exceeding that on natural foodstuffs, as has been commonly done
with domestic animals.  The author's opinion is that this can be
done best with more or less highly chemically-defined diets.  The
most ultimately rewarding technique will be to use artificial
diets, whose nutrient composition is known and manipulative, to
rear parasitoids directly without resort to a host insect.  Toward
this he has had some success in developing an artificial host;
that is, an encapsulated chemically-defined diet into which *I.
conquisitor* oviposited and in which egg eclosion occurred and

development of the larva progressed, but stopped just short of
pupation owing to technical difficulties (unpublished).  The
advantages of artificial diets, especially those sufficiently
chemically defined and alterable in composition to effect accordingly optimal nutrition and performance of a parasitoid, apply
quite as well for feeding the adult as for rearing the larva.  But
furthermore we should not overlook the cooperation of nutrition
and genetics and integrate them properly in our efforts to augment
natural enemies.

    We must recognize that laboratory rearing for the betterment
of parasitoids involves two prime considerations - genetics and
nutrition, as with any other kept animal.  From the standpoint of
genetics we can select for increased vigor, fecundity, or any other
desirable quality or performance.  Selection through the application
of principles of genetics was used to modify certain qualities or
traits advantageously in particular parasitoids (see: Wilkes 1942;
White *et al*. 1970).  To achieve somewhat the same end we recognize
hybridization as a common useful technique in plant and animal
breeding.  Heterosis, however, has been largely ignored in genetic
improvement programs with insects other than with silkworms and
honeybees (Hoy 1975).  From the standpoint of nutrition, when we
know the nutritional requirements of a particular parasitoid we
can set about to provide for its optimal nutrition to maximize its
activities and performance.  As mentioned above, Bracken (1965,
1966) showed that a probable fault in hymenopteran females incurred
in laboratory rearing could be avoided or corrected simply by feeding
certain essential nutrients to assure sufficient supplies for normal
fecundity.  Moreover it seems possible that diets could be devised
taking into account certain nutritional factors to enable a parasitoid to insure itself to adverse ambient temperature.  This might be
useful to preserve the insect during shipment or upon release in temperature to which the insect is unaccustomed, at least until naturally acclimatized and established (House *et al*. 1958; House 1966d).
We should expect to do something like that done with genetics and
nutrition in sericulture, as described by Yokoyama (1973).  The
point to be taken is that certainly insect parasitoids respond to
genetic and nutritional manipulations.  Nevertheless, we seem impractical in raising parasitoids; for if on the one hand in rearing
programs we have manipulated one of these factors, on the other hand,
we have ignored the other factor persistently.

    Fundamentally, the application of the principles of genetics to
increase the dominance of desirable qualities and traits and of nutrition to enable the organism to fulfil its potential capabilities
is the universally recognized way to improve animals.  Successful enterprises are built upon our will and ability to make changes to a
better way of doing things than before.  It is questionable whether
superficial means of augmenting natural enemies are sufficient.  However, are we in truth willing to attempt augmenting natural enemies

by means that are intrinsically founded on fundamental biological
bases and which are relatively undeveloped now?   Such an approach
to affect parasitoids is not the rule in applied economic entomology.
But should we attempt it, we have a bit of insight into some prereq-
uisites   as briefly pointed out in this discussion of the nutrition
of natural enemies.   All that appears necessary is to decide to do it
and begin.   By a rational breeding and feeding program that cooperates
and integrates genetic and nutritional principles, we most likely
will achieve best our objective to improve parasitoids.   But if one
is employed without the other or if either or both are applied
superficially, we surely will be short-changed.   This claim is made
because elsewhere animal husbandry (and including poultry husbandry), concerned with the production and care of domestic animals,
embraces scientific control and management in which genetics and
nutrition are principal factors brought together and applied advantageously with a definite purpose on a practical basis.   The
beneficial results can not be denied.   We need to engage our attention and to promote widely efforts to apply conjointly these principles of husbandry likewise to raise parasitoids of high quality.
This need is furthered by our inclination and increasing efforts
everywhere to raise all kinds of insects on unnatural foodstuff in
the laboratory.   Consequently, it may be that entomology needs a
new specialized discipline--insect husbandry.


CONCLUSION


     It would appear to date that there is nothing about the nu-
trition and qualitative nutritional requirements of parasitoids to
distinguish them from other kinds of insects.   This concept probably will hold because parasitoids are not highly specialized parasites.   Some are predators, and though they have a particular feeding habit they do not differ unusually in taxonomic position from
other insects:   i.e. some Diptera are parasitoids others are not.
Elsewhere with different species and different feeding habits among
mammals and among birds, respectively, nutritional requirements do
not significantly differ accordingly.   In fact, insect nutrition
may be regarded as a branch of animal nutrition because much the
same nutritional requirements and principles of nutrition are common to insects and higher animals.   With so little to distinguish
it, the subject of insect nutrition has little more than academic
interest in many quarters including some entomological.   It is unfortunate for the best interests of the subject that the nutrition
of insects has no obvious vulnerable point that can be exploited
readily in pest control.   Much entomology is preoccupied with killing insects directly, and nutrition can hardly be associated con-
textually with insecticides.   It is fair to say that the economic
importance of insect nutrition rarely has been fully realized in
any applied practice other than possibly in sericulture.   Work on
insect nutrition has meandered no doubt.   It has short comings;

some of which were underlined in this discussion where they concerned understanding of the nutrition of natural enemies.    Therefore, a reasoned judgment, or conclusion, might very well be simply
to depreciate the efforts and application, to question the value
of work on insect nutrition, or to deplore what seems to be a lack
of progress in it.    But to do so would lack discernment of the role
of nutrition.

Biological control with parasitoids provides a prime and somewhat unique opportunity to employ a means issueing from a physiological basis, nutrition, to attack pests; although indirectly.
We may conclude that we are in fair shape in many respects to do
this.    Some things have been learned in nutritional work on insects
in general that are very useful in understanding the nutrition of
natural enemies in particular.    The most insight into the nutritional requirements of a parasitoid and the role of nutritional factors
in various matters is centered on *A. housei*.    Perhaps as much or
probably more is understood about these in *A. housei* than in any
other insect except *D. melanogaster*.    According to Levinson (1955),
it was the first parasitoid to be reared axenically, or aseptically,
on a chemically well-defined, artificial, diet enabling the nutritional requirements of a parasitoid to be determined.    The emanated
work on *A. housei* since 1954 was described by Vanderzant (1974)
as notable among nutritional investigations.    Recently other similar
work on a few other parasitoids has issued.    This points up the fact
that despite their feeding habit parasitoids are amenable to nutritional investigation.    And we have some clear understanding of
the nutrition of parasitoids and the depth and range of understanding
are increasing on a sound basis.    It is reasonable to conclude that
the most promising nutritional means of augmenting natural enemies,
especially from the laboratory, will lie with the quantitative nutritional factor.    This is where the action is, particularly concerning
the proportional relations, or balance, of nutrients needed for optimal nutrition and the various effects when these relations are
suitably or unsuitably met.    We can conclude from the evidence in
hand that nutrition is a dynamic factor affecting insects in various
ways that have ecological significance.    We have mostly a number
of pioneering one-shot demonstrations concerning a miscellany of
matters involving nutrition.    What we need to do is take a long-
range view of nutrition and of augmenting natural enemies, and with
some appreciation of the role of nutrition undertake to use scientifically nutritional factors with competence and order.    That nutritional principles have never been used in this connexion with
parasitoids does not prove conclusively that such will not work.
For best results nutritional principles doubtlessly should be applied in close association with those of genetics.    The beneficial
effects of improved nutrition for any purpose express themselves
in a generation or so, those of genetics to develop or intensify
important traits in the parasitoid may take a little longer.    A
properly executed program using these principles will require some

time to maximize satisfactory results.  It might be questionable
therefore whether we can afford to do this.  For the usual methods
of importing parasitoids from afar and/or raising them on some
living host in the laboratory for release may not be entirely satisfactory, but it commonly is regarded widely as expediential, often politic, and it can be done immediately in hard-pressed programs.

We can not be satisfied with our overall efforts to date in
producing parasitoids, particularly in view of our wide-spread in-
difference to the role of nutrition and its various effects on them.
We may conclude that we have some understanding of some of the pre-
requisites for augmenting parasitoids, including the cooperation
and integration of genetic and nutritional principles, that would
enable us to expertly employ nutrition to produce superior parasitoids
in particular.  This author concludes, however, that we have some
and are gaining more understanding of parasitoid nutrition, and
that we could and perhaps now should undertake to apply it properly
to augment these allies in pest control.  His opinion is that our
efforts to rear superior parasitoids will be most advantageous
parasitoids were genetically selected for desirable traits and
reared directly on nutritionally well-designed artificial diets.
For resort to a host insect to rear parasitoids in the laboratory
is extravagant and parasitoid quality is hardly controllable.  But
furthermore, in order to advance much beyond our present state,
he suggests that the fundamental problems of producing superior
parasitoids are so manifold and complex as to necessitate much insight and training, which would be set out comprehensively if a
new specialized discipline--insect husbandry, were set up emulating
sound animal husbandry's preoccupation with instruction and practice in the essentials of good managements, breeding and nutrition.

## LITERATURE CITED

Albritton, E. C.  ed.  1954.  *In* Standard Values in Nutrition and
    Metabolism, pp. 21-30.  Philadelphia:  Saunders.  380 pp.
Altman, P. L., and D. S. Dittmer.  ed.  1968.  *In* Metabolism, pp.
    148-63.  Washington:  Fed. Am. Soc. Exp. Biol.  737 pp.
Altman, P. L., and D. S. Dittmer.  ed.  1974.  *In* Biology Data Book,
    vol. 111, pp. 1433-75.  Bethesda:  Fed. Am. Soc. Exp. Biol.
    690 pp.
Atallah, Y. H., and R. Killebrew.  1967.  Ecological and nutrition-
    al studies on *Coleomegilla maculata* (Coleoptera:Coccinellidae).
    IV.  Amino acid requirements of the adults determined by the
    use of $C^{14}$-labelled acetate.  Ann. Entomol. Soc. Am. 60:186-8.
Atwal, A. S., and S. L. Sethi.  1963.  Biochemical basis for the
    food preference of a predator beetle.  Current Sci. (India)
    11:511-2.
Beck, S. D.  1972.  Nutritional aspects of pest management.  *In*
    Insect and Mite Nutrition, ed. J. G. Rodriguez, pp. 555-6.

Amsterdam: North-Holland. 702 pp.

Beirne, B. P. 1962. Trends in applied biological control of insects. Ann. Rev. Entomol. 7:387-400.

Boller, E. 1972. Behavioral aspects of mass-rearing of insects.
Entomophaga 17:9-25.

Bracken, G. K. 1965. Effects of dietary components on fecundity
of the parasitoid *Exeristes comstockii* (Cress.) (Hymenoptera:
Ichneumonidae). Can. Entomol. 97:1037-41.

Bracken, G. K. 1966. Role of ten dietary vitamins on fecundity
of the parasitoid *Exeristes comstockii* (Cresson) (Hymenoptera:
Ichneumonidae). Can. Entomol. 98:918-22.

Brambell, M. R. 1972. Mammals: their nutrition and habitat.
*In* Biology of Nutrition, ed. R.N. T-W-Fiennes, 18:613-48.
International Encyclopaedia of Food and Nutrition. Oxford:
Pergamon. 681 pp.

Brues, C. T. 1946. Insect Dietary; an Account of the Food Habits
of Insects, Cambridge: Harvard. 466 pp.

Couch, J. R. 1967. Nutritional requirements (vertebrates other
than mammals). *In* The Encyclopedia of Biochemistry, ed. R.
J. Williams and E. M. Lansford, Jr., pp. 604-6. New York:
Reinhold. 876 pp.

Dadd, R. H. 1973. Insect nutrition: current developments and
metabolic implications. Ann. Rev. Entomol. 18:381-420.

Davis, G. R. F. 1966. Replacement of RNA in the diet of *Oryzaephilus surinamensis* (L.)(Coleoptera:Silvanidae) by purines,
pyrimidines, and ribose. Can. J. Zool. 44:781-5.

Davis, G. R. F. 1967. Effects of dietary lipids on survival and
development of the saw-toothed grain beetle, *Oryzaephilus surinamensis* (L.) (Coleoptera:Silvanidae). Revue Can. Biol. 26:
119-24.

Dethier, V. G. 1947. Chemical Insect Attractants and Repellents.
Philadelphia: Blakiston. 289 pp.

Dethier, V. G. 1970. Chemical interactions between plants and
insects. *In* Chemical Ecology, ed. E. Sondheimer and J. B.
Simeone, pp. 83-102. New York: Academic. 336 pp.

Evans, A. C. 1939. The utilization of food by certain lepidopterous larvae. Trans. Roy. Entomol. Soc. London (A) 89:13-22.

Gordon, H. T. 1959. Minimal nutritional requirements of the German
roach, *Blattella germanica* L. Ann. N.Y. Acad. Sci. 77:290-351.

Hagen, K. S. 1950. Fecundity of *Chrysopa californica* as affected
by synthetic foods. J. Econ. Entomol. 43:101-4.

Hagen, K. S., and R. Hale. 1974. Increasing natural enemies through
use of supplementary feeding and non-target prey. *In* Proceedings of the Summer Institute on Biological Control of Plant
Insects and Diseases, ed. F. G. Maxwell and F. A. Harris,
pp. 170-181. Jackson: Mississippi.

Hagen, K. S., and R. L. Tassan. 1966. The influence of protein
hydrolysates of yeast and chemically-defined diets upon the
fecundity of *Chrysopa carnea* Stephens (Neuroptera). Vestnik.
Cs. Spol. Zool. 30:219-27.

Hagen, K. S., and R. L. Tassan.  1970.  The influence of Food Wheast[R] and related *Saccharomyces fragilis* yeast products on the fecundity of *Chrysopa carnea* (Neuroptera:Chrysopidae).  Can. Entomol. 102:806-11.

Hagen, K. S., and R. L. Tassan.  1972.  Exploring nutritional roles of extra cellular symbiotes on the reproduction of honeydew feeding adult chrysopids and tephritids.  *In* Insect and Mite Nutrition, ed. J. G. Rodriguez, pp. 223-351.  Amsterdam: North-Holland.  702 pp.

Hagen, K. S., E. F. Sawall, Jr., and R. L. Tassan.  1971.  The use of food sprays to increase effectiveness of entomophagous insects.  Proc. Tall Timbers Conf. Ecol. Anim. Contr. Habitat Mange.  2:59-81 (Tall Timbers Res. Sta., Tallahassee, Fla.)

House, H. L.  1954a.  Nutritional studies with *Pseudosarcophaga affinis* (Fall.), a dipterous parasite of the spruce budworm, *Choristoneura fumiferana* (Clem.).  I. A chemically defined medium and aseptic-culture technique.  Can. J. Zool. 32:331-41.

House, H. L.  1954b.  Nutritional studies with *Pseudosarcophaga affinis* (Fall.), a dipterous parasite of the spruce budworm, *Choristoneura fumiferana* (Clem.).  III.  Effects of nineteen amino acids on growth.  Can. J. Zool.  32:351-7.

House, H. L.  1954c.  Nutritional studies with *Pseudosarcophaga affinis* (Fall.), a dipterous parasite of the spruce budworm, *Choristoneura fumiferana* (Clem.).  IV.  Effects of ribonucleic acid, glutathione, dextrose, a salt mixture, cholesterol, and fats.  Can. J. Zool.  32:358-65.

House, H. L.  1958.  Nutritional requirements of insects associated with animal parasitism.  Exp. Parasit.  7:555-609.

House, H. L.  1963.  Nutritional diseases.  *In* Insect Pathology, ed. E. A. Steinhaus, 1:133-60.  New York: Academic.  661 pp.

House, H. L. 1964.  Effects of dietetic nucleic acids and components on growth of *Agria affinis* (Fallen) (Diptera:Sarcophagidae).  Can. J. Zool. 42:801-6.

House, H. L.  1965a.  Effects of vitamin A acetate and structurally related substances on growth and reproduction of *Agria affinis* (Fallen) (Diptera:Sarcophagidae).  J. Insect Physiol. 11:1039-45.

House, H. L.  1965b.  Effects of low levels of the nutrient content of a food and of nutrient imbalance on the feeding and the nutrition of phytophgous larva, *Celerio euphorbiae* (Linnaeus)(Lepidoptera:Sphingidae).  Can. Entomol. 97:62-8.

House, H. L.  1966a.  The role of nutritional principles in biological control.  Can. Entomol.  98:1121-34.

House, H. L.  1966b.  Effects of varying the ratio between the amino acids and the other nutrients in conjuction with a salt mixture on the fly *Agria affinis* (Fall.).  J. Insect Physiol. 12:299-310.

House, H. L.  1966c.  Effects of vitamins E and A on growth and development, and the necessity of vitamin E for reproduction in the parasitoid *Agria affinis* (Fallen) (Diptera:Sarcophagidae). J. Insect Physiol. 12:409-17.

House, H. L. 1966d. Effect of temperature on the nutritional requirements of an insect, *Pseudosarcophaga affinis* Auct. nec Fallen (Diptera:Sarcophagidae), and its probably ecological significance. Ann. Entomol. Soc. Am. 59:1263-67.

House, H. L. 1967a. Artificial Diets for Insects: a Compilation of References with Abstracts. Inform. Bull. No. 5, Res. Inst., Can. Dept. Agr., Belleville, Ont. 163 pp.

House, H. L. 1967b. The role of nutritional factors in food selection and preference as related to larval nutrition of an insect, *Pseudosarcophaga affinis* (Diptera:Sarcophagidae), on synthetic diets. Can. Entomol. 99; 1310-21.

House, H. L. 1969. Effects of different proportions of nutrients on insects. Entomol. Exp. Appl. 12:651-69.

House, H. L. 1970. Choice of food by larvae of the fly, *Agria affinis*, related to dietary proportions of nutrients. J. Insect Physiol. 16:2041-50.

House, H. L. 1971a. Changes from initial food choice in a fly larva, *Agria affinis*, as related to dietary proportions of nutrients. J. Insect Physiol. 17:1051-59.

House, H. L. 1971b. Relations between dietary proportions of nutrients, growth rate, and choice of food in the fly larva *Agria affinis*. J. Insect Physiol. 17:1225-38.

House, H. L. 1972a. Insect nutrition. *In* Biology of Nutrition, ed. R.N. T-W-Fiennes, 18:513-73. International Encyclopaedia of Food and Nutrition. Oxford: Pergamon. 681 pp.

House, H. L. 1972b. Inversion in the order of food superiority between temperatures affected by nutrient balance in the fly larva *Agria housei* (Diptera:Sarcophagidae). Can. Entomol. 104:1559-64.

House, H. L. 1974. Nutrition. *In* The Physiology of Insecta, ed. M. Rockstein, 5:1-62. New York: Academic. 648 pp.

House, H. L. 1976. Interaction between amino acids and glucose in larval nutrition of the fly *Agria housei* Shewell at low temperature. Can. Entomol. In press.

House H. L., and J. S. Barlow. 1956. Nutritional studies with *Pseudosarcophaga affinis* (Fall.), a dipterous parasite of the spruce budworm, *Choristoneura fumiferana* (Clem.) V. Effects of various concentrations of the amino acid mixture, dextrose, potassium ion, the salt mixture, and lard on growth and development; and a substitute for lard. Can. J. Zool. 34:182-9.

House, H. L., and J. S. Barlow. 1960. Effects of oleic and other fatty acids on the growth rate of *Agria affinis* (Fall.) (Diptera:Sarcophagidae). J. Nutr. 72:409-14.

House, H. L., and J. S. Barlow. 1961. Effects of different diets of a host, *Agria affinis* (Fall.) (Diptera:Sarcophagidae), on the development of a parasitoid, *Aphaereta pallipes* (Say) (Hymenoptera:Braconidae). Can. Entomol. 93:1041-4.

House, H. L., and J. S. Barlow. 1965. Effects of a new salt mixture developed for *Agria affinis* (Fallén) (Diptera:Sarcophagidae) on the growth rate, body weight, and protein content of the larvae. J. Insect Physiol. 11:915-8.

House, H. L., D. F. Riordan, and J. S. Barlow. 1958. Effects of thermal conditioning and of degree of saturation of dietary lipids on resistance of an insect to a high temperature. Can. J. Zool. 36:629-32.

House, H. L., Pritam Singh, and W. W. Batsch. 1971. Artificial Diets: a Compilation of References with Abstracts. Inform. Bull. No. 7, Res. Inst., Can. Dept. Agr., Belleville, Ont. 156 pp.

Hoy, M. A. 1975. Improving the quality of laboratory-reared insects. J. N. Y. Entomol. Soc. 83: 276-7.

Kajita, H. 1973. Rearing of *Apanteles chilonis* Munakata on the rice stem borer, *Chilo suppressalis* Walker, bred on a semi-artificial diet. Jpn. J. Appl. Entomol. Zool. 17:5-9.

Legay, J. M. 1958. Recent advances in silkworm nutrition. Ann. Rev. Entomol. 3:75-86

Leius, K. 1960. Attractiveness of different foods and flowers to the adults of some hymenopterous parasites. Can. Entomol. 92:369-76.

Leius, K. 1961a. Influence of food on fecundity and longevity of adults of *Itoplectis conquisitor* (Say) (Hymenoptera: Ichneumonidae). Can. Entomol. 93:771-80.

Leius, K. 1961b. Influence of various foods on fecundity and longevity of adults of *Scambus buolianae* (Htg.) (Hymenoptera: Ichneumonidae). Can. Entomol. 93:1079-84.

Leius, K. 1967a. Influence of wild flowers on parasitism of tent caterpillar and codling moth. Can. Entomol. 99:444-6.

Leius, K. 1967b. Food sources and preferences of adults of a parasite, *Scambus buolianae* (Hym.:Echn.), and their consequences. Can. Entomol. 99:865-71

Levinson, Z. H. 1955. Nutritional requirements of insects. Riv. Parassitol. 16:113-138, 183-204.

Mackauer, M. 1976. Genetic problems in the production of biological control agents. Ann. Rev. Entomol. 21:369-85.

Mathai, S. 1972. Studies on the effect of host nutrition on *Bracon brevicornis* Wesmael. Agr. Res. J. Kerala (1971) 9:1-3.

Maynard, L. A., and J. K. Loosli. 1962. Animal Nutrition. New York: McGraw-Hill. 533 pp.

Morrison, F. B. 1941. Feeds and Feeding, Abridged. Ithaca: Morrison. 503 pp.

Pratt, J. J. Jr., H. L. House, and A. Mansingh. 1972. Insect control strategies based on nutritional principles: a prospectus. *In* Insect and Mite Nutrition, ed. J. G. Rodriguez, pp. 651-68. Amsterdam: North-Holland. 702 pp.

Sang, J. H. 1956a. The quantitative nutritional requirements of *Drosophila melanogaster*. J. Exp. Biol. 33:45-72.

Sang, J. H. 1956b. Differences in the nutritional requirements

of *D. melanogaster* and the relation to heterosis. Proc.
Int. Congr. Genet., 9th, 1953, Caryologia, Florence suppl.
(1954) 6:818-21.

Sang, J. H. 1957. Utilization of dietary purines and pyrimidines
by *Drosophila melanogaster*. Proc. Roy. Soc. Edinburgh
66:339-59.

Sang, J. H. 1959. Circumstances affecting the nutritional requirements of *Drosophila melanogaster*. Ann. N. Y. Acad.
Sci. 77:352-65.

Sang, J. H. 1962. Relationship between protein supplies and B-
vitamin requirements in axenically cultured *Drosophila*.
J. Nutr. 77:355-68.

Schoenheimer, R. 1942. The Dynamic State of the Body Constituents. Cambridge: Harvard. 78 pp.

Shteinberg, D. M. 1955. Some aspects of the problem of adoption
of entomophagous and phytophagous insects to their nutrition
[in Russian]. Trans. Zool. Inst. Acad. Sci. U.S.S.R. 21:36-
43. (Translation by E. R. Hope, Directorate of Scientific
Information Service, Defense Research Board, Ottawa, Canada).

Singh, Pritam. 1974. Artificial Diets for Insects: a Compilation
of References with Abstracts (1970-72). Bull. No. 214, N. Z.
Dept. Sci. Ind. Res., Entomol. Div., Auckland, N.Z.

Smith, B. C. 1965. Differences in *Anatis mali* Auct. and *Coleomegilla maculata lengi* Timberlake to changes in the quality
and quantity of the larval food (Coleoptera: Coccinellidae).
Can. Entomol. 97:1159-66.

Smith, J. M. 1957. Effects of the food plant of California red
scale, *Aonidiella aurantii* (Mask.) on reproduction of its
hymenopterous parasites. Can. Entomol. 89:219-30.

Snodgrass, R. E. 1961. The Caterpillar and the Butterfly.
Smithsonian Inst. Misc. Collect. 143:1-51. (publ. 4472).

Soo Hoo, C. F., and G. Fraenkel. 1966. The consumption, digestion,
and utilization of food plants by a polyphagous insect,
*Prodenia eridania* (Cramer). J. Insect Physiol. 12:711-30.

Tauber, M. J., and C. A. Tauber. 1974. Dietary influence on
reproduction in both sexes of five predaceous species
(Neuroptera). Can. Entomol. 106:921-5.

Thompson, S. N. 1975. Defined meridic and holidic diets and
aseptic feeding procedures for artificially rearing the
ectoparasitoid *Exeristes roborator* (Fabricius). Ann. Entomol.
Soc. Am. 68:220-6.

Thompson, S. N. 1976a. The amino acid requirements for larval
development of the hymenopterous parasitoid *Exeristes roborator* Fabricius (Hymenoptera:Ichneumonidae). Comp.
Biochem. Physiol. 53A:211-3.

Thompson, S. N. 1976b. Effects of dietary amino acid level and
nutritional balance on larval survival and development of
the parasite *Exeristes roborator*. Ann. Entomol. Soc. Am.
69:835-8.

Typpo, J. T., and G. M. Briggs. 1967. Nutritional requirements (mammals). *In* The Encyclopedia of Biochemistry, ed. R. J. Williams and E. M. Lansford, Jr., pp. 600-4. New York: Reinhold. 876 pp.

Uvarov, B. P. 1929. Insect nutrition and metabolism. Trans. Entomol. Soc. London 76(2):255-343.

Vanderzant, E. S. 1968. Synthetic diets: insects. *In* Metabolism, ed. P. L. Altman and D. S. Dittmer, pp. 163-7. Washington: Fed. Am. Soc. Exp. Biol. 737 pp.

Vanderzant, E. S. 1973. Improvements in the rearing diet for *Chrysopa carnea* and the amino acid requirements for growth. J. Econ. Entomol. 66:336-8.

Vanderzant, E. S. 1974. Development, significance, and application of artificial diets for insects. Ann. Rev. Entomol. 19: 139-60.

Waldbauer, G. P. 1964. The consumption, digestion and utilization of solanaceous and non-solanaceous plants by larvae of the tobacco hornworm, *Protoparce sexta* (Johan.) (Lepidoptera: Sphingidae). Entomol. Exp. Appl. 7:253-69.

Waldbauer, G. P. 1968. The consumption and utilization of foods by insects. Advan. Insect Physiol. 5:229-88.

White, E. B., P. DeBach, and M. J. Garber. 1970. Artificial selection to temperature extremes in *Aphytis lingnanensis* Compere (Hymenoptera:Aphelinidae). Hilgardia 40:161-92.

Wilkes, A. 1942. The influence of selection on the preferendum of a chalcid (*Microplectron fuscipennis* Zett.) and its significance in the biological control of an insect pest. Proc. Roy. Soc. London (B) 130:400-15.

Wilkes, A. 1947. The effects of selective breeding on the laboratory propagation of insect parasites. Proc. Roy. Soc. London (B) 134:227-45.

Yazgan, S. 1972. A chemically defined synthetic diet and larval nutritional requirements of the endoparasitoid *Itoplectis conquisitor* (Hymenoptera). J. Insect Physiol. 18:2123-41.

Yokoyama, T. 1973. The history of sericultural science in relation to industry. *In* History of Entomology ed. R. F. Smith, T. E. Mittler, and C. N. Smith, pp. 267-84. Palo Alto: Annual Reviews. 517 pp.

CHAPTER 6

MASS PRODUCTION OF NATURAL ENEMIES

R. K. Morrison and E. G. King

Agricultural Research Service
U. S. Department of Agriculture
College Station, Texas 77840, U.S.A.
Stoneville, Mississippi 38776, U.S.A.

The value of entomophagous arthropods (parasites and predators) in supressing pest populations has been well established.
However, sufficient numbers are not always present or available
to provide adequate control. Biological control by augmentation
of predators and parasites that have been mass produced in insectaries provides a means of insuring the presence of numbers that
are adequate to maintain pest populations at the desired level.
The feasibility of this approach to pest control has been documented by Ridgway et al. (1974), Stinner (1977) and Rabb et al.
(1976) and in other chapters of this book.

Production of parasites and predators for augmentation has
been of considerable interest to workers in biological control
since the beginning of the 20th century. Finney and Fisher (1964)
reviewed requirements necessary for successful culture of entomophagous arthropods and their hosts in the laboratory, and Smith
(1966) reviewed production techniques for a wide range of insects.
Significant advances have been made in recent years in mass rearing and in handling procedures for insects, but these have principally been nonentomophagous.

Although many research colonies of natural enemies have been
laboratory reared in meaningful numbers over several generations,
relatively few (10-20) have been produced in quantities that
approach or meet MacKauer's (1972) definition of mass production
(production of a million times the mean female offspring during
the time required for one generation cycle).

Beirne (1974) concluded that the chief obstacle to widespread
application of the augmentation method of pest control was the
difficulty in mass producing or mass collecting entomophagous
arthropods at economical costs.  He considered the lack of effective techniques for mass producing parasites and predators on artificial diet as a major limiting factor in utlization of this
approach.  Also, economic constraints and lack of techniques that
simulate natural conditions result in selection pressures that
lead to genetic deterioration of mass reared arthropods (MacKauer
1972).  Such deterioration can mean the loss of behavioral traits
basic to the effectiveness of released natural enemies (Boller
1972), thereby jeopardizing the success of a biological control
program.

## LOCATIONS AND CAPACITY FOR PRODUCTION OF
## ENTOMOPHAGOUS ARTHROPODS

Numerous centers throughout the world are presently producing
predators and/or parasites that can be used to augment entomophagous arthropod populations for control of arthropod pests.  These
centers may operate at the private, cooperative, state, or federal
level.  Some examples of parasites and predators that are produced
or collected in sizable numbers in the U.S.A. have been reported
by Ridgway et al. (1974).  DeBach (1974) also reported on sources.
However, the most extensive production centers for entomophagous
arthropods, particularly *Trichogramma* spp., are perhaps those located in the People's Republic of China and the U.S.S.R.  These
centers are operated at both the local and state level.  DeBach
(1974) reported that the U.S.S.R. Ministry of Agriculture annually
produces "thousands of millions" of *Trichogramma* spp. which are
used on a total area of about 4 million acres against more than 10
species of pests.  Other countries located in South America and
Europe that produce and distribute predators and parasites include
Venezuela (F. Ferrer, Intergrated Control Service, Acarigua
Venezuela, personal communication), Brazil (Teran 1976), Mexico
(Castillo Chacon 1967), West Germany (W. Stein 1960), and France
(Daumel, et al. 1975).  Where plants cultured under glass are
attacked by spider mites (Acarina: Tetranychidae) and whiteflies
(Hemiptera: Aleyrodidae), the use of predators (*Phytoseiulus
persimilus* Athias-Henriot) and parasites (*Encarsia formosa* Gahan),
respectively, in augmentation programs is gaining increasing credi-
bility as a control approach.  Studies to determine the feasibility
of the augmentation approach have required the development of large-
scale insect rearing operations.  Some recent examples where production figures are given or may be extrapolated include Bryan et
al. (1973), Starks et al. (1976), King et al. (1975b), Morrison
and Ridgway (1976), and Morrison et al. (1976).

## NATURAL AND UNNATURAL HOSTS

Finney and Fisher (1964) briefly reviewed the use of natural and unnatural hosts for rearing entomophagous arthropods. They defined a natural host as one that is usually attacked in nature by the parasite or predators. An unnatural host was not normally attacked, apparently because of some isolating mechanism, but it did serve as a suitable host in the insectary.

Numerous large-scale rearing programs have used unnatural hosts for production of predators and parasites. The Angoumois grain moth, *Sitotroga cerealella* (Olivier), has routinely been used for mass production of *Trichogramma* spp. (Flanders 1930, Morrison et al. 1976). Other examples include the following: (1) production of *Macrocentrus ancylivorus* Rohwer on the potato tuber-worm, *Phthorimaea operculella* (Zeller) for control of the oriental fruit moth, *Grapholitha molesta* (Busck) (Finney et al. 1947); (2) production of *Aphytis lingnanensis* Compere on oleander scale, *Aspidiotus nerii* Bouché, for control of California red scale, *Aonidiella aurantii* (Maskell) (DeBach and White 1960); (3) production of *Bracon kirkpatricki* (Wilkinson) on the beet army worm, *Spodoptera exigua* (Hubner), for control of the pink bollworm, *Pectinophora gossypiella* (Saunders) (Bryan et al. 1973); and (4) production of *Lixophaga diatraeae* (Townsend) on the greater wax moth, *Galleria mellonella* (L.). for control of the sugarcane borer, *Diatraea saccharalis* (F.) (E. G. King, unpublished data). In each instance, the factors leading to selection of the unnatural host were reduced production costs, convenience, and ease of handling.

Both Knipling (1966) and a panel on insect pest management and control (Anonymous 1969) have expressed concern that entomophagous arthropods reared on unnatural hosts may change their host preference as a result of preimaginal conditioning with the result that they will have reduced effectiveness when released against the natural host. However, experiments designed to test this phenomenon have failed to show reduced attraction to the natural host. Thorpe and Jones (1937) found that when *Venturia canescens* (Gravenharst) was reared on the unnatural host, *Achroia grisella* (F.), a conditioning response occurred that caused the parasite to respond more strongly to this host despite the presence of the natural host, *Anagasta kuehniella* (Zeller). However, in absence of the unnatural host, parasites reared on *A. grisella* responded as strongly to *A. kuehniella* as did parasites reared on *A. kuehniella*. Likewise, Salt (1935) demonstrated that when *Trichogramma evanescens* Westwood was reared exclusively on *S. cerealella* and *A. kuehniella* for 63 and 43 generations, respectively, no dependence on or preference for, the respective hosts was developed.

We have concluded that the preimaginal conditioning factor that results from rearing on unnatural hosts may not be as critical in host selection after field release as has been indicated. What does seem to be critical is reduced "vigor" of the parasite or predator because of exposure to laboratory hosts that supply inadequate nutrition for the attacker. Simmonds (1966) arrived at a similar conclusion and cited several examples where differences occurred between predators or parasites reared on natural or unnatural hosts. For example, *Trichogramma* spp. reared on unnatural hosts have been shown to have reduced size, fecundity, longevity, and general robustness (Boldt and Marston 1974, Lewis et al. 1976, and others). Further, host diet has also been shown to affect the suitability of a host for parasite development. Thus Etienne (1973) reported that *Lixophaga diatraeae* could not be continuously reared on greater wax moth larvae fed bees wax and pollen, unless the host diet was supplemented with vitamin E and/or wheat germ (Etienne 1974, Etienne, personal communication). In fact, problems in rearing *L. diatraeae* on greater wax moth larvae fed a diet similar to that reported by Dutky et al. (1962) were generally eliminated by substituting 120 g of wheat germ for 120 g of baby cereal per kilogram of diet, by using older larvae as hosts, by reducing the amount of superparasitization, and by reducing the density of flies in holding cages (E. G. King, unpublished data). In some instances entomophages may be of a higher quality than those produced on the normal or natural host. DeBach and White (1960) reported that the aphelinid, *Aphytis lingnanensis*, reared on an unnatural host, *Aspidiotus nerii*, resulted in the more uniform production of a larger, more robust parasite than when reared on the natural host, *Aonidiella aurantii*. Thus, suitability cannot be determined merely by screening hosts, but host nutrition and other factors must also be considered; and compromises between entomophagous arthropod quality and quantity may have to be made because of cost and the numbers required. However, in initial feasibility studies, every effort should be made to rear the parasite or predator on the target pest, and, if possible, the target pest should be fed the commodity that is to be protected.


## STORAGE OF NATURAL ENEMIES

The ability to store parasites and predators for relatively long periods will probably be a key factor in the successful development of the augmentation method of control. Stinner (1977) considered storage techniques critical to cost reduction. Additionally, labor would be simplified by continual rather than seasonal production and reserve quantities of entomophages would counteract periods of low production and/or high demand.

Since most parasites and predators will have to be produced and released as live organisms, conditions for storage will probably always be more complex than those for abiotic pesticides, which are routinely manufactured and stored for later use.

The use of low temperatures to reduce development rates has been the major component in almost all reported storage techniques. However, other methods in combination with low temperatures have been used.

Some examples where storage techniques have been developed are as follows: Stinner et al. (1974) reported that *Trichogramma pretiosum* was stored in the pharate adult stage (just prior to emergence) from 4 to 12 days without detrimental effects on emergence if they were allowed to develop for 8 days at 26.7°C and then held at 16.7°C. After 6 days at 16.7°C, the temperature was further reduced to 15°C. Continued storage past 12 days caused a significant decrease in emergence. Kinzer (1976) stored *Chrysopa carnea* for up to 14 days at 10°C and 75% RH without reduced viability. Aphid parasites have been successfully stored at low temperatures in the mummy stage. Starks et al. (1976) reported that *Lysiphlebus testaceipes* (Cresson) pupae acclimatized at 16°C for 12 hr prior to storage at 5°C emerged at a higher rate than those stored continually at 5°C. Nevertheless, storage was only for a few days. Stary (1970) reported that *Aphidius smithi* Sharma and Subba Rao could be stored in the mummy stage at 1 to 4°C for up to 2 months or more. Adults could be stored only 10 days at 10°C. Scopes (1968) concluded that as long as food was present, up to 80% survival of *Phytoseiulus persimilis* held in gelatin capsules could be expected for up to 12 weeks storage at 7.2°C. *Lixophaga diatraeae* puparia were stored for up to 14 days at 15.6°C without noticeable effects. Additionally, the maggot stage was also stored within the unnatural host, *Galleria mellonella,* at 13°C for 2 weeks before transfer to 26°C. This practice was validated in an ongoing experimental rearing and release program (E. G. King, unpublished data).

Host stages are also stored for later use in natural enemy production.

In a report on insect control in the Peoples' Republic of China (Anonymous 1977), infertile eggs are extracted from macerated *Samia cynthia* (Drury) virgin moths. These eggs, used for production of *Trichogramma* spp., may be stored for short periods. For longer storage periods, pupae are stored before emergence and subsequent maceration.

In another *Trichogramma* spp. production method, Voegele et al. (1974) was able to store the eggs of *Ephestia kuehniella* (Zeller) for up to 60 days before exposing them to the parasite by first sterilizing them with ultraviolet irradiation (before they reached the gastrulation stage) and then holding them at 4°C. Morrison

(1975) routinely uses *Sitotroga cerealella* eggs that have been
frozen up to 6 months for rearing *Chrysopa carnea*. *Galleria
mellonella* larvae in the cocoon stage can be held for several
months at 10-13°C and still be used for rearing *Lixophaga diatraeae*
(E. G. King, unpublished data).

    Reduction in viability and vigor of predators and parasites is
frequently induced by storage at low temperatures.  There was a
decline in adult emergence from puparia, percentage mating, adult
longevity, and fecundity when *L. diatraeae* puparia were held at
4.4°, 10°, or 15.6°C (E. G. King, unpublished data).  Similar results have been reported by others (Scopes and Biggerstaff 1971 -
*E. formosa*; Starks et al. 1976 - *L. testaceipes*).  Stinner (1977)
emphasized the importance of testing for detrimental effects due
to storage under conditions that to some degree simulated field
stresses.  Effects may not be apparent under nonstress environmental conditions in the laboratory, but when stressed, significant differences may occur between stored and nonstored individuals.  These differences may be magnified under field conditions.


                        MASS PRODUCTION TECHNIQUES


    In 1964, Finney and Fisher reviewed the culturing of entomo-
phagous arthropods and their hosts, the care of the host substrate,
and the information needed in developing a mass-production program.
Their report is still highly relevant; however, it does reflect
their experiences in rearing of Aphelinidae, Encyrtidae, and
*Macrocentrus ancylivorus* and their hosts.  We hope to both compli-
ment and supplement their review with additional recent examples
of mass production techniques.  In certain instances, examples we
select do not fit their concept of mass production, "a skillful
and highly refined processing of an entomophagous species and its
host and substrate through insectary procedures which result in
economical productions of millions of beneficial insects."  However, in reviewing these programs we will highlight certain techniques that could lead to mass production.  No attempt will be
made to cover all mass- or large-scale rearing programs of entomophagous arthropods that have been reported; instead, from one to
three examples will be selected for certain families.  Also, the
primary emphasis will be placed on the predator or parasite (host
and host food source will be considered only when it is an integral part of the program for rearing the predator or parasite),
but pertinent references on specific host rearing and care of the
host substrate (food) will be cited.  One family will not be covered in our review - Pteromalidae.  This family contains certain
parasites, e.g. *Spalangia endius* Walker, that attack muscoid flies.

## Chrysopidae

The first large-scale rearing of Chrysopidae was reported by Finney (1948, 1950). New methods and techniques for production of *Chrysopa carnea* Stephens were later reported by Ridgway et al. (1970), Hassan (1975), Morrison et al. (1975), Morrison and Ridgway (1976), and Morrison (1977). The development of inexpensive, highly acceptable adult diets that induce and maintain high rates of fecundity was accomplished by Hagen (1950) and Hagen and Tassan (1970).

The basic technique for production of adult chrysopids from eggs is to supply food to isolated or separated larvae since this prevents or reduces cannibalism. Techniques used for adult production by Finney (1948) consisted of the use of alternate layers of paper spread with prepared eggs and larvae of the potato tuber moth (food for the developing larvae). These layers were held in closed containers, and additional layers added as needed. Cannibalism was limited by the abundance of food and by spacial separation of the developing larvae.

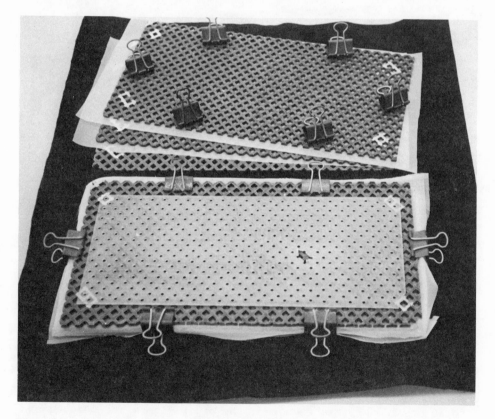

Fig. 1. *Chrysopa carnea* larval rearing unit and feeding plate.

Fig. 2.  Component parts of *Chrysopa carnea* oviposition and feeding unit.

More recent techniques involve isolating well-developed *C. carnea* eggs and providing an initial food supply of eggs of the Angoumois grain moth (first been frozen and then thawed) to small, organdy-covered, multi-cell, rearing units (Fig. 1).  Defined amounts of prepared eggs are then placed on the organdy at regular intervals, and the larvae feed on them through the cloth (Morrison et al. 1975).  After pupation the pupae are placed together for emergence.  As the adults emerge, they are collected daily, held in preoviposition units, and supplied food and water for 4 days. Thereafter, the units are lined with paper on which the adults lay their eggs.  This paper is coated with the adult diet, a combination of Food Wheast® (a yeast *Saccaromyces fragilis*, cultured on a whey substrate) and sugar (Hagen and Tassan 1970), which must be continuously available to maintain high fecundity (Fig. 2).  When 0 to 24 hr old eggs are required, the paper liners are removed and replaced daily.  Finney (1948) used $CO_2$ to immobilize the adults

Fig. 3.  Vacuum unit and holding belt used for handling and changing adult *Chrysopa carnea*.

during this change.  Morrison and Ridgway (1976) used a vacuum to
first immobilize the adults and to transfer them to clean units
(Fig. 3).

     Removal of *C. carnea* eggs from the oviposition paper was
accomplished by Finney (1948) with a solution of dilute sodium hy-
pochlorite.  The egg-coated paper was dipped into the solution for
a timed period and withdrawn for a timed period; then the eggs
washed off the paper with water into a catch basin, rinsed, spread
on drying cloths, and air dried before removal and subsequent use.
Ridgway et al. (1970) used a hand-held ball of nylon net to remove
eggs:  the ball was passed lightly over the paper oviposition
sheets, and the egg stalk was easily broken by the hard nylon
threads of the net.  More recently, Morrison (1977) reported that
a thin electrically heated wire will readily sever the egg stalk.
Vacuum is used in conjunction with the hot wire to extend the egg

Fig. 4.  Method and device for collection of adult *Cryptolaemus montrouzieri*.

to the limit of the stalk.  This provides sufficient distance between the egg and substrate to prevent heat damage and also immediately carries the disconnected egg away from the wire.

Morrison and Ridgway (1976) report that 16 man hours per day were required to produce about 350,000 eggs/day with their techniques.  Of this time, about 75% was devoted to handling the adults and to egg collection.

## Coccinellidae

Two coccinellids that have been reared or field collected and then released in large quantities are the convergent lady beetle, *Hippodamia convergens* Guérin-Méneville, and *Cryptolaemus montrouzieri* Mulsant.  DeBach and Hagen (1964) give an excellent

review on the collection of diapausing adult convergent lady
beetles from winter aggregations in the Sierra Nevada Mountains.
Smith and Armitage (1931) first reported a complete rearing procedure for *C. montrouzieri* for control of Pseudococcidae on citrus.
Until the early 1950's a large number of insectaries in California
were producing this beetle, but this approach was curtailed as a
result of the establishment of several mealybug parasites and the
advent of organic pesticides (Fisher 1963). Now the beetle is
reared primarily by one insectary--Associates Insectary at Santa
Paula, California.

Present day rearing procedures at Associates Insectary are
essentially the same as those reported by Fisher (1963). The
citrus mealybug, *Planococcus citri* (Risso), is mass reared on potato sprouts grown in subdued light in soil held in wooden trays.
After the sprouts are about 4.7 cm tall, mealybug crawlers are
removed from producing trays by allowing them to crawl onto freshly
cut leafy terminals of *Pittosporum undulatum* Vent. or *Schinus molle*
L., which are placed in trays for about 6 hours. These terminals
are then placed among the potato sprouts in darkened rooms and the
crawlers move onto the sprouts as the terminals begin to dry.
After 8 days, *C. montrouzieri* adults are placed in the trays where
they oviposit on the potato sprouts and trays for 12 days. Then
these adults are collected at opened windows screened with cotton
muslin, and released in the citrus orchards. Meanwhile, burlap
bands are attached to the front of the racks holding the trays as
a substrate for pupating beetle larvae, and the holding room is
again darkened for 6 days, the window shutter is once more raised,
and emerging beetles are collected for release at the cloth-covered
window opening (Fig. 4). Fisher (1963) presented the components of
the *C. montrouzieri* production system along with production data and
total cost per year for production. Associates Insectary maintained
an average production of 30 million per year with only about 2 men.

## Aphelinidae

The present rearing methods for 2 members of this group are
rather unique and worthy of note. The techniques used for the
first, *Aphytis melinus* DeBach, are as follows: Flanders (1951)
was the first to describe a large scale rearing program for *Aphytis
chrysomphali* (Mercet). The host used for production of the parasite was the California red scale, *Aonidiella aurantii*, grown on
potato tubers. However, the method developed by DeBach and White
(1960) for the production of the California red scale parasite,
*Aphytis lingnanensis* Compere, proved to be more satisfactory and
is still used by at least 2 commercial insectaries in California.
Subsequently an imported species, *Aphytis melinus*, proved to be
more satisfactory against California red scale for inland citrus
areas than *A. lingnanensis*. It is now reared in these commercial

Fig. 5.  *Aphytis* spp. oviposition and production cabinet used to
hold scale-infested squash.

concerns by the same techniques as follows:  The host is an alternate host, a uniparental strain of the oleander scale, *Aspidiotus
nerii* Bouché.  The banana squash, *Cucurbita maxima* Dene., is used
as host material for the scale.  Fresh squash is infested daily
with "crawlers" (the immature mobile stage of oleander scale)
collected from squash infested 58-73 days earlier.  Scale-infested
squash about 73 days old are offered to the parasite (after use as
"crawler" production material) in parasite oviposition-collection
units (Fig. 5) supplied with honey, the food for the adult parasites (Fig. 6).  A defined number of freshly emerged and collected

Fig. 6.  Technique used to supply adult food in *Aphytis* spp. production cabinet.

adult parasites are placed on the unparasitized scale-infested squash, and the unit is then closed.  After a 24 hr oviposition period, the parasites within the unit are anesthetized, removed, and prepared for field release.  The now parasitized scale-infested squash are rotated from the unit to room storage racks for parasite development and also to allow for full utilization of the oviposition-collection unit.  Two or 3 days before emergence of the adult parasites, the squash are placed back into a prepared unit, and the parasites then collected each 24 hr throughout the emergence cycle.  The newly emerged parasites are again used for oviposition for 24 hr and then released in the field.

The production system is rather unique in that, all the adult insects are utilized for field releases except for the use of produced parasites for a 24 hr oviposition period (only a small fraction of their reproductive potential).  Additionally, by first

using host material for production of reinfesting stock and then
for parasite production, the need for separate or secondary "brood"
cultures is eliminated.  Thus production efficiency (percent of
parasites produced that are utilized for field release) approaches
100 percent.

The economics of this production system were carefully analyzed and cost per thousand parasites was estimated to be $0.096
at that time (1960).  Further, it was determined that the cost of
the complete program of production and distribution compared favor-
ably with the cost of a typical annual insecticide program for control of California red scale.

The use of *Encarsia formosa* Gahan, a parasite of the greenhouse whitefly *Trialeurodes vaporariorum* (Westwood), in augmentative
releases is receiving increased attention as a control measure.

This method is principally used in Europe where whitefly problems occur in the production of high-value produce in glasshouses.
Two methods used to produce *E. formosa* are described below.

Scopes (1969) and Scopes and Biggerstaff (1971) described
rearing procedures for *E. formosa*.  Scopes (1969) described a production unit that would supply sufficient numbers to treat 80 ha
of glasshouse, and also presented detailed cost estimates.  The
natural host, *T. vaporariorum*, is produced on tobacco (*Nicotiana
tabacum* L.) exposed for 24 hr to whitefly adults.  Then the plants
are shaken and fumigated with dichlorvos (2,2-dichlorovinyl dimethyl
phosphate) to eliminate the adults; and the egg-infested plants are
held until the whitefly reaches the third instar.  At that time
adult parasites are introduced onto the plants, and ten days later,
the parasitized and unparasitized whitefly pupae are brushed from
the leaves into suitable containers and held for emergence of the
unparasitized whiteflies (they emerge before the parasites, thereby
providing a simple method of self-elimination from the parasite
culture).  The remaining culture of parasites is used for release
in commercial glasshouses and for the reproductive colony.

A more recent commercial production system of *E. formosa* in
the Netherlands (J. Woets, Proefstation, Naaldwijk, the Netherlands;
personal communication) uses cucumber plants grown vertically on
string as the host plant.  The plants are grown in "Venlo" type
glasshouses maintained at about 25°C (daytime temperature) and at
about 20°C at night.

Young plants are initially infested with about 100 *T. vaporariorum* adults.  Then about 2 weeks later, *E. formosa* is introduced as
mature pupae at a ratio of about 7 parasites:10 whiteflies.  After
an additional 2 weeks, harvest of the bottom leaves, containing

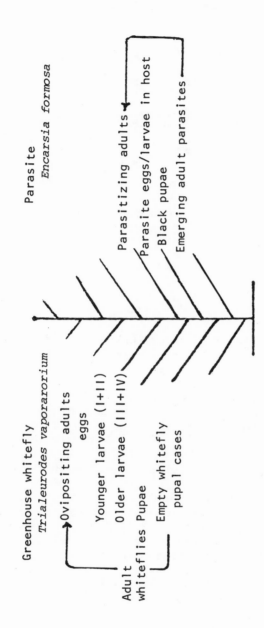

Fig. 7.   Schematic drawing of *Encarsia formosa* production on live cucumber plants.

only mature *E. formosa* pupae and emerging adults, begins after inspection insures that some parasite emergence has occurred.

The plant is now a self-sustaining production unit for several months with the bottom leaves being harvested regularly (Fig. 7). Both pest and parasite move up the plant to newly emerged or infested leaves, respectively, thus sustaining the cycle. If the rate of parasitism is out of balance, it can be increased or decreased by manipulating the glasshouse temperature ± 2-3°C.

Since surviving whitefly adults have all emerged from the lower leaves, the harvested leaves contain only mature parasite pupae and those leaves containing at least 200 parasite pupae are sold.

Woets states that during 1974, about 50 million parasites were produced to control whitefly in about 400 ha of glasshouse tomatoes. These parasites were produced on about 11,000 $m^2$ of glasshouse area. Grower cost, including consulting services, was about $.08/m$^2$. With the exception of initial planting, infesting, and harvest, this rather unique production system greatly reduces the labor used in previous production systems.

## Braconidae

The braconids *Aphidius* spp., *Lysiphlebus testaceipes* (Cresson), and *Bracon kirkpatricki* have been reared in large quantities for release in studies of feasibility. However, *Macrocentrus ancylivorus*, another braconid, was mass reared as part of area-wide programs to control the oriental fruit moth in New Jersey, California, and presently Colorado. These four species were all insectary reared, but Halfhill and Featherston (1967) mass reared the aphid parasites *Aphidius pulcher* Baker and *Aphidius smithi* Sharma and Subba Rao on their host in the field. The following is an outline of the rearing program for each of these parasites.

Finney et al. (1947) reported a complete rearing procedure for *Macrocentrus ancylivorus* on the potato tuberworm. Potato tubers were perforated and placed on a hardware cloth where they were infested with host larvae by covering them with the oviposition cloth containing host eggs. The young host larvae entered the punctures. Later they were exposed to the parasites by placing the trays in a closed compartment containing the parasites. Subsequently, the trays were placed in tiers above a shallow wooden container that retained the host larvae as they left the potatoes. Strips or ridges of finely crushed rock on waxed plywood or sheet-metal plates were placed within the containers to provide a place for the cocooning of the host larvae. The plates bearing the newly formed host cocoons were removed from the container daily. Parasite cocoons

and host pupae were freed of the host cocoons and then separated
so host-free parasites could be packed and shipped to the field.

Bryan et al. (1973) reported on the rearing program for *B.
kirkpatricki* on the beet armyworm:  Fifth-stage host larvae were
placed between 2 pieces of paper toweling resting on a sheet of
plastic.  Then the parasite cage, a plastic box covered with nylon
organdy, was arranged so the organdy cage top contacted the paper
covering the larvae.  The parasite probed with its ovipositor
through the organdy and toweling and injected a fluid into the host
that immobilized it.  Eggs were then deposited beside it.  The
toweling bearing the immobilized larvae was stacked and held until
the parasites completed development and had spun their cocoons.
Then it was placed in an emergence chamber where the adult parasites
were collected for field release.

Stary (1970) and Starks et al. (1976) reared aphid parasites
on aphids produced on live plants in the laboratory.  First the
aphids were established on the plants; then they were exposed to
the parasites.  Parasite mummies were removed from the plants and
shipped to the field where they were emerged and released.  Halfhill
and Featherston (1967) used field cages that produced a "greenhouse
effect" in that growth of early spring alfalfa was 4-6 weeks ahead
of growth in surrounding fields.  The host, the pea aphid,
*Acyrthosiphon pisum* (Harris), was introduced into the cage; the
parasite was introduced after the host was established.  The parasites escaped from the cage into the surrounding field through
temperature-controlled automatic screened vents in the cage roof
that retained the aphids and other relatively large arthropods.
Sufficient numbers of parasites remained in the cages to maintain
a favorable host-parasite ratio.

Finney et al. (1947) estimated that by his method it cost
about $0.78/1000 parasites to produce 19 million *Macrocentrus
ancylivorus* in any one season.  The cost per 1000 would probably
have to be multiplied by at least 4-fold to equal today's price.
Starks et al. (1976) estimated that it cost $19.00 to produce 1000
*Lysiphlebus testaceipes*, however, he also figured other costs and
compared them with other methods of control.  Cost estimates are
unavailable for the other braconid parasites though Bryan et al.
(1973) did present production figures.


Encyrtidae

*Metaphycus helvolus* was first successfully cultured on black
scale reared on potato sprouts by Flanders (1942).  A later method,
still in use at the Fillmore Citrus Protective District Insectary
at Fillmore, California (Lorbeer 1971), involves the use of potted
oleander plants as the black scale host instead of potato sprouts.

Oleander cuttings grown outside for two years are moved into a
6x8 m insectary room specially equipped with tightly-stretched
white muslin at the full-length windows on opposite sides of the
rooms.  The glass windows outside the muslin barrier can be opened
and closed for ventilation and temperature control; additionally,
they are shuttered for careful light and ventilation control.  Once
in the room, the potted oleanders are infested with immature, mobile, black scale crawlers, the room is darkened, and the scale are
allowed to distribute themselves uniformly over the plants before
permanent attachment occurs.  When the black scale develops to the
second instar, parasites are released in the rooms for oviposition.
Once parasite emergence begins, the glass windows are opened to
admit light, and the emerged parasites congregate on the muslin.
Each day the parasites are vacuum aspirated from the muslin into
honey-streaked (for adult food) clear plastic tubes, the tubes are
capped with cloth (for ventilation), and the parasites are released
in the field directly from the tubes.

Approximately 150,000 parasites are produced per room, a relatively low number.  This is caused by a high rate of host feeding
necessary if the parasite is to achieve successful parasitization.
However, since the released parasite also destroys large numbers
of hosts through host feeding, this low production is considered
acceptable.  Production costs for this system are unavailable.

## Trichogrammatidae

Flanders (1929, 1930) described the first known mass-production
system for *Trichogramma*.  Spencer et al. (1935) described a modified
system and also reported an analysis of production efficiency and
costs.  DeBach and Hagen (1964) reviewed production and various
aspects of the use of *Trichogramma*.  More recently, Lewis et al.
(1976) reported a mass production method of this parasite in which
gamma irradiated eggs of *Heliothis zea* (Boddie) were utilized, also
Morrison et al. (1976) described techniques used to produce about
15 million parasites per day.  Numerous recent reports [Dysart
(1973), Huffaker (Chapt. in text)] indicate extensive and successful
production and utilization of *Trichogramma* spp. in both the Soviet
Union and the People's Republic of China.

The basic method of producing *Trichogramma* is to confine the
adults and acceptable host eggs together in a restricted area,
generally for a limited period and in the presence of subdued light.
After the desired time, the adults are removed and the now parasitized host eggs are held for development of the parasite.  The modifications and variations of this scheme are numerous.  For example,
Flanders (1930) and Spencer et al. (1935) used closed petri dishes
as the production unit.  They introduced sufficient quantities of
eggs of the Angoumois grain moth, *Sitotroga cerealella*, glued to

paper circles the size of a petri dish and then added pharate adult
*Trichogramma* (retained from the previous generation); the dishes
were then held in subdued light for 4-6 days.  Thereafter, the parasitized eggs were separated from hatched *Sitotroga cerealella* larvae
with soft brushes and air currents, and the papers with parasitized
eggs were held for field release with a portion reserved for future
propagation.  Techniques used by Morrison et al. (1976) consisted
of preparing loose host eggs (*S. cerealella*) for parasitism by
attaching them uniformly to pieces of mucilage-dampened construction
paper (Fig. 8).  A mechanical, partially automated device developed
in the Soviet Union rapidly accomplishes this same basic task, presumably at levels limited only by the amount of host eggs available
(R. L. Ridgway, personal communication).

A unit used for parasitism of prepared host egg cards (Morrison
et al. 1976) allows 10 of these cards to be offered for parasitism
each 24 hr (Fig. 9).  Since the unit has removable glass faces on 2
opposite sides and the parasites are positively phototaxic, adults
can be readily moved to either of the glass faces by differential

Fig. 8.  Techniques used in preparation of host eggs for parasitization by *Trichogramma pretiosum*.

Fig. 9.  *Trichogramma pretiosum* production cabinet.

light intensity.  Egg sheets can thus be placed in and removed from
the unit by simply placing the face to be opened away from the
source of light.  Once the parasites have moved off that particular
glass front toward the light, it can be removed and the necessary
procedures performed.  Maintenance of the adult population in the
unit for oviposition is accomplished by placing pharate adults
attached to a paper substrate (portions of parasitized egg cards)
in the corners of the unit at the same time that the host egg cards
are changed.  Complete emergence of these adults occurs within a
4-day period, which allows expended material to be rotated out of
the cage and be replaced with fresh material each 24 hr on a 4-day
cycle.  Once the required changes are accomplished, the glass face
is replaced, which seals the unit, and the unit is then revolved to
face the light source.  The parasites then move towards the light
where they encounter the unparasitized host eggs.  However, light
intensity and diffusion must be adjusted so the parasite is

sufficiently attracted but not so strongly attracted that it remains
on the glass exclusively.

A newly developed production unit similar to the unit previously described uses a 2 hr cyclic photophase to continuously
draw the parasites back and forth over the host eggs (Morrison
1977).  Production efficiency with this unit has been increased 2-
fold over the previous one.  Additionally, the parasitized host
eggs are loose and unattached to a substrate, greatly increasing
the flexibility of parasite manipulation in field release programs.

Because the basic techniques used in producing *Trichogramma*
involve the use of pharate adults as oviposition stock and the
manipulation of the strong affinity of the parasite to light, we
will not describe further examples of production methods.

A cost analysis of *Trichogramma* production is difficult without
knowing the cost of the host egg being used in the production
scheme.  The unit described by Morrison et al. (1976) produced about
700,000 parasites per unit/day for field release and required about
1/2 man hr/day to operate.  Spencer et al. (1935) stated in his
analysis that total production costs per million parasites ranged
between $5 and 10.


                              Tachinidae

Numerous species of tachinids have been reared (small scale)
in the laboratory.  For example, Bennett (1969) reviewed the biology
and rearing of tachinids (6 genera) that parasitize Lepidopteran
sugarcane borers, and Bryan et al. (1969) reported laboratory rearing procedures for 7 species of tachinids that parasitize cotton
insect pests.  However, only *Lixophaga diatraeae* has been reared
on a large scale (King et al. 1975a).

Until 1969, laboratory rearing procedures for *L. diatraeae*
were those summarized by Bennett (1969).  Then King et al. (1975b)
reported procedures for large-scale production of *L. diatraeae* on
its natural host, *Diatraea saccharalis*; also, a mass-rearing program was developed for producing *L. diatraeae* on an unnatural host,
*Galleria mellonella* L (E. G. King, unpublished data).  Others have
reported successful rearing of *L. diatraeae* on *G. mellonella* larvae
fed diets similar to the one reported by Haydak (1936) and have
either suggested or been actively involved in augmentation programs
(Montes 1970; Summers et al. 1971; Guerra 1974; and Bennett, pers.
comm., 1975).  Problems associated with greater wax moth diets were
discussed earlier in the section on "Natural and Unnatural Hosts."

The following is an abridged outline of the *L. diatraeae* rearing program by King et al. (1975b) and King (unpublished data):  The

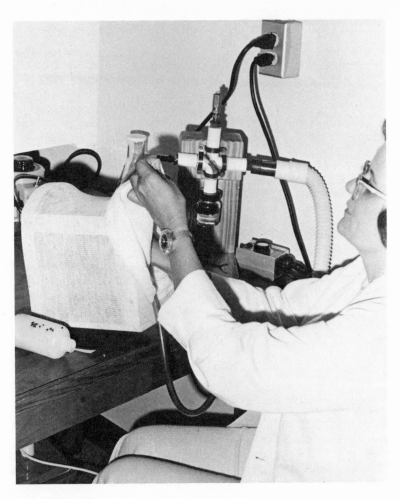

Fig. 10.  Aspiration device used to transfer *Lixophaga diatraeae*
adults from holding cage to 1% NaOCl solution.


adult flies are held in cages containing raw sugar and water until
they are 12-14 days old.  Then they are aspirated into a collection
jar containing 1% NaOCl (Fig. 10) (King et al. 1975b, Gantt et al.
1976).  After the NaOCl is rinsed from the flies, they are blended
at 8,500 rpm for 9 seconds in a 0.7% formalin solution to extract
the maggots (the solution serves to remove a bacterium from the
maggots that causes septicemia in parasitized host larvae).  After
5 minutes, the maggots are separated from the fly particles, rinsed,
suspended in a 0.15% agar-water solution, and placed in a gridded
petri plate where the total maggot number is determined.

Thereafter, additional agar-water solution can be added to obtain
the desired maggot density in a given solution.  When *L. diatraeae*
are reared on sugarcane borer larvae by this method, the solution
is metered (Gantt et al. 1976) into 30-ml cups containing early
5th-stage host larva (e) (Miles and King 1975, King et al. 1976)
and a soybean-wheat germ diet (Brewer 1976).  The maggots seek out
the host larvae and parasitize them.  When *L. diatraeae* are reared
on greater wax moth larvae by this method, hosts (selected just
before cocoon formation) are placed in trays (Fig. 11) containing
the parasite maggots, which have been applied by "pouring" or spraying with an air brush.  Then a screen is placed over the trays,
and after about 1 hr, corncob grits are added to the trays to
absorb excess moisture.  In either case, the maggots complete

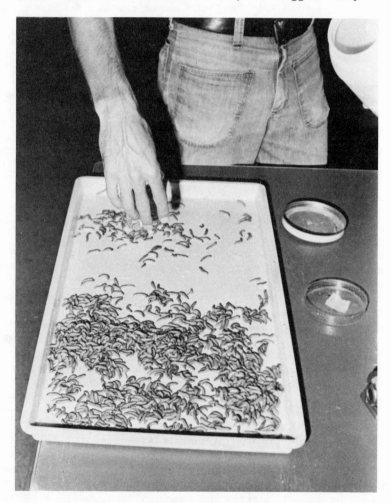

Fig. 11.  *Lixophaga diatraeae* maggots, suspended in 0.15% agar-
water solution on tray being offered *Galleria mellonella* larvae.

development in 6-9 days, and both hosts and parasite puparia are harvested on the eleventh day after parasitization. Puparia are removed by hand (with forceps) from cups containing sugarcane borers; puparia outside the greater wax moth cocoons are brushed from the trays, and those retained in the host cocoon are harvested by dissolving the cocoon in NaOCl and collection by flotation. Puparia to be shipped are packaged between layers of cotton in cartons and placed in styrofoam boxes that contain ice packs so low temperatures (18 to 24°C) will be maintained in transit.

During a 4 year period when tests were being conducted in Louisiana and Florida to determine the feasibility of using L. diatraeae in periodic releases for control of the sugarcane borer, a total of over 4.5 million parasites was produced. Each succeeding year production was either tripled or doubled as improvements were made in rearing techniques. Costs per 1000 L. diatraeae puparia (diet, labor, and rearing containers) were $51.00 and $9.80 for production on sugarcane borer and greater wax moth larvae, respectively. Thus, rearing on the unnatural host reduced costs 81%. Primary savings occurred in the area of labor and diet costs. With additional research, the cost of rearing L. diatraeae on the greater wax moth can probably be reduced one-half. However, additional research is needed concerning the relationship between nutrition of the greater wax moth and quality of the parasite.

## Phytoseiidae

Recent reports indicated that phytoseiids are mostly being reared and used on a commercial basis for control of Tetranychus urticae Koch in glasshouses in areas outside the United States (Hussey and Bravenboer 1971, Anonymous 1974). Procedures for rearing this prey and the predator mite have been reported primarily by McMurtry and Scriven (1965), Scriven and McMurtry (1971), McMurtry and Scriven (1975), and J. Woets (Proefstation, Naaldwijk, the Netherlands, personal communication). Scopes (1968) also reported on the rearing procedures of prey and predator, but his procedures were similar to those reported in the series of articles by McMurtry and Scriven. According to Woets (1974), Phytoseiulus persimilis is being reared on a large scale for control of T. urticae attacking tomatoes, cucumbers, and egg plant in glasshouses in the Netherlands.

Basically, the rearing procedures reported for the tetranychid and its predator are as follows: The prey is reared on bean plants and harvested by washing from the leaves in a specially designed washer. Then they are collected on a series of graduated fine mesh screens. The predator mites are held on various dry substrates (painted construction paper, fluted filter paper, or a polyurethane foam mat covered with metal tile) surrounded by moisture-bearing

material that prevents dispersal.  An established number of prey
(different stages) are sprinkled over the predator-infested surface.
Adult female predators are collected with a suction aspirator for
field release beginning after 14-16 days.  Scopes packaged them in
gelatin capsules with prey for shipment to release points.

A more recent production method used commercially in the
Netherlands (J. Woets, personal communication) utilizes the small,
rotating plots of bean plants grown in "Venlo" type glasshouses for
production of both host and predator.

The wire net-supported plants are infested with *T. urticae*
when they develop their first true leaves and the mites are allowed
to develop until the second true leaves appear.  *P. persimilis* is
now introduced and when it has virtually eliminated the prey, those
leaves containing about 20 adult *P. persimilis* and variable numbers
of eggs and immature stages are picked, placed in paper bags (not
plastic) and delivered to release areas.  The plot is then destroyed
and seeded for reuse.  This system eliminates the costly step of
host harvesting used in previous programs.

Woets states this method was used to produce about 15 million
*P. persimilis* in a 600 m$^2$ area in 1974.  They were used to control
spider mites on about 150 ha of glasshouse-cultured cucumbers and
bell peppers.

Scopes (1968) presented a detailed outline of costs of rearing sufficient *P. persimilis* in a central production unit to treat
about 80 ha of cucumbers under glass.  Also, Woets (1974) and
McMurtry and Scriven (1975) provided indirect cost estimates.
However, Woets based his figures on cost per phytoseiid for control
of *T. urticae* in tomatoes or cucumbers and included production and
consultant fees.  McMurtry and Scriven merely observed that their
production methods for *P. persimilis* were economically feasible
for control of *T. urticae* in glasshouse crops, but not in outdoor
strawberries.


## TIME AND MOTION STUDIES AND COST ANALYSIS

Many techniques have been developed for rearing parasitic and
predator arthropods, however, comprehensive production technology,
including time and motion studies, has been applied to only a
limited extent.  Also, detailed analyses of the cost of rearing
parasites and predators are almost nonexistent.

Mechanization of routine tasks has been an important factor
in reducing costs in successful mass-rearing projects because labor
is generally the major expense until that has been accomplished.
For example, the major cost of producing the screwworm, *Cochliomyia*

*hominivorax* (Coquerel), is now the larval diet (Anonymous 1977b);
and facility costs and expendable supplies are the major costs of
the boll weevil, *Anthonomus grandis* Boheman (Gast and Vardell 1963),
melon fly, *Dacus cucurbitae* Coquillett (Metchell and Steiner 1965),
and other successful insect mass-rearing projects (Burton 1969,
Mangum et al. 1969, and Schoenleger et al. 1970) now that labor has
been reduced to a secondary importance.

Reeves (1975) stated:

"An analysis for the development or improvement of
any integrated system such as insect production or distribution and release should begin with the development
of process operation flow charts (Apple 1963). This
approach allows the researcher to visualize and divide
the system into subsystems and steps so an analysis of
the labor supply, equipment, and space requirements of
the various operations can be made. Thus, a complete
description of the methods involved can be formulated
and priorities can be established for needed modifications of the systems. In this way, operations can be
mechanized and the organization of the work force can
be modified to improve the efficiency of the systems
(Eckman 1958, Nadler 1955, and Nieble 1967).

"Plant layout is a phase of system development
that assures efficient materials handling, space and
equipment utilization, and personal movement (Nieble
1967). Insect rearing, distribution and release
processes require a high degree of production quality
control, record keeping, and process coordination which
calls for effective departmental supervision; therefore,
plant layout should be a priority item in the design of
the pilot test production, distribution, and release
system."

He described in 9 operation flow charts the integrated systems
of insect production, distribtuion and field release for *Trichogramma pretiosum* Riley and *Chrysopa carnea*. His insect production
charts are shown in Figs. 12 and 13. These charts were then analyzed to establish engineering priority needs associated with a
large-scale field test of programmed releases of these 2 insects.
In addition, his charts were useful in the planning and construction of a production facility for these insects.

As the terminology for insect production improves, increased
importance will be placed on the monetary value of the product.
Although some analyses of costs have been conducted, there is little
uniformity in the reporting methods. Uniformity of reporting costs
of different production systems might, of course, be difficult, but

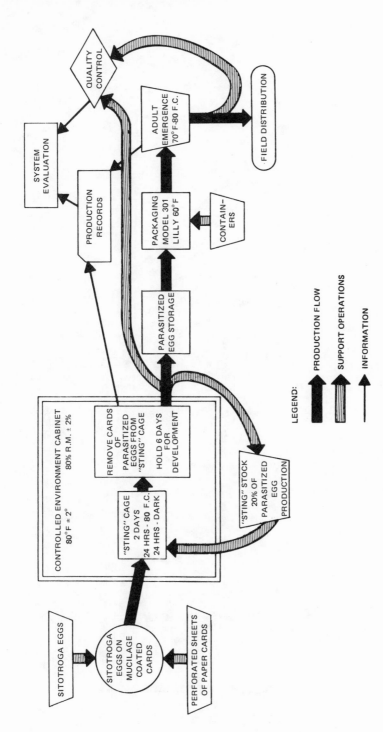

Fig. 12. Flow chart of *Trichogramma pretiosum* production system.

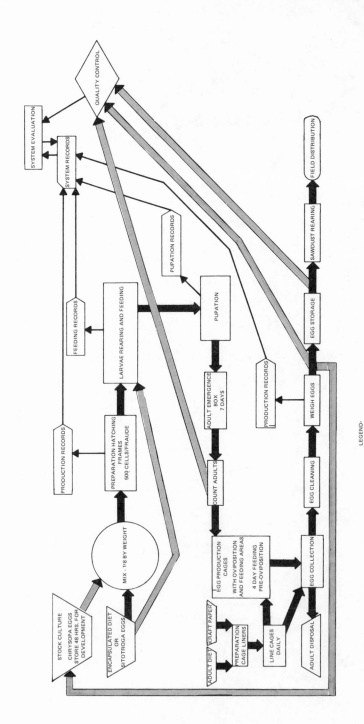

Fig. 13.  Flow chart of *Chrysopa carnea* production system.

such a standardized system would report only those factors necessary
to produce a given unit of the "finished product", which might allow
both present-day and future workers to rapidly estimate the costs
of a given production program despite changeable circumstances and
areas.   Such a factor analysis might also give additional insight
into cost reduction considerations of ongoing programs.

The first factor of such a system would be to define the
quality and quantity of man-hours required to produce a given number of parasites or predators, or the amount and technical level
of man-hours required per unit of production.

The second factor would be to state the units of host material
(natural or artificial) required per unit of production.   This
would require a separate analysis of host production (or preparation) but costs could easily be estimated if quality and quantity
requirements of the host material per unit of production were known.

The final and most difficult factor to be reported would be
the type or sophistication of housing facilities and equipment,
spacial requirements, and units of energy necessary per stated unit
of production.   It would be much simpler to evaluate such examples
as $m^2$, concrete-floored, kilowatt hr utilized, and 5000 BTU heating
system required to produce a given unit of insects rather than "a
building which cost $250/m^2$."

Admittedly we have oversimplified by omitting secondary factors
such as depreciation, support services and profit (where applicable)
but such a system, together with detailed cost analyses of each
factor, would introduce a measure of standardization and compre-
hension into an increasingly important but often neglected phase of
mass production.

## ACKNOWLEDGMENTS

We sincerely thank Dr. Paul DeBach, Univ. of Calif., Riverside,
for permission to use the photographs in Figs. 5 and 6; Dr. T. W.
Fisher, Univ. of Calif., Riverside, for permission to use the
photograph in Fig. 4; Dr. Jack Woets, Proefstation, Naaldwijk, the
Netherlands, for permission to use the drawing in Fig. 7; and Dr.
B. G. Reeves, USDA Exten. Serv., for permission to use the flow
charts shown in Figs. 12 and 13.

This is Technical Article 13271 from the Texas Agricultural
Experiment Station.

DISCLAIMER

Mention of a commercial (or proprietary) product in this paper does not constitute an endorsement of this product by the USDA.

LITERATURE CITED

Anonymous. 1969. Control by parasites, predators, and competitors. Pages 100-64 *in* Insect pest management and control. National Academy of Sciences Publication 1695, Washington, D. C.

Anonymous. 1974. Proc. of the International Organization of Biological Control. Conference on intergrated control in glasshouses. Report on the meeting held from 18 to 20 September 1973 at the Glasshouse Crops Research Station, Littlehampton, England. 73 p.

Anonymous. 1977a. Insect control in the People's Republic of China. National Academy of Sciences. Washington, D. C. 218 p.

Anonymous. 1977b. The sterile screwworm fly production plant. Mission, Texas. USDA, APHIS, U. S. Govt. Printing Off. 1977-772-925/272/7. 6 p.

Apple, J. M. 1963. Plant layout and materials handling 2nd edition. The Ronald Press Co., New York.

Beirne, B. P. 1974. Status of biological control procedures that involve parasites and predators. Pages 69-76 *in* Proceedings of the summer institute on biological control of plant insects and diseases (F. G. Maxwell and F. A. Harris, Eds.). Univ. Press of Mississippi, Jackson.

Bennett, F. D. 1969. Tachinid flies as biological control agents for sugarcane moth borers. Pages 117-48 *in* Pests of sugarcane (J. R. Williams, J. R. Metcalf, R. W. Muntgomery, and R. Mathes, Eds.). Elsevier Publ. Co., New York.

Boldt, P. E., and N. Marston. 1974. Eggs of the greater wax moth as a host for *Trichogramma*. Environ. Entomol. 3: 545-8.

Boller, E. 1972. Behavioral aspects of mass-rearing of insects. Entomophaga 17: 9-25.

Brewer, F. D. 1976. Development of the sugarcane borer on different artificial diets. USDA, ARS-S-116. 6 p.

Bryan, D. E., C. G. Jackson, and A. Stoner. 1969. Rearing cotton insect parasites in the laboratory. USDA Prod. Res. Rep. 109. 13 p.

Bryan, D. E., R. E. Fye, C. G. Jackson, and R. Patana. 1973. Releases of parasites for suppression of pink bollworms in Arizona. USDA, ARS W-7. 8 p.

Burton, R. L. 1969. Mass rearing of corn earworm in the laboratory. USDA, ARS 33-134. 8 p.

Castillo Chacon, I. A. R. 1967. Comportamiento de *Heliothis* spp. al quinto ano de liberaciones masivas del parasito *Trichogramma* en la Comarca Lagu nera y posibilidades de control biologico

del gusano rosado del algodonero. XV Reunion del personal
tecnico de low programas cooperativos. Mexico - Estados
Unisos, para la prevencion y combate de las plagas de las
plantas. Torreon, Coahuila, 20 de diciembre de 1967.

Daumel, J., J. Voegele, et P. Brun. 1975. Les Trichogrammes II.
Unite de production massive et qualidienne d'un hote de
substitution *Ephestia kuhniella* Zell. (Lepidoptera, Pyralidae).
Ann. Zool. Ecol. Anim. 7, 1, 45-59.

DeBach, P. (ed.). 1974. Biological control by natural enemies.
Cambridge Univ. Press, New York. 323 p.

DeBach, P., and K. S. Hagen. 1964. Manipulation of entomophagous
species. Pages 429-58 *in* Biological control of insect pests
and weeds (P. DeBach, Ed.). Chapman and Hall, Ltd., London.

DeBach, P., and E. B. White. 1960. Commercial mass culture of the
California red scale parasite *Aphytis lingnanensis*. Calif.
Agric. Exp. Stn. Bull. 770. 58 p.

Dutky, S. R., J. V. Thompson, and G. E. Cantwell. 1962. A technique for mass rearing the greater wax moth (Lepidoptera:
Galleriidae). Proc. Entomol. Soc. Wash. 64: 56-8.

Dysart, F. J. 1973. The use of *Trichogramma* in the USSR Proc.
Tall Timbers Conf. Ecol. Anim. Control Habitat Manage. 4:
165-73.

Eckman, D. P. 1958. Industrial instrumentation. 6th printing.
John Wiley and Sons, Inc., New York.

Etienne, J. 1973. Consequences de l'elevage continu de *Lixophaga
diatraeae* [Dipt. Tachinidae] sur l'hote de remplacement:
*Galleria mellonella* [Lep. Galleriidae]. Entomophaga 18:
193-203.

Etienne, J. 1974. Tachinidae: *Lixophaga diatraeae*. Pages 45-51.
Institut de Recherches Agronomiques Tropicales et des cultures
Virrieres (*in* Irat Reunion Rapport). St. Denis, Isle de la
Réunion.

Finney, G. L. 1948. Culturing *Chrysopa californica* and obtaining
eggs for field distribution. J. Econ. Entomol. 41: 719-21.

Finney, G. L. 1950. Mass culturing *Chrysopa californica* to obtain
eggs for field distribution. Ibid. 43: 97-100.

Finney, G. L., and T. W. Fisher. 1964. Culture of entomophagous
insects and their hosts. Pages 329-55 *in* Biological control
of insect pests and weeds (P. DeBach, Ed.). Chapman and Hill,
Ltd., London.

Finney, G. L., S. E. Flanders, and H. S. Smith. 1947. Mass culture
of *Macrocentrus ancylivorus* and its host the potato tuber moth.
Hilgardia 17: 437-83.

Fisher, T. W. 1963. Mass culture of *Cryptolaemus* and *Leptomastix* -
natural enemies of citrus mealybug. Calif. Agric. Exp. Stn.
Bull. 797. 39 p.

Flanders, S. E. 1929. The mass production of *Trichogramma minutum*
Riley and observations on the natural and artificial parasitism
of the codling moth egg. Trans. 4th Int. Congr. Entomol. 2:
110-30.

Flanders, S. E.   1930.   Mass production of egg parasites of the
    genus *Trichogramma*.   Hilgardia   4:   465-501.

Flanders, S. E.   1942.   Propagation of black scale on potato
    sprouts.   J. Econ. Entomol.   35:   687-9.

Flanders, S. E.   1951.   Mass culture of California red scale and its
    golden chalcid parasites.   Hilgardia   21:   1-12.

Gantt, C. W., E. G. King, and D. F. Martin.   1976.   New machines for
    use in a biological insect-control program.   Trans. Amer. Soc.
    Agric. Eng.   19:   242-3.

Gast, R. T., and H. Vardell.   1963.   Mechanical devices to expedite
    boll weevil rearing in the laboratory.   USDA, ARS 33-89.

Guerra, M. de Souza.   1974.   Methods and recommendations for mass
    rearing of the natural enemies of the sugarcane borer
    (*Diatraea* spp.) (Lepidoptera: Crambidae).   Pages 397-406 *in*
    Proceedings of the international society of sugarcane tech-
    nology, 15th congress, entomology section.

Hagen, K. S.   1950.   Fecundity of *Chrysopa californica* as affected
    by synthetic foods.   J. Econ. Entomol.   43:   101-4.

Hagen, K. S., and R. L. Tassen.   1970.   The influence of Food Wheast
    and related *Saccharomyles fragilis* yeast products on the fecun-
    dity of *Chrysopa carnea* (Neuroptera: Chrysopidae).   Can.
    Entomol.   102:   806-11.

Halfhill, J. E., and P. E. Featherston.   1967.   Propagation of
    braconid parasites of the pea aphid.   J. Econ. Entomol.   60:
    1756.

Hassan, Van S. A.   1975.   Uber die massenzucht von *Chrysopa carnea*
    Steph. (Neuroptera, Chrysopidae).   Z. Angus. Entomol.   79:
    310-5.

Haydak, M. H.   1936.   A food for rearing laboratory insects.   J.
    Econ. Entomol.   29:   1026.

Hussey, N. W., and L. Bravenboer.   1971.   Control of pests in glass-
    house culture by the introduction of natural enemies.   Pages
    195-216 *in* Biological control (C. B. Huffaker, Ed.).   Plenum
    Press, New York.

King, E. G., J. V. Bell, and D. F. Martin.   1975a.   Control of the
    bacterium *Serratia marcescens* in an insect-host-parasite rear-
    ing program.   J. Invertbr. Pathol.   26:   35-40.

King, E. G., D. F. Martin, and L. R. Miles.   1975b.   Advances in
    rearing of *Lixophaga diatraeae* [Dipt.: Tachinidae].   Ento-
    mophaga   20:   307-11.

King, E. G., L. R. Miles, and D. F. Martin.   1976.   Some effects
    of superparasitism by *Lixophaga diatraeae* of sugarcane borer
    larvae in the laboratory.   Entomol. Exp. App.   20:   261-9.

Kinzer, R. E.   1976.   Ph. D. Dissertation, Texas A&M Univ. Library,
    College Station.

Knipling, E. F.   1966.   Introduction.   Pages 2-12 *in* Insect coloni-
    zation and mass production (C. N. Smith, Ed.).   Academic Press,
    New York and London.

Lewis, W. J., D. A. Nordlund, H. R. Gross, Jr., W. D. Perkins, E.
    F. Knipling, and J. Voegele.   1976.   Production and performance

of *Trichogramma* reared on eggs of *Heliothis zea* and other
    hosts. Environ. Entomol. 5: 449-52.
Lorbeer, H. 1971. From cyanide to parasites. Fillmore Herald,
    January 2.
Mackauer, M. 1972. Genetic aspects of insect control. Entomophaga
    17: 27-48.
Mangum, C. C., W. O. Ridgway, and J. R. Brazzel. 1969. Large-scale
    laboratory production of the pink bollworm for sterilization
    programs. USDA, ARS 81-85.
McMurtry, J. A., and G. T. Scriven. 1965. Insectary production of
    phytoseiid mites. J. Econ. Entomol. 58: 282-4.
McMurtry, J. A., and G. T. Scriven. 1975. Population increase of
    *Phytoseiiulus persimilis* on different insectary feeding pro-
    grams. Ibid. 68: 219-21.
Metchell, S. N. T., and L. F. Steiner. 1965. Methods of mass
    culturing melon flies and oriental and mediterranean fruit
    flies. USDA, ARS 33-104.
Miles, L. R., and E. G. King. 1975. Development of the tachinid
    parasite, *Lixophaga diatraeae*, on various developmental stages
    of the sugarcane borer in the laboratory. Environ. Entomol.
    4: 811-4.
Montes, M. 1970. Estudios solve bionomis y la biometri de
    *Lixophaga diatraeae* (Towns.) (Diptera, Tachinidae). Ciencias
    Biolo. 4: 1-60.
Morrison, R. K. 1977. New developments in mass production of
    *Trichogramma* and *Chrysopa* spp. *in* Proceedings of the 1977
    beltwide cotton production research conferences. In Press.
Morrison, R. K., and R. L. Ridgway. 1976. Improvements in tech-
    niques and equipment for production of a common green lacewing,
    *Chrysopa carnea*. USDA, ARS-S-143. 5 p.
Morrison, R. K., V. S. House, and R. L. Ridgway. 1975. An improved
    rearing unit for larvae of a common green lacewing. J. Econ.
    Entomol. 68: 821-2.
Morrison, R. K., R. E. Stinner, and R. L. Ridgway. 1976. Mass
    production of *Trichogramma pretiosum* on eggs of the Angoumois
    grain moth. Southwestern Entomol. 1: 74-80.
Nadler, G. 1955. Motion and time study. McGraw-Hill Co., New
    York.
Nieble, B. W. 1967. Motion and time study, 4th ed. Irwin Publ.
    Co., Homewood, Illinois.
Rabb, R. L., R. E. Stinner, R. van den Bosch. 1976. Conservation
    and augmentation of natural enemies. Pages 233-49 *in* Theory
    and practice of biological control (C. B. Huffaker and P. S.
    Messenger, Eds.). Academic Press, New York.
Reeves, B. G. 1975. Design and evaluation of facilities and equip-
    ment for mass production and field release of an insect para-
    site and an insect predator. Ph. D. Dissertation. Texas A&M
    Univ. Library, College Station.
Ridgway, R. L., R. E. Kinzer, and R. K. Morrison. 1974. Production
    and supplemental releases of parasites and predators for

control of insect and spider mite pests of crops. Pages 110-6 *in* Proceedings of the summer institute on biological control of plant insects and diseases (F. G. Maxwell and F. A. Harris, Eds.). Univ. Press of Mississippi, Jackson.

Ridgway, R. L., R. K. Morrison, and M. Badgley. 1970. Mass-rearing a green lacewing. J. Econ. Entomol. 63: 834-6.

Salt, G. 1935. Experimental studies in insect parasitism. III. Host selection. Proc. R. Soc. London [Ser. B] Biol. Sci. 117: 413-35.

Schoenleger, L. G., B. A. Butt, and L. D. White. 1970. Equipment and methods for sorting insects by sex. USDA, ARS 42-166.

Scopes, N. E. A. 1968. Mass-rearing of *Phytoseiulus riegeli* Dosse, for use in commercial horticulture. Plant Pathol. 17: 123-6.

Scopes, N. E. A. 1969. The economics of mass rearing *Encarsia formosa*, a parasite of the whitefly, *Trialeurodes vaporariorum*, for use in commercial horticulture. Ibid. 18: 130-2.

Scopes, N. E. A., and S. M. Biggerstaff. 1971. The production, handling, and distribution of the whitefly *Trialeurodes vaporariorum* and its parasite *Encarsia formosa* for use in biological control programmes in glasshouses. Ibid. 20: 111-6.

Scriven, G. T., and J. A. McMurtry. 1971. Quantitative production and processing of tetranychid mites for large-scale testing of predator production. J. Econ. Entomol. 64: 1255-7.

Simmonds, F. J. 1966. Insect parasites and predators. Pages 489-99 *in* Insect colonization and mass production (C. N. Smith, Ed.). Academic Press, New York.

Smith, C. N. (ed.). 1966. Insect colonization and mass production. Academic Press, New York and London. 618 p.

Smith, H. S., and H. M. Armitage. 1931. The biological control of mealybugs attacking citrus. Univ. Calif. Bull. 509. 74 p.

Spencer, H., L. Brown, and A. M. Phillips. 1935. New equipment for obtaining host material for mass production of *Trichogramma minutum*, an egg parasite of various insect pests. USDA Circ. 376. 17 p.

Stary, P. 1970. Methods of mass-rearing, collection and release of *Aphidius smithi* (Hymenoptera, Aphidiidae) in Czechoslovakia. Acta Entomol. Bohemoslov. 67: 339-46.

Starks, K. J., R. L. Burton, G. L. Leetes, and E. A. Wood. 1976. Release of parasitoids to control greenbugs on sorghum. USDA, ARS-S-91. 12 p.

Stein, W. 1960. Versuche zur biologischen bekampfung des apfelwicklers [*Carpocapsa pomonella* (L.)] durch ein parasiten der gattung *Trichogramma*. Entomophaga 5: 237-59.

Stinner, R. E. 1977. Efficacy of inundative releases. Annu. Rev. Entomol. 22: 515-31.

Stinner, R. E., R. L. Ridgway, and R. E. Kinzer. 1974. Storage, manipulation of emergence and estimation of numbers of *Trichogramma pretiosum*. Environ. Entomol. 3: 505-7.

Summers, T. E., M. G. Bell, and F. D. Bennett. 1971. *Galleria mellonella*: An alternate laboratory host for mass rearing tachinid parasites of sugarcane borers, *Diatraeae* spp. Proc. Am. Soc. Sugarcane Tech. 1: 63-54.

Teran, F. O. 1976. Perspectivas do controle biologico da broca da cana-de-acucar. Boletin Tecnico Copersucar. 2: 5-9.

Thorpe, W. H., and F. G. W. Jones. 1937. Olfactory conditioning in a parsitic insect and its relation to the problem of host selection. Proc. R. Soc. London (Ser. B). 124: 56-81.

Voegele, J., J. Daumel, Ph. Brun, and J. Onillon. 1974. Action du traitement au froid aux ultraviolets de l'ceuf d'*Ephestia kuehniella* (Pyralidae) sur le touh de multiplication de *Trichogramma evanescens* et *T. brasiliensis* (Hymenoptera: Trichogrammatidae). Entomophaga 19: 341-8.

Woets, J. 1974. Integrated control in vegetables under glass in the Netherlands *in* Proc. of the international organization of biological control conference on intergrated control in glasshouses, Littlehampton, England. 73 p.

# CHAPTER 7

QUALITY ASPECTS OF MASS-REARED INSECTS

E. F. Boller and D. L. Chambers

Swiss Federal Research Station, CH-8820 Wädenswil,
Switzerland, and Insect Attractants, Behavior and Basic
Biology Research Laboratory, Agricultural Research Service
USDA, Gainesville, Florida 32604 U.S.A.

Control of the quality of mass-produced industrial goods is
a matter of survival for any industrial enterprise that has to
compete on the market. A constant survey of the market situation
provides feedback to the factory where quality, design, or price
levels have to be adjusted in order for the company to remain in
business. The impression that quality control of mass-produced
insects has received only scant attention until recently indicates
that these basic rules of the marketplace apparently do not apply
to this special category of mass production. That those in charge
of insect rearing may lack this concern of industrial product
management may be caused by several factors. First, in most cases
there is no market and there is no producer-competitor-customer
situation enforcing a constant improvement of the end product,
because mass-reared insects are usually produced and used by the
same institution--often by the same individuals. Secondly, the
definition, monitoring, and manipulation of quality in mass-reared
insects is a relatively difficult matter; and it is heavily influenced by personal judgment and not least by a general lack of
knowledge with regard to the characteristics that enable the insect
to perform its intended role in pest control. This second aspect
has probably been the major block to the implementation of quality
control procedures in our insectaries. However, it is encouraging
to perceive that quality is receiving wider attention wherever
insects are reared in large numbers, and that an increasing number
of specialists and research institutions are devoting considerable
time and effort to developing and applying quality control concepts
and techniques to the particular organisms being mass produced.
Furthermore, we can recognize a positive trend toward increasing
interdisciplinary and international cooperation aiming toward the
development of general concepts and standards in quality control

that could be valid beyond the specific requirements of the individual insectaries.  This period of active research and development is generating a wealth of new ideas, hypotheses, models, and methods.  Some of these will endure and become part of a generally accepted concept and methodology.

It cannot be the aim of this paper to cover all the pertinent conceptual and technical aspects of quality control, and we refer to several comprehensive reviews and basic articles of recent origin that provide an introduction into the complex topic (Boller 1972, Chambers 1975, 1977, Huettel 1976, Mackauer 1972, 1976).  We can mention only briefly the importance of the contributions made by several disciplines of science that have a direct bearing on the understanding and solution of the quality problem.  These are, in particular, the fields of behavior, genetics, behavioral genetics, physiology, ecology, and bioengineering.

Simplicity and clarity of concepts and methods will be important criteria for successful implementation of quality control procedures despite the inherent complexity of this interdisciplinary operation. As applied entomologists at the ultimate level of application of the procedures in the laboratory and field we must accept certain short-cuts and simplifications to achieve progress and to establish feasible methods for day-to-day application.  This paper tries to put emphasis on the applied aspects of quality control that might be valid beyond the specific needs of individual rearing programs. Most of the thoughts expressed in this contribution derive from experience gained in research carried out on agricultural pests, but we shall refer to the situation in entomophagous species wherever feasible.

## DEFINITIONS

*Mass rearing* has been defined recently by Chambers (1977), reflecting the terminology of several other authors, as follows: "Mass rearing is the production of insects competent to achieve program goals with an acceptable cost/benefit ratio and in numbers exceeding ten-thousand to one-million times the mean productivity of the native population female."

*Quality control* is in more critical need of definition.  Again, Chambers' interpretation is adopted, as it analyzes and describes the complex nature of the term quality:  "*Quality* has, by dictionary definition, three components--skill, relativity, and reference. That is, quality is the degree of excellence in some trait(s) or skill(s) relative to a reference.  Thus, skills are the performance requirements for achieving the objective; relativity has to do with ranking (comparing) the degree of excellence of the performance of the skills; and the references are the standards, needs and con-

straints against which the skills are compared.  The definition of
*control* includes the notion of checking, verification, comparison,
regulation, periodicity, and restraint.  In the context with
quality there are also implications of input and feedback mechanisms, decision-making, and the development of protocols for the
imposition of control upon quality" (Chambers 1977).


## THE STEPS IN QUALITY CONTROL

The chronological steps in the development of a quality
control program—namely objectives, standards, monitoring and
action—are logical and expected to be recognized as major elements
in quality control.


### Objectives

The purposes of mass rearing insects are manifold and range
from the production of pathogens to the inundative release of the
reared insects against a target population.  The objectives to be
achieved by the organisms will provide the general framework for
the definition of those characteristics that must be present for
performing the intended role.  It is generally assumed that the
strictest quality requirements must be anticipated in those cases
where the mass-reared insects must have an immediate impact on the
target population either as predators and parasites (in interspecific
action) or as carriers of a genetic load (intraspecific action).
The nature of the intended impact on the target is *the* important
criterion for the definition of standards.  For further discussion
of this aspect we refer to the literature (Boller 1972, Chambers
1975, 1977, Huettel 1976, Mackauer 1972, 1976, Vinson 1976).


### Standards

The standards provide the yardstick for measuring quality,
and they may be considered as the very basis for the development
and application of quality control procedures.  It is therefore
surprising that this crucial element of any comparative investigation of quality components is often obscure and vague.  Standards
should, by definition, be described as precisely as possible, including not only average performance levels, but also their toler-
able band-widths, in order to indicate the acceptable plasticity
or required rigidity of the trait.  Vague definition of standards
is partly caused by our lack of essential background information
about the organism, especially with respect to its behavior,
ecology and genetics.  This general lack of comprehensive knowledge
suggests that the definition of standards in absolute terms (as is
applied in industrial production) is not yet feasible and therefore

its expectation may be unrealistic. One possible solution to this problem is the proper selection of a standard or reference strain against which our mass-reared organisms are compared. Where laboratory-reared insects are released in a genetic control program or in other intraspecific operations against their wild counterparts, it has become a widely accepted custom to take the wild target population as the best available standard.

This situation may differ for entomophagous species used exclusively in interspecific operations. There, one might apply the original concept of establishing standards based on theoretical considerations or empirical knowledge about the desirable attributes required to achieve the given objective. Off-hand, one might conclude that standards for entomophagous insects should be more flexible in principle than those for insects used in autocidal programs (Vinson 1976). This seems to be indeed the case. DeBach (1965) reached the conclusion that in 225 cases of successful or partly successful biological control projects there were no out-standing peculiar or common performance characteristics. Obviously, the biotic mortality agents were well adapted to their environment and usually highly specific to their insect host. It is generally considered that they must have a high searching ability, i.e., capacity to find hosts when these are scarce. Complete precoloni-zation adaptation to the new environment seems to be the rule (Messenger and van den Bosch 1971). Several authors (e.g., Doutt 1964, Mackauer 1976, Ridgway et al. 1974) mention the following essential characteristics of successful parasitoids: ability to find the host; the capacity to increase in numbers as fast or faster than the pest; a high degree of specificity to and synchron-ization with the pest; and adaptability to varying environmental conditions. For discussion of the specific aspect of host selection by parasites we refer to the recent review by Vinson (1976).

Thus, Mackauer and van den Bosch (1973) and Mackauer (1976) conclude that there may not exist precise and consistent attributes that would enable the identification of biotic agents suitable for control in a particular situation. But without knowing the neces-sary performance attributes and the conditions of the specific biological control program it is difficult to establish general standards and to develop procedures for monitoring parasitoid quality. At present it seems possible to do no more than monitor conditions that can be expected to lead to undesirable genetic changes or general genetic decay in laboratory populations.

In general the widely used concept of the *internal standard* (Chambers 1975, 1977; Huettel 1976) can be applied wherever quality control is a matter of the detection of deleterious effects caused by intrinsic and extrinsic factors such as disease, diet, marking, chilling, sterilization, handling, etc. In these cases untreated samples of the laboratory strain are used as an internal standard.

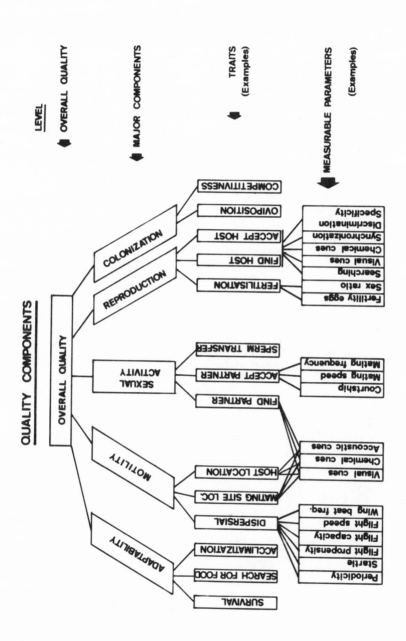

Fig. 1. Schematic presentation of a possible hierarchy of quality components with examples of second and third order interrelationships.

What, specifically, are these standards and how can they be established?  Again, we have to point out that the objectives of a program will largely determine the complexity of the standards. Furthermore, it becomes more and more evident that there is no one quality of an insect population but a quality that is composed of a multitude of components.  These components have confused the issue in many discussions, and it seems to be necessary to rank the various bits and pieces of quality in a hierarchy.  A model of such a quality hierarchy is shown in Fig. 1.

The *overall quality* (Huettel 1976) of a laboratory population is measured in terms of how well it functions in its intended role; that is, how effective it will be on the target after it has been released.  Suboptimum success in a release program may be indicative of lowered insect quality, but it tells us nothing about the real causes of failure.  It is certainly true that a successful field program is an indication of logistically adequate insect quality, but there is need to develop and to apply test procedures that assess quantitatively the overall quality at the very top of the hierarchy shown in Fig. 1.  An assessment of integral field performance may seem adequate as long as no major problems are in sight.  The limitations of this approach become evident as soon as problems start to appear.  These problems might arise very rapidly because apparent success in the field may mask concealed problems or gradual deterioration of quality.  Then when unfavorable conditions cause a sudden breakdown of the performance of the insect and explanations must become available for the causes, it will become evident that descriptive overall quality measurement lacks the analytical power to uncover the causes of the failure.  Thus, quality has to be divided into components amenable to analysis.

One such division is shown in Fig. 1 where quality is divided into five *major components* covering the aspects of adaptability, motility, sexual activities (courtship and mating), reproduction, and colonization.  This division (or similar attempts to divide quality into major components) is necessary to sort out those components that might be of major importance for the observed failure or are known to be especially sensitive to alteration. The selection of a few major quality components is a matter of personal judgment, but it is anticipated that a generally acceptable consensus will emerge in the near future.  The most difficult problem at this level is the determination of the relative impor- tance of the individual major components for needed program achieve- ment.  An example of such an attempt to weight the components in relation to the objective or purpose of mass rearing is given in Fig. 2.  Placement of the priorities depends largely on the charac- teristics of the individual program, and generalizations at this level will certainly have a limited value.  However, there is general agreement that in genetic control, aspects associated with

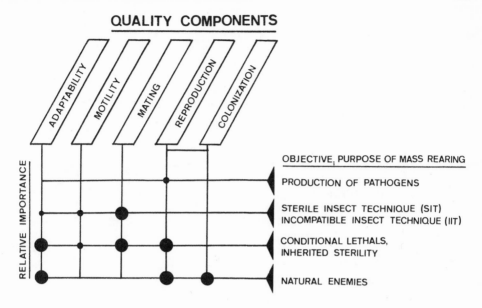

Fig. 2. Relative importance of major quality components according to the objectives of a rearing program.

sexual activities deserve special attention.  The development of isolating mechanisms (Dobzhansky 1937) between wild and laboratory-adapted strains is a major problem wherever the function of the released insect is interference with the wild population of the same species.  Entomophagous organisms have quite different re-quirements, and those characteristics connected with adaptability, reproduction and colonization might play the central roles in a quality control program.  The evolution of reproductively isolated laboratory strains of parasitoids with altered biological charac-teristics might not necessarily cause the failure of a control program (as it would in genetic control) and could even have advantageous aspects.  As Doutt and DeBach (1964) point out, it is the behavior of the individual entomophagous insect that deter-mines its usefulness in biological control.  They concluded accordingly that if any species has a strain or race exhibiting advantageous characters, that strain should be introduced.

The opinions about the overall importance of motility for program achievement are less pronounced.  Several authors (e.g., Coluzzi 1971, Boller 1972, Bush et al. 1976) stress the importance of normal dispersive behavior, but it is also proposed (e.g., Chambers 1977) that deteriorated flight abilities of laboratory-reared insects might be compensated for, at least to a certain

extent, by logistic procedures. This is unlikely where released
insects must exhibit cruising ranges of several kilometers but
might not be a problem at all with very small parasites that can
be distributed adequately in the immediate vicinity of suitable
hosts (DeBach and Bartlett 1964). Certain quality components are
irrelevant for the establishment of standards (e.g., colonizing
properties in the field in releases of sterile insects in an
autocidal program).

These few major components are of no great help in the analysis
of problems unless they are subdivided into *individual quality
traits* containing measurable parameters. This level of the hierarchy shows by necessity a great variety of items, as many traits
will reflect the peculiarities of the species under consideration.
Certain traits will find broader application, and some of these
are listed in Fig. 1. Again, this check list of quality traits
should be made as complete as possible in the planning stage of
quality control and then those singled out that might have a major
impact on overall performance.

The actual design of the quality tests will call for an even
further division of the traits into *individual parameters* amenable
to direct measurement. One example is the subdivision of the
trait embracing the activities involved in the location of the
mating site, the sexual partner, or the host. Because both physical
and chemical stimuli might be involved in directing the insect to
the proper site (such as color, odor, shape, or sound) different
techniques are required to measure and handle visual, olfactory
and acoustical processes. This is the final level where a variety
of techniques is to be developed for measuring and monitoring
quality, where a wealth of data will be produced, and where it must
be determined what techniques will produce information relevant for
the events in the field.

The relatively complex structure of quality we have presented
brings us back to the statement in the introduction that quality
control procedures should be relatively simple in order to be
applied widely and routinely. This breakdown of quality into
innumerable parameters might indeed lead to the wrong conclusion,
i.e., that quality control is so complex and sophisticated that it
becomes the privilege (or pleasure) of a few specialists. This is
not the case. Thanks to the rapidly accumulating experience,
improved and simplified methods and devices, and not the least to
the services provided by specialized facilities, entomologists
should soon be able to analyze their problems and select or develop
the proper key elements for biotests that monitor the quality of
the insects they have to produce.

## Monitoring

Figure 3 shows schematically the steps and activities occurring in quality control.  The actual assessment of quality follows basic patterns that can be observed in nearly any manufacturing process.  One activity deals with the testing of the end product by both the manufacturer and consumer.  The other activity concerns the internal monitoring of the factory production line.

*Testing the performance* of mass-reared organisms prior to and sometimes after their release is the principal activity of a quality control program that can provide input and feedback data for curative action.  The routine monitoring of essential traits should provide an early warning that latent problems are at hand whenever significant deviations from the adopted standard are observed.  These early indications of deteriorating quality components should set in motion more elaborate investigations in the field and in the laboratory, aimed at a proper analysis and correction of the particular problem.  The causes of significant changes have to be found by comparing the data of performance tests with those derived from the second type of monitoring, the *internal monitoring of the production line*.  Only the combination of insight gained by monitoring both performance and production will provide the necessary information needed to plan and carry out effective curative action.

## MONITORING

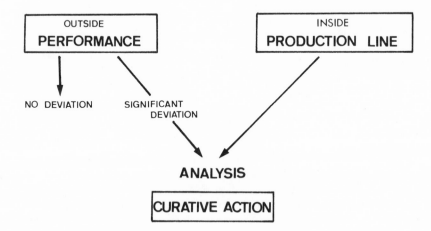

Fig. 3.  Schematic presentation of the monitoring activities in a quality control program leading to curative action.

## Monitoring Performance

It is beyond the scope of this paper to mention and to discuss
the technical details of all methods that have been developed to
measure specific quality traits.  We refer to a recent publication
by Huettel (1976) for a more complete coverage and discussion of
methodologies.

The question still arises whether laboratory tests for monitoring insect quality can adequately assess traits that have to
function under field conditions.  Criticism focuses in particular
on the validity of laboratory tests that measure behavior.  There
is now, however, a general consensus that both field and laboratory
tests should be developed that complement each other because it
has become obvious that both approaches have their definite advantages and shortcomings.

Field tests or tests in larger field cages have the advantage
that certain components such as dispersal, orientation mechanisms,
survival rates and production of viable offspring can be studied
under the direct influence of the complex environment of the target
area.  However, the traits amenable to direct field assessment are
limited, and several major components such as sexual activity,
especially mating frequency, or the early detection of gradual
changes in threshold values are very difficult to cope with.  Unless
carried out in a highly effective manner, field experiments often
produce data that are influenced by a multitude of variables that
cannot be identified precisely; hence, cause and effect are obscured
and analysis is difficult (Parker and Pinnell 1974).  Therefore,
it is our personal opinion that these cost- and labor-intensive
field tests are the logical followup to preliminary laboratory
experiments that provide the first indications of when and in what
direction research has to take in the field.

Testing Overall Quality.  Two tests have been designed to study
the integrated action of various quality components.  One of these
is the widely used ratio or "competitiveness" test (Steiner and
Christenson 1956, Fried 1971).  This test measures the effect of
sterilization on the fertility of eggs produced by a mixed population of fertile and sterile insects.  It can be carried out in the
laboratory or in field cages and can be considered a precursor of
the final field performance test.  The advantage is a defined experimental setup; the weakness is again the lack of analytical power
since deviations from the expected fertility rates usually have
multiple causation such as differential mating activity of fertile
and sterile insects, differential fecundity, or potential distortions
of the results caused by sperm precedence.

The second test, similar but geared to the comparison of two
fertile strains, is the "genetic integration technique" (Huettel

1976). The overall quality of laboratory-reared insects is assessed by monitoring the success of fertile insects released into a wild population. With morphological or genetic markers such as polymorphic enzymes, samples of $F_1$ progeny can be tested, and the recovery of the marker in the expected ratio indicates that the reared insects compete successfully with their wild counterparts.

Both methods suffer from the shortcoming that the interpretation of significant deviations from the expected values depends largely on input from additional tests that measure individual traits. These difficulties have also stimulated the search for sensitive laboratory tests.

The Variation Problem. Since we are dealing with phenotypes and phenotypic expressions of quality traits, one realizes that observed variation in data collected from a large number of individuals is the product of genotypic or inherited variation and environmental influences. If the performance of two strains has to be compared or the impact of certain treatments has to be assessed by comparison with an internal standard, it becomes a necessity to reduce the experimental variables to the greatest possible extent. Ideally, the comparison should be done with two identical strains that differ only in the characteristic under investigation. This can never be achieved because of the inherent variability of our test material. Therefore, in work with phenotypes exhibiting genetic and environmental components of variability, it is of prime interest to identify the environmental factors influencing our data, and to reduce the overall variation by standardization or elimination of those influences that can be manipulated. An example of such a reduction of variation has been described by Remund et al. (1976) for flight data obtained with the olive fly, *Dacus oleae* (Gmelin). The flies were flown to exhaustion on a multirotor flight mill in the laboratory. After identification and standardization of the variables—sex, age, body size, temperature, and food (energy level) —flight data were obtained that fluctuated within a greatly reduced range around the population mean, a fluctuation probably caused to a large extent by the inherent variability of the genotype and by a residual environmental variation caused by unidentified factors. Therein lies one of the major advantages of laboratory tests: They can be carried out under defined experimental conditions that allow the detection of details that would be overlooked in the field. The importance of the heritability of behavioral traits and of the methodologies of behavioral genetics can only be mentioned in this context. Several comprehensive textbooks (such as Hirsch 1967, Parsons 1967, Van Abeelen 1974) will provide a stimulating introduction to this topic.

The Laboratory Performance Test. What does a comparative laboratory test tell us? What does it mean when we observe that our laboratory strain shows significant deviations from the wild stan-

dard?  First, we can state that the two strains produced significantly different results when tested under identical experimental
conditions.  Nothing can be said *a priori* about the relevance of the
measured indices and the significance of the observed differences to
field performance.  However, we have an indication that our lab
strain has apparently undergone certain modifications during the
rearing process that have altered its performance patterns or threshold values.  Such data should serve as warning that changes are
occurring.  Then by comparing consecutive generations, one should be
able to assess speed and intensity of this change.  Evidence is
accumulating that adaptation to laboratory conditions occurs relatively fast and tends to reach a plateau after some 5-7 generations
(Boller 1972, Bush 1975, Herzog and Phillips 1974, Leppla et al.
1976, Raulston 1975, Rössler 1975).  Secondly, the warning signal
should stimulate additional studies, especially in the field, to
determine the relevance of the observed deviations to field performance.  Examples are increasing wherein deficiencies detected in the
laboratory were confirmed in subsequent field studies (Lewis et al.
1976, Remund et al. 1976, Stinner et al. 1974).

    One particular case of an unexpected discovery was an apparent
increase of mating activity observed in the laboratory when European
cherry fruit flies, *Rhagoletis cerasi* L., were topically marked with
a yellow fluorescent spray.  Field cage tests carried out to study
this phenomenon under outdoor conditions not only confirmed the
tendency observed in the laboratory but showed this stimulation to
be even more pronounced than anticipated (Boller et al. unpublished
data).  The observation could never have been made in the field that
mating speed and mating frequency of *R. cerasi* are correlated and
that the portions of the population that are fast, intermediate,
and slow in mating is apparently well balanced in nature.

    It will be the challenging task of those individuals responsible
for the quality aspects of insects to review the existing techniques
and to sort out those that might be applicable to their particular
species, to modify them according to local needs, possibilities,
and limitations, and to participate in the international effort to
pool and disseminate key information and experience for the benefit
of all concerned.


## Monitoring Production

    The second set of data having a bearing on quality control is
derived from the internal monitoring of the production line.  Life-
table studies have been of great value for the identification and
quantification of bottlenecks in the life cycle of laboratory
colonies (Hathaway et al. 1973, Leppla et al. 1976, Tamaki et al.
1972).  Briefly, life-table records (Deevey 1947, Southwood 1966)
are obtained by sampling a defined number of adults from a given

laboratory generation and following their development, measuring
such values as:  survival, fecundity, fertility of the eggs, larval
and pupal development, emergence rates, and sex ratio of the new
adults.  Observations can be added *ad libitum*, but important information is lost when the protocol is reduced without urgent reasons.

Some crucial events during adaptation to the laboratory environment occur very early as when wild adults are brought to the
laboratory and forced to reproduce under artificial conditions.
This situation is shown in a simplified manner in Fig. 4.

Two *major bottlenecks* often occur when all life stages are
reared exclusively on artificial substrates: one that occurs in the
adult stage reduces drastically the reproducing portion of the
founder population.  The second one occurs during the larval stage,
and its effect is often more conspicuous because the pupal yield in
the early phase of mass rearing on artificial diets is usually very
low.  The impact of selection during the early stages of colonization may have more subtle effects on the quality as defined by the
standard.  Mortality occurring during the larval stage will eliminate those individuals that cannot survive on the new diet and
selection will favor those exhibiting the appropriate physiological
disposition.  Its effect on important behavioral traits will not
*appear* to be dramatic, as it will influence mostly characteristics
that rank very low in the quality hierarchy outlined in Fig. 1.  The
impact of bottlenecks that affect the adult stage are more severe,
as such selection acts directly on those characteristics of adult
behavior that have an immediate relationship with performance
quality.  Of special concern are major quality components such as
sexual and reproductive behavior and, to a certain extent, motility.

It is surprising that records dating back to the first introduction of wild material to the rearing facility are often not
available in those laboratories that have gradually built a major
rearing program from a smaller founder population.  The production
figures available are often highly impressive, indication that the
laboratory strain has reached the upper limits of reproductive
capacity.  There is no doubt that the removal of natural regulating
factors is a benefit for the laboratory colony, but this very fact
too often obscures the tremendous loss suffered at the very beginning of laboratory propagation.  It seems that the problem of
deteriorating quality could be tackled with greater success if
attention were directed to the removal of the first bottleneck in
the adult stage.  Sporadic introduction of wild material into
existing colonies probably is of little value as such material
stands a good chance of being eliminated by the same selective
forces that affected the colony before.  Such considerations might
be helpful where mass rearing is still in the planning phase or has
retained a certain flexibility.  We realize that it is extremely
hard to change rearing procedures once the rearing plant has reached

**STAGE**

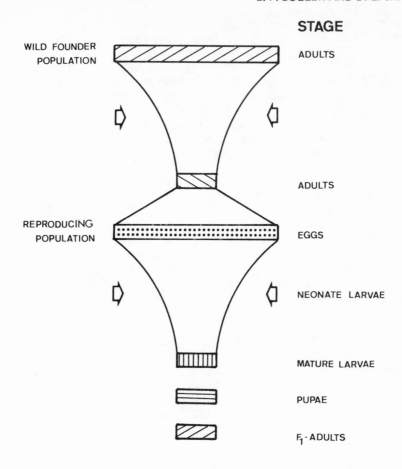

Fig. 4.  Two of the major bottlenecks acting upon wild insects
transferred to artificial laboratory conditions for propagation.


dimensions where even minor modifications in the established routine
create major logistic or organizational problems.

      Another method of monitoring production line continuity, the
allozyme technique, has received increased attention.  Loss of
variation at specific genetic loci can be assessed by electrophoretic separation and identification of polymorphic enzymes of individual insects sampled from the test population (Bush 1975, Bush
and Huettel 1976, Bush et al. 1976, Wagner and Selander 1974).  The
advantages of such a technique include the objectivity of a chemical
method, a relatively simple procedure that can be carried out at
moderate cost, and the fact that frozen specimens can be stored and
shipped to any destination where such equipment and service are

available.  However, no method is without disadvantages.  The
allozyme technique suffers from the same shortcomings as other tests
measuring overall quality.  It can provide a description of the
*status quo* but has no direct analytical power.  Progress has been
made in correlating changes at the enzyme level with changes in
certain quality traits (e.g., enzymes that are important in flight
metabolism, Bush et al. 1976).  But until the data provided by
allozyme analysis can be related directly to specific quality traits,
the technique will remain what it is at the moment--a very valuable
auxiliary tool in the monitoring of genetic changes occurring in
insect colonies.


## CONCLUSIONS

Quality control has made substantial progress within the last
few years, and the emphasis of discussions on the topic has shifted
from the question of whether quality control is necessary at all to
the problem of the optimum implementation of quality control procedures in on-going rearing programs.  Despite a fruitful collabora-
tion between laboratories that devote considerable time and effort
to research aimed at a better understanding of the basic and applied
aspects of insect quality, we have not yet reached the point where
widely applicable principles or a manual of available techniques are
at hand.  However, it can be anticipated that such mutual efforts
will continue to increase and to produce results.

Quality control research combines the interests of specialists
involved in basic research at the academic level with those of
applied field workers that know the problems at the farm and laboratory levels.  There is excellent opportunity for a multidisciplin-
ary approach involving all levels of research and application.
Results produced would not just add more bulk to the immense professional literature but could lead to tangible action.  One step in
this direction could be planned workshops on quality control problems
at the location of major rearing plants now in operation.  These
could provide the necessary direct contact between individuals
involved in basic and applied research, including those that have to
carry the burden and responsibility for front line action.


## REFERENCES CITED

Boller, E.  1972.  Behavioral aspects of mass-rearing of insects.
    Entomophaga 17: 9-25.
Bush, G. L.  1975.  Genetic variation in natural insect populations
    and its bearing on mass-rearing programmes.  p. 9-17 In *Con-
    trolling Fruit Flies by the Sterile-Insect Technique*. Vienna:
    IAEA/FAO Panel Proc. Ser.

Bush, G. L., and M. D. Huettel. 1976. Population and ecological
    genetics. p. 43-9 In V. L. Delucci, ed. *Studies in Biological
    Control*. International Biological Programme, 9. Cambridge
    University Press, Cambridge, England.
Bush, G. L., R. W. Neck, and G. B. Kitto. 1976. Screwworm eradi-
    cation: Inadvertent selection for noncompetitive ecotypes during
    mass rearing. Science 193: 491-3.
Chambers, D. L. 1975. Quality in mass-produced insects. p. 19-32
    In *Controlling Fruit Flies by the Sterile-Insect Technique*.
    Vienna: IAEA/FAO Panel Proc. Ser.
Chambers, D. L. 1977. Quality control in mass rearing. Annu. Rev.
    Entomol. 22: 289-308.
Coluzzi, M. 1971. Problémes théoriques et pratiques liés à l'
    élevage et à la production de masse des Culicides. Ann.
    Parasitol. 46: 91-101.
DeBach, P. 1965. Some biological and ecological phenomena associ-
    ated with colonizing entomophagous insects. p. 287-303 In
    E. G. Baker and G. L. Stebbins, eds. *The Genetics of Colonizing
    Species*. Academic Press, N.Y. 588 p.
DeBach, P., and B. R. Bartlett. 1964. Methods of colonization,
    recovery and evaluation. p. 402-426 In P. DeBach, ed. *Bio-
    logical Control of Insect Pests and Weeds*. Chapman and Hall,
    London.
Deevey, E. S. 1947. Life tables for natural populations of animals.
    Quart. Rev. Biol. 22: 283-314.
Dobzhansky, T. 1937. *Genetics and the Origin of Species*. Columbia
    Univ. Press. 364 pp.
Doutt, R. L. 1964. Biological characteristics of entomophagous
    adults. p. 145-67 In P. DeBach, ed. *Biological Control of
    Insect Pests and Weeds*. Chapman & Hall, London.
Doutt, R. L., and P. DeBach. 1964. Some biological control concepts
    and questions. Ibid. p. 118-42.
Fried, M. 1971. Determination of sterile-insect competitiveness.
    J. Econ. Entomol. 64: 869-72.
Hathaway, D. O., L. V. Lydin, B. A. Butt, and L. J. Morton. 1973.
    Monitoring mass rearing of the codling moth. J. Econ. Entomol.
    66: 390-3.
Herzog, G. A., and J. R. Phillips. 1974. Selection for a non-
    diapause strain of the bollworm, *Heliothis zea* (Lepidoptera:
    Noctuidae). Environ. Entomol. 3: 525-7.
Hirsch, J. (ed.) 1967. *Behavior-Genetic Analysis*. McGraw-Hill,
    N.Y. 522 pp.
Huettel, M. D. 1976. Monitoring the quality of laboratory-reared
    insects: A biological and behavioral perspective. Environ.
    Entomol. 5: 807-14.
Leppla, N. C., M. D. Huettel, D. L. Chambers, and W. K. Turner.
    1976. Comparative life history and respiratory activity of
    "wild" and colonized Caribbean fruit flies. Entomophaga 21:
    353-7.

Lewis, W. J., Donald A. Nordlund, H. R. Gross, Jr., W. D. Perkins,
E. F. Knipling, and J. Voegele. 1976. Production and performance of *Trichogramma* reared on eggs of *Heliothis zea* and
other hosts. Environ. Entomol. 5: 449-52.

Mackauer, M. 1972. Genetic aspects of insect production. Entomophaga 17: 27-48.

Mackauer, M. 1976. Genetic problems in the production of biological
control agents. Annu. Rev. Entomol. 21: 369-85.

Mackauer, M., and R. van den Bosch. 1973. General applicability of
evaluation results. J. Appl. Ecol. 10: 330-5.

Messenger, P. S., and R. van den Bosch. 1971. The adaptability of
introduced biological control agents. p. 68-92 In C. B.
Huffaker, ed. *Biological Control*. Plenum Press, N.Y.

Parker, F. D., and R. E. Pinnell. 1974. Effectiveness of *Trichogramma* spp. in parasitizing eggs of *Pieris rapae* and *Trichoplusia ni* in the laboratory. Environ. Entomol. 3: 935-8.

Parsons, P. A. 1967. *The Genetic Analysis of Behaviour*. Methuen's
Monographs on Biological Subjects. Methuen, London. 174 pp.

Raulston, J. R. 1975. Tobacco budworm: Observations on the laboratory adaptation of a wild strain. Ann. Entomol. Soc. Am. 68:
139-42.

Remund, U., E. F. Boller, A. P. Economopoulos, and J. A. Tsitsipis.
1976. Flight performance of *Dacus oleae* reared on olives and
artificial diet. Z. ang. Entomol. 81:

Ridgway, R. L., R. E. Kinzer, and R. K. Morrison. 1974. Production
and supplemental releases of parasites and predators for control
of insect and spider mite pests of crops. Proc. Summer Institute on Biological Control of Plant Insects and Diseases. Univ.
Press, Jackson, MS. p. 110-116.

Rössler, Y. 1975. Reproductive differences between laboratory-
reared and field-collected populations of the Mediterranean
fruit fly *Ceratitis capitata*. Ann. Entomol. Soc. Am. 68:
987-91.

Steiner, L. F., and L. D. Christenson. 1956. Potential usefulness
of the sterile fly release method in fruit fly eradication
programs. Proc. Hawaii Acad. Sci. 3: 17-8.

Stinner, R. E., R. L. Ridgway, and R. K. Morrison. 1974. Longevity,
fecundity, and searching ability of *Trichogramma pretiosum*
reared by three methods. Environ. Entomol. 3: 558-9.

Southwood, T. R. E. 1966. *Ecological Methods*. Methuen, London.
391 pp.

Tamaki, G., J. E. Turner, and R. L. Wallis. 1972. Life tables for
evaluating the rearing of the zebra caterpillar. J. Econ.
Entomol. 65: 1024-7.

Van Abeelen, J. H. F. 1974. The genetics of behavior. *Frontiers
of Biology*. Vol. 38, North Holland Publ. Co., Amsterdam. 450 pp.

Vinson, S. B. 1976. Host selection by insect parasitoids. Annu.
Rev. Entomol. 21: 109-33.

Wagner, R. P., and R. K. Selander. 1974. Isozymes in insects and
their significance. Annu. Rev. Entomol. 19: 117-38.

CHAPTER 8

BEHAVIORAL CHEMICALS IN THE AUGMENTATION

OF NATURAL ENEMIES

S. Bradleigh Vinson

Texas A&M University
Department of Entomology
College Station, Texas 77843    U.S.A.

The behavioral manipulation of natural enemies of pest insects
has been a long standing dream among the biological control practi-
tioners.  Recent work concerning the behavior of some natural
enemies of pest insects suggests that this dream may become a
practical reality.  Wright (1964, 1965) recognized the potential
for pest management by the behavioral modification of insects
through the use of non-toxic agents that modify the insects' behavior.  As stated by Wright (1965) these chemicals may effect
control of a pest by inhibiting a correct response or eliciting
an incorrect one.  He coined the term "metarchon" for the introduced
stimulus which includes chemical repellents and attractants as
well as physical factors such as light and sound which interfere
with the insects ability to communicate.  While much effort is
underway to develop methods of manipulating the pest insect,
a different approach is needed for the manipulation of natural
enemies.  Instead of attempting to elicit an incorrect response
or inhibit a correct one, the goal of entomophagous insect manipulation is to redirect or stimulate the response.  The behavioral
modification of certain arthropod enemies of insects through
the use of chemicals has opened up new opportunities for the
manipulation of these arthropods for the benefit of man.  However,
such an approach requires a working understanding of the behavior
of entomophagous arthropods and the role played by chemicals
in their behavioral patterns.

Entomophagous arthropods can be divided into 3 major groups,
the parasites, parasitoids and predators.  My comments will be
restricted to the use of parasitoids and predators in the management
of insect pests of agronomic and veterinary or medical importance.
Much of the work discussed is derived from the parasitoid Hymenoptera

although many of the concepts are valid for the parasitoid Diptera
and some predators.

A great deal has been written concerning the behavior of parasitoid Hymenoptera in recent years (Matthews 1974, Vinson 1975a,
1976, Jones et al. 1976, Lewis et al. 1976) and a detailed review
will not be attempted here. But before the potential of behavioral
chemicals for use in the augmentation of parasitoids and predators
can be determined, the biology and behavior of the entomophagous
species in the context of the host selection process must be understood. Such an analysis can pinpoint those areas where chemicals
may play a role in the behavior of the entomophagous-host relationship
and may give some clues to the potential of the compound for
use in a pest management program. This analysis is also necessary
for the development of specific bioassays that are essential
for the isolation and identification of behavioral chemicals.


## HOST SELECTION STRATEGY

There are several strategies that are employed for host
location and selection by different parasitoid and predator species.
The strategies that have evolved appear to be related in part
to the specificity of the parasitoid-host or predator-prey association. In the more opportunistic relationships, host encounter
may be only random (Fig. 1a) while in the more restricted relationships host encounter may be directed more by the innate behavior
of the predator or parasitoid in response to various stimuli
eminating from the host or the host's environment.

Some authors consider that host encounters are random (Cushman
1926, Clausen 1940, Fleschner 1950, Rogers 1972). Indeed, even
those organisms that appear to be directed to the host from some
distance by orientating to stimuli produced by or associated
with the host presumably must initially perceive the stimuli
by random encounter. In fact, Salt (1934) and Ullyett (1947)
suggested that the searching behavior of most parasitoids was
somewhere between that of a random search and a systematic one.
However, the evolution of host selection based entirely upon
random encounter is probably rare. Also, there would appear
to be little potential for the chemical manipulation of those
parasitoids or predators where host selection is entirely random.

At the other extreme is the concept that the parasitoid
or predator is led directly to the host by a single stimulus
(Fig. 1b). Cues eminating from a host can be important in orientating
a parasitoid to a distant host. These cues can act as a "trail"
or "beacon" directing a parasitoid to its host. In such parasitoid-
host relationships the parasitoid may locate the "trails" by random
encounter and then is directed to the host. There are several

examples where only a single stimulus appears capable of directing
a parasitoid to its host.  Sound produced by the cricket, *Gryllus
integer*, has been found to be important in attracting its habitual
tachinid parasitoid, *Euphasiopteryx ochracea* (Cade 1975).  Sound
also has been reported to direct a sarcophagid parasitoid to its
cicada host (Soper et al. 1976) although sight appeared to play
a role in host selection once the parasitoid was near the potential
host.

     In other cases the trails may be due to air-borne odors and
are therefore less directional than host sounds.  These odor trails
may act to orient the parasitoid to its host similar to the sex
pheromones orienting one mate to the other (Farkas and Shorey
1974).  In fact, there are several reports where the pheromone
communication system between individuals of a species also orient
and attract some of their parasitoids (Rice 1969, Mitchell and
Man 1970, Sternlicht 1973).

     Such trails may be ground based as are the terrestrial pheromone
trails used by ants and termites (Blum 1974, Moore 1974).  These
trails can serve to lead a potential parasitoid or predator to
the host or host nest.  Doutt (1957) found that *Solenotus begini*
(Ashmead) will follow the serpentine leaf mine of her agromyzid
host, *Phytomyza atricornis* Meigen.  As she progresses along the
length of the mine, the female parasitoid swings her antennae
back and forth across the mine in search of the larva and will
do so even in the absence of a host in the mine.  Weseloh (1977)
has found that *Apanteles melanoscelus* (Ratzeberg), a larval parasitoid
of the gypsy moth, *Lymantria dispar* L., responds to the silk or
webbing produced by its host and thus may be capable of following
the silk to the host.  The habitual parasitoid of *Heliothis virescens*
(F), *Cardiochiles nigriceps* Vierick, will follow a trail of the
mandibular gland contents of its host (Vinson 1968).

     In contrast to the case for random host contact, the potential
exists for the manipulation of the parasitoid responding to a
single cue leading to the host.  However, such manipulation may
be counter-productive.  The application of a compound to an area
that orients a beneficial arthropod directly to its host may lead
to confusion.  It is just this concept which has been proposed
for the pheromonal control of pest insects.  The application to
and contamination of an area with a pheromone which leads or orien-
tates one sex to the other may lead to the disruption of the insect's
sexual behavior (Tetle 1974).  Thus, the application or release
of a stimulus that leads a parasitoid or predator to its host
(See Fig. 1b), could disrupt the beneficial's host or prey seeking
behavior.

     However, most parasitoids appear to locate their hosts through
a sequence of steps (Fig. 1c,d) rather than a completely random

encounter or by being led directly to a host by a single stimulus.
Based on the observations of Salt (1935) and Flanders (1953),
Doutt (1964) divided the process leading to host location and
oviposition into 3 steps:  1) host habitat location, 2) host location
and 3) host acceptance.  These three steps can be combined under
the broad heading of host selection (Vinson 1976).

While the division of host selection into a series of steps
is useful in terms of the sources and types of the stimuli involved,
and the type of behavior that is elicited, it still gives the
impression that the parastoid is led to the host (Fig. 1e).  This
may or may not be the case as described below.  Certainly some
parts of the host selection process of some parasitoids may involve
a sequence of cues.  These cues can follow one another, each leading
the parasitoid closer to the host.  For example, *C. nigriceps*
is first attracted to the host plant of its host.  Once in the
proper plant habitat females orient to the host plants and fly
from plant to plant, searching each in turn by flying 2-5 cm from
the stems, leaves and buds.  Females alight briefly and antennate

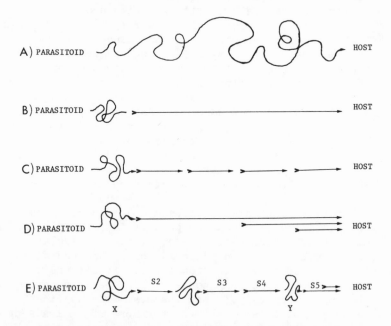

Figure 1.  Host selection strategy of insect parasitoids.
     a.  A completely random encounter.  b .  Random encounter of
a cue that directs or leads a parasitoid to a host.  c.  Random
encounter of a series of cues that progressively lead to a host.
d.  An initial encounter of a hierarchy of cues leading to a host.
e.  A random encounter of a series of cues with points of random
searching between cues.

damaged plant tissue and may then resume their search.  However,
if the damage was inflicted by a potential host, the female parasitoid crawls over the plant and rubs the surface with her antennae.
This latter behavior is elicited by a chemical deposited on the
plant by the feeding activity of the host (Vinson 1968).

Although, one cue may lead to another and bring the parasitoid
closer to the host at each step, any given cue may elicit the
proper response only in the presence of essential preceeding cues
or stimulus.  Thus the parasitoid may be led to a host through
a hierarchy of cues eminating from the host's immediate environment
(Fig. 1d).  Like the sex pheromonal system of insects (Shorey
1973), additional stimuli are perceived as the female parasitoid
approaches the host.  Different stimuli or different concentrations
of a single stimulus can be involved (Schwinch 1955, Traynier
1968).  Whether the female parasitoid responds to a series of
independent cues or a hierarchy of cues, each step serves to reduce
the distance between the parasitoid and host, thereby increasing
the potential for encounter.  The application of these stimuli
may have utility for entomophagous insect manipulation.  But,
like the single stimulus, the application of only one stimulus
in a chain could lead to confusion.

While some parasitoids are capable of following a trail leading
to a host, a continuous trail is not necessary.  In fact, the
occurence of continuous trail of a stimulus or series of stimuli
leading to a host in nature is probably uncommon.  Thus, a female
must be capable of following a non-continuous trail.  When a female
encounters one cue she searches for the next but within the context
of the present or last stimulus.  Some stimuli release an innate
searching behavioral pattern in certain parasitoid species.  Females
encountering a chemical contaminated surface will search not only
the contaminated surface but also adjacent areas.  A good example
is provided by *Microplitis croceipes* (Cresson) which, upon contacting
the frass of its host *Heliothis zea* (Boddie) placed on a leaf,
will crawl over the leaf while antennating the surface.  The female
may return to the frass or search the underside of the leaf and
adjacent stem before leaving (personal observation).  Lewis et
al. (1975b) reported that parasitism by *Trichogramma* and *M. croceipes*
was greater in areas adjacent to those treated with chemicals
isolated from the host.  Such phenomena support the view that
host and host-damage related chemicals act as a searching stimulant
as they release an innate host-searching behavior in the female.
Searching stimulants appear most promising for parasitoid manipulation
for control.

The division of the host selection process into host habitat
location, host location and host acceptance often gives the impression
that a parasitoid is led to its host through a series of cues,
one leading to the next.  Such a view appears to be supported

by examples such as described for *C. nigriceps* above, but such
views are often too simplistic. As shown in Fig. 1e, host selection
may involve a sequence of steps leading to a host but a certain
element of random searching often occurs at several points in
the process. Thus, a parasitoid is led to a certain point and
then must search for the next cue. At each step the search for
the next cue is more restricted than the preceding search. As
described earlier, once *C. nigriceps* reaches a habitat containing
a population of its host's food plants, the parasitoid will proceed
to search the plants. The plants actually searched appear to
be selected at random. However, this is a more restricted random
search, confined to the patch of plants and probably under the
influence of plant- or habitat-directed stimuli. The next cue
may elicit another random search of an even more restricted area.
In fact the mandibular secretion from *H. virescens* was described
as a searching stimulant (Vinson 1968).

Initially, parasitoids may be located in a habitat devoid
of hosts. For example, a female parasitoid may ecdyse to imago
in a habitat no longer suitable for hosts or the host population
may be at a density too low to maintain the parasitoid population.
Thus a female must find a new site for host exploitation. Since
many suitable host habitats are discontinuous, having a random
or clumped spatial dispersion (Southwood 1966), a female must
respond to those cues which are capable of orientating her at
a relatively long distance (note Fig. 1e stimuli S2 and S3).
Prior to contacting such long distance cues a female may randomly
fly, crawl or be carried down wind until encountering an orientating
stimulus (Fig. 1e "X"). These initial stimuli may eminate from
the host food plant, the host insect or organisms associated with
the host insect. These stimuli serve to orient the parasitoid
to the proper habitat where the female may then begin a random
search for the next sequential cue. However, the random search
is now much more restricted in space thus increasing the chance
of locating the next cue. One of the potentials of behavioral
chemicals is their ability to permit a female engaged in the initial
semi-random searching of a wide area to a semi-random searching
of a more restricted area where potential hosts may exist. Essentially, some behavioral chemicals cause a female parasitoid located
at point "x" to move to point "y" (Figure 1e), thus increasing
the chance for parasitism.


## ROLE OF CHEMICALS IN HOST OR PREY SELECTION

With regard to host selection, the various types of stimuli
that may elicit a behavioral response in a entomophagous arthropod
are given in Fig. 2. Although factors such as sound, color, electromagnetic radiation or shape have been found to be important in
host selection (Vinson 1976), they do not appear to show great

promise in the augmentation of entomophagous arthropods.  In
contrast, chemicals appear to play the major role in releasing
behavioral patterns involved in host selection, particularly
in the parasitoid Hymenoptera.  Furthermore, chemicals show
great promise in the augmentation of entomophagous arthropods.
The concept of using behavioral chemicals to augment natural
enemies has arisen from an increased understanding of the role
of behavioral chemicals in parasitoid behavior, particularly
their role in the process of host selection.

Chemicals involved in host selection and prey location
have been defined on the basis of their origin and the behavioral
response which they elicit (Brown et al. 1970).  A kairomone
is a chemical produced by an individual of one species that
releases a behavioral response in a second species and the second
species benefits from the response (Nordland and Lewis 1976).
An allomone is a chemical produced by an individual of one species
which elicits a behavioral response in a second species.  However,
the producer is benefited (Nordland and Lewis 1976).  Both allomones
and kairomones appear to have potential in the manipulation of
predators and parasitoids.  Pheromones are compounds released
by an individual that affect the behavior of other members of
the same species (Brown et al. 1970) and may also have a place
in the manipulation of beneficial insects.

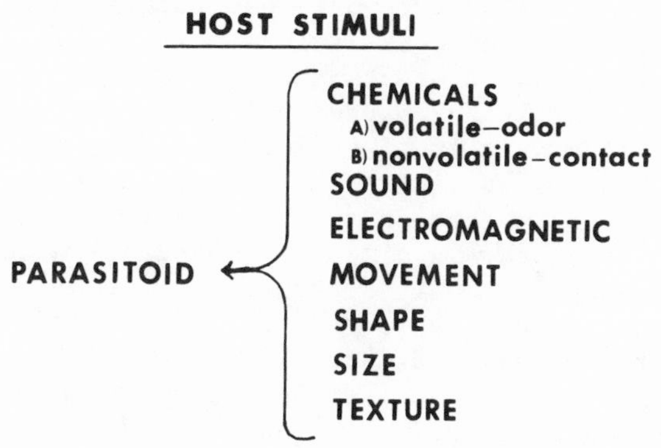

**HOST STIMULI**

PARASITOID ⟵ {
CHEMICALS
A) volatile—odor
B) nonvolatile—contact
SOUND
ELECTROMAGNETIC
MOVEMENT
SHAPE
SIZE
TEXTURE

Figure 2.  Types of cues that serve to orientate a parasitoid to
its host.

A number of chemicals have been shown to release specific
behavioral patterns in parasitoids and those which have been
isolated and identified are presented in Table 1.  A number of
other chemicals have been implicated in the parasitoid host selection
process and some have been extracted (Table 2) but have not yet
been identified.  Though relatively few chemicals have presently
been identified, tables 1 and 2 show that such chemicals are
diverse.  Furthermore, the types of behavioral responses elicited
by these chemicals differ considerably.  The varied behavioral
responses to chemicals observed reflect the different aspects
or levels of the host selection process.

## Host Habitat Location

As previously noted, females usually must first locate a proper
host habitat prior to actual host location.  Chemicals play an important role in host habitat location.  To be effective in orienting
a predator or parasitoid, such chemicals would have to possess
a degree of volatility.  Although the existence of such chemicals
has been indicated in the literature, only a few have been isolated
and identified (Table 1, S2 type stimuli).  The potential source
of such chemical cues which also applies to the nonchemical cues,
are the host plant, the host insect or non-host organisms associated
with the host habitat.

Many parasitoids orient and are attracted to a habitat regardless of the presence or absence of hosts.  Such entomophagous
insects probably orient to the plant or other food material of
the potential host.  Zwölfer and Kraus (1957) found that the
ichneumonid, *Apecthis rufata* Gmelin, attacked pupae of the fir
budworm, *Choristoneura murinana* Hubner, when such pupae were
placed in oak leaves.  Such attacks occurred on oak despite the
fact *A. rufata* usually attacks only oak tortricids.  The authors
concluded the parasitoid cues first on the plant.  They found
that the parasitoid would attack the fir budworm pupae in oak
leaves but would not attack pupae in fir trees in the same area
although such pupae were 10 times more abundant in fir.  In another
example, *C. nigriceps* was attracted to and searched tobacco in
the absence of host insects on such plants (Vinson 1975a).  Taylor
(1932) found that *Heliothis armigera* (Hübner), which feeds on
a variety of plants, was only attacked by *Microbracon brevicornis*
Wesman in South Africa when it fed on *Antirrhinum*.  Stary (1966)
observed that the parasitoids attacking *Aphis fabae* on the spindle
tree, *Enonymus europaeus*, in a forest community were different
than the parasitoids attacking the same aphid on the beet, *Beta
vulgaris*, located in the Steppe community of Czechoslovakia.
His results not only support the importance of the host plant
but also the habitat which may influence the types of parasitoids
that attack a host.  As Salt (1935) concluded, parasitoids are

| Family | Parasitoid | Host | Stimulus Type | Source | Solubility | Reference |
|---|---|---|---|---|---|---|
| Braconidae | Aphidius (Diaeretiella) rapae (Curtis) | Myzus persicae Sulzer | $S_2$ | Host Plant | Allyl isothiocyanate | Reed et al. 1970 |
| Braconidae | Microplitis croceipes (Cresson) | Heliothis zea (Boddie) | $S_3$ | Host frass | 13-Methyl Hentria-contane | Jones et al. 1971 |
| Braconidae | Cardiochiles nigriceps Vierick | Heliothis zea (Boddie) | $S_3$ | Mandibular gland | Misture of methyl esters of hentria-, dotria-, and tritria-contanes | Vinson et al. 1975 |
| Braconidae | Orgilus lepidus Muesebeck | Phthorimaea operculella (Zeller) | $S_5$ | Host frass | n-heptanoic acid | Hendry et al. 1973 |
| Braconidae | Bracon mellitor (Say) | Anthonomus grandis Boheman | $S_5$ | Host frass | cholesterol linenolate | Henson et al. 1977 |
| Braconidae | Biosteres (Opius) longicaudatus (Ashmead) | Anastrepha suspensa (Loew) | $S_3$ | Host associated fungi | Acedlaldehyde | Greany et al. 1977 |
| Ichneumonidae | Itoplectis conquisitor Say | Galleria mellonella (L.) | $S_6$ | Hemolymph | Serine, lysine, Leucine, MgCl | Authur et al. 1969; Hegdekar & Authur 1973 |
| Pteromalidae | Heydenia unica Cook and Davis | Dendroctonus frontalis Zimmerman | $S_2$ | Host tree | α-pinene | Camors & Payne 1971 |
| Trichogrammatidae | Trichogramma evanesens Westwood | eggs of wide range of lepido-pterous hosts | $S_3$ | ♀ moth scales | Triconsane | Jones et al. 1973 |
| Tachinidae | Cyzenis albicans (Fall.) | Operoptera brumata (L.) | $S_3$ | Damaged host plant | Sucrose & Fructose | Hassel 1968 |
| Tachinidae | Archytas marmoratus (Zeller) | Heliothis virescens (F.) | $S_6$ | Frass | Protein | Nettles & Burks 1975 |

Table 1. Parasitoids and their hosts for which chemicals involved in some aspect of host selection have been identified. See figure 4 with regard to the stimulus type. (Stimuli $S_3$ may also serve as $S_4$.

| Family | Parasitoid | Host | Stimulus Type | Source | Solubility | Reference |
|---|---|---|---|---|---|---|
| Aphelinidae | Aphytis coheni DeBach | Aonidiella aurantii (Maskell) | $S_4$ | Scale cuticle | $H_2O$ | Quednau & Hübsch 1964 |
| Braconidae | Perilitus coccinellae (Schrank) | Coccinellidae | $S_2$ | Coxal secretion | ? | Richerson & DeLoach 1972 |
| Braconidae | Apanteles melanoscelus (Ratzeburg) | Lymantria dispar L. | $S_4$ | larval cuticle | Hexane | Leonard et al. 1975 Weseloh 1974 |
| Braconidae | Brachymeria intermedia (Nees) | Lymantria dispar L. | $S_4$ | pupal cuticle | Hexane | Leonard et al. 1975 Weseloh 1974 |
| Braconidae | Peristenus pseudopallipes (Loan) | Lygus lineolaris Palisot deBeauvois | $S_2$ | flower odor | ? | Shahjahan 1974 |
| Encyrtidae | Cheiloneurus noxius Compere | Coccus hesperidum L. | $S_4$ | Scale cuticle | $H_2O$ | Weseloh & Bartlett 1971 |
| Ibaliidae | Ibalia leucospoides (Hochenwarth) | Sirex noctilia F. | $S_{3+4}$ | associated fungus | ? | Madden 1968 Spradberry 1970 |
| Ibaliidae | Ibalia drewseni Borries | Sirex noctilia F. | $S_{3+4}$ | associated fungus | ? | Spradberry 1970 |
| Ichneumonidae | Venturia canescens Gravenhorst | Anagasta Kuehniella (Zeller) | $S_2$ | Host odor | ? | Williams 1951 |
| Ichneumonidae | Venturia canescens Gravenhorst | Anagasta Kuehniella (Zeller) | $S_4$ | Mandibular gland | Hexane | Corbet 1971 Mudd & Corbet 1973 |
| Ichneumonidae | Campolitis sonorensis (Cameron) | Heliothis zea (Boddie) | $S_4$ | Cuticle | Hexane | Schmidt 1974 Wilson et al. 1974 |
| Ichneumonidae | Phaeogenes cynarea Bragg | Platyptilia carduidactyla (Riley) | $S_3$ | odor plant | ? | Bragg 1974 |
| Ichneumonidae | Pimpla instigator F. | Pieris brassicae (L.) | $S_2$ | Host odor | ? | Carton 1974 |
| Tachinidae | Trichopoda pennipes (F) | Nezara viridula (L.) | $S_2$ | sex pheromone | ? | Mitchell & Mann 1970 |
| Tachinidae | Lixophaga diatraeae (Townsend) | Heliothis zea (Boddie) and Diatraea grandiosella (Dyar). | $S_6$ | Frass | Methanol | Roth 1976 |

Table 2. Parasitoids and their hosts for which chemicals appear to play a role in host selection.

first attracted to a particular environment or habitat.  Although the female may be attracted to a habitat devoid of hosts, she may not be retained in the habitat in the absence of hosts.

While the presence of host odor is not necessary for host habitat location, the presence of odors from the habitat has been shown to be important.  Arthur (1962) showed that *Itoplectus conquisitor* (Say) was attracted to Scotts pine, *Pinus resinosa* Vit., but not red pine, *Pinus sylvestris* L., although hosts were present on both.  Using an olfactometer Arthur (1962) found that *I. conquisitor* would respond to the odor of Scotts pine but not red pine.  Several dipterous parasitoids have been found to respond to odors of the food plants of their host (Monteith 1955, 1956, Herrebout and van den Veer 1969).

Some entomophagous insects cue on organisms associated with their hosts.  The production of acetylaldehyde by the fungus, *Monolinia fructicola*, in rotting peaches occupied by tephritid fruit fly larvae was shown by Greany et al. (1977) to attract *Biosteres* (*Opius*) *longicaldatus* (Ashmead), a parasitoid of the fly, to the host vicinity.  The parasitoid was attracted to rotting fruit irrespective of the presence or absence of host larvae.  Similarly, Madden (1968) demonstrated that a chemical from a symbiotic fungus associated with *Sirex noctilio* F. stimulated its parasitoids although the fungus apparently stimulated ovipositor probing rather than attracting the parasitoid to the host site.

Chemicals that orient a parasitoid or predator to a particular habitat may be the result of a host insect-plant interaction.  Camors and Payne (1971) showed that following a bark beetle attack, host tree terpenes were released which attracted *Heydenia unica* Cook & Davis, a parasitoid of southern pine beetle larvae.  As the infestation develops bark beetle pheromones are released attracting additional species of parasitoids and predators.  As brood development occurred, the complex of parasitoids and predators continued to change in response to changes in the tree's physiological state and to changes in associated organisms and bark beetle development (Camors and Payne 1973).  A similar situation may exist in the complex succession of parasitoids and predators of a developing gall infestation (Askew 1966).  Monteith (1955) found that the tachinid, *Drino bohemica* Mesnill, was preferentially attracted to the unhealthy food plant of its host.  The tachinid, *Cyzenis albicans* (Fall.), was found to cue on sugars released from the plant due to the feeding activity of its host (Hassell 1968).

## Host Location

Once a female parasitoid reaches the proper habitat she

must begin a search for the host. Depending upon the parasitoid
species, the female may randomly search entire plants within the
habitat or the female may respond by searching only certain parts
of the plant. An example is provided by *Cardiochiles abdominalis*
Guenéé which searches the flower heads of wild aster, *Aster pilosus*,
and parasitizes the larvae of *Schinia arcigera* Guenéé that occur
there. Host larvae occurring on the leaves and stems were not
observed to be attacked (personal observation). However, the individual plants that were searched appeared to be randomly selected
by this species.

Various stimuli from the host usually serve as the next step
in the host selection process (Vinson 1976). Chemicals again appear
to play a major role. These compounds may be odors acting at a
short distance (Hendry et al. 1973, Carton 1974, Schmidt 1974)
or they may elicit a response only upon contact (Quednau 1967,
Vinson 1968). Females contacting such chemicals may orient and
be led to the host, but many of the chemicals act as searching
stimulants. These compounds release a searching behavior of a
more restricted area than the previous cues (Vinson and Lewis 1965,
Vinson 1975b).

## Host Acceptance

The final process in parasitoid selection of hosts is host
acceptance by the female parasitoid. A number of factors are
involved in the acceptance of hosts. In some cases ovipositor
probing is elicited by host movements or vibrations (Labeyrie 1958,
Walker 1961, Ryan and Rudinsky 1962, Baier 1964). In other instances
movements are not important and acceptance appears to depend more
on chemical mediators (Hays and Vinson 1971, Schmidt 1974). Texture
and shape also have been found to influence acceptance (Martin
1946, Weseloh 1970, 1974, Carton 1974), although odors or chemicals
in combination with the physical factors also play an important
part in host acceptance (Slobodchikoff 1973, Wilson et al. 1974).

Chemicals have been found to stimulate ovipositor thrusting
and probing (Jones et al. 1971, Schmidt 1974, Henson et al. 1977)
and as Picard (1922) pointed out, host odors may result in a reflex
action of ovipositor piercing or probing. Actual egg release may
depend on yet other host chemicals. As an example, *I. conquisitor*
was found to oviposit into simulated hosts containing host hemolymph
or several amino acids and salts (Arthur et al. 1969, Hegdekar
and Arthur 1973). Similar results were reported by Rajendram and
Hagen (1974) although in both cases the factors initiating the
prior step of ovipositor thrusting were not investigated thoroughly.

## Success Motivated Searching

Parasitoids often exhibit an intense searching behavior after ovipositing (Edwards 1954, Chabora 1967, Gerling and Schwartz 1974). Croze (1970) referred to similar behavior as area restricted search i.e., as a food item is discovered the search behavior is intensified. In high host densities or clumped host dispersion patterns, female parasitoids intensely searching the area around a host would be a reasonable strategy. Kairomones seem to release and perpetuate this same behavior (Gross et al. 1975, Vinson 1975a). In a sense the presence of a kairomone indicates to a female that a host is or has been in the area. While kairomones or allomones concerned with host habitat location may orient and retain the female parasitoid in the habitat for a short period of time, kairomones associated with the host and associated success motivated searching may retain the parasitoid longer. However, if no hosts are encountered, such searching is eventually abandoned.

## Host Preference

The species of potential host attacked or host preference also depends on chemicals. It has been shown that an unacceptable host may be rendered acceptable after contamination by an acceptable host (Thorpe and Jones 1937, Bartlett 1953, Spradberry 1968). For example *C. nigriceps* would not attack the fall armyworm, *Spodoptera frugiperda* (Smith) or granulated cutworm, *Feltia subterranea* (F.) but would oviposit in both hosts after they were treated with searching stimulant extracted from its habitual host, *H. virescens* (Vinson 1975a). Thus, use of these chemicals may have some importance in altering host preference when applied to fields. What effect altering host preferences might have from an ecological or practical point of view is unknown.

## Prey Selection

The role played by chemicals in predator behavior towards prey is not as well established as that for parasitoids. From an evolutionary view, the predatory strategy emphasizes ambush or search and attack. While sight or movement (Corbet et al. 1960, Pritchard 1965, Alcock 1973), or sound or vibration (Murphey and Mendenhall 1973), most often have been described as the major sensory input into predatory attack behavior. However, a role for chemicals has been implicated more recently.

Wood et al. (1968) found 2 species of predators attracted to the sex pheromone of their prey, *Ips confusus* (LeConte). Wilbert (1974) concluded that the aphid predator, *Aphidoletes aphidimyza* (Rond.), located its prey in part by odor over short distances.

Vité and Williamson (1970) found that both sexes of the predator, *Thanasimus dubius* (F.), responded to frontalin (1,5-dimethyl-6,8-dioxybicyclo [3.2.1] octane), a pheromone released by its prey. Host location by a staphylinid myrmecophilic beetle was found by Hölldobler (1969) to involve odor. He found that young beetles were attracted by the odor of *Myrmica* but that this response changed with the beetles age. After hibernating, the staphylinid beetle oriented to the odor of *Formica polyctena* Forest (Hölldobler 1969).

As pointed out by Thompson (1951) many predators are more prey specific than commonly thought. Many predators prey only on a selected number of prey species although they may accept other species of prey when starved. Thus the predators may resemble the parasitoids in specificity, some having a wide range of prey species and others being more prey specific. In the future it may be found that chemicals play a more significant role in prey location and selection by predators than here-to-fore suspected, particularly in the more prey specific species.

## BEHAVIORAL RESPONSES OF ENTOMOPHAGOUS INSECTS

There are a number of different types of behavioral patterns that may be elicited by allomones or kairomones depending on the species of parasitoid and host involved and the level of the host selection process. Thus a description of the various behavioral responses is extremely important when concerned with the several levels of the host selection process. Whether a particular allomone or kairomone will prove useful in a parasitoid or predator manipulation program will ultimately depend upon the type of response elicited by the compound.

Some compounds may elicit activity in the parasitoid stimulating it to fly or crawl upwind or only to orientate to the odor source. Such compounds would have potential in attracting the entomophagous insects to a local area.

Antennation in response to contact chemicals is a common feature elicited by host kairomones in the Hymenoptera (Williams 1951, Moran et al. 1969, Gordh 1973). Usually the female's antennae alternately move up and down, occasionally touching the substrate as the female moves about. Upon contacting a host or traces of a host (i.e. anything that has been contaminated by the presence of a host), the behavior of the female changes. The antennae are brought into contact with the host-contaminated material and rub or palpate the object. The kairomone may elicit antennation of the contaminated substrate followed by examination of the immediate area around the contamination. In some cases the kairomones elicit antennal drumming which is common in the Chalcidoidea, due in part to their "elbowed" antennae (Edwards 1954). The antennal response is often followed by ovipositor probing. For example, in *Bracon mellitor* Say, the female is stimulated by esters of cholesterol to probe filter paper discs to which the ester is applied (Henson et al. 1977). However, this behavior is usually preceeded by antennation of the contaminated filter paper spot. Probing behavior often consumes a great deal of time. Probing activity by a single *B. mellitor* female has been observed to occur for over an hour in response to cholesterol esters (personal observation).

A kairomone resident in the honeydew of soft scales, elicits turning behavior in *Microterys flavus* (Howard) with females often running sideways and even backwards upon contacting the water soluble material (Harland and Vinson, unpublished data). In contrast, a female of *Euplectrus plathypenae* (Howard) mounts up on the host dorsum and remains motionless until the host molts at which time the parasitoid oviposits (personal observation).

Oviposition is another response observed in parasitoids exposed to certain kairomones. An interesting case was reported by Nettles and Burks (1975). These authors found that the tachinid, *Archytas marmoratus* (Zeller), was stimulated to larviposit on contact with host frass. Roth (1976) has found that *Lixophaga diatraeae* (Townsend), another tachinid, will also larviposit on contacting host frass.

Several species of parasitoid Hymenoptera have been found to oviposit upon ovipositor contact with several amino acids and salts (Hegdekar and Arthur 1973, Rajendram and Hagen 1974). In such cases the response was ellicited by solutions in containers. The importance of these containers is not known but may be necessary for an ovipositional response. The response reported by these authors is at the level of host acceptance and may only occur after females have been stimulated by other kairomones or other factors such as shape or texture. Whether these oviposition stimulants could be used in augmentation is difficult to envision, however, they may have a place in mass rearing parasitoids. One problem encountered in artificial rearing is obtaining an abundance of viable, fertilized eggs. The development of artificial hosts or diets with the proper chemicals to elicit oviposition may become an important prerequisite to innundative releases of laboratory reared parasitoids.

Hymenopteran antennae contain many sensory receptors utilized in host detection (Slifer 1969, Miller 1972, Norton and Vinson 1974). The ovipositor also may be involved in host detection (Vinson 1975a) but is more important in host acceptance (Gutierrez 1970). The sense organs of the ovipositor of parasitoid hymenoptera have been described by Hawke et al. (1973). The location of receptors involved in host detection on predators and dipterous parasitoids

as well as the behavioral responses elicited by kairomones in
these two groups is generally unknown. However, a recent investigation by Roth (1976) showed that the tachinid, *L. diatraeae*, would
larvaposit on tarsal contact with the host's frass. Documentation
of the role of kairomones on the behavior of dipterous parasitoids
and the predators needs to be undertaken, as does work regarding
the location of sensory receptors involved and types of behavioral
patterns elicited before definite statements about their behavior
can be made.

## SOURCES OF HOST KAIROMONES

There are several sources from the host where kairomone activity
has been located. Potential sources are diagrammatically presented
in Figure 3 and those which have been found are listed in Tables
1 and 2. The source as well as the chemical nature of a kairomone
may differ for different parasitoids of the same host. The larva
of the tobacco budworm, *H. virescens*, serves as a host for several
parasitoid species. The tachinid, *A. marmoratus* is stimulated
by a protein in the host frass to larviposit (Nettles and Burks
1975), whereas *M. crociepes* is stimulated to search by a hydrocarbon
in the frass (Jones et al. 1971). *C. nigriceps*, on the other
hand, is stimulated to search by secretions from the mandibular
gland of its host (Vinson 1968) and *Campoletis sonorensis* (Cameron)
is stimulated by frass and the host cuticle (Wilson et al. 1974,
Schmidt 1974). Parasitoids attacking the eggs of *H. virescens*
are influenced by chemicals from the female moth and the egg chorion
(Lewis et al. 1972a, Vinson 1975b). Host frass has been shown
to contain a number of factors which elicit host searching or
other host related behavior (Jones et al. 1971, Hendry et al.
1973, Nettles and Burks 1975, Henson et al. 1977) and appears
to be a common source of cues utilized by parasitoids. The host
cuticle also has been found to influence the behavior of a number
of parasitoids (Wilson et al. 1974, Schmidt 1974, Leonard et al.
1975, Weseloh and Bartlett 1971).

The kairomone that releases searching behavior in *Trichogramma
evanescens* (Westwood) is a hydrocarbon present on adult tobacco
budworm scales (Jones et al. 1973). These scales are deposited
near the egg during the act of oviposition (Lewis et al. 1972).
*Chelonus texanus* Cresson, a hymenopterous parasitoid which attacks
eggs of some of the same host species as *Trichogramma*, is stimulated
to search by a hexane fraction of the moth scales and *C. texanus*
also probes with its ovipositor when stimulated by a factor present
on the egg chorion (Vinson 1975b).

Compounds which orient a predator or parasitoid to the host
would not appear to be advantagous for the host and would be selected
against. However, such compounds are probably important to the

host and serve as pheromones, allomones or chemicals important in the host's physiology. For example, the mandibular gland secretion of *Anagasta* (*Ephestia*) *kuehniella* (Zeller) serves as a kairomone for its habitual parasitoid, *Venturia* (*Nemeritis*) *canescens* (Gravenhorst) while serving as a dispersal pheromone for the host larvae (Corbet 1971, Mudd and Corbet 1973).

Several cases have been reported where the host's sex pheromones attract their predators and parasitoids (Rice 1969, Mitchell and Man 1970, Vité and Williamson 1970, Sternlicht 1973). An interesting case is provided by R. Prokopy (personal communication) concerning the apple maggot adult, *Rhagoletis pomonella* (Walsh) which attacks cherries in Wisconsin. While ovipositing an egg in a cherry the fly deposits a marking pheromone on the fruit. The marking pheromone reduces the attractiveness of the marked fruit as an oviposition site for female *Rhagoletis*. The marked fruit, however, is attractive to the apple maggot parasitoid, *Opius lectus*, and elicits ovipositor probing of the marked fruit by the female parasitoid.

## Kairomone Isolation

Knowledge of the sources of the kairomone is not necessary for the extraction and purification, although knowledge of the source may aid in these efforts. As previously mentioned, the frass or webbing associated with the host is often active in eliciting some components of host selection (Vinson 1968, Jones et al. 1971, Hendry et al. 1973, Wilson et al. 1974, Weseloh 1977). Whether a kairomone is located in the frass, a gland or host cuticle, the material can be removed by soaking or macerating in the appropriate solvent. Since a wide range of chemicals have been found to act as kairomones (Table 1), a series of chemicals of increasing polarities should be employed to extract active compounds.

The methodologies used for isolation, purification and identification of kairomones and allomones are similar to those used for pheromones (Jacobson 1972, Tumlinson and Heath 1976) and will not be covered. Most of the kairomones thus far isolated and identified are relatively nonvolatile and occur in nanogram quantities. This has made isolation and identification less difficult. The nanogram range is the optimum concentration that will elicit a positive response in those parasitoids for which a kairomone has been identified (Jones et al. 1971, Hendry et al. 1973, Vinson et al. 1975, Henson et al. 1977). This has further aided in their isolation. However, some species may well respond to lower concentrations in the fentogram range, concentrations at which some pheromones elicit behavioral responses.

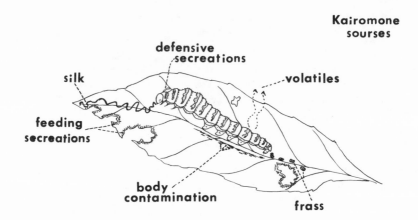

Figure 3.  Potential sources of kairomones that can influence host selection in insect parasitoids.

Bioassays

The type of bioassay conducted depends on the nature of the host-parasitoid or predator-prey relationship under study.  An olfactometer of some type would be the best choice for those compounds that attract the parasitoid or predator over a distance of several cm or more.  However, many kairomones isolated thus far do not possess much volatility and only elicit a response over extremely short distances.  These chemicals have been referred to as contact chemicals (Quednau 1967, Vinson 1968) and require a different type bioassay than that employed for volatile chemicals.  A simple Petri dish bioassay has often been employed where the chemical in question is applied as a small spot to a suitable substrate such as filter paper (Jones et al. 1973, Vinson et al. 1976). A female parasitoid or predator is introduced and its behavioral response upon contacting the chemical is recorded as positive or negative or is rated on a subjective scale.  In some cases surro-gate hosts have been employed (Richerson and DeLoach 1972, Schmidt 1974).

Many behavioral criteria have been used to interpret the

bioassays.  These include antennation of the chemical (Vinson
and Lewis 1965, Lewis and Jones 1971, Wilson et al. 1974), ovipositor
probing (Hendry et al. 1973, Vinson 1975a, Vinson et al. 1976,
Henson et al. 1977), increased turning (Harland and Vinson, unpublished data) and oviposition or larviposition (Arthur et al. 1969,
1972, Hegdekar and Arthur 1973, Rajendram and Hagen 1974, Nettles
and Burks 1975).  While such a bioassay has proven useful, there
are a number of factors that can influence the results.  Some of
the most important ones are enumerated below.  A)  The female
may have a preoviposition period and fail to search for hosts
or prey until she becomes reproductively mature (DeBach 1964).
B)  The ovipositional experience of the female may be important.
Samson-Boshuizen et al. (1974) have shown that oviposition in
*Pseudeucolia bochei* Weld is a matter of experience.  C)  The female
must be in the proper physiological condition and the bioassays
should be conducted under specific environmental conditions of
temperature, light and humidity (Edwards 1955, Orphanides and
Gonzales 1970, Lewis and Jones 1971, Vinson et al. 1973).  D)
Optimum concentrations of the chemical may be important.  Several
parasitoids fail to respond to low or high concentrations of a
kairomone (Jones et al. 1971, Hendry et al. 1973, Vinson et al.
1976).  E)  A proper mixture was found to be important (Vinson
et al. 1975) as has been shown for many pheromones (Payne 1974).
F)  A number of parasitoids mark their hosts (Arthur et al. 1964,
Bakker et al. 1967, Greany and Oatman 1971a,b) or the areas they
have searched (Price 1972).  Host marking can present a special
problem as the presence of marking pheromones may reduce the response
of females to the kairomone and necessitate using new females and
arenas for each exposure.  G)  Another problem that may arise is
that of associative learning.  Vinson et al. (1976) isolated methyl
p-hydroxy benzoate from laboratory reared boll weevil larvae that
elicited ovipositor probing in the parasitoid B. *mellitor*.  However,
the failure to detect methyl p-hydroxy benzoate in field boll weevil
larvae and the isolation of esters of cholesterol from field boll
weevils that released ovipositor probing (Henson et al. 1977) suggested
the females were capable of learning to respond to the contaminant
(Vinson et al. 1977).  The possibility that some parasitoids can
associatively learn to respond to certain cues (Arthur 1966, 1971,
Robacker et al. 1976) may hinder efforts to isolate and identify
the innate cues involved in host selection.  Since associative
learning is a possibility, inexperienced female response to the
isolated chemicals should be determined.  H)  Females may become
habituated to the chemical soon after exposure.  This may be a
particular problem when high concentrations are used.  Such females
may respond once or twice and then fail to respond further as long
as they are confined with the kairomone.  I)  The proper physical
cues such as shape, movement, color or tactal cues may be necessary
before a response to a chemical will occur (Weseloh 1970, Slobodchikoff
1973).  There are many other considerations with regard to bioassays
and the reader should refer to Young and Silverstein (1976).

## BEHAVIORAL CHEMICALS IN AUGMENTATION

Allomones and kairomones involved in host location by parasitoids and predators offer new concepts and potential new methods for the regulation of pest populations. A simplified, diagrammatic representation of the stimuli releasing the various behavioral patterns at each phase of the host selection process may be helpful. Such a diagram has been developed by Lewis et al. (1976). These authors refer to the essential behavioral patterns involved in host selection as the "find and attack cycle". A modified version of this cycle is presented in Figure 4. The solid lines represent the behavior that occurs if the necessary stimulus is provided at the next step. The dotted lines show the alternative behavioral response that occurs in the absence of a proper stimulus. As developed by Lewis et at. (1976) the stimuli ($S_1$ to $S_{10}$) provided at each step result in a transition ($T_1$ to $T_{10}$) of the beneficial insect from one behavioral pattern to the next. For example, the transition ($T_1$) from inactivity to initial random movement is initiated by an innate appetitive drive together with prevailing environmental conditions and the current physiological state of the parasitoid (Lewis et al. 1976).

As already pointed out, parasitoids may make a transition from a random movement to a host habitat orientation ($T_2$) by orienting to odors ($S_2$) eminating from the host insect, host plant or organisms associated with the host insect (Fig. 1). The identification of these chemicals may have some practical use in natural enemy augmentation. The application of these compounds could be useful in attracting more entomophagous insects to a host infestation. They also may be important in attracting or orientating entomophagous insects to those crops or habitats which lack proper orientation cues. While hosts for *I. conquisitor* occur in red and Scotts Pine, the parasitoid is attracted to and only attacks hosts in Scotts Pine (Arthur 1962). The application of the odor of Scotts pine to red pine could attract the parasitoid to red pine and could result in the parasitism of the hosts present. A similar observation that *C. nigriceps* attacks *Heliothis* on cotton or tobacco but not peanuts or corn suggests that a plant factor necessary to orientate the female to the proper habitat is lacking in the latter 2 crops (Vinson 1975a). Doutt (1964) described a similar case with *Cyrtorhinus mundulus* (Breddin) an egg predator which was more effective in sugar cane than corn, although potential prey occurred in both. In principle, the provision of an attractant or long distance orientation factor to peanuts or corn by either selective plant breeding or application of an attractant could result in the attack of *Heliothis* on these crops by otherwise ineffective entomophagous insects. Further, in the development of pest resistant plant varieties, the elimination of factors necessary for the orientation of a parasitoid could have detrimental consequences. If a crop variety was developed that reduced the pest popula-

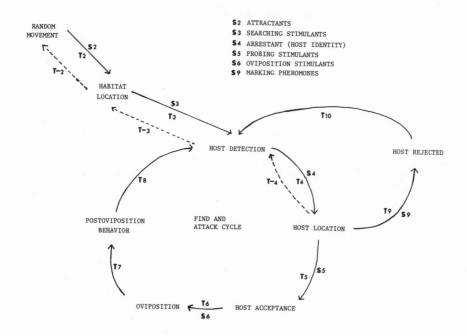

Figure 4. Diagrammatic representation of the parasitoid find and attack cycle. (Adopted from Lewis et al. 1976). See text for explanation.

tion by 50% but also reduced the effectiveness of the entomophagous organisms by 50% or more, the net effect could be an increase in the pest infestation.

It is the $T_3$ transition released by stimuli indicating the presence of hosts within the habitat ($S_3$) that has received the most intense study from a behavioral and chemical standpoint as well as from a pest management point of view. The $S_3$ stimuli or searching stimulants release an intensive searching behavior in certain female parasitoids. While continual contact with the $S_3$ stimuli is not necessary, contact with the mediator must occur periodically to reinforce and maintain the process. Lack of reinforcement results in a $T_{-3}$ transition to general habitat scanning and possible loss of the parasitoid from the area (Lewis et al. 1976).

The application of the $S_3$ stimuli to a target area in sufficient quantity and of a proper distribution serves to reinforce and maintain the parasite in the $T_3$ transition. In effect, such an application results in increasing the apparent host density.

This serves to retain the parasitoid in the $T_3$ transition increasing
the chances that the parasitoid will contact a host.  It is the
capacity of the host searching stimulants ($S_3$ - stimuli) to retain
parasitoids in the $T_3$ transition that offers an opportunity to
manipulate the parasitoids behavior for the effective use in pest
control programs.

W. J. Lewis and associates (Lewis et al. 1976, Jones et al.
1976) have conducted a number of studies demonstrating the potential
of the $S_3$ stimuli in the manipulation of parasitoids for pest
management.  Lewis et al. (1972a) released *T. evanescens* reared
from *Cadra cautella* (Walker) eggs into cotton fields containing
plants with alternating leaves treated with hexane (control) or
hexane extracts of *H. zea* moth scales (the $S_3$ stimuli).  *H. zea*
eggs were placed on control and treated leaves for 3 hr on 5 subsequent days.  The eggs were then collected and the degree of parasitism determined.  The results showed that about twice as many
eggs were parasitized on the treated compared to the control leaves
(Lewis et al. 1972a).  In another study, crimson clover plots
were sprayed with as little as 150 mg of tricosane ($S_3$ stimuli)
or hexane (control) per acre.  Eggs of *H. zea* were placed on plants
in each plot and subjected to natural *Trichogramma* populations
for 24 hr.  Parasitization by all *Trichogramma* spp. was significantly
higher in eggs from the $S_3$ treated plots than the control (Lewis
et al. 1975a).  Similar results also were obtained with released
*Trichogramma* spp. on tricosane versus control treated plots (Jones
et al. 1973, Lewis et al. 1975a).

The mechanisms that result in increased parasitism observed
in field studies have been investigated by Lewis et al. (1975b).
These authors divided the process leading to increased parasitism
by the $S_3$ stimuli into 3 behavioral mechanisms:  1) activation,
2) retention and 3) egg distribution.

## Activation

A test to determine the effect of complete kairomone coverage
compared to partial or no coverage was conducted by Lewis et al.
(1975b).  These authors found in greenhouse studies that the percent
parasitism by *Trichogramma* of *H. zea* eggs placed on pans of pea
plants uniformly treated with tricosane was greater ($\overline{X}$ = 71%) than
control pea plants treated with hexane ($\overline{X}$ = 23%).  Parasitism of
partially treated pans of pea plants was intermediate between
completely treated and untreated plants.  Analysis of the parasitism
of eggs on the treated and untreated pea plants in partially treated
pans revealed that egg parasitism on treated plants was higher
($\overline{X}$ = 57%) compared to eggs on the untreated plants in the same
pan ($\overline{X}$ = 47%).  However, the parasitism of eggs in the treated
area of partially treated pans was lower than the parasitism of

completely treated pans. Also the parasitism of eggs on the untreated
plants in partially treated pans was higher than eggs in untreated
pans.

These results suggest the parasitoid is stimulated by the
kairomone to search rather than guiding the parasitoid to the
host egg. In completely treated pans of pea plants the female
would be stimulated to search and receive continual reinforcement
retaining her in the search and attack cycle. In untreated pans
the parasitoid would only receive stimulation upon random contact
with host eggs and would fail to receive reinforcement resulting
in poor parasitism. In the partially treated pan the females
were activated to search by the kairomone. They not only searched
the treated area but were even stimulated to search the adjoining
untreated area as well. The stimulants thus appear to cause the
parasitoids to search not only areas contaminated by the kairomone
but areas surrounding the kairomone-contaminated area.


## Retention

Kairomones also appear to retain the parasitoids in contaminated
areas. Lewis et al. (1975b) found that released *M. croceipes*
females were retained on pea plants which contained hosts as well
as pea plants lacking larvae but contaminated by frass and feeding
damage. On the other hand released parasitoids deserted the pea
seedlings that were not contaminated by host frass (Lewis et al.
1975b).

Retention of females on a group of plants may not depend
on whether or not the plants are infested but the degree of infestation. When the host density is low, the percent of each plant
contaminated by the pest is low and the chance a female parasitoid
will contact contaminated areas is decreased. Although a few
hosts may be located on the plant only a small percent of the
female parasitoids are stimulated to search these plants. Also,
because the level of contamination is low the parasitoid may not
be retained on the plant for a very long period. The net effect
is that a large percentage of the hosts are not located and attacked.

At high host densities much of the plant is contaminated
and thus a larger percentage of parasitoids contacting the plant
are stimulated to search. The presence of hosts, stimulates parasitoids to search and because they search longer (i.e. are retained
in the area longer) a greater percentage of the hosts are detected
and parasitized. By applying kairomones to plants having a low
density of hosts, the parasitoids are activated and retained on
the treated plants. This retention and activation increases the
chance for host contact and results in an increased percentage
of hosts parasitized. This can make a difference in the number

of hosts available for the next generation and the potential for
economic damage lowered.

## Egg Distribution

One of the effects of applying a kairomone to field plots
is an improved distribution of eggs by the female parasitoid (Lewis
et al. 1975b). An explanation of this phenomenon can be provided
by the concept of host marking. Salt (1937) was the first to
describe host marking behavior in parasitoids. He showed that
*Trichogramma* females deposited a substance on an egg either during
or immediately after oviposition which reduce the acceptance of
the egg by another female. Since his classic work, host marking
by female parasitoids has been shown to be a widespread phenomenon
(van Lenteren 1976).

Although superparasitism may occur, many parasitoids are
solitary. When 2 or more eggs are deposited within a host, the
competitors are usually destroyed and only a single adult emerges
(Fisher 1971, Vinson 1972). Host marking and the discrimination
by subsequent female parasitoids appears to be important in the
avoidance of parasitized hosts and therefore improves the efficiency
of the parasitoid by reducing the competition for hosts and the
wastage of eggs. However, host marking may be of even greater
importance. As shown by Vinson (1972), *C. nigriceps* not only
marks the host but also marks the substrate around the host during
intensive searching thus avoiding previously searched areas.
Price (1972) has shown that some species avoid previously searched
areas. This behavior results in the dispersal of the parasitoids
into non-searched areas (Fig. 4, T9 transition). As discussed
by Price (1975, see p. 243) the avoidance of searched areas leads
to a dispersal of the parasitoid population and thus reduces an
effective increase in the parasitoids density in response to high
host density.

While host marking may be important in dispersal and reducing
superparasitism by other females, host marking also may be important
in reducing repeated oviposition by the same female. As discussed
previously, many parasitoids exhibit success-motivated searching,
i.e. they will intensively search the adjacent area around a host
after successful oviposition. In a low host density situation,
a female's success motivated search would be more restricted due
to the restricted kairomone contaminated area. Thus, a female's
chance of contacting an already parasitized host is increased
and with it an increased probability that a marked host will be
accepted. In fact, while host marking pheromones are important
in reducing superparasitism, many parasitoids may still parasitize
a host several times depending on host availability and other
conditions. An example is provided by *C. nigriceps*. When given

a choice between a nonparasitized host and one that has been parasitized only once, the parasitoid accepts both. If no choice is provided a female will parasitize the same host many times. If, however, a choice is provided between a nonparasitized larvae and one that has been parasitized several times the female parasitoid displays a distinct preference for the nonparasitized host (Vinson and Guillot 1972). Thus when host density is low there is an increased chance a host will be attacked several times. In a high host density and increased kairomone distribution, the host marking pheromone would act as a dispersal agent leading the parasitoid away from the host just attacked therefore increasing the chance a different host will be encountered.

From the above discussion it should be expected that augmentation involving a blanket coverage by a kairomone might lead to improved egg distribution. In the presence of a blanket treatment of kairomone the female would be stimulated to search and would more likely be led away from a host that was just attacked. Such a situation has been reported by Lewis et al. (1975b). In plots treated with kairomones that trigger searching behavior in *Trichogramma*, it was found that host eggs in the kairomone treated plots contained fewer *Trichogramma* eggs per host egg than in control plots (Lewis et al. 1975a). These authors concluded that the presence of the kairomone on the plants probably induce the parasitoids to depart the host egg more rapidly and to disperse leading to improved and more efficient parasitoid egg distribution.

## Release Manipulation

While the application of certain kairomones to crops has potential in increasing the efficiency of egg distribution, retention of the parasitoid and activation to a searching pattern, there are other possible uses of these compounds. One of the problems encountered in the release of natural or exotic entomophagous insects is retaining them in the area of release. In many instances, upon release the insects rapidly disperse, being carried by the wind or flying from the release site. Many of the failures in the establishment of parasitoids may be due in part to this phenomenon. Gross et al. (1975) used kairomones or sign stimuli (releasers) to place the parasitoid in the "host find and attack cycle", thus reducing the tendency of the parasitoid to enter the escape and dispersal behavior phase upon its release. These authors placed frass and host larvae reared on pea plants near *M. croceipes*. Thus positioned the parasitoid on release would contact the host or its frass. The released females showed little tendency to disperse after host contact. Instead these "prestimulated" females assumed a host-seeking pattern that resulted in the orientation by flight to host infested pea plants located nearby (Gross et al. 1975). In the laboratory unstimulated *M. croceipes* exhibited

a positive phototaxis on release by flying to the ceiling of the
greenhouse where they remained without initiating search.

The findings of Gross et al. (1975) are of particular importance.
The isolation and identification of the kairomones involved in
the prestimulation concept may prove to be a highly significant
factor relative to the introduction and release of exotic parasitoids
in biological control programs.

Another aspect of parasitoid behavior that may play an important
role in the manipulation of these beneficial insects is the concept
of associative learning. Learning by certain insects has been
well demonstrated (Alloway 1973, Quinn et al. 1974) and documented
in the higher Hymenoptera (Hoefer and Lindauer 1975, 1976). Learning
has also been observed in certain aspects of the parasitoids behavior
Robacker et al. 1976). The demonstration that some parasitoids
can associatively learn certain cues involved with host location
(Arthur 1971, Vinson et al. 1977) may have further importance
with regard to the prestimulation concept.

It has been found that B. mellitor, a parasitoid with a wide
host range (Cross and Chesnut 1971) will respond to some chemicals
only after the female has associated the chemical with a host
(Vinson et al. 1977). The demonstration that B. mellitor can
associatively learn to use certain chemicals as cues for ovipositor
probing has important adaptive implications. B. mellitor is
a polyphagous parasitoid and certain potential hosts may be abundant
only at a particular time and place. A female after having found
and parasitized a few hosts of one species, may learn to associate
additional chemical cues with that particular host species. Such
a female would thus become more selective and efficient in locating
its host.

The demonstration that some parasitoids can associatively
learn and the prestimulation concept suggest that parasitoids
such as B. mellitor could be prestimulated not only to search
but to search for a particular host. This concept may be particularly applicable to inundative releases of parasitoids which are
reared in the laboratory on one species of host but released for
the control of another.


Predator Manipulation

The concepts concerning the use of kairomones and allomones
for augmentation of parasitoids also apply to many predators.
This may be particularly true of the host specific species which
may be strongly influenced by chemical cues. However, certain
aspects of predatory behavior may be more amenable to manipulation.

Many adult predators require the consumption of a number
of prey as adults for egg production and oviposition though others
require pollen or nectar (Clausen et al. 1933, Finney 1948, Schneider
1948, Sundby 1967). The need for supplemental food coupled with
the transient and homogenous agronomic environments offers some
additional approaches for the augmentation of certain predators.
The potential for parasitoid and predator augmentation by supplemental release of hosts or prey, the provision of alternate hosts
or prey and provision of plants that will supply the necessary
pollen and nectar is beyond the scope of this chapter. These
considerations are discussed by DeBach and Hagen (1964) and van
den Bosch and Telford (1964).

Some predators require the presence of supplemental food
to attract and retain them as well as to provide the stimulus
necessary for egg production. Hagen et al. (1970) noted that
it is usually the adult stage of most predators that bridge the
gaps of prey shortage, for adults can subsist on alternate foods
such as honeydew, plant secretions and pollen and often it is
in the adult stage that predators aestivate or hibernate. Hagen
et al. (1970) found that an artificial honeydew preparation (Hagen
1950, Hagen and Tassan 1970) applied to crops attracted, retained
and stimulated egg production in the green lacewing, *Chrysopa
carnea* Stephens. Hagen et al. (1970) also found that artificial
honeydew attracted and sustained syrphid adults. Aphid juice
and artificial honeydews were found to attract several species
of predators to potato plants (Saad and Bishop 1976). The amino
acid tryptophan has been identified as the source of the attractant
for the adults of *C. carnea* (Hagen et al. 1976). Hagen et al.
(1970) found that the application of food sprays to cotton and
alfalfa increased the number of predators and resulted in a decrease
in the number of pests in the field. The application of attractants
to fields to augment predator populations and to provide supplemental
food necessary to sustain and stimulate egg production prior to
prey outbreak appear to be promising approaches to control.

Kairomones may also stimulate predator searching. Larvae
of *C. carnea* respond to kairomones extracted from the scales of
adult female *Heliothis* (Lewis et al. 1977). This extract increased
the rate of predation of *H. zea* eggs by *C. carnea* presumably by
stimulating larval searching.


PHEROMONES


Pheromones also offer several theoretically important approaches
for the augmentation of natural enemies. Of particular importance
are the host marking pheromones first reported by Salt (1937).
Since his report, host-marking pheromones have been reported for
a number of parasitoids (Fisher 1971, van Lenteren 1976, Vinson

1976). Host-marking pheromones are those where a female parasitoid externally marks the host. The host is perceived as marked by the same individual or other members of the same parasitoid species prior to ovipositor insertion (Lloyd 1942, Hsiao et al. 1966, Vinson and Guillot 1972).

In some cases a parasitized host is attacked but rejected after ovipositor insertion. Numerous examples have been described of previously parasitized hosts being rejected after ovipositor insertion (Fisher 1961, Wylie 1965, Greany and Oatman 1971a,b, Griffiths 1971). In fact, Guillot and Vinson (1972) found that *C. sonorensis* injects a fluid along with the egg that results in host rejection after host attack but requires several hours to become effective. Although, van Lenteren (1966) suggests that *P. bochei* injects a factor that prevents superparasitism prior to ovipositor insertion.

However, the host marking pheromones of interest in pest management are those that are applied externally and reduce parasitism prior to ovipositor insertion. The source and nature of these marking pheromones have only been investigated in a few species. In *C. nigriceps, M. croceipes* and *C. sonorensis* Dufours gland produces a host marking pheromone (Guillot and Vinson 1972, Vinson and Guillot 1972). The responsible factor for *C. nigriceps* appears to be a hydrocarbon component of the gland (Guillot et al. 1974). A topical application of the host marking pheromone to nonparasitized larvae rendered them unacceptable to searching female parasitoids. If isolation, identification and synthesis of host marking pheromones is realized they may offer some unique approaches to parasitoid augmentation.

The two most promising areas for the use of marking pheromones in parasitoid augmentation include the biological control of weeds and manipulation of hyperparasitoids. The biological control of weeds has been compared to the inverse of the biological control of insects (Tillyard 1930, Sweetman 1958). However, as pointed out by DeBach (1964) many of the principles are similar. One of the problems encountered in the biological control of weeds is the presence of parasitoids or predators of the biological weed control agent. Pettey (1948) cites an example where the coccinellid, *Cryptolaemus montrouzieri* Mulsant, introduced into South Africa for the control of mealybugs fed on *Dactylopius tomentosus* (Lamarck) which had been introduced for the control of prickly pear. Dodd (1953) reported that *Opius tryoni* (Cameron) introduced to control the fruit fly, *Ceratitis capitata* (Wiedemann) in Hawaii also attacked *Procecidochares utilis* (Stone), a tephritied gall formerly used to control the pamakani weed, *Eupatorium glandulosum* Sprengel. The effectiveness of the introduced weed control agents was reduced by the introduction of its parasitoids and predators.

In situations of high weed density the application of a marking
pheromone could free the biological weed control agent from attack
by its parasitoids thus increasing its effectiveness.  The same
reasoning could also apply to indigenous weeds which have evolved
with a complex of phytophagous insects and their parasitoids.
By freeing the phytophagous insect from its parasitoids, the phytophagous insect might be effective in the reduction of the weed in
local areas.

The impact of hyperparasitoids on the effectiveness of primary
parasitoids has yet to be clearly demonstrated.  From a theoretical
point of view, hyperparasitoids may be important in preventing
the establishment or buildup of the primary parasitoid (Muesebeck
and Dohanion 1927).  In an effort to demonstrate the effectiveness
of the hyperparasitoid *Lygocerus* sp. in reducing the effectiveness
of *Anarhopus sydneyensis* Timberlake, a mealybug parasitoid, DeBach
(1949) applied DDT which had little effect on the primary parasitoid
but retarded the hyperparasitoid.  The results suggested that
the hyperparasitoid was effective in reducing the control potential
of the primary parasitoid.  Whether hyperparasitoids are a detriment
in biological control or a benefit through damping oscillations
in parasitoid populations, the ability to manipulate these insects
may be of importance.

It would be surprising if hyperparasitoids did not mark their
hosts as do the primary parasitoids.  The demonstration, isolation
and identification of such factors needs serious attention.  Such
compounds could be effective in reducing hyperparasitism by dispers
ing the hyperparasitoids from areas where pest populations are
increasing thus freeing the primary parasitoid from their attack.

## PROBLEMS IN THE USE OF BEHAVIORAL CHEMICALS

The chemicals that attract the parasitoid to the proper habitat
($S_2$) and the searching stimulants that release the female into
the search and attack cycle ($S_3$) both show promise in augmentation
of natural enemies.  However there are a number of potential problems
associated with the use of kairomones and allomones that must
be considered.

### Reduced Searching

Although no data is available at present, one can speculate
on the potential problems with some of the kairomones.  Kairomones
that release ovipositor probing might possibly reduce parasitoid
effectiveness.  For example, Henson et al. (1977) found that contact
with cholesterol esters elicited ovipositor probing behavior in
*B. mellitor*.  This probing activity may last for 10 minutes or

more.  Application of such a material in the field could result
in stimulating female *B. mellitor* to probe indiscriminately thereby
reducing the time a female spends searching or probing potentially
productive sites.  A similar reduction in fecundity and efficiency
might be expected by the kairomone isolated from eggs of *Heliothis*
that stimulate ovipositor probing by *C. texanus* (Vinson 1975b).
Those kairomones that result in arrested movement or extensive
antennal drumming would be expected to reduce the parasitoids
effectiveness.


## Constant Stimulation

In some insects constant stimulation often can result in
death.  This death has been attributed to depletion of energy
reserves and the build-up of a neuro-toxic factor (Granett and
Leeling 1971).  Prolonged exposure to a kairomone could reduce
the longevity of the parasitoid due to continued stimulation of
the female.  However, Nordland et al. (1976) found that constant
exposure of *Trichogramma pretiosum* Riley to a kairomone actually
improved longevity.  Ashley et al. (1974) similarly found that
*T. pretiosum* females lived longer when provided with naturally
deposited *H. zea* eggs rather than *Trichoplusia ni* (Hübner) eggs
and suggested the difference was due to the different stimuli
perceived by the females.  The work of Nordland et al. (1976)
also suggested that continued exposure of *T. pretiosum* to the
kairomone did not result in habituation and that total productivity
was greater in females exposed to the kairomone and provided host
eggs daily.  The effects of kairomones on longevity, productivity
and habituation in other host-parasitoid relationships remains
to be determined.


## Host Density Influence

There are several approaches to the use of kairomones in
the augmentation of parasitoids and predators each with its own
specific problems.  One is the repeated application of the kairomone
to attract and act as a releaser to the search and attack cycle
in the parasitoid female.  Such a releaser would result in both
activation and retention of the parasitoid in the treated area.
When relying on the natural parasitoid population, the constant
attraction and retention of parasitoids in the target area may
result in the depletion of these insects from surrounding areas.
This could have several deleterious effects.  The loss of parasitoids
from untreated crops could result in aggravated pest problems
in those areas "robbed" of their normal parasitoid complement.
Secondly, the retention of parasitoids in an area with limited
host resources could result in a reduced number of parasitoids
in the subsequent generation and a lower natality.  Although Nord-

land et al. (1976) reported increased parasitoid fecundity, his
results are based on a continual host supply which may not occur
under field conditions.  These potential problems suggest a need
for judicial use of kairomones in pest management research schemes.

While the application of a kairomone to a low density host
population could result in reduced parasitoid natality, kairomone
application to a high density host population may not have much
effect on the parasitoids due to the elevated level of natural
kairomone.  Kairomones could be used during a host build-up when
the host population would insure reasonable parasitoid production
and at the same time reduce the host population to noneconomic
levels.  As theorized by Knipling (Chapter 3, this treatise) the
host population must reach certain levels before the release of
parasitoids has a beneficial effect.  Application of a kairomone
might be expected to have the greatest benefit just prior to and
after this optimal period.  The careful use of kairomones to attract
and stimulate entomophagous insects as the pest population begins
to increase, but before it reaches the economic threshold or before
the pest population is sufficiently dense to attract and stimulate
parasitoids on its own, appears to show considerable promise as
a pest management tactic.

A second approach to the use of kairomones is through a release
program coupled with behavioral manipulations.  While rearing
of parasitoids and predators is possible (see Morrison and King,
Chapter 6, this treatise) it is often prohibitively expensive.
However, as the development of artificial rearing techniques progresses it may be possible to economically rear large numbers of natural
enemies.  Many of the techniques in this approach are directly
applicable to the importation and release of exotic parasitoids
and predators as well as natural enemies.

The release of parasitoids has often not been effective in
suppressing pest damage.  The lack of success has been partly
the result of failure to retain the parasitoids in the release
and target areas.  The use of kairomones as releasers to activate
parasitoid host-seeking behavior during their release, as described
by Gross et al. (1975) would appear to offer a solution to this
problem.  Even more intriguing is the possibility that the parasitoids
can be directed toward specific crops and hosts.  As suggested
by Arthur (1966) and Vinson et al. (1977), a parasitoid, having
found and parasitized a few hosts of one species, may learn to
associate additional chemical cues with that particular host species.
The result is a more selective and efficient host selection process.
Gross et al. (1975) suggested that plant components in addition
to the host seeking kairomone produce the oriented searching pattern
as observed upon stimulation of *M. croceipes* by larval frass.
The use of kairomones or associated chemicals may not only stimulate
the released parasitoid to remain in the area but the use of chemicals

from the host plant and host may aid in orientating the released
female toward the particular host in the particular crop as required.


### ACCESSORY USES OF KAIROMONES, ALLOMONES AND PHEROMONES

Kairomones may have some use in traps to survey for presence
and abundance of certain species of parasitoids or predators.
Kairomones that act over long distances would appear particularly
useful.  Another possibility is the use of pheromone traps for
parasitoids and predators.  Lewis et al. (1970) found that caging
female *C. nigriceps* in the field resulted in the trapping of both
sexes.

Kairomones may be of great importance in the development
of methods for rearing of entomophagous insects on artificial
host diets.  One of the problems in the artificial rearing of
parasitoids, even if an adequate diet is available (see House,
Chapter 5, this treatise), is the collection of large numbers
of viable parasitoid eggs.  The use of kairomones offers a means
of readily obtaining the eggs necessary for a mass artificial
rearing facility.  Nettles and Burks (1975) found that a protein
from the frass of *H. zea* stimulated its parasitoid, *A. marmoratus*,
to larviposit.  In studies with *I. conquisitor*, a larval parasitoid
of *Galleria mellonella* (L.), Arthur et al. (1972) and Hegdekar
and Arthur (1973) found that several amino acids and $CaCl_2$ from
the hosts hemolymph would result in oviposition.  Rajendram and
Hagen (1974) using small wax droplets containing several amino
acids were able to induce oviposition by *Trichogramma californicum*
Nafaraja and Nagarkatti.  The ability to release oviposition behavior
in some parasitoids may be of added significance.  As shown by
Went and Krause (1973) eggs dissected from the ovary of the female
parasitoid *Pimpla turionellae* L, will only hatch when activated
by pressure induced deformation as the eggs pass down the egg
canal of the ovipositor.  By using kairomones to induce oviposition
it may be possible to obtain large numbers of viable eggs required
in an artificial mass rearing program.

Kairomones also play a role in host specificity and preference.
Some species of parasitoids readily attack certain insect hosts
when encountered even though these do not serve as natural hosts
for the parasitoid (Lewis and Vinson 1971, Lingren and Noble 1971,
Calvert 1973).  It has been shown that the contamination of an
otherwise unacceptable host by the odor or kairomones of a preferred
host may result in the attack of the unacceptable host (Thorpe
and Jones 1937, Bartlett 1953, Tawfik 1957).  For example, *C.
nigriceps* will normally not parasitize *G. mellonella* or *S. frugiperda* but will do so following the application of the kairomones
from *H. virescens*, its natural host (Vinson 1975a).  In some cases
the novel host has been found to be suitable for the development

of the parasitoids progeny (Thorpe and Jones, 1937, Spradberry 1968). The application of kairomones and the resultant contamination of pest species normally not attacked by a particular parasitoid might cause such hosts to become susceptible to the parasitoid. While such a situation might rarely occur it could offer some unique opportunities. Some host insects are cheaper and easier to rear. If by kairomone manipulation, non-hosts could be converted to a acceptable host in a mass rearing program it might cut costs. This approach also could be used to rear parasitoids in areas where the natural host is quarantined.

## REFERENCES CITED

Alcock, J. 1973. Cues used in searching for food by red-winged blackbirds (*Agelaius phoeniceus*). Behaviour 46:174-88.

Alloway, T. M. 1973. Learning in insects except apoidea. pp. 131-71. In Invertebrate Learning. Vol. 1. [W. C. Corning; J. A. Dayal and A. O. D. Willows, Eds.] Plenum Press, New York.

Arthur, A. P. 1962. Influence of host tree on abundance of *Itoplectis conquisitor* (Say), a polyphagous parasite of the European shoot moth *Rhyacionia buoliana* (Schiff.). Can. Entomol. 94:337-47.

Arthur, A. P. 1966. Associative Learning in *Itoplectis conquisitor* (Say). Can. Entomol. 98:213-23.

Arthur, A. P. 1967. Influences of position and size of host on host searching by *Itoplectis conquisitor*. Can. Entomol. 99:877-86.

Arthur, A. P. 1971. Associate learning by *Nemeritis canescens*. Can. Entomol. 103:1137-41.

Arthur, A. P., B. M. Hegdekar, and C. Rollins. 1969. Component of the host hemolymph that induces oviposition in a parasitic insect. Nature 223:966-67.

Arthur, A. P., B. M. Hegdekar, and W. W. Batsch. 1972. A chemically defined synthetic medium that induces oviposition in the parasite *Itoplectis conquisitor*. Can. Entomol. 104:1251-58.

Arthur, A. P., J. E. R. Stainer, and A. L. Turnbull. 1964. The interaction between *Orgilus obscurator* (Nees) and *Temelucha* (Grav.), parasites of the pine shoot moth *Rhyacionia buoliana* (Schiff.). Can. Entomol. 96:1030-34.

Ashley, T. R., J. C. Allen, and D. Gonzales. 1974. Successful parasitization of *Heliothis zea* and *Trichoplusia ni*. eggs by *Trichogramma*. Environ. Entomol. 3:319-22.

Askew, R. R. 1966. On the biology of the inhabitants of oak galls of Cynipidae (Hymenoptera) in Britain. Trans. Soc. Br. Ent. 14:237-68.

Baier, M. 1964. Zer Biologie and Grodologie des Sattelmucke *Haplodiplosis equestris* Wagener. Z. Angew. Ent. 53:217-73.

Bakker, K., S. N. Bagchee, W. R. Van Swet, and E. Meelis. 1967.
    Host discrimination in *Pseudeucoila bochei*. Entomol. Exp.
    Appl. 10:295-311.
Bartlett, B. R. 1953. A tactile ovipositional stimulus to culture
    *Macrocentrus ancylivorus* on an unnatural host. J. Econ. Ento-
    mol. 46:525.
Blum, M. S. 1974. Pheromonal sociality in the hymenoptera.
    pp. 222-49. *In* Pheromones [M. C. Birch, Ed.]. American
    Elsevier Pub., New York, 495 pp.
Bosch, van den, R., and A. D. Telford. 1964. Environmental modifica-
    tion and biological control. pp. 459-88. *In* Biological Con-
    trol of Insect Pests and Weeds. [P. DeBach, Ed]. Chapman
    and Hall, London. 844 pp.
Bragg, D. 1974. Ecological and behavioral studies of *Phaeogenes
    cynarae*: Ecology; host specificity; search and oviposition;
    and avoidance of super-parasitism. Ann. Entomol. Soc. Am.
    67:931-36.
Brown, W. L., Jr., T. Eisner and R. H. Whittaker. 1970. Allomones
    and kairomones: Transpecific chemical messengers. Bioscience
    20:21-2.
Cade, W. 1975. Acoustically orienting parasitoids: Fly phono-
    taxis to cricket song. Science 190:1312-3.
Calvert, D. 1973. Experimental host Preference of *Monoctonus
    paulensis*, including a hypothetical scheme of host selection.
    Ann. Entomol. Soc. Am 66:28-33.
Camors, F. B. Jr., and T. L. Payne. 1971. Response in *Heydenia
    unica* to *Dendroctonus frontalis* pheromones and a host-tree
    terpene. Ann. Entomol. Soc. Am. 65:31-3.
Camors, F. B., Jr., and T. L. Payne. 1973. Sequence of arrival of
    entomophagous insects to trees infested with the southern pine
    beetle. Environ. Entomol. 2:267-70.
Carton, Y. 1974. Biologie de Pimpla instigator III Analyse
    experimentale du processus de reconnaissance de l'hotechyr-
    solida. Entomol. Exp. Appl. 17:265-78.
Chabora, P. C. 1967. Hereditary behavior variation in oviposition
    patterns in the parasite *Nasonia vitripennis*. Can. Entomol.
    99:763-65.
Clausen, C. P. 1940. Entomophagous Insects. McGraw-Hill:New York.
    688 pp.
Clausen, C. P., H. A. Jaynes and T. R. Gardner. 1933. Further
    investigations of the parasites of *Popillia japonica* in the
    far east. U.S. Dept. of Agric. Tech. Bull. 366. 58 pp.
Corbet, S. A. 1971. Mandibular gland secretion of larvae of the
    flour moth, *Anagasta kuehniella*, contains an epidemic pheromone
    and elicits oviposition movements in a hymenopteran parasite.
    Nature 232:481-84.
Corbet, P. S., C. Longfield and N. W. Moore. 1960. Dragonflies.
    Collins, London.
Cross, W. W. and T. L. Chesnut. 1971. Arthropod parasites of the
    boll weevil, *Anthonomus grandis*: 1. An annotated list. Ann.

Entomol. Soc. Amer. 64:516-27.

Croze, H. 1970. Searching Image in Carrion Crows. Parey, Berlin. 86 pp.

Cushman, R. A. 1926. Location of individual hosts versus systematic relation of host species as a determing factor in parasite attack. Proc. Entomol. Soc. Wash. 28:5-6.

DeBach, P. 1949. Population studies of the long-tailed mealybug and its natural enemies on citrus trees in southern California, 1949. Ecology 30:14-25.

DeBach, P. 1964. Biological Control of Insect Pests and Weeds. Chapman and Hall, Ltd. London. 844 pp.

DeBach, P. and K. S. Hagen. 1964. Manipulation of entomophagous species. pp. 429-58. In Biological Control of Insect Pests and Weeds. [P. DeBach, Ed]. Chapman and Hall, London. 844 pp.

Dodd, A. P. 1953. Observations on the stem gall-fly of Pamekani, Eupatorium glandulosum. Proc. Hawaiian Entomol. Soc. 15:41-4.

Doutt, R. L. 1957. Biology of Solenotus begini (Ashmead). J. Econ. Entomol. 50:373-4.

Doutt, R. L. 1964. Biological characteristics of entomophagous adults. pp. 145-67. In Biological Control of Insect Pests and Weeds [P. DeBach, Ed.]. Chapman and Hall, London. 844 pp.

Edwards, R. L. 1954. The host-finding and oviposition behavior of Mormoniella vitripennis (Walker), a parasite of muscoid flies. Behavior 7:88-112.

Edwards, R. L. 1955. How the hymenopteran parasite Mormoniella vitripennis (Walker) finds its host. Bri. J. Anim. Behav. 3:37-8.

Farkas, S. R. and H. H. Shorey. 1974. Mechanisms of orientation to a distant pheromone source. pp. 81-95. In Pheromones. [M. C. Birch, Ed]. American Elsevier Pub. Co. N.Y. 495 pp.

Fleschner, C. A. 1950. Studies on searching capacity of the larvae of three predators of the citrus red mite. Hilgardia 20:233-65.

Finney, G. L. 1948. Culturing Chrysopa californica and obtaining eggs for field distribution. J. Econ. Entomol. 41:719-21.

Fisher, R. C. 1961. A study in insect multiparasitism. I. Host selection and oviposition. J. Exp. Biol. 38:267-75.

Fisher, R. C. 1971. Aspects of the physiology of endoparasitic hymenoptera. Cambridge Phil. Soc. Biol. Rev. 46:243-78.

Flanders, S. E. 1953. Variations in susceptibility of citrus-infesting coccids to parasitization. J. Econ. Entomol. 46:266-69.

Gerling, D. and A. Schwartz. 1974. Host selection by Telonomus remus, a parasite of Spodoptera littoralis eggs. Entomol. Exp. Appl. 17:391-96.

Gordh, G. 1973. Biological investigations on Comperia merceti (Compere), an encyritid parasite of the cockroach Supella longepalpa (Serville). J. Econ. Entomol. Ser. A 47:115-27.

Greany, P. D. and E. R. Oatman. 1971a. Demonstration of host

discrimination in the parasite *Orgilus lepidus*. Ann. Entomol.
Soc. Am. 65:375-6.

Greany, P. D. and E. R. Oatman. 1971b. Analysis of host discrimination in the parasite *Orgilus lepidus*. Ann. Entomol. Soc.
Am. 65:377-83.

Greany, P. D., J. H. Tumlinson, D. L. Chambers and G. M. Boush.
1977. Chemically-mediated host finding by *Biosteres* (*Opius*)
*longicaudatus*, a parasitoid of tephritid fruit fly larvae.
J. Chem. Ecol. 3:189-95.

Granett, J. and N. C. Leeling. 1971. A hyperglycemic agent in the
serum of DDT - prostrate American cockroaches, *Periplaneta
americana*. Ann. Ent. Soc. Amer. 65:299-302.

Griffiths, K. J. 1971. Discrimination between parasitized and
unparasitized hosts by *Pleolophus basizonus*. Proc. Entomol.
Soc. Ont. 102:83-91.

Gross, H. R., W. J. Lewis, and R. L. Jones. 1975. Kairomones and
their use in the management of entomophagous insects. III.
Stimulation of *Trichogramma achaeae*, *T. pretiosum* and *Microplitis croceipes* with host-seeking stimuli at time of release to
improve their efficiency. J. Chem. Ecology. 1:431-8.

Guillot, F. S. and S. B. Vinson. 1972. Sources of substances which
elicit a behavioral response from the insect parasitoid, *Campoletis perdistinctus*. Nature 235:169-70.

Guillot, F. S., R. L. Joiner, and S. B. Vinson. 1974. Host discrimination: isolation of hydrocarbons from Dufour's gland of a
braconid parasitoid. Ann. Entomol. Soc. Am. 67:720-1.

Gutierrez, A. P. 1970. Studies on host selection and host specificity of the aphid hyperparasite *Charips victrix*. 6. Description of sensory sturctures and a synopsis of host selection
and host specificity. Ann. Entomol. Soc. Am. 63:1705-9.

Hagen, K. S. 1950. Fecundity of *Chrysopa californica* as affected
by synthetic foods. J. Econ. Entomol. 43:101-4.

Hagen, K. S. and R. L. Tassan. 1970. The influence of food Wheast[R]
and related *Saccharomyces fragilis* yeast products in the fecundity of *Chrysopa carnea*. Can. Entomol. 102:806-11.

Hagen, K. S., E. F. Sawall, Jr., and R. L. Tassan. 1970. The use of
food sprays to increase effectiveness of entomophagous insects.
Proc. Tall Timb. Conf. Ecol. An. Hab. Mat. 2:59-81.

Hagen, K. S., P. Greany, E. F. Sawall, Jr. and R. L. Tassan. 1976.
Tryptophan in artificial honeydews or a source of an attractant
for adult *Chrysopa carnea*. Environ. Entomol. (in press).

Hassell, M. P. 1968. The behavioral response of a tachinid fly
[*Cyzenis albicans* (Fall.)] to its host, the winter moth
[*Operophtera brumata* (L.)]. J. Anim. Ecol. 37:627-39.

Hawke, S. D., R. D. Farley, and P. D. Greany. 1973. The fine
structure of sense organs in the ovipositor of the parasitic
wasp. Tissue Cell 5:171-84.

Hays, D. B. and S. B. Vinson. 1971. Host acceptance by the parasite *Cardiochiles nigriceps* Viereck. Anim. Behav. 19:344-52.

Hegdekar, B. M. and A. P. Arthur. 1973. Host hemolymph chemicals that induce oviposition in the parasite *Itoplectis conquisitor*. Can. Entomol. 105:787-93.

Hendry, L. B., P. D. Greany, and R. J. Gill. 1973. Kairomone mediated host-finding behavior in the parasitic wasp *Orgilus lepidus*. Entomol. Exp. Appl. 16:471-77.

Henson, R. D., S. B. Vinson, and C. S. Barfield. 1977. Ovipositional behavior of *Bracon mellitor* Say, a parasitoid of the boll weevil. III. Isolation and identification of natural releases of oviposition probing. J. Chem. Ecol. 3:151-8.

Herrebout, W. M. and J. van den Veer. 1969. Habitat selection in *Eucarcelis retilla* Vill. III. Preliminary results of olfactometer experiments with females of known age. Z. Angew. Entomol. 64:55-61.

Hölldobler, B. 1969. Host finding of odor in the myrmercophilis beetle *Atemeles pubicollis* Bris. (Coleoptera; Staphylinidae). Science 166:757-8.

Hoefer, I. and M. Lindauer. 1975. Das lernverhalten zwier bienenrassen unter veränderten orientierungsbedingungen. J. Comp. Physiol. 99:119-38.

Hoefer, I. and M. Lindauer. 1976. Der einfluss einer vordressur auf das lernverhalten der honigbiene. J. Comp. Physiol. 109:249-64.

Hsiao, T., F. G. Holdaway, and C. H. Chang. 1966. Ecological and physiological adaptations in insect parasitism. Entomol. Exp. A. pp. 9:113-23.

Jacobson, M. 1972. Insect Sex Pheromones. Academic Press, New York 382 pp.

Jones, R. L., W. J. Lewis, M. C. Bowman, M. Beroza, and B. A. Bierl. 1971. Host-seeking stimulant for parasite of corn earworm: isolation identification, and synthesis. Science 173:842-3.

Jones, R. L., W. J. Lewis, M. Beroza, B. A. Bierl, and A. N. Sparks. 1973. Host-seeking stimulants (kairomones) for the egg parasite *Trichogramma evanescens*. Environ. Entomol. 2:593-6.

Jones, R. L., W. J. Lewis, H. R. Gross, Jr., and D. A. Nordland. 1976. Use of kairomones to promote action by beneficial insect parasites. pp. 119-34. *In* Pest Management With Insect Sex Attractants. [M. Beroza, Ed] Acs symposium series. 23.

Labeyrie, V. 1958. Facteurs conditionnant la ponte de *Microgaster globatus* Nees. Bull. Soc. Entomol. Fr. 63:62-6.

Lenteren, J. C. van. 1976. The development of host discrimination and the prevention of superparasitism in the parasite *Pseudeucoila bochei* Weld. (Hym.:Cynipidae). Neth. J. Zool. 26:1-83.

Lewis, W. J. and S. B. Vinson. 1971. Suitability of certain *Heliothis* as hosts the parasite *Cardiochiles nigriceps*. Ann. Entomol. Soc. Am. 64:970-2.

Lewis, W. J., J. W. Snow and R. L. Jones. 1970. A pheromone trap for studying populations of *Cardiochiles nigriceps*, a parasite of *Heliothis virescens*. J. Econ. Entomol.

     64:1417-21.
Lewis, W. J. and R. L. Jones. 1971. Substance that stimulates
     host-seeking by *Microplitis croceipes*, a parasite of *Heliothis*
     species. Ann. Entomol. Soc. Am. 64:471-3.
Lewis, W. J., R. L. Jones, and A. N. Sparks. 1972. A host-
     seeking stimulant for the egg parasite *Trichogramma evanescens*;
     its source and a demonstration of its laboratory and field
     activity. Ann. Entomol. Soc. Am. 65:1087-9.
Lewis, W. J., R. L. Jones, D. A. Nordland and A. N. Sparks. 1975a.
     Kairomones and their use for management of entomophagous in-
     sects:  I.  Evaluation for increasing rate of parasitization
     by *Trichogramma* spp. in the field. J. Chem. Ecol. 1:343-7.
Lewis, W. J., R. L. Jones, D. A. Nordland and H. R. Gross, Jr.
     1975b. Kairomones and their use for management of entomopha-
     gous insects:  II.  Mechanisms causing increase in rate of
     parasitization by *Trichogramma* spp. J. Chem. Ecol. 1:349-60.
Lewis,  W. J., R. L. Jones, H. R. Gross, Jr., and D. A. Nordland.
     1976.  The role of kairomones and other behavioral chemicals in
     host finding by parasitic insects. Behavioral Biology 16:267-89.
Lewis, W. J., D. A. Nordland, H. R. Gross, Jr., R. L. Jones and S.
     L. Jones. 1977. Kairomones and their use for management of
     entomophagous insects:  V. Moth scales as a stimulus for pred-
     ators of *Heliothis zea* (Boddie) eggs by *Chrysopa carnea*
     Stephens larvae. J. Chem. Ecol. (In press).
Leonard, D. E., B. A. Bierl and M. Beroza. 1975. Gypsy moth kairo-
     mones influencing behavior of the parasitoids *Brachymeria inter-
     media* and *Apanteles melanoscelus*. Environ. Entomol. 4:929-30.
Lingren, P. D. and L. W. Noble. 1971. Preference of *Campoletis*
     *perdistinctus* for certain noctuid larvae. J. Econ. Entomol.
     65:104-7.
Lloyd, D. C. 1942. Further experiments on host selection by
     hymenopterous parasites of the moth, *Plutella maculipennis*
     Curtis. Rev. Can. Biol. 1:633-45.
Madden J. 1968. Behavioral responses of parasites to symbiotic
     fungus associated with *Sirex noctilio* F. Nature (London)
     218:189-90.
Martin, C. H. 1946. Influence of characteristics of the puncture
     in potatoes on parasitism of potato tuber moth by *Macrocen-
     turs ancylivorus*. J. Econ. Entomol. 39:516-21.
Matthews, R. W. 1974. Biology of Braconidae. Ann. Rev. Entomol.
     19:15-32.
Miller, M. C. 1972. Scanning electron microscope studies of the
     flagellar sense receptors of *Paridesmia discus* and *Nasonia*
     *vitripennis*. Ann. Entomol. Soc. Am. 65:1119-23.
Mitchell, W. C. and R. F. L. Man. 1970. Response of the female
     southern green stink bug and its parasite *Trichopods pennipes*
     to male stink bug pheromones. J. Econ. Entomol. 64:856-9.
Monteith, L. G. 1955. Host preferences of *Drino bohemica* Mesn.
     with particular reference to olfactory responses. Can. Ento-
     mol. 87:509-30.

Monteith, L. G. 1956. Influence of host movement on selection of hosts by *Drino bohemica* Mesn. as determined in an olfacto-meter. Can. Entomol. 88:583-6.

Moore, B. P. 1974. Pheromones in the termite societies. pp. 250-6. *In* Pheromones [M. C. Birch, Ed]. American Elsevier Pub. New York, 495 pp.

Moran, V. C., D. J. Brothers, and J. J. Case. 1969. Observations on the biology of *Tetrastichus flavigaster* Brothers and Moran parasitic on psyllid nymphs. Trans. Roy. Entomol. Soc. London 121:41-58.

Mudd, A. and S. A. Corbet. 1973. Mandibular gland secretion of larvae of the stored products pests *Anagasta kuehniella*, *Ephestia cautella*, *Plodia interpunctella* and *Ephestia elutella*. Entomol. Exp. App. 16:291-3.

Muesebeck, C. F. W. and S. M. Dohanian. 1927. A study of hyper-parasitism with particular reference to the parasites of *Apanteles melanoscelus* (Ratzeburg). U.S. Dept. Agric. Bull. 1487. 35 pp.

Murphey, R. K. and B. Mendenhall. 1973. Localization of receptors controlling orientation to prey by the back swimmer *Notonecta undulata*. J. Comp. Physiol. 84:19-30.

Nettles, W. C. and M. L. Burks. 1975. A substance from *Heliothis virescens* larvae stimulating larviposition by females of the tachinid, *Archytas marmoratus*. J. Insect Physiol. 21:965-78.

Nordland, D. A. and W. J. Lewis. 1976. Terminology of chemical releasing stimuli in intraspecific and interspecific inter-actions. J. Chem. Ecol. 2:211-20.

Nordland, D. A., W. J. Lewis, R. L. Jones and H. R. Gross. 1976. Kairomones and their use for management of entomophagous insects: IV. Effect of kairomones on productivity and longevity of *Trichogramma pretiosum* Riley. J. Chem. Ecol. 2:67-72.

Norton, W. N. and S. B. Vinson. 1974. A comparative ultrastructural and behavioral study of the antennal sensory sensilla of the parasitoid, *Cardiochiles nigriceps*. J. Morphol. 142:329-50.

Orphanides, G. M. and D. Gonzales. 1970. Effects of adhesive materials and host location on parasitization by uniparental race of *Trichogramma pretiosum*. J. Econ. Entomol. 63:1891-8.

Payne, T. L. 1974. Pheromone perception. pp. 35-61. *In* Phero-mones [M. C. Birch, Ed]. American Elsevier Pub. N.Y., 495 pp.

Pettey, F. W. 1948. The biological control of prickly pears in South Africa. Union S. Africa Dept. Agric. Sci. Bull. 271:163 pp.

Picard, F. 1922. Parasites de Pieris brassicae L. Bull. Biol. Fr. Belg. 56:54.

Price, P. W. 1972. Behaviour of the parasitoid *Pleolphus basi-zonus* in response to changes in host and parasitoid density. Can. Entomol. 104:129-40.

Price, P. W. 1975. Insect Ecology. John Wiley and Sons, New York. 514 pp.

Pritchard, G. 1965. Prey capture in dragonfly larvae (Odonata:

Anisoptera).  Can. J. Zool.  43:271–90.

Quednau, F. W.  1967.  Notes on mating behaviour and oviposition
of *Chrysocharis laricinellae*, a parasitoid of the larch
casebearer.  Can. Entomol.  99:326–31.

Quednau, F. W. and H. M. Hübsch.  1964.  Factors influencing the
host-finding and host acceptance pattern in some *Aphytis*
species.  S. Afr. J. Agric. Sci.  7:543–54.

Quinn, W. G., W. A. Harris and S. Benzer.  1974.  Conditioned behavior in *Drosophila melanogaster*.  Proc. Nat. Acad. Sci.
71:708–12.

Rajendram, G. F. and K. S. Hagen.  1974.  *Trichogramma* oviposition
into artificial substrates.  Environ. Entomol.  3:399–401.

Read, D. P., P. P. Feeney, and R. B. Root.  1970.  Habitat selection
by the aphid parasite *Diaretiella rapae*.  Can. Entomol.
102:1567–78.

Rice, R. E.  1969.  Response of some predators and parasites of *Ips
confusus* (LeC.) to olfactory attractants.  Contrib. Boyce
Thompson Inst.  23:189–94.

Richerson, J. V. and C. J. DeLoach.  1972.  Some aspects of host
selection by *Perilitus coccinellae*.  Ann. Entomol. Soc. Am.
65:834–9.

Robacker, D. C., K. M. Weaver and L. B. Hendry.  1976.  Sexual
communication and associative learning in the parasitic wasp,
*Itoplectis conquisitor* (Say).  J. Chem. Ecol.  2:39–48.

Roth, J. P.  1976.  Host habitat location and larviposition stimulation response of the tachinid, *Lixophaga diatraeae* (Townsend).  Dissertation, Mississippi State University, Mississippi State, MS.  68 pp.

Rogers, D.  1972.  Random search and insect population models.
J. Anim. Ecol.  41:369–83.

Ryan, R. E. and J. A. Rudinsky.  1962.  Biology and habits of the
douglas fir beetle parasite, *Coeloides brunneri* Viereck.
Can. Entomol.  94:748–63.

Saad, A. A. B. and G. W. Bishop.  1976.  Attraction of insects
to potato plants through the use of artificial honeydews and
aphid juice.  Entomophaga.  21:49–57.

Salt, G.  1934.  Experimental studies in parasitism. II.  Superparasitism.  Proc. Roy. Soc. London Ser. B.  114:455–76.

Salt, G.  1935.  Experimental studies in insect parasitism. III.
Host selection.  Proc. Roy. Soc. B.  117:413–35.

Salt, G.  1937.  Experimental studies in insect parasitism. V.
The sense used by *Trichogramma* to distinguish between para-
sitized and unparasitized hosts.  Proc. Roy. Soc. London
Ser. B.  122:57–75.

Samson-Boshuizen, M., J. C. van Lenteren, and K. Bakker.  1974.
Success of parasitization of *Pseudeucoila bochei* Weld a
matter of experience.  Neth. J. Zool.  24:67–85.

Schmidt, G. T.  1974.  Host-acceptance behaviour of *Campoletis sonorensis* toward *Heliothis zea*.  Ann. Entomol. Soc. Am. 67:835–44.

Schneider, F.  1948.  Beitrag zer Kenntnis der Generationsverhäll-

nisse und Diapause räuberrscher Schwebefliegen (Syrphidae).
Mitteil., Schweiz. Ent. Ges. 21:249-85.

Schwinch, I. 1955. Weitere Untersuchungen zur Farge Geuichsori-
entierung der Nochtschmetterlinge: Partielle Fühleramputa-
tion bei Spinnermännchen, insbesondere am Slidenspinner
*Bomblyx mori*. L. Z. Vergl. Physiol. 37:439-58.

Shahjahan, M. 1974. *Erigeron* flowers as food and attractive odor
source for *Peristenus pseudopallipes*, a braconid parasitoid
of the tarnished plant bug. Environ. Entomol. 3:69-72.

Shorey, H. H. 1973. Behavioral responses to insect pheromones.
Ann. Rev. Entomol. 18:349-80.

Slifer, E. H. 1969. Sense organs on the antennae of a parasitic
wasp, *Nasonia vitripennis*. Biol. Bull. 136:253-63.

Slobodchikoff, C. N. 1973. Behavioral studies of three morphotypes
of *Therion circumflexum*. Pan-Pac. Entomol. 49:197-206.

Soper, R. S., G. E. Shewell and D. Tyrrell. 1976. *Colcondamyia
auditrix* New Sp. Diptera:Saracophagidae), a parasite which is
attracted by the mating song of its host, *Okanagana rimosa*
(Homoptera: Cicadidae) Can. Ent. 108:61-8.

Southwood, T. R. E. 1966. Ecological Methods. Methuen and Co.
London. 391 pp.

Spradberry, J. P. 1968. A technique for artificially culturing
ichneumonid parasites of wood wasps. Entomol. Exp. Appl.
11:257-60.

Spradberry, J. P. 1970. The biology of *Ibalia drewseni* Borries
a parasite of siricid woodwasps. Proc. Roy. Entomol. Soc.
London Ser. B 45:104-13.

Stary, P. 1966. Aphid parasites of Czechoslovakia. A review of the
Czechoslovak Aphidiidae (Hymenoptera). The Hague: Junk. 242 pp.

Sternlicht, M. 1973. Parasitic wasps attracted by the sex pheromone
of the coccid host. Entomophaga 18:339-42.

Sundby, R. A. 1967. Influence of food on fecundity of *Chrysopa
carnea* Stephens. Entomophaga 12:475-9.

Sweetman, H. L. 1958. The Principles of Biological Control. Wm.
C. Brown Co., Dubuque, Iowa. 560 pp.

Taylor, J. S. 1932. Report on cotton insects and disease investigations. Pt. II. Notes on the American bollworm (*Heliothis
obsoleta* F.) on cotton and on its parasites (*Microbracon brevicornis* Wesm). Sci. Bull. Dept. Agric. For. Un. S. Afr. No.
113. 18 pp.

Tawfik, M. F. S. 1957. Host parasite specificity in a braconid
*Apanteles glomeratus* L. Nature 179:1031-2.

Tetle, J. P. 1974. Pheromones in insect population management.
pp. 399-410. *In* Pheromones [M. C. Birch, Ed.]. American
Elsevier Pub. New York. 495 pp.

Thorpe, W. H. and F. G. W. Jones. 1937. Olfactory conditioning
in a parasite insect and its relation to the problem of host
selection. Proc. Roy. Entomol. Soc. London Ser. B. 124:56-81.

Thompson, W. R. 1951. The specificity of host relations in predaceous insects. Can. Ent. 83:262-9.

Tillyard, R. J. 1930. The biological control of noxious weeds. Trans. 4th Internatl. Congr. Ent. 2:4-9 (1928).

Traynier, R. M. M. 1968. Sex attraction in the Mediterranean flower moth, *Anagasta kuehniella*: location of the female by the male. Can. Entomol. 100:5-10.

Tumlinson, J. H. and R. R. Heath. 1976. Structure elucidation of insect pheromones by microanalytical methods. J. Chem. Ecol. 2:87-99.

Ullyett, G. C. 1947. Mortality factors in populations of *Plutella maculipennis* Curtis (Lepidoptera: Tineidae), and their relation to the problems of control. Union S. Africa, Dept. Agric. Ent. Mem. 2:77-202.

Vinson, S. B. 1968. Source of a substance in *Heliothis virescens* that elicits a searching response in its habitual parasite, *Cardiochiles nigriceps*. Ann. Entomol. Soc. Am. 61:8-10.

Vinson, S. B. 1972. Competition and host discrimination between two species of tobacco budworm parasitoids. Ann. Entomol. Soc. Am. 65;229-36.

Vinson, S. B. 1975a. Biochemical coevolution between parasitoids and their hosts. pp. 14-48. *In* Evolutionary Strategies of Parasitic Insects and Mites. [P. W. Price, Ed]. Plenum:New York. 225 pp.

Vinson, S. B. 1975b. Source of material in the tobacco budworm involved in host recognition by the egg-larval parasitoid, *Chelonus texanus*. Ann. Entomol. Soc. Am. 68:381-4.

Vinson, S. B. 1976. Host selection by insect parasitoids. Ann. Rev. Entomol. 21:109-33.

Vinson, S. B. and W. J. Lewis. 1965. A method of host selection by *Cardiochiles nigriceps*. J. Econ. Entomol. 58:869-71.

Vinson, S. B. and F. S. Guillot. 1972. Host-marking: source of a substance that results in host discrimination in insect parasitoids. Entomophaga 17:241-5.

Vinson, S. B., F. S. Guillot and D. B. Hays. 1973. Rearing of *Cardiochiles nigriceps* in the laboratory with *Heliothis virescens* as hosts. Ann. Entomol. Soc. Am. 66:1170-72.

Vinson, S. B., R. L. Jones, P. Sonnet, B. A. Beirl, and M. Beroza. 1975. Isolation, identification and synthesis of host-seeking stimulants for *Cardiochiles nigriceps*, a parasitoid of the tobacco budworm. Entomol. Exp. App. 18:443-50.

Vinson, S. B., R. D. Henson and C. S. Barfield. 1976. Ovipositional behavior of *Bracon mellitor* Say (Hymenoptera: Braconidae), a parasitoid of boll weevil (*Anthonomus grandis* Boh.) I. Isolation and identification of a synthetic releaser of ovipositor probing. J. Chem. Ecol. 2:431-40.

Vinson, S. B., C. S. Barfield and R. D. Henson. 1977. Oviposition behavior of *Bracon mellitor*, a parasitoid of the boll weevil (*Anthonomus grandis*) II. Associative learning. Physiological Entomology (In Press).

Vité, J. P. and D. L. Williamson. 1970. *Thanasimus dubius*: Prey preception. J. Insect Physiol. 16:233-7.

Walker, M. F. 1961. Some observations on the biology of the lady-bird parasite, *Perilitus coccinellae* (Schrank) with special reference to host selection and recognition. Entomol. Mon. Mag. 97:240-4.

Went, D. F. and G. Krause. 1973. Normal development of mechanically activated, unlaid eggs of an endoparasitic hymenoptern. Nature 244:454-6.

Weseloh, R. M. 1970. Influence of host deprivation and physical host characteristics on host selection behavior of the hyperparasitic *Cheiloneurus noxius*. Ann. Entomol. Soc. Am. 64:580-6.

Weseloh, R. M. 1974. Host recognition by the gypsy moth larval parasitoid, *Apanteles melanoscelus*. Ann. Entomol. Soc. Am. 67:585-7.

Weseloh, R. M. 1977. Behavioral responses of the parasite, *Apanteles melanoscelus*, to Gypsy moth silk. Environmental Entomol. 5:1128-32.

Weseloh, R. M. and B. R. Bartlett. 1971. Influence of chemical characteristics of the secondary scale host on host selection behavior of the hyperparasite *Cheiloneurus noxius*. Ann. Entomol. Soc. Am. 64:1259-64.

Williams, J. R. 1951. The factors which promote and influence the oviposition of *Nemeritis canescens* Gravs. Proc. Roy. Entomol. Soc. London Ser. A 26:49-58.

Wilbert, H. 1974. Die Wahrnehmung von Beute durch die Eilarven von *Aphidoletes aphidimyza* (Cedidomyiidae). Entomophaga 19:173-81.

Wilson, D. D., R. L. Ridgway, and S. B. Vinson. 1974. Host acceptance and oviposition behavior of the parasitoid *Campoletis sonorensis*. Ann. Entomol. Soc. Am. 67:271-4.

Wood, D. L., L. E. Brown, W. D. Bedard, P. E. Tilden, R. M. Silverstein and J. O. Rodin. 1968. Response of *Ips confusus* to synthetic sex pheromones in nature. Science 59:1373-74.

Wright, R. H. 1964. Metarchon: A new term for a class of non-toxic pest control agents. Nature (London) 204:603-4.

Wright, R. H. 1965. Finding metarchons for pest control. Nature (London) 207:103-4.

Wylie, H. G. 1965. Discrimination between parasitized and unparasitized house fly pupae by females of *Nasonia vitripennis* Walk. Can. Entomol. 97:279-86.

Young, J. C. and R. M. Silverstein. 1976. Biological and chemical methodology in the study of insect communication. pp. 75-161. *In* Methods in Olfactory Research. [Moulton, D. G., A. Turk and J. W. Johnston, Jr., Ed]. Academic Press, N. Y. pp. 497.

Zwölfer, H. and M. Kraus. 1957. Biocoenotic studies on the parasites of two fir and two oak-tortricids. Entomophaga 2:173-96.

# Section III

# EXPERIMENTAL AND PRACTICAL

# APPLICATIONS OF AUGMENTATION

CHAPTER 9

SEASONAL COLONIZATION OF ENTOMOPHAGES IN THE U.S.S.R.

G. A. Beglyarov and A. I. Smetnik

All-Union Scientific Research Institute of
Phytopathology, Moscow, U.S.S.R., and Central Research
Laboratory for Plant Quarantine, Moscow, U.S.S.R.

The first attempts to use entomophages by means of mass
releases were undertaken in Russia at the beginning of the current
century (Vasil'ev 1906, Radetskii 1911). However, the systematic
development of this approach to biological control was not started
until 1930 after the organization of the Biomethods Laboratory of
the All-Union Institute of Plant Protection (AIPP). From 1930 to
1950 mass releases of the egg parasites, *Telenomus* spp., were
widely used against *Eurygaster integriceps* Puton. Since 1934, and
especially after the organization of a network of mass-rearing
laboratories in the Ukraine in 1944, and later in other republics,
egg parasites, *Trichogramma* spp., began to be used widely against
many harmful Lepidoptera. Because of the labor and expense of
using entomophages and the successes of chemical methods in the
1950's, work was conducted on a small scale by individual researchers and small production laboratories.

A new, vigorous stage in the development of methods of mass
releases of entomophages began in the early 1960's. At the present
time, this approach to biological control in the Soviet Union consists in releasing more than 15 species of predators and parasites
against many species of Lepidoptera that damage grain, deciduous
fruit, citrus, mulberry, sugar beet, and vegetable crops in glasshouses and in the field. In 1976 entomophages were used on more
than 10 million hectares.

In recent years, special attention has been devoted to the
development of methods for commercial rearing and mechanized

283

dispersal of entomophages. More than 10 biological factories are working on the commercial rearing of *Trichogramma*; the first examples of apparatus for the dispersal of this entomophage have already been produced and are being tested; production lines for commercial rearing of green lacewings, *Chrysopa* spp., are close to completion; and production lines for commercial rearing of *Aleochara* are being developed.

We present below the principal results of work on the mass release of entomophages in the U.S.S.R.

## TRICHOGRAMMA

The history of the use of *Trichogramma* in Russia began with preliminary experiments on transporting the effective strains of *Trichogramma evanescens* Westwood from the land along the Volga (Astrakhan) to Central Asia for the purpose of combatting the codling moth (Vasil'ev 1913, Radetskii 1911). The next stage of development consisted in repeated experiments on the study and testing of *Trichogramma* in various countries which were made possible by the development in the U.S.A. (Flanders 1928) of available means of year-round rearing of the parasite in eggs of the Angoumois grain moth, *Sitotroga cerealella* (Oliver).

In the U.S.S.R., extensive investigations on the study and commercial testing of *Trichogramma* were begun in 1937 after the organization of the Biomethods Laboratory of the AIPP under the leadership of N. F. Meier. Shortly thereafter, other scientific research institutions and mass-rearing laboratories were included in the work.

However, in the initial period of use of *Trichogramma*, it was not possible to avoid mistakes and even failures. First, it was supposed that only one species, *Trichogramma evanescens*, lived in the U.S.S.R. This polyphagous species parasitized the eggs of almost 200 species of 5 orders of insects. Consequently, it was used against many pests on different field crops as well as in orchards and forests. In this process, the same species was used everywhere, and was reared on cabbage from eggs of the cabbage moth, [*Mamestra* (=*Barathra*) *brassicae* (L.)], at a station in the Slavyanskii Krasnodar region. To obtain maximum production, the *Trichogramma* was reared under constant conditions in Leningrad and sent throughout the whole country. The wide use of the oophage began when there had been insufficient study of its biology and ecology; therefore, the results were contradictory. In control of the codling moth, the effectiveness of this form of *Trichogramma* varied from 0 to 95% (Lapina 1936, Afanes'eva and Alekseeva 1935, Afanas'eva 1936, 1937, Alekseeva 1936, Dirsh 1937, Koroleva, 1938,

Meier, 1937, 1941, Sidorovkina 1936, Tyumeneva, 1936, Shchepetil'-nikova 1939).

Further study of the ecology of *Trichogramma* reared from eggs of the cabbage moth showed that it has a clearly expressed selective capability in relation to the eggs of noctuids and is extremely hygrophilous (Shchepetil'nikova 1939, 1940, 1962, Meier 1940, 1941).

The necessity arose for investigating the intraspecies differentiation and seeking specialized forms (species or races) of *Trichogramma* which would be adapted to one or another pest and would be less sensitive to unfavorable conditions. In addition to *T. evanescens*, other species of the genus *Trichogramma* and ecologically distinct geographic races were detected by the investigations of N. F. Meier (1940, 1941). Of these, the yellow *Trichogramma* (*T. cacoeciae pallida* Meyer) proved to be more adapted to the codling moth, and *T. cacoeciae pini* Meyer more adapted to the pine moth, and consequently to the habitats of these pests. It was further found that *T. evanescens* prefers the eggs of noctuids and is closely linked with the field biotypes.

Investigations on *Trichogramma* in the prewar period showed that not only the species and intraspecies differentiation of the parasite, but also the conditions of rearing the oophage are important for success. Prolonged rearing of *Trichogramma* for many generations under constant conditions of temperature and humidity that are not natural to it had a negative effect on the subsequent selection of microhabitats by the oophage. For example, the natural *Trichogramma* is active within wide limits of temperature (17-30°C) and humidity (55-95%). On the contrary, a "constant" population that had been reared at 25°C and 75-80% humidity retained its activity only within narrow limits of temperature (22-27°C) and humidity (75-85%). When released in the field, this *Trichogramma*, in contrast to those developing naturally, infested eggs only on the lower side of the leaves, where it was damp, avoiding the drier upper surfaces, where the cutworm, [*Agrotis* (=*Euxoa*) *segetum* *Schiffermuller*], also deposits its eggs.

In the postwar period investigations were intensified on the biological basis of the usefulness of *Trichogramma* and other entomophages. It was found that the regulating role of polyphagous entomophages is limited by noncoincidence of the periods of their development with those of their host; therefore, the number of polyphages at the beginning of the development of a generation of a pest usually is small (Shchepetil'nikova 1954, 1957). This fact brought about the necessity of developing the method of seasonal colonization of entomophages. This method envisages yearly releases of the entomophages at the beginning of development of

the pest and is dependent on further independent multiplication of
the parasite in nature during the warm season of the year. The
effectiveness of use of *Trichogramma* by the method of seasonal
colonization depends on a number of factors. The most important
factors to be considered are ability to accumulate rapidly, ecological pliability, and ability to quickly seek out insect hosts
and select habitats.

Along with many-sided ecological investigations, work was
carried out on the study of the taxonomy of the genus *Trichogramma*.
As a result of his taxonomic investigations, I. A. Telenga (1959)
introduced changes into the existing concept of the species of this
genus and established that *T. cacoeciae pallida* and *T. cacoeciae
pini* are subspecies of *T. cacoeciae* Marchal, but retained *T.
embryophagum* (Hartig) and *T. evanescens* Westwood as independent
species.

Investigations by the All-Union and Ukrainian Institutes of
Plant Protection, the Belorussian Institute of Fruit, Vegetable
and Potato Culture, and other institutions on the ecology of
*Trichogramma* species and intraspecies differentiation, and on
methods of rearing and use were prerequisite to the regionalizing
of different species and intraspecies forms of *Trichogramma* by
zones, crops, and pests (Telenga and Shchepetil'nikova 1949,
Volkov 1954, 1959, Kovaleva 1954, Telenga 1956, 1959,
Shchepetil'nikova 1962, Telenga and Kiseleva 1959, Kolmakova 1962,
1965, Shchepetil'nikova and Fedorinchik 1968, Kartavtsev 1968).

In the Soviet Union there are 5 species (4 domestic and 1
introduced) and numerous intraspecies forms of *Trichogramma*. They
differ substantially from one another in their relationship with
hosts, adaptability to ecological conditions and habitats, and
existence in specific biocenoses. Each species has its own group
of hosts.

*Trichogramma evanescens* is a species that multiplies predominantly on pests populating field habitats. It is adapted
mainly to life near the ground. It is used in the U.S.S.R. against
cutworm, *Agrotis segetun Schif* Fermuller, cabbage moth, *Mamestra
brassica* L., gamma moth, [*Autographa* (=*Phytometra*) *gamma* L.] and a
whole complex of noctuid species that accompany them (10 species in
all); against corn borer, [*Ostrinia* (=*Pyrausta*) *nubilalis* Hubner],
and pea moth, *Laspeyresia nigricana* (Stephens), on a total area of
about 6 million hectares.

*Trichogramma cacoeciae pallida* Meyer and *T. embryophagum* Hartig
parasitize the eggs of pests of orchards or forests and are constant
inhabitants of these biotopes. These *Trichogramma* are confined to
conditions of life in the tree crown. They are rarely encountered

in field habitats; then they are found only in fields bordering on wooded areas (Telenga and Shchepetil'nikova 1949, Volkov 1959, Telenga 1959, Shchepetil'nikova et al. 1968).  According to data reported by the above-mentioned authors, *T. c. pallida* prefers crowded, humid orchards located in low places, in bottom lands of rivers, mainly in the forest-steppe zone; whereas, *T. embryophagum* predominantly populates dry localities.

*Trichogramma minutum* Riley was introduced into the U.S.S.R. from the U.S.A. in 1963.  This population was taken from an arid area in the southwestern part of California from eggs of a Heliothis moth.  All the domestic species of *Trichogramma* are widely distributed in the European part of the U.S.S.R., from the Moscow region to the Black Sea coast of the Caucasus, and in Western and Eastern Siberia and in Central Asia.

The regions where *Trichogramma* is consistently effective are characterized by a hydrothermal coefficient for the period of egg deposition of the insect hosts from 0.9 to 1.2, i.e. regions of optimal humidity.  In these regions, 8-9 years out of 10 are favorable for releases and further activity of the parasite.  The zone embraces the main regions of damage by cutworms and corn borers.  This zone includes the Ukraine, mainly the forest-steppe; the northern Caucasus, primarily the foothills; the forest-steppe zone of the land along the Volga; and the southern part of the central black earth zone of the Russian Soviet Federative Socialist Republic.  In these regions effective results have been obtained from the use of local forms of *Trichogramma* over a period of many years.

In regions that do not lie in this zone, where the weather is rainy and cold at the beginning of the releases, seasonal colonization with *Trichogramma* is inconsistent and insufficient (Moscow, Gorkov, and the north of the Voronezh district).  In fact, in these regions, only the direct action  of *Trichogramma* released in the period of mass oviposition by the pests, is effective.

All the species of *Trichogramma* are generally polyphages, are not synchronized with the development of the insect hosts, and have limited searching ability.  Because of this, the regulating ability of natural populations is limited.  In contrast with specialized species that are synchronized with the development of their hosts, are efficient   searchers, and are effective regulators of the numbers of a pest at any density of its population, polyphages require supplementary hosts in addition to the main host.  In this case, many different acceptable hosts should ensure survival of the polyphages.

In order to compensate for the biological properties associated with polyphages, the method of seasonal colonization is used.  The

main factors determining the effectivesss of season colonization
are the following:

(1) selection of species and forms for the given pest,
crop, and specific climatic and microclimatic conditions;

(2) time of release, rate, number, and point of releases
in relation to the phenology of the pest, its population density,
the leaf surface of the plants, and physical factors;

(3) combination with microbiological, chemical, and agrotechnical measures;

(4) conditions of rearing the *Trichogramma*.

*T. evanescens* is most widely and successfully used in the
U.S.S.R. in controlling 15 species of field pests. This species
has 4 races that are distinguished by their adaptation to the host
(noctuid, pierid, corn [borer], fruit moth) and about 20 ecological
forms. Field tests with artificial placement of eggs on plants of
cabbage, potatoes, and corn showed that when the noctuid race was
released about 90% of the total number of infested eggs of different hosts proved to be eggs of noctuids. The percentage of infestation of eggs of the corn borer by this race was 3 times less than
of the favorite noctuid hosts; grain moth eggs were infested still
less. All populations of *T. evanescens* reared from eggs of cabbage
and other noctuid moths in different parts of the Ukraine, the
center of the U.S.S.R., the Caucasus, and Central Asia preferred
the eggs of different species of noctuids to eggs of the corn borer,
codling moth, and tineids. They ignored the eggs of pierids,
scarcely infesting them even on the same leaves of cabbage. The
productiveness of the progeny of this race after parasitization of
pierid eggs decreased 50-67% (Shchepetil'nikova 1974).

All these data indicate the necessity, when mass rearing in
eggs of the grain moth (a host foreign to the *Trichogramma*), of
carrying out yearly passage of 1-2 generations of the parent culture
through eggs of noctuids, tortricids, and others. The fecundity
when these *Trichogramma* are again reared on moth eggs is twice as
high as previously.

*Trichogramma* used against harmful noctuids on sugar beets,
fallow fields, perennial grasses, cabbage, and potatoes in many
regions are highly effective on large areas with a decrease in the
numbers of a pest by 60-90% and saving of about 10% of the crop
(Meier 1951, Telenga and Shchepetil'nikova 1949, Shchepetil'nikova
1968, 1974). The rate of release in the period of development of
each generation, depending on the density of the pest, was 20-60
thousand individuals/ha. The percentage of parasitization of eggs

of cutworm and cabbage moth on sugar beets in the areas where
*Trichogramma* were released was 77–85%, whereas the parasitization
of eggs in the control did not exceed 9%.  *Trichogramma* releases
against cutworm on occupied and clear fallows under winter crops
resulted in parasitization of 60–88% of eggs, whereas in the control 7–25% were parasitized; releases of *Trichogramma* on cabbage
against cabbage moth resulted in 74–93.6% of the eggs being parasitized, yet only 3–12% were parasitized in the control.

The pierid race readily parasitizes the eggs of pierids.
When releases are made against cabbage white butterfly and other
pierids, only this race is effective.  It has high productivity
and prefers warm places.  However, the pierid race is rare.

The corn borer race is encountered in the northern Caucasus,
in the Stavropol and Krasnodar regions.  Its population parasitizes
eggs of the corn borer from year to year.  The second generation
of the borer often has 60–75% of its eggs parasitized by *Trichogramma*.

In experiments on the study of the selective ability of the
corn borer race of *Trichogramma*, half of the total number of eggs
infested by this race proved to be eggs of the corn borer; the
rest were divided between noctuids, tortricids and tineids
(Kartavtsev 1968).  These data indicate that the specialization in
the corn borer race is not as sharply expressed as in the noctuid
and pierid races.  The eggs of noctuids are attractive to it in
some degree.  Apparently the adaptations of this race are still
being formed and the race itself is in the process of formation.
In contrast to the noctuid forms living in the same area on cabbage,
the corn borer race, which is adapted to the biotope of a corn
field, is tolerant of low humidity.  The rates of release of
*Trichogramma* of this race are 30,000–50,000 individuals per hectare;
with this treatment the damage to stalks is decreased by 50% and
that to ears by 60% (Telenga 1965, Dyadechko and Tron' 1969).

In the structure of the species, *T. evanescens*, a special
place is occupied by the tortricid race, which is a component of
the agrobiocenosis of a fruit orchard.  It selects tortricid eggs.
It inhabits the top part of the tree crown, in contrast with the
noctuid race, which upon release onto a tree accumulates in the
lower stratum.  In test releases against noctuids in fields, the
tortricid race flies up and disperses.  These data emphasize that
the use of the tortricid race on fields and of the noctuid race in
orchards is unpromising.  The tortricid race is also a geographic
race, since it is distributed locally (still found nowhere except
in Astrakhan).  This race is tolerant of dryness.  It has normal
fecundity when the humidity of the air is 25% and the temperature
is 25–27°C.  With respect to effectiveness of use against tortricids,

the Astrakhan tortricid race is not inferior to the orchard species
*T. c. pallida* and *T. embryophagum* (Ismailov and Shchichenkov 1940).

The orchard species of *Trichogramma*, *T. c. pallida* and *T.
embryophagum*, are recommended for use in controlling tortricids in
the eastern and western parts of the zone with one generation of
codling moth.  The rate of release is 1500–2000 individuals per
tree 2–3 times.  The first release is carried out at the beginning of oviposition, the second at the beginning of mass deposition
of the moth eggs, and the third (when necessary) 5–7 days after
the second (Shchepetil'nikova 1968).

The most important question in the technology of use of
*Trichogramma* is adjustment of the rates and the number of releases
of the oophage depending on the density of the pest population.
At a high density the frequency of encounters of the parasite with
the host increases and stimulates the parasitization of the latter.
In *Trichogramma* which are polyphagous the hunting ability is not
high.  Only after encountering the first eggs of the host does the
activity of the parasite intensify and the searching efficiency
increase.  This is especially important when eggs of the host are
deposited individually; whereas, when eggs are deposited in
groups, the discovery of the host is facilitated.  With single egg
deposits the low density of the pest, which hinders searching,
limits the effectiveness of parasitization and the percentage of
parasitized eggs is not sufficiently high.

For pests with group deposition of eggs (cabbage moth), when
the density of eggs deposits increases, the rates of release of
*Trichogramma* also are increased in order to maintain a 1:20 ratio
(Shchepetil'nikova 1968, 1971).  In recent years field experiments
have been carried out to study parasitization of single deposits of
cutworm eggs, codling moth, and oriental fruit moth.  It was
established that at a 1:3 ratio of parasite and host eggs the
parasitization was 75–85%; with an increase in density of the pest
(1:5 ratio) the parasitization was 80–85%, and at 1:10 the parasitization was up to 90% (Kalashnikova 1971).

When there is a high density of eggs, in spite of a high level
of parasitization, some caterpillars still hatch and damage the
planting.  Therefore, an increase or decrease in the rate of release
directly proportional to the number of eggs of the host does not
give the desired effect.  On the basis of an analysis of the available information for a zone of consistent effectiveness, it is
recommended that 30,000 *Trichogramma* be released per hectare which
at a density of 30 cutworm eggs per square meter is equal to a
ratio of 1:10.  The same rate of release is maintained even when
the numbers of the pest decrease because of the reduction in
searching efficiency.  When there is an increase in density above

50 eggs per square meter the rates of release are increased.
Furthermore, the number of releases is increased up to 3. In
regions with low humidity, like Kirghizia, the release is increased
to 80,000-100,000 individuals per hectare (Shchepetil'nikova 1974).

Use of *Trichogramma* by the method of seasonal colonization in
the U.S.S.R. has substantially increased in recent years and
exceeded 7.5 million hectares in 1976.

An important achievement of the last decade has been the
development by coworkers at the All-Union Institute of Plant
Protection of mass-rearing technology designed for mechanization
and automation of the principal production processes. On the basis
of this technology the world's first experimental automated factory
for mass-rearing *Trichogramma* was constructed.

At the present time more than 10 biofactories have been built
in various regions of the country with a productive capacity of
more than 50 billion oophages/season, and the transition from out-
dated technology to modern commercial technology is being accomplished. In the arsenal of biological agents used in production,
*Trichogramma* has great relative significance. According to data
of the Ministry of Agriculture, in 1975-76 about 70% of the area
of use of the biomethod was devoted to this insect.

Hence it is clear what an important problem the mechanization
of the distribution process is. This operation is accomplished by
hand at the present time. In recent years the All-Union Scientific
Research Institute of Biological Methods of Plant Protection
(Kishinev) has constructed and is testing a special apparatus for
scattering paper capsules with *Trichogramma* packaged in them.


### TELENOMUS

Of the more than 40 species of entomophages of the harmful
stinkbug, *Eurygaster integriceps* Puton, known in the U.S.S.R. the
most important are the *Telenomus* oophages and the phasiidine flies
*Diptera, Tachinidae* that are parasites of the adults (Shchepetil'
nikova and Gusev 1970).

The first attempts to use the *Telenomus* oophages in controlling
the harmful stinkbug were undertaken in 1903 by I. V. Vasil'ev.
This investigator brought from Central Asia to the former Kharkov
province and released in nature onto plots infested by the harmful
stinkbug 12,000 oophages belonging to the species *Trissolcus
vassilievi* (Mayr). On the basis of calculations and observations
that had been carried out in the course of 1 year, the author
reported that on the plots where the release was made 57% of the

eggs of the pest were parasitized by the introduced parasite. At
the same time he speculated that the activity of *T. vassilievi*
might be even more effective if local species were not present
(Vasil'ev 1906). However, information now available indicates that
this introduction was unsuccessful. What was accepted by
I. V. Vasil'ev as the species *T. vassilievi* apparently was a
complex of species.

According to reliable data of recent years, in nature the
eggs of the harmful stinkbug are parasitized by a whole series of
species: *Trissolcus grandis* (Thomson), *T. simoni* (Mayr), *T.
scutellaris* (Thomson), *T. pseudoturesis* Rjachovsky, and *Telenomus
chloropus* Thomson. The oophages collected by I. V. Vasil'ev in
the Turkmen republic and released in the Ukraine were not determined before release. If *T. vassilievi* was among the colonized
telenomins, then it did not acclimatize in the Ukraine. This
species has not been found in the European part of the U.S.S.R.
Reports of the distribution of the species in the European part of
the U.S.S.R. are apparently the result of incorrect determinations
(Kozlov 1971).

Numerous investigations carried out in various zones of damage
by the stinkbug in the U.S.S.R. have shown that natural populations
of parasites usually do not suppress the multiplication of the
harmful stinkbug, although they noticeably limit the increase in
its numbers. At the beginning of oviposition of the stinkbug the
infestation of eggs by parasites is slight, increasing with the
development of the generations of oophages, which have a high rate
of multiplication. Only toward the end of oviposition does the
infestation by the oophages reach large proportions. Thus, infestation by the oophages lags with respect to the rate of oviposition
by the harmful stinkbug. This characteristic appears in various
degree depending on the agroclimatic conditions. In a great part
of their geographic range the oophages do not provide a sufficiently
high effect, a fact that has stimulated repeated attempts to use
*Telenomus* by means of seasonal colonization or flooding releases.

In the period from 1938 to 1943, during the massive multipli-
cation of the harmful stinkbug in the U.S.S.R., broad working-
scale tests were carried out on the seasonal colonization of
*Telenomus* (Kulakov 1940, Matovskaya 1940, Talitskii 1940,
Shchepetil'nikova 1942). In some regions the use of oophage
*Telenomus* has entered into the practice of exterminatory work.
Thus, in 1940 in the Ukraine in many specially constructed kolkhoz
laboratories 277.3 million oophages were bred and released on an
area of 50 thousand hectares. In the Dzhambul district of
Kazakhstan alone, where work was carried out under the direction of
Professor N. F. Meier in 1942 on an area of 1140 hectares, more
than 4.9 million oophages were colonized, and in 1943 about 9

million were colonized on an area of 1400 hectares.  *Telenomus*
was also used on a wide scale in the lower Volga area and in some
other regions (Shchepetil'nikova 1949, Kozlov 1971).

For carrying out commercial tests and for practical use
*Telenomus* have been reared on the eggs of the principal host, the
harmful stinkbug (Matkovskii 1940, Talitskii 1940, Shchepetil'
nikova 1938).  The adult stinkbugs that were necessary for
obtaining the eggs were collected in places of overwintering in
the early spring period.  The considerable losses from storage of
the collected insects and the low fecundity of the surviving
stinkbugs (5 eggs per female on the average) led to great expenditures of labor.  For obtaining 1 million oophages it was necessary
to collect about 40 kilograms of stinkbugs (Shchepetil'nikova 1958,
Popov 1971).  Nevertheless, this method has been widely used in
Iran, as well as the U.S.S.R. (1945-1963).

The extreme laboriousness and awkwardness of the technology
of breeding the oophages and also the vigorous development of the
chemical method of controlling the harmful stinkbug, were the
reasons for stopping the work on biological control of this pest
in the postwar period.  However, investigations were continued as
directed toward perfection of the method for mass-rearing of eggs
of the harmful stinkbug.  Especially good results were obtained
as a result of joint investigations by entomologists of the
U.S.S.R., Iran, and France (Alexandrov 1947-1949, Remaudiere 1960,
1961, Zommorodi 1959, Popov 1971).

Along with the use of eggs of the harmful stinkbug for rearing
telenomins, beginning with the second-half of the 1960's much
attention was given to elucidation of the possibility of using
other species of insects for this purpose.  Specialists of the All
Union Institute of Plant Protection carried out an evaluation of
the possibility of mass-rearing of more than 10 species of
Pentatomidae: *Aelia acuminata* (L.), *A. fieberi* Scott, *Palomena
prasina* (L.), *Dolycoris baccarum* (L.), *Graphosoma italica* (Mueller),
*Graphosoma semipunctata* (F.), *Carpcoris* sp., and others.  The most
suitable for mass-rearing are acknowledged to be *G. semipunctata*
and *D. baccarum* (Shchepetil'nikova and Gusev 1970, Gusev 1974).

At the end of the 1960's and the beginning of the 1970's the
All-Union Institute of Plant Protection and other institutions for
4 years again carried out broad experiments on the mass release of
*Telenomus*.  The tests, which were carried out on a total area of
about 3000 hectares in the Krasnodar area and the Voronezh, Kharkov,
and Kirovograd districts, did not give consistently positive results.
The parasitization of eggs of the stinkbug in the tests was
increased by 5-70% (10-20% average) in comparison with the parasitization on fields where the oophages were not released.  It was

determined that the reason for the ineffectiveness was the weak
hunting ability of the colonized *Telenomus*, which appeared
especially noticeable when there were low numbers of the pest
(Kartavtsev et al. 1971, Voronin 1974, Gusev 1974, Kartavtsev et
al. 1975).

At the present time it is acknowledged that the use of
telenomins by means of seasonal colonization is not suitable for
control of the harmful stinkbug.  Furthermore, definite recommendations have been developed with respect to the preservation of
natural populations of entomophages of the harmful stinkbug in a
system of integrated protection of wheat.  For some zones criteria
also have been proposed for ratios of the numbers of the harmful
stinkbug and of its oophages that make it possible to determine
the necessity of carrying out chemical treatments or the possibility
of discontinuing them.


*APHELINUS*

*Aphelinus* [*Aphelinus mali* (Haldeman)] was first brought into
our country in 1926 by Doctor Radzhabli from Italy for controlling
one of the most dangerous pests of apple, the woolly apple aphid
*Eriosoma lanigerum* (Hausmann).  The high effectiveness of *Aphelinus*
became apparent in the very first year in some orchards on the
territory of the Azerbaidzhan S.S.R.  After repeated introduction
beginning in the spring of 1931 under the direction of Professor
N. F. Meier, broad work was carried out on the dispersal and use
of *Aphelinus*.  At present the parasite is distributed practically
throughout the entire geographic range of the woolly apple aphid
(Telenga 1937).

Under the conditions of the Crimea and the Caucasus 6-9 generations of *Aphelinus* develop per year, and in the Transcaucasus
up to 12 (Goryunova 1967).  In September-October, when temperatures
of $13^{\circ}$ and lower set in, the larvae fall into diapause and stay in
the mummies of the aphids to overwinter.  Usually the flight of the
adult parasites ceases at the end of October.  The activity of the
*Aphelinus* depends on the weather conditions.  Therefore, in places
with cold damp springs and cloudy summers *Aphelinus* is ineffective.
Each female deposits 60-140 eggs.  The parasite weakly infests
aphids on young nursery plants and on trees with thinned out
crowns and aphids developing on the root collar.  Therefore, under
natural conditions there constantly are reserves of the pest from
which it can spread (Boldyreva 1970).

The economically unfavorable biological characteristics of
*Aphelinus* that have been mentioned, however, do not exert a determining influence on its effectiveness under the conditions of the

southern fruit orchards.  Not infrequently in the first year of
release of *Aphelinus*, up to 95-98% of the pest already is destroyed.
In almost the entire geographic range of the woolly apple aphid no
other control measures are required.

The short periods of development, high fertility, predominance
of females in the progeny, and capacity for parthenogenic reproduction guarantee high rates of increase of *Aphelinus*.  The parasite is specialized with respect to the woolly apple aphid and only
rarely infests other species of insects.  It is a valuable circumstance that the period of development of the *Aphelinus* coincides
with the period of development of the woolly apple aphid.  Thanks
to the enumerated characteristics and the high resistance to cold
of the overwintering larvae, which withstand a temperature drop to
$-30^{\circ}$ and below, the *Aphelinus* has firmly acclimatized, and in the
course of many years it has kept the numbers of the woolly apple
aphid everywhere at an economically inappreciable, low level.

The method of using *Aphelinus* consists in moving it from
places where it has spread to freshly arising centers of infestation of the woolly apple aphid.  For this purpose, in late autumn
before the onset of frosts, small branches and nourishing shoots
and sprouts covered with infested aphids are stored up in orchards
with *Aphelinus*.  During the winter the branches and shoots bound
in bundles are kept in a dry, well ventilated cold room.  In the
spring when the temperature rises to $+8^{\circ}$ the branches are transferred to a colder place.  After dry, warm weather is established,
the branches and shoots are hung on trees in the orchard.

Approximately 1000 *Aphelinus* individuals are required for the
treatment of 1 hectare.

Taking into consideration that *Aphelinus* spreads rather slowly
and independently (100-400 meters from the place of release per
season), it is necessary to hang branches and shoots on 8-10 trees
in each hectare.  During the summer period, in case of necessity
*Aphelinus* is transferred from old centers of infestation of the
pest, where the parasite is present, to newly arising infestations.

## PSEUDAPHYCUS

The most effective results of biological control on large
areas and in different zones of the country have been obtained from
the use of *Pseudaphycus*.  *Pseudaphycus malinus* Gahan was brought
into the U.S.S.R. from the U.S.A. in 1954 for control of Comstock
mealybug [*Pseudococcus comstocki* (Kuwana)].  The length of life of
the adult *Pseudaphycus* is 3-12 days.  In the summertime development
of one generation is completed in 17-21 days.

*Pseudaphycus* has been introduced into the Uzbek S.S.R. and the Tadzhik S.S.R. Under the conditions of these republics 6-8 generations develop per summer. On the average 2 generations of *Pseudaphycus* develop in the time of development of one generation of Comstock mealybug. Thanks to this fact the occupation by *Pseudaphycus* of plants infested with the mealybug makes it possible to sharply decrease the numbers of the pest during one season.

In the first years after the introduction of *Pseudaphycus* for resettlement to new centers of infestation of the Comstock mealybug, the mummified mealybugs were collected in places where they concentrated. For the collection of the mummies in nature, baits were placed on infested trees to serve as sheltering places for the mealybug. Subsequently, laboratories were constructed in which the *Pseudaphycus* was raised on colonies of the mealybug developed on etiolated potato sprouts or the fruits of some varieties of squash.

The mummies collected during the winter were kept in gauze bags (20-30 thousand in each) or in cardboard boxes at a temperature from +6 to -3°. Two to 3 days before release the mummies were packaged 1.5-2 thousand in a test tube. The parasite was released at a calculated rate of 100-150 mummies to each 5 trees. When individual trees were treated the mummies were placed on each one.

*Pseudaphycus* is a highly effective parasite; it suppresses the mealybug by 85-95%.

At present *Pseudaphycus* has successfully acclimatized in the entire geographic range of the Comstock mealybug and keeps the numbers of the pest at an economically inappreciable, low level. In 1963, out of 9 million infested trees in Uzbekistan multiplication of the mealybug on 8 million was suppressed by *Pseudaphycus* and chemical treatments were cancelled. The total yield of cocoons of silkworm, *Bombyx mori* (L.), was increased by 25%, and the cost of expenditures on one tree when the biological method was used amounted to 1.5-2 kopecs (Yasnosh 1974).

Use of *Pseudaphycus* consists in its further planned settlement in newly arising and reestablished centers of infestation of the Comstock mealybug; chemical treatments are carried out on 10% of the infested area, predominately in new centers of infestation of the pest.

Inasmuch as *Pseudaphycus* still does not always and everywhere finally cope with the mealybug, in 1962 the parasites *Allotropa convexifrons* Muesebeck, and *A. burrelli* Muesebeck were brought into the Soviet Union from Korea. These entomophages have acclimatized

in Central Asia and the Caucasus, but still are of secondary
importance in limiting the numbers of the pest.


*CRYPTOLAEMUS*


    The predatory beetle *Cryptolaemus montrouziere* Mulsant was
brought into the U.S.S.R. from Egypt in 1933 for controlling
mealybugs and soft scales.  In practice *Cryptolaemus* most often
is used against the grape mealybug, but the beetle is used with
considerable success also against citrus mealybug, and maritime
mealybug, cottony scale (*Pulvinaria aurantii*), and against *Icerya
purchasi* Maskell.

    Attempts to acclimatize *Cryptolaemus* in the Soviet subtropics
have not been successful.  The predator endures poorly a decrease
in temperature even to $-1^{o}C$, and at $-6^{o}C$ mass death occurs.  The
situation is aggravated by the fact that the beetles become active
during short warm periods in the winter and die from lack of food.
However, in connection with the fact that the activity of
*Cryptolaemus* is rather effective, during the growing period it is
used by the method of seasonal colonization or yearly releases
(Stepanov 1935, Fedorov 1935).

    A temperature of 20-26° in combination with a high relative
humidity of the air (70-85%) is favorable for the development of
*Cryptolaemus*.  In hot, dry weather a suppression of reproduction
of the predator is observed.  Therefore it is not possible to use
it in controlling the grape mealybug in Azerbaidzhan and Dagestan
and against the Comstock mealybug in the republics of Central Asia.

    The adult beetles live up to 3-7 months.  The fertility of the
females is 200-500 eggs.  In the Abkhazian A.S.S.R. 3-4 generations
of *Cryptolaemus* develop per year and are difficult to delimit
because of the protracted period of oviposition and development.
In a warm period all stages of development of the parasite are
encountered simultaneously.

    The technical simplicity and rapidity of rearing and the
economic profitability of using *Cryptolaemus* are its main advantages.  However, the necessity for constant renewal of the
*Cryptolaemus* by artificial propagation and use of it by the
methods of inundation and seasonal colonization at the same time
is a negative aspect of its use.

    Mass propagation and use of *Cryptolaemus* is carried out by
production laboratories.  These laboratories yearly rear and
release hundreds of thousands of the beetles on infested plantations of citrus, grapes, tea and other plants in the Abkhazian

A.S.S.R., the Adzharian A.S.S.R. and the Black Sea subtropics of
the Russian Soviet Federative Socialist Republic.  On plots treated
with *Cryptolaemus* the mealybugs and soft scales are suppressed in
the course of a season by 90-95% (Stepanov 1955, Kolotova 1938,
Gaprindashvili and Chochia 1944).

The rate of release is 10 beetles per tree for citrus or 3
beetles per plant for grapes.

During the warm time of year the *Cryptolaemus* is raised on
mealybugs that are breeding on specially grown plants of soy, corn,
sunflower, and horseradish.  In winter and early spring the mealybugs and *Cryptolaemus* are reared in the laboratory on etiolated
sprouts of potato or squash fruits.

In recent years intensive investigations have been carried
out on the development of artificial nutrient media for *Cryptolaemus*.
Media have been obtained (Sagoyan 1969) that are fully suitable for
rearing *Crypotlaemus*.  The larvae reared on these media do not lag
behind the controls in growth.  The yield of adults is 95%.  At
present, work is directed toward simplifying and cheapening the
media.

Introduction into the U.S.S.R. of cold-resistant forms of
*Cryptolaemus* from various regions of its earlier acclimatization,
e.g. Menton (France), is promising (Rubtsov 1947).


## LINDORUS

The predatory beetle *Lindorus lophantae* Blaisdell was introduced in 1947 from Italy for control of diaspid scales.  Although
it proved to be more resistant to cold than *Cryptolaemus*, because
it does not have a fixed winter diapause it dies where there are
prolonged low air temperatures in all the plantations of the Adjar
and Abkhazian republics.  Because of this the *Lindorus* is transferred
from sanctuaries under natural conditions, or releases are made of
beetles artificially reared in the laboratory, where they are raised
from potato tubers infested with oleander scale, *Aspidiotus nerii*
Bouche.

In 1954-1962 in the Batum biolaboratory of the Georgian Institute of Plant Protection attempts were undertaken to increase the
cold-resistance of *Lindorus* (Gaprindashvili 1963).  By means of
directed selection during 33 generations it was possible to select
a winter-hardy form of *Lindorus*.  These forms since 1956 have overwintered yearly in plantations in the Adjar A.S.S.R., Abkhazian
A.S.S.R., and western Georgia, although the temperature falls for a
long time to -6, -9°.  In the Adjar A.S.S.R. in different years not

not more than 52% of the beetles died in the winter; not more than
11% of the larvae and pupae died.  The cold-resistant forms, when
they were later moved, did not lose any of their acquired property
of cold-resistance.  The present geographic range of *Lindorus*,
where it multiplies openly in nature, testifies to its considerable
cold-resistance; the beetles live in regions where the temperature
falls in winter to -10° and lower.

## CHILOCORUS

In orchards on the Black Sea coast the numbers of San Jose
scale are appreciably decreased by the predators *Chilocorus
reinpustulatus* (Scriba) and *C. bipustalatus* (L.).  On the coast
these species develop 3 generations.  The average fertility varies
between 110 and 140 eggs.  The beetles live 33-39 days and overwinter in orchards under fallen leaves.

In places where chemical methods of protection are not used
and the predators are not destroyed, they almost fully suppress
the numbers of the scale.  Because of the high effectiveness of
the 2 species of *Chilocorus*, in a number of regions in winter or
early spring only selective sprayings are carried out and only in
case the infestation exceeds degree II (28 scales per square centi-
meter of bark) (Popova 1971).

## RODOLIA

*Rodolia cardinalis* (Muslant) was brought into the U.S.S.R.
from Egypt in 1931.  The beetles initially were reared in greenhouses in Leningrad (Meier 1937).  In the spring of 1932 the first
group of beetles was sent to the Abkhazian A.S.S.R. where at that
time preparatory work had been carried out; insectaries had been
built, and territory infested with *Icerya* had been found.  During
the summer of 1932 the beetle spread and successfully multiplied
in the Abkhazian A.S.S.R.  Already in the first year of acclimatization the beneficial action of the beetle was shown in individual
centers of *Icerya* infestation.

At present the entomophage has acclimatized and contends successfully with *Icerya* fluted scale in all regions where the pest
has spread (Abkhazian A.S.S.R., Adjarian A.S.S.R., subtropics of
the Krasnodar area).  The full cycle of development of the beetle
is completed in 20-40 days, depending on the temperature and
humidity.  During a year under the conditions of the Abkhazian
region 4 generations of  *odolia* develop.  In the autumn the beetles
get in under the leaves or dry plant residues, where they overwinter.
Under natural conditions, development of *Icerya* begins long before

the overwintered beetles of the predator start ovipositing. During
the first half of the year the beetle succeeds in producing 6 generations under the conditions of the subtropics of the U.S.S.R.,
but the *Icerya* produces only 3 generations (Rubtsov 1954). The
imperfect coincidence of the times of development of the predator
and its victim, and also the ease of spread of the latter, are the
main reasons for a periodic increase in the numbers of *Icerya* in
local centers of infestation in the beginning period of the year.
Furthermore, even when there are a small number of overwintered
beetles, the centers of *Icerya* infestation are invariably suppressed
by *Rodolia*, which spreads independently from other centers or is
artifically transferred (Stepanov 1940). Release of 10-20 beetles
per center is sufficient to suppress the pest for a number of years.

Rodolia is a narrowly specialized predator, and although it
sometimes feeds on other coccids, its acclimatization in regions
where *Icerya* is absent is not successful. Sometimes *Rodolia* dies
in a locality because of the complete extermination of *Icerya*.

The technique of practical use of *Rodolia* is relatively simple
and amounts to moving the beetles from old centers of infestation
of the pest to new ones, if the predator has not succeeded in
moving there independently. Because in some more northern sub-
tropical regions *Rodolia* does not always overwinter. It is necessary to periodically renew its natural reserves from more southern
regions.

Icerya is losing its importance as a pest; following it with
the beetle is becoming rare, but both species as a rule continue
to exist in a given locality. In the spring they sometimes still
appear in noticeable quantities, but by the middle of summer the
trees infested by *Rodolia* already are clean.

Icerya may show flareups of multiplication in new, isolated
centers of infestation where *Rodolia* is not present. In such cases
supplementary colonization of the beetle is necessary. For this
purpose in the Soviet subtropics there are biolaboratories and
insectaries where the beetle is either kipt in the winter months
in special greenhouses or is additionally collected in regions
where it has survived, and from which it is transferred in the
spring and summer to centers of *Icerya* infestation.

Because of the danger of possible death of the *Rodolia* (in
case of a severe winter) under natural conditions, it is reared in
special insectaries on plants infested with *Icerya*. The development
and reproduction of *Rodolia* in the greenhouse takes place in the
winter in contrast to natural conditions, where there is a winter
pause in reproduction of the beneficial insect. Similar seasonal
colonization of the beetle at the rate of 10-20 beetles per

infestation center is sufficient to suppress the pest for a number
of years (Meier 1937).


## PROSPALTELLA

In our country 2 geographically separate populations of
*Prospaltella* occur that are morphologically indistinguishable but
differ with respect to the presence or absence of males in their
populations.  In the primary center of the pest (in the Far East)
it is parasitized by the bisexual form of *Prospaltella*, but in the
western center (southern regions of the European part of the
U.S.S.R.) it is parasitized by the maleless form, which reproduces
parthenogenetically (Chumakova and Goryunova 1963).  The work of
A. Khuba (Niva 1957, 1958) has shown that populations of *Prospaltella*
of different origin (Chinese, Caucasian) differ from each other
with respect to heat and to speed of development.

*Prospaltella perniciosi* Tower in our country parasitizes only
San Jose scale, *Quadraspidiotus perniciosus* (Comstock), infesting
the females and the nymphal stage of the pest.  The entire development of the parasite from egg to adult insect takes place within
the body of the host.  The *Prospaltella* is widely adapted to this
species, and 2-3 generations of the parasite develop in one generation of the host.  In different zones of the Soviet Union
*Prospaltella* gives from 3 to 6-8 generations per year.

In conformity with the nature of the differences existing in
populations of *Prospaltella*, they correspond most to the idea of
ecotypes.  For the Caucasian population 3 ecotypes have been distinguished: steppe, foothill and maritime (Goryunova 1966).  It
has been proposed that the series of Caucasian populations be considered as a subspecies (or geographic race), differing from the
subspecies inhabiting the Far East.  The Far Eastern *Prospaltella*
has reproduced successfully under high temperature conditions (25-
27°), but low humidity of the air causes a depression in its
development.

Experiments on acclimatization of the Far Eastern *Prospaltella*
were carried out in the 1960's under conditions of the steppe and
foothill zones of the northern Caucasus.  Already in the first year
of acclimatization under the conditions of the foothill zone, when
10,000 parasites were released per tree, the degree of parasitization in the scales reached 52% by the end of the season in the test
orchard.  In the control the parasitization of the pest by local
*Prospaltella* did not exceed 6-9%.

The effectiveness of *Prospaltella* in controlling San Jose scale
is especially high when there is a weak infestation of plants by the

pest.  In some years it is able even on strongly infested plants
to suppress the numbers of the host to a minimum at which losses
in yield from the scale are not economically significant.

Such effectiveness has served as a basis for the development
of a method of using the parasite in seasonal colonization (Popova
1969).  For this purpose a method of mass-rearing *Prospaltella*
(*Prospaltella perniciosi*) was developed.  By systematic observations of the behavior of *Prospaltella* in nature it was established
that the most favorable time for releases of the parasite in fruit
plantings is in the spring, in April-May (Popova 1962).  For this
time massive amounts of material are prepared and kept at low
temperatures.

## APHYTIS

*Aphytis proclia* (Walker), an effective native parasite of the
San Jose scale, is common in orchards of the northern Caucasus.
With respect to fertility and time of development it is capable of
suppressing the pest.  However, successful development of its
generations depends on the presence in nature of females of the
scale.  Out of 5 generations of the parasite (the fifth is incomplete) 2 hatch in periods when there are very few of the adult
pests, and its population consists mainly of larvae of the I-II
instars which the parasite does not infest.  Usually under these
circumstances the parasite dies, having left very few progeny.

Efforts were undertaken to eliminate the lack of synchronism
of development of the *Aphytis* with that of the pest (Chumakova
1961).  Sowings of nectar-bearing plants, facelia, and dill, as a
source of nectaral nutriment for the parasites raised the parasitization of the scale from 1.0 to 72.2%.  Sowing of nectar bearers
between the orchard rows is proposed.

Because of poor overwintering in regions of the northern
Caucasus of the *Aphytis* introduced from Pakistan and other adjacent
countries, it was necessary to make additional releases of *Aphytis*,
especially in the spring period when it was very low in numbers
(Goryunova 1965).  Thus it appeared necessary to develop
methods for mass-rearing *Aphytis* for the purpose of periodic
seasonal colonization.  At present, mass-rearing of the parasite
is being started in a number of laboratories in the northern
Caucasus.

## COCCOPHAGUS

In 1960 *Coccophagus gurneyi* Compere, a specialized parasite of
citrus mealybug, was brought into the U.S.S.R.  In the period 1960

to 1963 the *Coccophagus* was successfully acclimatized in the
Abkhazian A.S.S.R., and at present it is the main natural regulator
of the numbers of the citrus mealybug.

In the subtropics the *Coccophagus* overwinters in the body of
the host in the third instar larval stage and the pupa.  The adult
parasites appear from the end of April through May.  The parasite is
capable of both sexual and parthenogenetic reproduction.  The
females most willingly oviposit in the third-instar larvae,
although they parasitize scales of all stages.  The only exception
is the young larvae newly hatched from the eggs, which the parasite
usually does not infest.

The fertility of the females is 45-60 eggs.  The length of
life is 2-3 weeks or more.  Complete development of a generation
is accomplished in approximately a month, but may vary with temperature.  In the course of a year in the Abkhazian A.S.S.R. 5-6
generations of *Coccophagus* develop.

Under laboratory conditions for rearing *Coccophagus*, the
mealybugs are raised on etiolated sprouts of potato, on squash,
and on citrus fruits and plants.  For this purpose, under natural
conditions in the autumn small cuttings of branches of various
plants infested with the mealybug are prepared.  Then females of
the mealybug are placed in desiccators, where infestation of tubers
and by crawlers takes place.  At the appropriate time, material
with the mealybug is transferred to another room for parasitization
by *Coccophagus*.  Thanks to the activity of the artifically distributed parasite, the numbers of the mealybug are sharply decreased.


*PERILLUS*

One of the principal, most effective destroyers of the Colorado
potato beetle in the U.S.A. and Canada is the bug *Perilloides
bioculatus* (F.).

In the Soviet Union investigations directed to acclimatization
of the predator have been carried on since 1961 by a number of
laboratories of the All-Union Institute of Plant Protection and
the Central Scientific Research Laboratory.  *Perillus* develops more
rapidly than the Colorado potato beetle; in the lowland regions of
the Transcarpathian area 3-4 generations develop to 2 generations
of the Colorado potato beetle.  The fertility of the bug is 500-
800 eggs.  Its voracity has great importance in the effectiveness
of the predator.  Each pair of insects during only the period of
oviposition destroy an average of 2760 eggs of the beetle.

In the Institute of Evolutionary Morphology and Ecology of
Animals of the Academy of Sciences of the U.S.S.R., studies have
been made of the effect of various temperatures on the insect and
of the state of winter rest. The lower threshold of feeding of
larvae and adults varied within the limits 5-8°C and the upper
50-54°C; the lower threshold of torpor was 3-6°C and the upper
58-61.5°C. It was established that upon passage into the resting
state an increase in resistance to cold was observed. On the basis
of the ecological characteristics of the predator that were
established, the northern limit of the potential geographic range
was determined to be in the European part of the U.S.S.R. (Shagov
1967).

A number of authors think that practical investigation of
P. bioculatus as an entomophage of the Colorado potato beetle under
the conditions of the U.S.S.R. may be made both by means of its
acclimatization and by seasonal colonization (Gusev et al. 1971).

To increase the profitability of the method a considerable
improvement is necessary in the existing methods of mass-rearing
of the insect which is being raised exclusively on the Colorado
potato beetle (Methodological Directions for Laboratory and
Laboratory-Field Rearing of *Perillus* 1971).

Acclimatization of the predator may be achieved as a result
of natural selection taking place when there is a one-time seasonal
colonization in one region.

At present cases have been noted of overwintering of the insect
under natural conditions in the Odessa district and on the coast of
the northern Caucasus. It is possible that selection of a winter-
hardy form of the predator has occurred, which will permit gradually
widening the limits of its present experimental use, extending them
to the zone of potato planting.


*GREEN LACEWINGS*


In the U.S.S.R. attention was first attracted to the beneficial
activity of green lacewings, particularly *Crysopa carnea* Stephens,
because of the spread and massive multiplication of the Comstock
mealybug (*Pseudococcus comstocki*). Initially the possibility of
adaptation of *C. carnea* to feeding on this pest, which was new to
the fauna of the U.S.S.R., was investigated (Meier and Meier 1946).
Later data were obtained on the high effectiveness of the lacewing
in suppressing the multiplication of the mealybug on mulberry and
catalpa in Central Asia (Luppova 1949, 1950, 1952).

Information on the fauna of the *Chrysopidae* of the Soviet Union is presented mainly in small publications containing a list of species from Central Asia (Luppova 1966), Azerbaidzhan (Kurbanov 1971), Georgia (Vashadze 1959, Tvaradze 1969), Moldavia (Zeleny and Talitskii 1955), and eastern Siberia (Pleshanov et al. 1965). Recently a list has been published of 40 species recorded on the territory of the U.S.S.R. (Shuvakhina 1974).

At present it is known that on the territory of the U.S.S.R. *C. carnea* is the most widely distributed and numerous species. Also widely distributed are *C. formosa* Brauer, *C. septempunctata* Wesmael, *C. perla* (L.), *C. phyllochroma* Wesmael, *C. ventralis* Curtis (= *C. prasina* Burmeister), *C. abbreviata* Curtis, *C. albolineata* Killington, *C. (Nineta) flava* (Scopoli), *C. (Nineta) vittata* Wesmael, and some others.

Investigations carried out in various regions of the country have established that the lacewings are often primarily important in the natural regulation of the numbers of aphids, mealybugs, spider mites, and other pests on cotton, fruit crops, citrus, and other plants (Uspenskii 1951, Luppova 1955, 1966, Kurbanov 1966, Beglyarov 1957, Gaprindashvili 1964, Andreeva 1967, Yuzbyashýan et al. 1970). Unfortunately, under natural conditions this very active and voracious emtomophage usually has an appreciable effect on the numbers of the pests too late, when the plants already have been definitely damaged. Therefore, the idea originated of utilizing lacewings by means of seasonal colonization or flooding releases.

In domestic literature there is repeated mention of the idea of the possibility of using the common lacewing by this method against aphids, mealybugs, Colorado potato beetle, and other pests (Shengelaya 1953, Luppova 1955, Bashadze 1959, Shuvakhina et al. 1959).

The first experiments on direct use of the entomophage in the U.S.S.R. were carried out rather recently (Beglyarov et al. 1970), Nyaes 1970, Bondarenko and Moiseev 1971, 1972, Beglyarov and Ushchekov 1972). Wide possibilities for testing emerged after the development in detail of a simple method for mass-rearing the lacewing, at the basis of which is feeding of the predator larvae on eggs of the grain moth, *Sitotroga cerealella* (Oliver) (Beglyarov et al. 1972).

In recent years the specialists of a number of scientific research institutes and mass-rearing laboratories have been occupied with the study of the effectiveness of using the common lacewing on various crops in different regions of the country. We shall present the principal results of this work.

Testing in the Field

Tests carried out in Moldavia on eggplants in field plantings
have established the possibility of effective destruction of the
Colorado potato beetle in the egg or young larval stage by colonization of eggs and larvae of *C. carnea*. Thus, when the eggs were
colonized (1:1 ratio of predator to host), the population of the
host was decreased by 74%. First-instar larvae at a starting ratio
from 1:1 to 1:5 destroyed 86-91% of the eggs of the Colorado potato
beetle. Larvae of the second and third instars colonized at the
same ratios destroyed 87-100% of the eggs in a comparatively short
time (5-10 days) (Adashkevich and Kuzina 1971, Adashkevich et al.
1972, Adashkevic and Kuzina 1974). The authors also report that
similar results were obtained from the use of the lacewing on
potatoes. Investigations on the determination of the effective
rates and times of using the lacewing for control of the Colorado
potato beetle on potatoes have been carried out also by other
authors. Thus, field tests by E. Ya. Shuvakhina (1974, 1975)
established that only release of larvae of the predator in the
second and the beginning of the third instar decreases the numbers
of the Colorado potato beetle to an acceptable level when there is
a low initial density of the pest (up to 2 egg masses per plant)
and the predator-pest ratio is from 1:10 to 1:20.

Data on the effectiveness of seasonal colonization of the
lacewing for control of aphids in vegetable crops are of practical
interest. The results of plot tests in Moldavia showed that against
most species of aphids colonization of second-instar larvae at a
ratio of not less than 1:5 is the most effective. The technical
effectiveness of using the predator against the green peach aphid,
*Myzus Lachmus persicae* (Sulzer) on tomatoes was 72%, and on eggplants 43-97%.

A rather stable and high effect was obtained when second-instar
larvae were colonized to control the green peach aphid and the
alfalfa aphid, *Aphis craccivora* Koch (= *Aphis laburni* Kaltenbach),
on peppers. Thus, the numbers of aphids on peppers were decreased
by 94-98% 6 days after colonization. The increase in yield of
pepper fruits from biological control was 13%.

In contrast to other crops, high effectiveness of the lacewing
against the pea aphid, *Acyrthosiphon pisum* (Harris), on peas was
achieved only when a very large number of larvae was colonized
(predator-host ratio 1.5:1). Lower rates of colonization proved to
be ineffective (Kuzina 1974).

Concurrent with the positive results, in a number of experiments the use of the lacewing turned out to be completely or
partially ineffective. Thus, for example, on tomatoes the

colonization of second-instar larvae at the rate of 1:5 destroyed
only 72% of the aphids.  On fall cabbage the same rate and an even
higher rate (1:1) of colonization of second- and third-instar
larvae gave a decrease of not more than 74% in the numbers of
aphids (Adashkevich and Kuzina 1974).

Experiments and observations by Tsintsadze (1975) have shown
that seasonal colonization of *C. carnea* is not promising for use
in controlling whitefly, aphids, and spider mites on citrus in
the subtropics of Georgia.

## Principal Results of Testing in Glasshouses

Many years of investigations carried out in the Moscow district (Beglyarov et al. 1977) have established that the effectiveness of *C. carnea* when it is used against aphids on the main crop
on enclosed ground, cucumbers, is not stable and depends on many
factors that affect the behavior of the larvae (type of protected
ground, age, and size of the protected plants, temperature,
illumination, etc.).

Under the conditions of a mild climate (Moscow and Leningrad
district, Estonia) inundative releases of second-instar larvae of
*C. carnea* give effective control of the multiplication of aphids
on cucumbers in glass-enclosed winter hothouses at the beginning
(up to May) and the end (September-October) of growth of the crop.
During the entire period of growth an effect is obtained only on
a trailing crop of cucumbers under glass, on low-growing green
vegetables (lettuce, celery, parsley, dill, coriander) and on
flowering plants (carnations, Tulips) in hothouses (Bondarenko et al.
1971, 1972, Nyaes 1971, Beglyarov et al. 1972, Beglyarov and
Ushchekov 1972, 1974, 1977).

Lacewings (*C. carnea* and *C. Septempunctata*) in hothouses do
not settle in nor reproduce independently.  Therefore, at present
they can be used effectively only as a "living insecticide," i.e.
without counting on multiplication of the colonized individuals.
Use of these entomophages by means of inundative releases is too
laborious and not always a profitable measure.  Actually, when
*C. septempunctata* is used, about 5-10 times fewer larvae are
required than when *C. carnea* is used.  However, the method of mass-
rearing of the first species is still too complicated and more
laborious than that of the second.  Therefore, at present much
attention is being given both to the improvement of the methods of
mass-rearing lacewings and to the elucidation of the possibilities
of decreasing the rates of colonization of the predators.

It has been established that the rates of release of the
sevenspotted lacewing can be lowered by colonization of fertilized
females.  They are capable of independently seeking out infested
plants and depositing eggs in a colony of aphids.  The larvae that
hatch out in rather high numbers effectively repress multiplication
of the pest.  However, a direct correlation exists between the
density of the aphid population, the number of eggs deposited by
the females of the entomophage, and the effectiveness of colonization.  Thus, in tests carried out in the region near Moscow in 1976
in spring film-enclosed hothouses on the background of an artificial infestation, with a high density of the aphid population
(5000 per fruit-bearing cucumber plant), colonization of the fertilized females (at a predator-host ratio of 1:500) gave a 97%
decrease in the numbers of the pest by the end of the third week.
In variants with low density of the host (200 and 1000 aphids per
plant) colonization of the females at the same rate (1:500) proved
to be ineffective, and a continuous increase was noted in the
numbers of the pest.

At present the possibilities of obtaining a high effect by
increasing the rate and number of releases of the predator when
there is a threshold number of aphids (not more than 1000 per plant)
are being studied.  Investigations show that real possibilities
also exist for decreasing the total expenditure of larvae of *C.
carnea* on cucumbers by combining the activity of the entomophage
with chemical treatments.  Three years of testing (1974-1976) in
spring film-enclosed hothouses in the Moscow district have given
2 methods of eliminating 1-2 sprayings and postponing the time of
chemical treatments without any damage:

(1) inundative releases of second-instar larvae of *C.
carnea* in an aphid-breeding area at the rate of 1:5, and

(2) spraying of carbophos (malathion) only when the aphid
population exceeds 10,000 on one or more plants in the area.  In
1976, on an experimental plot where the lacewing was used, it was
necessary to carry out only one spraying with carbophos, while on
the control (without the predator) 3 sprayings were necessary.

The outlook for practical use of lacewings for controlling
aphids on cucumbers is directly dependent on how quickly economically justified and reasonable methods of mechanized mass-rearing
and distributing these entomophages can be developed.  It is
entirely obvious also that it is necessary to select other aphidophages capable of giving stable effectiveness on different types
of enclosed ground.

Development of the biological method of controlling aphids is
especially important for green vegetable crops, since the use of

chemical pesticides for their protection is not recommended in general.

In recent years tests in the Leningrad district have shown the possibility in principle of using *C. carnea* against aphids on green crops (Bondarenko and Moiseev 1971, 1972, Moiseev et al. 1972, Moiseev 1973). Tests carried out in 1974-75 in commercial hothouses of the Moscow district have made it possible to recommend for practical use the method of flooding releases of second-instar larvae. The numbers of aphids on green crops usually are significantly less than on cucumbers, and the microclimate here is more favorable for the activity of the larvae of the entomophage. Therefore, significantly lower rates of release of the lacewing are required for the protection of green crops. The expenditures associated with the rearing and use of the entomophage are small and are fully justified by the increase in quality of the production (Beglyarov and Ushchekov 1977).

In 1976 the biological method of controlling aphids was used in the Moscow district for protecting green crops on a total area of 21,000 square meters (Belov 1977).

In conclusion we should like to emphasize once more that the possibilities of widening the scale of use of lacewings at present is directly dependent on the cost of the entomophages, and on the mechanization and automation of the processes of mass-rearing and distributing them. At present, great attention is being devoted to the development of these problems.

## CECIDOMYIID

### *Aphidoletes aphidimyza (Rondani)*

At the beginning of the 1930's it was established that the appearance in hothouses of the predatory cecidomyiids *A. aphidimyza* and *Arthrocnodax* sp. at the very start of reproduction of the green peach aphid [*Myzus* (= *Myzodes*) *persicae* (Sulzer)] and spider mite (*Tetranychus urticae* Koch) on cucumbers gives good results in controlling these pests (Zorin 1934a, 1934b). Upon coming to this conclusion, the author undertook in hothouses near Leningrad for the first time in the U.S.S.R. attempts to use cecidomyiids for protecting plants on enclosed ground. P. V. Zorin started from the fact that "the use of predatory insects for controlling pests of cucumbers under the conditions of enclosed ground can be important only if we are in a position to destroy by this means all 3 of the most important pests of enclosed ground: spider mite, green peach aphid, and tobacco thrips, *Frankliniella fusca* (Hinds)." Although the experiments on utilization of gall midges gave encouraging

results, P. V. Zorin acknowledged that these predators were not
very well suited for practical use, since" ... neither one nor
the other of them affects the tobacco thrips" and furthermore
" ... in nature the larvae of the gall midges suffer severely from
parasitic hymenopterous insects." The investigations of P. V.
Zorin were further developed in the work of N. V. Bondarenko and
B. I. Asyakin (1974, 1975a, 1975b), which led to the conclusion
that of the aphidophages studied up to that time the use of the
cecidomyiid *A. aphidimyza* gave the best results in controlling
*Aphis gossypii* Glover on cucumbers in hothouses.

At present, methods already have been proposed for using this
aphidophage (Bondarenko and Asyakin 1975a, 1975b). In 1976,
commercial verification of the effectiveness of cecidomyiid in
hothouses of farms in the Leningrad and Moscow districts was
successfully carried out on an area of 230,000 square meters
(Belov 1977).

Besides *C. carnea*, *C. septempunctata*, and *A. aphidimyza*,
entomologists in the U.S.S.R. have studied to some extent the
possibilities of using a series of other chrysopid, coccinellid,
and syrphid parasites of the Aphididae for control of aphids on
various crops on open and enclosed ground. Work with these entomophages has not yet given practical results, and we consider it
permissible not to dwell on a discussion of the published information (Adashkevich 1971, 1974, 1975, Beglyarov et al. 1972,
Bondarenko and Asyakin 1972, Bondarenko et al. 1973).


*PHYTOSEIULUS*


*Phytoseiulus persimilis* Athias-Henroit

Investigation of the biology of the predatory mite, *Phytoseiulus persimilis* Athias-Henroit and development of methods for
rearing and using it were begun in the U.S.S.R. in 1963 (Beglyarov
et al. 1964). The positive results of the first commercial tests,
carried out in 1964 in the Moscow district, made it possible even
in 1965 to start testing in various regions of the country (Moscow,
Leningrad, and Sverdlovsk districts, Armenia, Uzbekistan, Georgia,
and others).

Intensive investigations carried out in the following 10-15
years in the U.S.S.R. and abroad have permitted practical resolution of the problem of biological control of spider mites, a
scourge of the main crop in glasshouses, cucumbers, and a
dangerous pest of many other plants. The essence of the method,
which has received wide recognition, is the carrying out of regular
inspections and the release of the predator *Phytoseiulus* onto the

infested plants. Uncomplicated methods of mass rearing and use
of *Phytoseiulus* have been described in detail in a number of publications (Beglyarov et al. 1967, Beglyarov 1968, 1970, 1976). In
many cases the use of *Phytoseiulus* makes it possible to entirely
avoid the use of acaricides. It also has been found that for
obtaining the maximum effect (with the least expenditure of
predators) from *Phytoseiulus* it is necessary to make a release
into the breeding grounds of spider mites at the very beginning of
their appearance. Under commercial conditions determination of
the rate of release of the acariphage by estimation of the numbers
of the pest proved to be unacceptable. Because of this there was
tested and then recommended for adoption a method that provides
for populating each infested plant with the predator on the basis
of a visual evaluation of the degree of damage (weak, medium, and
strong). The results of commercial tests made it possible to
establish that in those hothouses where biological control is
carried out 2 years in a row and more for protection of the crop,
during the season approximately half as many predators are required
as on areas where *Phytoseiulus* is used for the first time. In the
spring, in the second year of use of *Phytoseiulus*, mass development
of the harmful mites starts 3-4 weeks later. This is explained by
the fact that the use of the predator permits practically avoiding
completely the formation of an overwintering stock of diapausing
females of the pest.

It has been noted that in hothouses where *Phytoseiulus* is used,
because of the exclusion of a large number of chemical treatments,
a more intensive activity of bees is observed and consequently
better pollination of a large number of [plant] ovaries. As a consequence of the sharp curtailment of the number of chemical treatments, the growing and fruit-bearing periods of the plants are
lengthened.

Many years of commercial testing carried out in various
regions of the country have shown that timely use of *P. persimilis*
guarantees protection of a crop from the harmful activity of spider
mites. The number of plants infested with spider mites is insignificant, rarely reaching 10% of the total number; the degree of
damage of leaves usually is evaluated as grade one (not more than
25% of the leaf surface area damaged). It is known that this
degree of damage does not affect the yield. On farms of the Moscow
and Leningrad districts, the use of *Phytoseiulus* gives an increase
in yield of 4.7-6.9 kg. of cucumbers per square meter of those
grown on winter ground, 3.3 kg. of those grown hydroponically, and
2.7-4.5 kg. in spring hothouses in comparison with hothouses in
which from 7 to 27 sprayings have been carried out with chemical
agents; thiophos [parathion], kelthane [dicofol], etc. Because of
the additional yield obtained, each ruble spent on biological control of mites on cucumbers is repaid 10-15 times and more

(Beglyarov et al. 1967, 1968, Bondarenko 1969, Bushchik and
Plotnikov 1967, Chalkov 1968).

During the years of the ninth five-year plan the scale of
use of *Phytoseiulus* on enclosed ground increased by more than 8
times (1.2 million square meters in 1971, more than 10 million in
1975). According to the data of the Ministry of Agriculture of
the U.S.S.R., in 1976 *Phytoseiulus* was used on a total area of
about 13 million square meters.

In the process of wide use of *Phytoseiulus* it has been found
that the elimination of repeated chemical treatments against mites
promotes the relatively unhindered multiplication of another group
of dangerous pests, the aphids. The use of chemical agents against
aphids, in turn (because of the toxicity of the preparations for
*Phytoseiulus*), creates definite difficulties for the practical
accomplishments of biological control of mites. Therefore, in
recent years many specialists have devoted much attention to the
development of a biological method of controlling aphids. The
main results of this work are discussed within the sections of
the present article that are devoted to the use of lacewings and
aphidimyza cecidomyiids. Here we shall note only that besides
this is a toxicological evaluation of all the pesticides used on
enclosed ground has been carried out for the purpose of selecting
aphicides that are least dangerous for *Phytoseiulus* (Beglyarov et
al. 1966, Anikina et al. 1971, Zhuravleva et al. 1970), and also
that investigations have been started with a race of *Phytoseiulus*
that is resistant to organophosphorus compounds (Beglyarov et al.
In press).

The use of pesticide-resistant acariphages and entomophages
is in principle a new direction for biological control. The
resolution of this problem may lead to a considerable broadening
of the volume of use of *Phytoseiulus* by means of its release in
hothouses where it still is necessary to carry out chemical treatments against aphids. The first tests in commercial hothouses near
Moscow gave encouraging results.

## *ENCARSIA FORMOSA* GAHAN

In connection with the growth and intensification of hothouse
production in recent years on many farms the damage by the greenhouse whitefly [*Trialeurodes vaporariorum* (Westwood)] has increased
considerably. At present, in the U.S.S.R. definite attention is
being given to the development of a biological method of controlling
this pest. Studies and tests are being made of the parasite *E.
formosa*, which is used in some countries, and also of the entomopathogenic fungi *Aschersonia* spp., which were introduced previously

for control of citrus whitefly *Dialeurodes citri* (Ashmead). Tests
with *Encarsia* give encouraging results, although they are only of
a preliminary nature.

## ALEOCHARA BILINEATA GYLLENHAL

The beneficial activity of *Aleochara*, a widely distributed
parasite and predator of the cabbage maggot, *Hylemia brassicae*
(Wiedemann), onion fly, *H. antiqua* (Meigen), and other flies,
was noted in the U.S.S.R. in the 1920's (Zorin 1923, Vodinskaya
1928). At that same time the first, though unsuccessful, attempts
were undertaken for laboratory rearing of this entomophage (Zorin
1927). In the last decade interest in the study of *Aleochara* has
intensified. Methods were developed for its mass-rearing in the
laboratory (Adashkevich 1970, Adashkevich and Perekrest 1973),
which made it possible to carry out field tests to elucidate the
possibility and effectiveness of using *Aleochara* by means of
inundative releases or seasonal colonization against the cabbage
maggot and onion fly.

Four years of testing in Moldavia showed that release of
20,000-40,000 beetles per hectare at 2-3 dates, beginning from the
first egg laying by the pest, gives full protection of cabbage from
the cabbage maggot. In controlling the onion fly (using 33,000
beetles per hectare) *Aleochara* proved to be less effective; on the
third day after release 76% of the eggs of the pest were destroyed
(Adashkevich and Perekrest 1974, Adashkevich 1975).

At present, specialists of the Institute of Biological Methods
for Plant Protection (in Kishinev and Moldavia) are developing
methods for mechanized rearing and distribution of *Aleochara*. This
will make it possible in the next few years to carry out wide
testing under working conditions in various regions of the country
and to determine the prospects for successful use of the entomophage
in protection of plants from cabbage maggot and onion fly.

Besides the work that has been enumerated, in the Soviet Union
investigations are being carried out to explore the possibilities
of practical use of a number of other entomophages and acariphages
for controlling pests of cotton (*Bracon hebetor* Say., *Apanteles
kazak* Telenga, *Trichogramma* spp., and others against *Heliothis* spp.,
*Agrotis* (= Euxoa*) segetum* Schiffermuller, and other noctuids), and
of grain (*Spalangia drosophilae* Ashmead against *Oscinella* spp.),and of
fruit, berry, and citrus crops (phytoseiid mites). At present,
detailed studies have been made of the main biological and ecological characteristics of the most promising species, methods have
been developed for laboratory rearing, and laboratory-field testing
has been carried out. However, at this stage of the investigations
it is not possible to draw final conclusions.

REFERENCES CITED

Adashkevich, B. P. 1970. Razvedenie *Aleochara bilineata*
    (Coleoptera, Staphylinidae) v laboratorii [Rearing *Aleochara*
    *bilineata* (Coleoptera: Staphylinidae) in the laboratory].
    Zool. Zh. 49 (7): 1081-5.
Adashkevich, B. P. 1971. Perspektivnye vidy parazitov dlya bor'by
    s gorokhovoi tlei [Promising species of parasites for con-
    trolling pea aphids].  Pages 3-4 In Biologicheskaya
    zashchita plodovykh i ovoshchnykh kul'tur [Biological pro-
    tection of fruit and vegetable crops]. Kishinev.
Adashkevich, B. P. 1974. Konkuriruyushchaya sposobnost' parazitov
    gorokhovoi tli (*Acyrthosiphon pisum* Harris) [Competing
    ability of parasites of the pea aphid (*Acyrthosiphon pisum*
    Harris)].  Pages 3-4 In Entomofagi, fitosagi i mikroorganizma
    v zashchite rastenii [Entomophages, phytosages and micro-
    organisms in the protection of plants].  Publishing House
    Shtiintsa, Kishinev.
Adashkevich, B. P. 1975. Entomofagi i akarifagi v bor'be s
    vreditelyami ovoshchnykh kul'tur v SSSR [Entomophages and
    acariphages in controlling pests of vegetable crops in the
    USSR].  Pages 7-12 In Doklady na sektsii VIII Mezhdunarodnogo
    kongressa po zashchite rastenii. Vol. 3. Moscow.
Adashkevich, B. P. and N. P. Kuzina. 1971. Khrizopa protiv
    koloradskogo zhuka [Chrysopa against the Colorado potato
    beetle].  Zashch. Rast. 12: 23.
Adashkevich, B. P. and N. P. Kuzina. 1974. Zlatoglazka na
    ovoshchnykh kul'turakh [Lacewing on vegetable crops].   Ibid.
    9: 28-9.
Adashkevich, B. P., N. P. Kuzina, and Shiiko. 1972. Razvedenie,
    khranenie i primenenie zlatoglazki *Chrysopa carnea* Steph.
    [Rearing, storage and use of the lacewing *Chrysopa carnea*
    Steph.].  Pages 15-18 In Voprosy biologicheskoi zashchity
    rastenii. Kishinev.
Adashkevich, B. P. and O. N. Perekrest. 1973. Massovoe razvedenie
    *Aleochara bilineata* (Coleoptera, Staphylinidae) v laboratorii
    [Mass-rearing of *Aleochara bilineata* (Coleoptera:
    Staphylinidae) in the laboratory]. Zool. Zh. 52 (11): 1705-9.
Adashkevich, B. P. and O. N. Perekrest. 1974. Primenenie *Aleochara*
    *bilineata* Gill. (Coleoptera, Staphylinidae) v bor'be s
    kapustnoi i lukovoi mukhami [Use of *Aleochara bilineata* Gill.
    (Coleoptera: Staphylinidae) in controlling cabbage maggot and
    onion fly].  Pages 9-16 In Entomofagi, fitofagi i mikro-
    organizmy v zashchite rastenii. Publishing House Shtiintsa,
    Kishinev.

Afanas'eva, O. V. 1936. Primenenie trikhogrammy v bor'be s yablonnoi plodozhorkoi i kukuruznym motyl'kom v Moldavskoi SSR [Use of *Trichogramma* in controlling codling moth and corn borer in the Moldavian SSR]. Itogi nauchno-issledovatel'-skikh rabot Vsesoyuznogo Instituta Zashchita Rastenii, 1936.

Afanas'eva, O. V. 1937. Primenenie trikhogrammy v plodovom khozyaistve [Use of *Trichogramma* in fruit farming]. Trudy VASKhNIL.

Afanas'eva, O. V. and Ya. A. Alekseev. 1935. Dostizhenie biometoda v bor'be s vreditelyami plodovodstva [Achievement of the bio-method in controlling pests of fruit growing]. Na zashchitu urozhaya. Vol. 1

Alekseev, Ya. A. 1936. Primenenie trikhogrammy (*Trichogramma evanescens* Westw.) v bor'be s yablonnoi plodozhorkoi v usloviyakh Leningradskoi oblasti [Use of *Trichogramma* (*Trichogramma evanescens* Westw.) in controlling codling moth under the conditions of the Leningrad district]. Itogi nauchno-issledo-vatel'skikh rabot Vsesoyuznogo Instituta Zashchity Rastenii, 1936.

Alexandrov, N. 1947-1949. *Eurygaster integriceps* Put. a varamine et ses parasites. Entomol. Phytopathol. Appl. 5: 28-41, 6-7: 28-47, and 8: 16-52.

Andreeva, L. A. 1967. Fauna i biologiya nekotorykh khishchnykh nasekomykh, pitayushchikhsya plodovymi kleshchami [Fauna and biology of some predatory insects feeding on fruit mites]. Vest. Skh. Nauki 1: 98-104.

Anikina, N. E., G. A. Beglyarov, T. N. Bushchik, and V. F. Plotnikov. 1971. O toksichnosti pestitsidov dlya khishchnogo kleshcha fitoseiulyusa (*Phytoseiulus persimilis* A.-H.) [The toxicity of pesticides for the predatory mite phytoseiulus (*Phytoseiulus persimilis* A.-H.)]. Page 121 In Trudy XIII Mezhdunarodnogo entomologicheskogo kongressa. Moskva, 2-9 avgusta 1968 g. Publishing House Nauka, Leningrad.

Beglyarov, G. A. 1937. Vliyanie DDT na chislennost' tetranik-hovykh kleshchei i ikh khishchnikov [Effect of DDT on the numbers of tetranychid mites and their predators]. Entomol. Obozre. 36 (2): 370-85.

Beglyarov, G. A. 1968. Page 20 In Metodicheskie ukazaniya po massovomu razvedeniyu i primeneniyu khishchnogo kleshcha fitoseiulyusa dlya bor'by s pautinnymi kleshchami v zashchishchennom grunte na ogurtsakh [Methodological directions for mass-rearing and use of the predatory mite *Phytoseiulus* for controlling spider mites on cucumbers on enclosed ground]. Kolos, Moscow.

Beglyarov, G. A.   1970.   Page 7 In Biologicheskii metod bor'by
     s pautinnymi kleshchami v zashchishchennom grunte [Biologi-
     cal method of controlling spider mites on enclosed ground].
     Kolos, Moscow.
Beglyarov, G. A.   1976. Page 25 In Metodicheskie ukazaniya po
     massovomu razvedeniyu i primeneniyu khishchnogo kleshcha
     fitoseiulyusa dlya bor'by s pautinnymi kleshchami v
     zashchishchennom grunte na ogurtsakh [Methodological
     directions for mass-rearing and use of the predatory mite
     Phytoseiulus for controlling spider mites on cucumbers on
     enclosed ground].   (2nd edition, corrected and supplemented).
     Kolos, Moscow.
Beglyarov, G. A., T. N. Bushchik, and D. Tsedev.  1966. O
     vozmozhnosti ispol'zovaniya fitoseiulyusa (Phytoseiulus
     persimilis A.-N., Phytoseiidae) protiv pautinnogo kleshcha
     v zashchishchennom grunte. [The possibility of using
     Phytoseiulus (Phytoseiulus persimilis A.-N., Phytoseiidae)
     against spider mite on enclosed ground].   Pages 24-25 In
     Pervoe akarologicheskoe soveshchanie.  Tezisy doklada.
     Moscow-Leningrad.
Beglyarov, G. A., Uy. I. Kuznetsove, and A. T. Ushchekov.  1972.
     Page 32 In Metodicheskie ukazaniya po massovomu razvedeniyu
     i ispytanniyam effektivnosti zlatoglazki obyknovennoi
     (Methodological directions for mass-rearing and testing of
     the effectiveness of the common lacewing].  Kolos, Moscow.
Beglyarov, G. A., and A. T. Ushchekov.  1972.  O vozmozhnosti
     ispol'zovaniya zlatoglazki obyknovennoi dlya bor'by s tlyami
     v zashchishchennom grunte [The possibility of utilizing the
     common lacewing for controlling aphids on enclosed ground].
     Pages 33-43 In Biologicheskii metod bor'by s vreditelyami
     ovoshchnykh kul'tur [Biological method of controlling pests
     of vegetable crops].  Kolos, Moscow.
Beglyarov, G. A., and A. T. Ushchekov.  1974.  Opyt i perspektivy
     ispol'zovaniya zlatoglazki [Experiment and prospects of using
     lacewing].  Zashch. rast. 9: 25-27.
Beglyarov, G. A., and A. T. Ushchekov.  1977. Biologicheskaya
     bor'ba s tlyami na zelennykh kul'turakh [Biological control
     of aphids on greens crops].  Ibid. 2.
Beglyarov, G. A., A. T. Ushchekov, and T. A. Kozlova.  1972.  Opyt
     i perspektivy ispol'zovaniya parazita diarielly dlya bor'by
     s tlyami v zashchishchennom grunte [Experiment and prospects
     of use of the parasite Diareliella for control of aphids on
     enclosed ground].   Pages 5-11 In Biologicheskii metod bor'by
     s vreditelyami ovoshchnykh kul'tur [Biological method of con-
     trolling pests of vegetable crops].  Kolos. Moscow.

Beglyarov, G. A., A. T. Ushchekov, and I. A. Ponomareva. 1970.
Biologicheskaya bor'ba s tlyami v teplitsakh [Biological
control of aphids in hothouses]. Pages 498-9 In Tezisy
dokladov VII mezhdunarodnyi kongress po zashchite rastenii.
Parizh, 21-25 sentyabr 1970. (In English)

Beglyarov, G. A., R. A. Vasil'ev, and R. I. Khloptseva. 1967.
Page 23 In Metodicheskie ukazaniya po massovomu razvedehiyu
khishchnogo kleshcha fitoseiulyusa i ispytaniya effektivnosti
v bor'be s pautinnymi kleshchami v zashchishchennom grunte na
ogurtsakh [Methodological directions for mass-rearing of
the predatory mite Phytoseiulus and testing its effectiveness
in controlling spider mites on cucumbers on enclosed ground].
Publishing House of the Ministry of Agriculture, USSR
(Rotaprint).

Beglyarov, G. A., R. A. Vasil'ev, R. I. Khloshcheva, and R. A.
Listkova. 1964. Razrabotka biologicheskogo metoda bor'by s
pautinnym kleshchom v teplitsakh. [Development of a biological
method of controlling spider mite in hothouses]. Pages 119-
22 In Issledovaniya po biologicheskomu metodu bor'by s
vreditelyami sel'skogo i lesnogo khozyaistva. Doklady k
simpoziumu. [Investigations on the biological method of controlling pests of agriculture and forestry. Reports to
symposium]. Novosibirsk.

Beglyarov, G. A., R. A. Vasil'ev, I. V. Korol', A. I. Batov, and
R. V. Kokina. 1968. Biologicheskaya bor'ba s pautinnym
kleshchom [Biological control of spider mite]. Zashch.
rast. 7: 26-29.

Belov, V. K. 1977. Biometod na kazhdyi chetvertyi gektar
[Biomethod on every fourth hectar]. Zashch. rast. 1: 12-4.

Boldyreva, E. P. 1970. Ekologiya Aphelinus mali Hal (Hymenoptera,
Aphelinidae)--parazit krovyanoi tli v tadzhikistane [Ecology
of Aphelinus mali Hal (Hymenoptera: Aphelinidae)--a parasite
of the woolly apple aphid in Tadzhikistan. Entomol. obozr.
49 (4): 744-8.

Bondarenko, N. V., and B. P. Asyakin. 1972. Rol' mukh sirfid
kak afidofagov v zashchishchennom grunte Liningradskoi
oblasti [Role of syrphid flies as aphidophages on enclosed
ground in the Leningrad district]. Zap. Leningr. Skh. Inst.
190.

Bondarenko, N. V., and B. P. Asyakin. 1973. Otsenka effektivnosti
vechnogo sirfa (Syrphus corollae) v podavlenii razmnozheniya
tlei na ogurtsakh v teplitsakh [Evaluation of the effectiveness of the everlasting sirphus (Syrphus corollae) in suppressing the multiplication of aphids on cucumbers in hothouses].
Ibid.: 212.

Bondarenko, N. V. and B. P. Asyakin. 1974. Otsenka razlichnykh
    sposobov primeneniya khishchnoi gallitsy afidimizy
    Aphidoletes aphidimyza Rond. v bor'be s tlyami na ogurtsakh
    v teplitsakh [Evaluation of various means of using the pred-
    atory aphidimyza gall midge Aphidoletes aphidimyza Rond. in
    controlling aphids on cucumbers in hothouses]. Ibid. 239:
    3-12.
Bondarenko, N. V. and B. P. Asyakin. 1975. Metodika massovogo
    razvedeniya khishchnoi gallitsy afidimizy [Methods for mass
    rearing of the predatory aphidimyza]. Zashch. rast. 8: 42-3.
Bondarenko, N. V. and B. P. Asyakin. 1975. Effektivnost'
    gallitsy Aphidoletes aphidimyza Rond. v bor'be s tlyami v
    zimnykh teplitsakh [Effectivenss of cecidomyiid Aphidoletes
    aphidimyza Rond. in controlling aphids in winter hothouses].
    Zap. Leningr. Skh. Inst. 270: 29-33.
Bondarenko, N. V., A. A. Chalkov, and A. A. Elokhin. 1969.
    Primenenie fitoseiulyusa [Use of Phytoseiulus]. Ibid. 6:
    8-9.
Bondarenko, N. V., and E. G. Moiseev. 1971. Effektivnost'
    zlatoglazki obyknovennoi v bor'be s tiyami no ovoshchnykh i
    dekorativnykh kul'turakh v teplitsakhL éningradskoi oblasti
    [Effectiveness of the common lacewing in controlling aphids
    on vegetable and ornamental crops in hothouses of the
    Leningrad district]. Pages 16-17 In Biologicheskaya
    zashchita plodovykh i ovoshchnykh kul'tur [Biological pro-
    tection of fruit and vegetable crops]. Kishinev.
Bondarenko, N. V. and E. G. Moiseev. 1972. Otsenka effektivnosti
    zlatoglazki v bor'be s tlyami [Evaluation of the effective-
    ness of lacewing in controlling aphids]. Zashch. rast.
    2: 19-20.
Bushchik, T. I., and V. F. Plotnikov. 1967. Khishchnyi klop
    protiv vreditelya [A predatory bug against a pest]. Sel'
    skokhozyaistvennoe proizvodstvo nechernozemnoi zony. 1: 12.
Chalkov, A. A. 1968. Biologicheskaya bor'ba s pautinnym kleshchom
    [Biological control of spider mite]. Zashch. Rast. 7: 29.
Chumakova, B. M. 1961. Afitis i kaliforniiskaya shchitovka
    [Aphytis and San Jose scale]. Ibid. 2: 47.
Dirsh, V. M. 1937. Opyt ispol'zovaniya Trichogramma evanescens
    Westw. protiv yablonnoi plodozhorki v Krymu [Experiment on
    the use of Trichogramma evanescens Westw. against the codling
    moth in the Crimea]. Zashch. rast. 16.
Flanders, S. E. 1928. The mass production of Trichogramma minutum
    Riley and observation of the natural and artificial parasitism
    of the codling moth eggs. Pages 110-30 In 4th International
    Congress of Entomology. Vol. 2: Transactions. Naumburg/
    Saale, 1929.

Gaprindashvili, N. K. 1963. O napravlennom povyshenii kholodostoikosti entomofaga lindorusa [Directed augmentation of the cold-resistance of the entomophage lindorus]. Agrobiologiya 6 (144): 836-943.

Gaprindashvili, N. K. 1964. Prisposoblenie mestnykh entomofagov i akarifagov k pitaniyu tsitrusovoi belokrylkoi [Adaptation of local entomophages and acariphages to feeding on citrus whitefly]. Pages 131-3 In Issledovaniya po biologicheskomu metodu bor'by s vreditelyami sel'skogo i lesnogo khozyaistva. Publishing House of the Siberian Division of the Academy of Sciences of the USSR. Novosibirsk.

Goryunova, Z. S. 1965. Biologiya parazita kaliforniiskoi shchitovki afitisa korot-kobakhramchatogo (Aphytis proclia Wek.) i puti ego ispol'zovaniya [Biology of a parasite of San Jose scale, the short-fringed aphytis (Aphytis proclia Wek.) and ways of using it]. Tr. Vses. Inst. Zashch. Rast. 24: 211-6.

Goryunova, Z. S. 1967. Krovyanaya tlya i afelinus [Woolly apple aphid and Aphelinus]. Zashch. rast. 10: 39.

Goryunova, Z. S. Biologiya Prospaltella perniciosi Tow. i Aphytis proclia Wlk. parazitov kaliforniiskoi shchitovki razlichnykh lokal'nykh populyatsii [Biology of Prospaltella perniciosi Tow. and Aphytis proclia Wlk., parasites of San Jose scale of various local populations].

Gusev, G. B. 1974. Ispol'zovanie entomofagov v bor'be s vrednoi cherepashkoi [Utilization of entomophages in controlling the harmful stinkbug]. Pages 104-13 In Biologicheskie sredstva zashchity rastenii [Biological agents for plant protection]. Kolos, Moscow.

Gusev, G. B., E. M. Shagov, and Yu. V. Zayats. 1971. Vosmozhnosti prakticheskogo ispol'zovaniya perillyusa v usloviyakh SSSR [Possibilities of practical utilization of Perillus under the conditions of the USSR]. Byull. Vses. Nauchno-Issled. Inst. Zashch. Rast. 18: 7-11.

Ismailov, Ya. I., and P. I. Shchichenko. 1940. Nablyudeniya nad povedeniem yaitseeda Trichogramma evanescens Westw. v krone plodovogo dereva [Observations on the behavior of the oophage, Trichogramma evanescens Westw., in the crown of a fruit tree]. Vestn. Zashch. Rast. 4: 78-80.

Kalashnikova, G. I. 1971. Biologicheskie osobennosti nekotorykh vidov i form trikhogrammy v usloviyakh Chuiskoi doliny Kirgizii [Biological characteristics of some species and forms of Trichogramma under the conditions of the Chuiskii Valley of Kirghizia]. Pages 41-2 In Biologicheskie metody zashchity plodovykh i ovoshchnykh kul'tur ot vreditelei, boleznei i sornyakov kak osnovy integriro vannykh sistem [Biological methods of protecting fruit and vegetable crops from pests, diseases and weeds as the basis for integrated systems]. Kishinev.

Kartavtsev, N. I.   1968. Biologicheskie ocobennosti vidov i
    vnutrividovykh form trikhogrammy i effektivnost' ikh
    primeneniya v bor'be s kukuruznym motylkom [Biological
    characteristics of species and interspecies forms of
    Trichogramma and the effectiveness of their use in control-
    ling corn borer].   Tr. Vses. Inst. Zashch. Rast. 31: 63-85.
Kartavtsev, N. I., K. E. Voronin, A. F. Sumaroka, and Z. A.
    Dzyuba.   1971. Primenenie introdutsirovannykh vidov
    telenomin v bor'be s vrednoi cherepashkoi v usloviyakh
    stepnoi zony Kubani [Use of introduced and local species
    of Telenomus in controlling the harmful stinkbug under condi-
    tions of the steppe zone of the Kuban].   Pages 51-66 In
    Tezisy dokladov k soveshchaniyu po priemam biologicheskoi
    bor'by s vrednoi cherepashkoi v integriovannykh sistemakh
    zashchity zernovykh kul'tur [Summaries of reports to con-
    ference on means of biological control of the harmful stink-
    bug in integrated systems of protecting grain crops].
    Leningrad.
Kartavtsev, N. I., K. E. Voronin, A. F. Sumaroka, Z. A. Dzyuba,
    and G. A. Pukinskaya.   1975. Opyty mnogoletnykh issledovanii
    po sezonnoi kolonizatsii telenomin v bor'be s vrednoi
    cherepashkoi v Krasnodarskom krae [Experiments in many years
    investigations on the seasonal colonization of Telenomus in
    controlling the harmful stinkbug in the Krasnodar area].
    Tr. Vses. Inst. Zashch. Rast. 44:83-90.
Kolmakova, V. D.   1962. O primenenii belorusskikh form trikhogrammy
    v sadakh Buryatii [Use of Belorussian form of Trichogramma
    in the orchards of Buryatia].   In Vrediteli lesa i plodovykh
    kul'tur Zabaikal'ya [Pests of the forest and of fruit crops
    of the Transbaikal region].   Ulan-Ude.
Kolmakova, V. D.   1965. Primenenie mestnoi formy trikhogrammy
    (Trichogramma embryophagum) v sadakh Zabaikal'ya [Use of a
    local form of Trichogramma (Trichogramma embryophagum) in
    the orchards of the Transbaikal region.] Tr. Vses. Inst.
    Zashch. Rast. 24: 203-7.
Koroleva, N. I.   1938. Primenenie trikhogrammy v sadakh protiv
    yablonnoi plodozhorki [Use of Trichogramma in orchards
    against the codling moth].   "Za Michur." Plodov. 86.
Kovaleva, M. F.   1954. Puti povysheniya effektivnosti trikhogrammy
    v bor'be s vreditelyami sel'skokhozyaistvennykh kul'tur [Ways
    of increasing the effectiveness of Trichogramma in controlling
    pests of agricultural crops].   Zool. Zhu.   33  (1).

Kozlov, M. A. 1971. Primenenie yaitseedov-telenomin v bor'be s vrednoi cherepashkoi [Use of telenomin oophages in controlling the harmful stinkbug]. Pages 59-60 In Kratkie tezisy dokladov k soveshchaniyu po priemam biologicheskoi bor'by s vrednoi cherepashkoi v integrirovannoi sisteme zashchity zernovykh kul'tur [Short summaries of reports to conference on means of biological control of the harmful stinkbug in an integrated system of protecting grain crops]. Voronezh, 23-26 November 1971. Leningrad.

Kulakov, M. F. 1940. Bor'ba s vrednoi cherepashkoi (*Eurygaster integriceps* Put.) pri pomoshchi yaitseeda telenomusa (*Microphanurus* Nees) [Control of the harmful stinkbug (*Eurygaster integriceps* Put.) with the aid of the oophage *Telenomus (Microphanurus* Nees)]. Sb. Rab. Azovo-Chernomorskogo Skh. Inst. 11: 103-35.

Kurbanov, G. G. 1966. Materialy k izucheniyu zlatoglazki v Nakhichevanskoi ASSR [Information on the study of lacewing in the Nakhichevanskii ASSR]. Pages 294-7 In Materialy sessii Zakavkazskogo soveta po koordimatsii nauchno-issledovateli'skikh rabot po zashchite rastenii [Information for the session of the Transcaucasian council on the coordination of scientific research work on plant protection]. Publishing House of the Academy of Sciences of the Azerbaidzhan SSR, Baku.

Kurbanov, G. G. 1971. K izucheniyu khishchnykh zlatoglazok, rasprostranennykh na klopkovykh i khlebnykh polyakh v Nakhichevanskoi ASSR [Study of predatory lacewings distributed on cotton and grain fields in the Nakhichevanskii ASSR]. Izv. Akad. Nauk Az. SSR Ser. Biol. 4: 91-5.

Lapina, V. F. 1936. Rezul'taty primeneniya trikhogrammy (*Trichogramma evanescens* Westw.) protiv yablonnoi plodozhorki v Leningradskoi oblasti v 1929-1934 gg. [Results of the use of *Trichogramma (Trichogramma evanescens* Westw.) against the codling moth in the Leningrad district in 1929-1934]. Xth All-Union Meeting of the Lenin Komsomol.

Luppova, E. P. 1949. Biologiya chervetsa Komstoka v Tadzhikistane i biologicheskii metod bor'by s nim [Biology of the Comstock mealybug and a biological method of controlling it]. Pages 65-68 In Tezisy dokladov XIX plenuma sektsii zashchity rastenii VASKhNIL. Vol. 3.

Luppova, E. P. 1950. Chervets Komstoka v Tadzhikistane [The Comstock mealybug in Tadzhikistan]. Soobshch. Tadzh. Fil. Akad. Nauk. SSSR. 24: 27-30.

Luppova, E. P. 1952. Khimicheskii i biologocheskii metody bor'by s chervetsom Komstoka [Chemical and biological methods of controlling the Comstock mealybug]. Dokl. Akad. Nauk. Tadz. SSR. 2: 25-9.

Luppova, E. P.  1955.  Poleznye nasekomye v bor'be s pautinnym
    kleshchom [Beneficial insects in controlling spider mite].
    Sel'sk. khoz. Tadzh. 7: 34-6.
Luppova, E. P.  1966.  Itogi izucheniya setchatokrylykh (Neurop-
    tera) Srednei Azii [Results of study of neuroptera (Neurop-
    tera) of Central Asia].  Pages 245-52 In Fauna i zoogeografiya
    nasekomykh Srednei Azii [Fauna and zoogeography of insects of
    Central Asia].  Dushanbe.
Matkovskii, S. G.  1940.  Primenenie telenomusa dlya bor'y s
    klompom-cherepashkoi v kolkhoznykh usloviyakh [Use of
    telenomus for controlling stinkbug under kolkhoz conditions].
    Sov. agron. 8-9: 73-8.
Meier, N. F.  1937.  Biologicheskii metod bor'by s vrednymi
    nasekomymi [Biological method of controlling harmful insects].
    Sel'khozgiz.
Meier, N. F.  1940.  Vidy i rasy trikhogrammy (Trichogramma
    evanescens Westw.) [Species and races of Trichogramma
    (Trichogramma evanescens Westw.)].  Vestn. zashch. rast. 4.
Meier, N. F.  1941.  Trikhogramma [Trichogramma]. Sel'kohozgiz.
Meier, N. F. and Z. A. Meier.  1946.  Ob obrazovanii biologiche-
    skikh form u Chrysopa vulgaris Schr. (Neuroptera, Chrysopidae)
    [Formation of biological forms in Chrysopa vulgaris Schr.
    (Neuroptera: Chrysopidae)].  Zool. zh. 25 (2): 115-20.
Meier, N. F. and N. A. Telenga.  1932.  O biologicheskom metode
    bor'by s krovyanoi tlei (Eriosoma lanigerum Hausm.) pri
    pomoshchi ee parazite (Aphelinus mali Held.) v SSSR
    [Biological method of controlling the woolly apple aphid
    (Eriosoma lanigerum Hausm.) with the aid of its parasite
    (Aphelinus mali Held.) in the USSR].  Zashch. rast. 3: 15-24.
Moiseev, E. G.  1973.  Effektivnost' zlatoglazki v bor'be s tlyami
    na ovoshchnykh kul'tur v teplitsakh [Effectiveness of lacewing
    in controlling aphids on vegetable crops in hothouses].
    Avtoreferat dissertatsii na soiskanie uchenoi stepeni
    kandidate biologicheskikh nauk [Author's abstract of disserta-
    tion in competition for the scientific degree of candidate
    of biological sciences].  Leningrad-Pushkin.
Moiseev, E. G., N. V. Bondarenko, and Yu. V. Storozhkov.  1972.
    Slatoglazka v sakrytom grunte [Lacewing on enclosed ground].
    Zashch. Rast. 11: 30-1.
Nyaes, A.  1970.  Ispol'zovanie zlatoglazki protiv tlei v
    zakrytom grunte [Use of lacewing against aphids on enclosed
    ground].  Materialy VII pribaltiiskogo soveshchaniya po
    zashchite rastenii [Information for the VIIth Pribaltic meet-
    ing on plant protection].  Pages 38-39 In Vrediteli sel'-
    skokhozyaistvennykh i lesnykh rastenii i mery bor'by s nimi
    [Pests of agricultural and forest plants and measures for
    controlling them].  Part I, Elgava.

Pleshanov, A. S.  1965.  Ekologo-geograficheskii analiz fauny
    flernits (Chrysopidae) kak afidofagov v lesakh yuga Vostochnoi
    Sibiri [Ecologico-geographic analysis of the fauna of the
    green lacewings (Chrysopidae) as aphidophages in the forests
    of the south of Eastern Siberia].  Izv. Vost.-Sib. Otd. Geogr.
    Ova. SSSR, 64: 60-2.
Popov, G. A. 1971.  Razvedenie klopov-shchitnikov (Hemiptera,
    Pentatomidae), dlya yaitseedov vrednoi cherepashki [Rearing
    shield bugs (Hemiptera: Pentatomidae) for oophages of the
    harmful stinkbug].  Byull. Vses. Inst. Zashch. Rast. 19: 3-9.
Popova, A. I.  1962. Biologiya *Prospaltella perniciosis* Tow.
    parazita kaliforniiskoi shchitovki i ispol'zovanie ego na
    Chernomorskom poberezh'e Krasnodarskogo kraya [Biology of
    *Prospaltella perniciosis* Tow., a parasite of San Jose scale,
    and its use on the Black Sea Coast of the Krasnodar area].
    In Biologicheskii metod bor'by s vreditelyami i boleznyami
    sel'skkhozyaistvennykh kul'tur [Biological method of con-
    trolling pests and diseases of agricultural crops].
    Sel'khozizdat.
Popova, A. I.  1964.  K metodike massovogo razmnozheniya parazita
    kaliforniiskoi shchitovka [Method of mass rearing a parasite
    of San Jose scale].  Tru. Vses. Nauchno-Issled. Inst. Zashch.
    Rast. 20 (I): 61-4.
Popova, A. I.  1971.  Khilokorucy v bor'be s kaliforniiskoi
    shchitovkoi [Chilocoruses in controlling San Jose scale].
    Zashch. Rast.
Radetskii, A. F.  1911.  O parazitarnom metode bor'by s yablonnoi
    plodozhorkoi (*Carpocapsa pomonella* L.) [Parasite method of
    controlling the codling moth (*Carpocapsa pomonella* L.)].
    Tashketn, p. 19.
Remaudiere, G. 1960.  Sunn pest in the Middle East.  Development
    of Research- first practical application.  FAO SUNN Pest
    Inform. Document Centre, Paris 19 p.
Remaudiere, G. 1961.  Sunn pest in Middle East.  Draft report at
    termination of mission.  Ibid. 15 p.
Rubtsov, I. A.  1947.  Novye effektivnye biologicheskie agenty v
    bor'be s kaliforniiskoi shchitovkoi [New effective biological
    agents in controlling the San Jose scale].  Priroda 2: 61-3.
Rubtsov, I. A.  1947.  Akklimatizatsiya kriptolemusa na
    Chernomorskom poberezhe [Acclimatization of cryptolaemus on
    the Black Sea Coast].  Ibid. 2: 63-4.
Shagov, E. M.  1967.  Ob akklimatizatsii perillusa [The acclimati-
    zation of perillus].  Ibid. 10: 49-50.
Shchepetil'nikova, V. A.  1939.  O roli temperatury i vlazhnosti
    v biologii azovo-chernomorskoi rasy trikhogrammy [The role
    of temperature and humidity in the biology of the Azov-Black
    Sea race of *Trichogramma*].  Zashch. Rast. 19: 56-66.

Shchepetil'nikova, V. A.   1940.   K polevoi ekologii azovo-
    chernomorskoi rasy trikhogrammy (*Trichogramma evanescens*
    Westw.) [The field ecology of the Azov-Black Sea race of
    *Trichogramma* (*Trichogramma evanescens* Westw.)].   Vestn.
    Zashch. Rast. 1-2: 140.
Shchepetil'nikova, V. A.   1942.   Rasprostranenie, biologiya i
    primenenie razlichnykh vidov yaitseedov cherepashki v
    raznykh usloviyakh sushchestvovaniya [Distribution, biology
    and use of various species of oophages of stinkbug under
    different conditions of existence].   Dokl. Vses. Akad. Skh.
    Nauk. 5-6: 20-8.
Shchepetil'nikova, V. A.   1949.   Effektivnost' yaitseedov
    vrednoi cherepashki i faktory, ee obuslavlivayushchie
    [Effectiveness of oophages of the harmful stinkbug and
    factors that are responsible for it].   Tr. Vses. Inst.
    Zashch. Rast. 9: 243-84.
Shchepetil'nikova, V. A.   1954.   Faktory, opredelyayushchie
    narastanie chislennosti entomofagov raznoi stepeni
    spetsializatsii, i ikh znachenie v biologicheskom metode
    [Factors determining the increase in numbers of entomophages
    of different degrees of specialization and their importance
    in the biological method].   Pages 311-17 In Tret'ya
    ekologicheskaya konferentsiya [Third ecological conference].
    Tezisy dokladov. Kiev.
Shchepetil'nikova, V. A.   1957.   Zakonomernosti, opredelyayushchie
    effektivnost' entomofagov [Rules determining the effectiveness
    of entomophages].   Obshch. Biol. 13: 381-94.
Shchepetil'nikova, V. A.   1958.   Pages 30-34, 80-86 In Parazity
    klopa-cherepashki i ispol'zovanie yaitseedov.   Bor'ba s
    klopomi [sic] cherepashkoi [Parasites of stinkbug and use
    of oophages.   Control of stinkbug].   Sel'khozgiz, Moscow-
    Leningrad.
Shchepetil'nikova, V. A.   1962.   Vnutrividovye formy *Trichogramma
    evanescens* Westw. i faktory, opredelyayushchie ikh effekti-
    nost' [Intraspecies forms of *Trichogramma evanescens* Westw.
    and factors that determine their effectiveness].   Pages 39-67
    In Biologicheskii metod bor'by s vreditelyami i boleznyami
    sel'skokhozyaistvennykh kul'tur [Biological method of con-
    trolling pests and diseases of agricultural crops].
    Sek'khozizdat.
Shchepetil'nikova, V. A.   1971.   Effektivnost' trikhogrammy v
    sadakh i na ovoshchnykh kul'turakh [Effectiveness of
    *Trichogramma* in orchards and on agricultural crops].   Pages
    114-7 In Biologicheskie metody zashchity plodovykh i
    ovoshchnykh kul'tur ot vreditelei, boleznei i sornyakov kak
    osnovy integrirovannykh sistem [Biological methods of pro-
    tecting fruit and vegetable crops from pests, diseases and
    weeds as the basis of integrated systems].   Kishinev.

Shchepetil'nikova, V. A. 1974. Primenenie trikhogrammy v SSSR
    [Use of *Trichogramma* in the USSR]. Pages 138-58 In
    Biologicheskie sredstva zashchity rastenii [Biological
    agents for plant protection]. Kolos, Moscow.
Shchepetil'nikova, V. A. and N. S. Fedorinchik. 1968. Pages 1-
    112 In Biologicheskii metod bor'by s vreditelyami sel'sko-
    khozyaistvennykh kul'tur [Biological method of controlling
    pests of agricultural crops]. Kolos, Moscow.
Shchepetil'nikova, V. A. and G. V. Gusev. 1970. Izuchenie
    metodov biologicheskoi bor'by protiv khlebnogo klopa
    (*Eurygaster integriceps* Put.) [Study of methods of biologi-
    cally controlling a grain bug (*Eurygaster integriceps* Put.)].
    Pages 95-101 In Doklady sovetsko-frantsuzskogo simposiuma po
    primeneniyu entomofagov [Reports of Soviet-French Symposium
    on the Use of Entomophages]. Antibes, 13-18 May 1968.
Shchepetil'nikova, V. A., N. S. Fedorinchi, and O. V. Kapustina.
    1968. Kompleks priemov biologicheskoi bor'by kak osnova
    sistemy zashchity plodovogo sada ot vreditelei v zone s
    odnim pokoleniem yablonnoi plodozhorki [A complex of means
    of biological control as the basis of a system of protecting
    a fruit orchard from pests in a zone with one generation of
    codling moth]. Tr. Vses. Zashch. Rast. 21-62.
Shengeliya, E. S. 1953. Zlatoglazki (Chrysopidae) [Lacewings
    (Chrysopidae)]. Tr. Inst. Zool. Akad. Nauk. Gruz. SSR
    11: 141-5.
Shuvakhina, E. Ya. 1969. Zlatoglazka--khishchnik koloradskogo
    zhuka [Lacewing--a predator of the Colorado potato beetle].
    Zashch. Rast. 6: 51.
Shuvakhina, E. Ya. 1974. Zlatoglazki i ikh ispol'zovanie v
    bor'be s vreditelyami sel'skokhozyaistvennykh kul'tur
    [Lacewings and their use in controlling pests of agricultural
    crops]. Pages 185-99 In Biologicheskie sredstva zashchity
    rastenii. [Biological agents for plant protection]. Kolos,
    Moscow.
Shuvakhina, E. Ya. 1975. Obyknovennaya zlatoglazka kak
    khishchnik koloradskogo zhuka i vozmozhnosti ee primeneniya
    na kartofele [The common lacewing as a predator of the
    Colorado potato beetle and the possibilities of its use on
    potatoes]. Biologicheskii metod zashchity rastenii
    [Biological method of plant protection]. Tr. Vses. Zashch.
    Rast. 44: 154-61.
Sidorovkina, E. P. 1936. Trikhogramma v sadakh Azerbaidzhana
    [*Trichogramma* in the orchards of Azerbaidzhan]. Itogi
    nauchno-issledovatel'skikh rabot Vsesayuznogo Instituta
    Zashchita Rastenii. Leningrad.
Sogoyan, L. N. 1969. Iskusstvennye sredy dlya kriptolemusa
    [Artificial media for cryptolaemus]. Zashch. Rast. 7: 53.

Talitskii, V. I.   1940. Razmnozhenie i primenenie telenomusa
    dlya bor'by s klopomcherepashkoi [Rearing and use of telenomus
    for controlling stinkbug].   Kiev-Khar'kov. p. 72.
Telenga, I. A.   1937.   Rezul'taty akklimatizatsii parazita
    krovyanoi tli v SSSR.   Biologicheskii metod bor'by s
    vreditelyami sel'skokhozyaistvennykh kul'tur [Results of
    acclimatization of a parasite of the woolly apple aphid in
    the USSR.   Biological method of controlling pests of agri-
    cultural crops].   Tr. VASKhNIL. 10: 56-67.
Telenga, I. A.   1956.   Issledovaniya *Trichogramma evanescens* Westw.
    i *T. pallida* Meyer (Hymenoptera, Trichogrammatidae) i ikh
    primenenie dlya bor'by s vrednymi nasekomymi v SSSR [Inves-
    tigations of *Trichogramma evanescens* Westw. i *T. pallida*
    Meyer (Hymenoptera: Trichogrammatidae) and their use for
    controlling harmful insects in the USSR].   Entomol. Obozr.
    35 (3): 599-610.
Telenga, I. A.   1959.   Taksonomicheskaya i ekologicheskaya
    kharakteristika vidov roda *Trichogramma* (Hymenoptera,
    Trichogrammatidae) [Taxonomic and ecological characteristics
    of species of the genus *Trichogramma* (Hymenoptera:
    Trichogrammatidae)].   Vol. 8. Nauchnye trudy Ukrainskii
    Institut Zaschita Rastenii. Kiev.
Telenga, I. A. and O. M. Kiseleva.   1959.   Ispol'zovanie
    trikhogrammy dlya bor'by s vreditelyami sel'skokhozya-
    istvennykh kul'tur [Use of *Trichogramma* for controlling
    pests of agricultural crops].   Vestn. sel'skogospodarskogo
    nauka. Kiev.
Telenga, I. A. and V. A. Shchepetil'nikova.   1949.   Rukovodstvo
    po razmnozheniyu i primeneniyu trikhogrammy dlya bor'by s
    vreditelyami sel'skokhozyaistvennykh kul'tur [Handbook on
    rearing and using *Trichogramma* for controlling pests of
    agricultural crops].   Kiev.
Tsintsadze, K. V.   1975.   Pages 132-4 In Pishchevaya spetsializa-
    tsiya osnovykh vidov zlatoglazok, rasprostranennykh v
    Adzharii.   Subtropicheskie kul'tury [Food specialization of
    the main species of lacewings distributed in Adzharia.
    Subtropical crops].   Vol. 4. (139). Publishing House of the
    Institute of Tea and Subtropical Crops.
Tvaradze, M. S.   1969.   K. faune zlatoglazki (Neuroptera,
    Chrysopidae) Borzhomskogo ushchel'ya [Lacewing fauna
    (Neuroptera, Chrysopidae) of the Borzhomskii Canyon]. Tr.
    Gruz. Inst. Sashch. Rast. 21: 245-7.
Tyumeneva, V. A. 1936.  Biologicheskii metod bor'by s yablonnoi
    plodozhorkoi posredstvom yaitseeda trikhogrammy (Biological
    method of controlling the codling moth by means of the
    oophage *Trichogramma*].   Itogi nauchno-issledovatel'skikh
    rabot Vsesoyuznogo Instituta Zashchita Rastenii.
    Leningrad. 1936.

Uspenskii, F. M.  1951.  Vidovoi sostav i znachenie khishchnikov pautinnogo kleshcha [Species composition and significance of predators of spider mite].  Sbornik nauchnykh rabot STAZR Soyuz NIKhI. 44-8.

Vashadze.  1959.  Nekotorye dannye k voprosu biologicheskogo metody bor'by s yaponskoi voskovoi lozhnoshchitovkoi [Some data on the question of the biological control of Japanese wax scale].  Tr. Sukhum. Bot. Sada 12: 359-64.

Vasil'ev, I. V.  1906.  Vrednaya cherepashka (*Eurygaster integriceps* Osch) i novye metody bor'by s nei pri pomoshchi parazitov iz mira nasekomykh (The harmful stinkbug (*Eurygaster integriceps* Osch) and new methods of controlling it with the aid of parasites from the insect world].  Tr. Byuro Entomol. 4 (11): 28-55.

Vasil'ev I. V.  1913.  Vrednaya cherepashka (*Eurygaster integriceps* (Osch.) Put.) i novye metody bor'by s nei pri pomoshchi parazitov is mira nasekomykh [The harmful stinkbug (*Eurygaster integriceps* (Osch.) Put.) and new methods of controlling it with the aid of parasites from the insect world].  Ibid. 4 (1): 1-13.

Vodinskaya, K. I.  1928.  Materially po biologii i ekologii kapustnykh mukh *Hylemis brassicae* Bch. i *H. floralis* Fall. [Information on the biology and ecology of the cabbage maggots *Hylemia brassicae* Bch. and *H. floralis* Fall.].  Izv. Otd. Prikl. Entomol. 3 (2): 229-49.

Yolkov, V. F.  1954.  K voprosu ob otsenke effektivnosti primeneniya obyknovennoi i zheltoi trikhogrammy (*Trichogramma evanescens* Westw. i *T. pallida*) v sadakh v bor'be s yablonnoi plodozhorkoi i nekotorymi listovertkami [The problem of the evaluation of the effectiveness of using the common and the yellow *Trichogramma* (*Trichogramma evanescens* Westw. and *T. pallida*) in orchards in controlling the codling moth and some leaf rollers].  Page 323 In Biological metod bor'by s vrednymi nasekomymi.  Vol. 5.  Publishing House of the Academy of Sciences of the Ukrainian SSR.  Kiev.

Volkov, V. F.  1959.  K ekologii zheltoi trikhogrammy (*Trichogramma cacoecia pallida* Meyer) -- parazita yablonnoi plodozhorki i listovertok v sadakh USSR [The ecology of the yellow *Trichogramma* (*Trichogramma cacoecia pallida* Meyer) - a parasite of the codling moth and leaf rollers in orchards of the Ukrainian SSR.  Nauchn. Tru. Ukr. Nauchno-Issled. Inst. Zashch. Rast. 8: 137-63.

Voronin, K. E.  1974.  Prakticheskoe znachenie povedeniya telenomin [Practical significance of the behavior of telenomins].  Pages 114-28 In Biologicheskie sredstva zashchity eastenii [Biological agents for plant protection].  Kolos. Moscow.

Yasnosh, V. A. 1974. Biologicheskaya bor'ba s koktsidami v SSSR
    [Biological control of coccids in the USSR]. Pages 247-51 In
    Biologicheskie sredstva zashchity rastenii [Biological agents
    for plant protection]. Kolos, Moscow.
Yuzbash'yan, O. Sh. 1970. Zlatoglazka obyknovennaya (Chrysopa
    carnea Steph.) i ee rol' v ogranichenii chislennosti
    sosushchikh vreditelei khlopchatnika [The common lacewing
    (Chrysopa carnea Steph.) and its role in limiting the numbers
    of sucking pests of cotton]. Tashkentskii sel'skokhozyaistvennyi
    institut. Avtorreferat dissertatsii na soiskanie uchenoi
    stepeni kandidata nauk [Tashkent Agricultural Institute.
    Author's abstract of dissertation in competition for the
    scientific degree of candidate of Sciences]. Tashkent. 30 p.
Zeleny, I., and V. I. Talitskii. 1966. K poznaniyu fauny verbly-
    udok (Rhaphidioptera), setchatokrylykh (Neuroptera) i
    skorpionovykh mukh (Mecoptera) v Moldavii [Toward the know-
    ledge of the fauna of snakeflies (Rhaphidioptera), Neuroptera
    (Neuroptera) and scorpion flies (Mecoptera) in Moldavia].
    Tr. Mold. Nauchno-Issled. Inst. Sadovod. Vinograd. Vinodel.
    13: 85-91.
Zhuravleva, A. M., and I. V. Zil'bermints. 1970. K izucheniyu
    vozmozhnostei sochetaniya khimicheskogo i biologicheskogo
    metodov bor'by s pautinnym kleshchom [Study of the possibili-
    ties of combining the chemical and biological methods of
    controlling spider mite. Pages 213-4 In Tezisy dokl. vtoroe
    akarologicheskoe soveshchanie. Part 1. Publishing House
    Naukova dumka. Kiev.
Zomorrodi, A. 1959. La lutte biologique contre la punaise du
    ble Eurygaster intergriceps Put. par Microphanurus
    semistriatus Nees en Iran. Rev. Pathol. Veg. Entomol.
    Agric. Fr. 38 (3): 167-74.
Zorin, P. V. 1923. Vestnik oblastnoi stantsii. Zashch. Rast.
    Vred. 4: 6-7.
Zorin, P. V. 1927. Nablyudeniya nad zhukom Aleochara bilineata
    Gyll. [Observations on the beetle Aleochara bilineata Gyll.].
    Ibid. 4: 9-12.
Zorin, P. V. 1934a. Bor'ba s vreditelyami orgutsov v teplitsakh
    i parnikakh [Control of pests of cucumbers in hothouses and
    hotbeds]. Vses. Inst. Zashch. Rast. 8: 61-63.
Zorin, P. V. 1934b. Is-pod yuzhnogo neba v severnye teplitsy
    (From under the southern sky into northern hothouses].
    Ibid. 8: 177-8.

EDITORS' NOTE

    The reference citations for this chapter are reproduced as
received from the author. However, some references cited in the
text are not listed above and some listed references are incomplete.

CHAPTER 10

AUGMENTATION OF NATURAL ENEMIES

IN THE PEOPLE'S REPUBLIC OF CHINA

Carl B. Huffaker[1]

Division of Biological Control, Department of
Entomological Sciences, University of California, Berkeley
Albany, California 94706    U.S.A.

Biological control has received effective emphasis in China.
Since the Cultural Revolution government policy has decreed that
scientists channel their research toward practical assistance of
the communes.  But importation of natural enemies from foreign lands
has received almost no attention.  Biological control has been used
widely, although in limited ways.  It has not been taken as the
central feature of their integrated control approach.  Cultural
control occupies this role.  Undoubtedly, greater use than was
apparent to us is being made of indigenous natural enemies because
much attention is given to selective pesticide technology to protect
them.  The development of pest resistant crops as a central or high
priority component has had little active or organized support, although varieties of some crop species resistant to a variety
of pests were reported (NAS, 1977).

As in other areas of entomology, there appeared to be no highly
structured national organization of biological control.  The emphasis
given to particular areas of biological control appeared to be
decided rather independently at the Province and lesser political
levels.  In Shanghai, we were told that there is no central or
national agency that determines advisability of any pest control
measure and then encourages or directs its development and use
in all provinces.  Within the political areas, it seems to be de-

---

[1] Member, U.S. Insect Control Delegation of the Committee on
Scholarly Communication with the People's Republic of China,
sponsored jointly by the American Council of Learned Societies, the
National Academy of Sciences and the Social Sciences Research
Council (August 4-29, 1975).

pendent, as in the U.S.A., on the interest, competence and aggressiveness of particular scientists who have promoted particular
areas. National direction is not however, entirely lacking.  They
do have national conferences and scientific journals at which,
or in which, research and biological control programs are reported.
A scientist knowledgeable in a given program may go to another
region or province to assist in its development.  There is national
support and coordination of some areas from Peking.  For example,
the Government has sponsored, and has shown in various provinces,
the excellent Chinese movie produced in Kwangtung Province, to
popularize use of *Trichogramma*.

The big advantage that China has over most of the world in using
specific approaches to biological control is its low-cost manpower.
But to utilize this manpower in these programs, each area requires
its own corps of scientists, and its own research to relate findings in other areas to their own conditions.

The vigor of China's biological control programs is seen in
several areas--for example in the widely developed and reportedly
quite effective large-scale augmentations of several species of
*Trichogramma* and, more restrictedly, of *Anastatus*, and the distribution of *Bacillus thuringiensis* Berliner and *Beauveria bassiana*
(Bals.) V.  These methods have been used against a number of major
pests.  They also use ducks for both insect and weed control in
rice.  The widescale use of *Trichogramma* and *B. thuringiensis*
have been a major stimulus to other efforts.  Establishment of
biological control is an important part of integrated pest control
in China, along with, but as yet secondary to, cultural controls.
These tactics have been used along with chemicals.

This report is restricted to those efforts our committee
observed or discussed with Chinese scientists, and is further
restricted to the parasite and predator efforts, as appropriate
to this treatise.

It is possibly significant to their biological control approaches, that Chinese scientists had a close association with
scientists in the U.S.S.R. during the 1950's.  The Chinese emphasis placed on *Trichogramma* spp. and *Beauveria bassiana*,
in particular, may stem from that association, as well as from
internal political emphasis suggesting "...the use of natural
forces".  The emphasis on *B. thurigniensis* is derived, in
part at least, from the work of E. A. Steinhaus and colleagues
in California and the subsequent development of this bacterium
for use in several countries.  The extensive research on, and utilization of *Trichogramma* in the U.S.S.R. is a matter of record
[e.g., Klassen 1973, Maria Tuteovitch, personal communication
(reported in trip report of the author to W. G. Tweedy following
a visit to Moldavia in 1974)].  Yet the People's Republic of

China has also made innovative accomplishments of its own in
biological control (e.g. Collaborative Research Group on Biological Control of Rice Pests, Kwangtung Province 1974).

We are concerned in this book primarily with augmentation of
natural enemies for biological control, particularly parasites and
predators.  Three distinct efforts of this nature have been undertaken in the People's Republic of  China.  These are the extensive
releases of *Trichogramma* egg parasites, of *Anastatus* for control of
lichee stink bugs, and of both *Bracon greeni* Ashmead and its pest
host, *Eublemma amabilis* Moore, for control of the latter as a
pest of lac production.


## USE OF *TRICHOGRAMMA*

Several species of *Trichogramma*, *T. dendrolimi*, *T. australicum*,
*T. ostrineae* and *T. japonicum*, among 12 reported in China (Pang and
Chen 1974), are mass produced in the communes of various provinces
and are released extensively for control of the European corn borer,
pine caterpillars, rice leaf roller, sugar cane borer, and other
Lepidoptera, as in cotton, for example.  The host eggs used for
rearing are the giant silkworms *Samia* (=*Philosamia*) *cynthia* (Drury)
(or *S. ricini*) and *Antherea perniyi* (Gnérin-Mádneville), and the
rice grain moth *Corcyra cephalonica*.  *Sitotroga cereallella* Oliver
is the host used in most other areas of the world.  *Trichogramma*
has been released in some 1,700,000 acres of cotton yearly.

Under certain conditions they are released in rice at about
300,000 per acre, at 30 points per acre, and at five different
times for a single rice crop if densities of target insects (e.g.
rice leaf roller, *Cnaphalocrocis medinalis* Guenee) are high.
Lower numbers are used to control low density infestations.  The
Chinese report getting ca 80% parasitization or more, and this
is adequate.  Costs are reported to be about half that of chemical
control.  For corn borer, *Trichogramma*  are released no more than
three times, at about 130,000 per acre, with parasitization averaging
75%, with the cost in a one-generation area being about $0.50
to 0.70 per acre.  The labor of distribution costs about $0.52
to cover about 8 acres.


### *Trichogramma* Use in Kirin Province

*Trichogramma* for Corn Borer Control in Kirin.  The European
cornborer is a pest mainly of corn, millet, and sorghum in Kirin.
Prior to the Cultural Revolution, mainly chemicals were used for
its control, but with emphasis since then put on integrated control and use of natural forces, biological control has been developed and "popularized" for use in the countryside.  There has been

a definite increase in the biological control in Kirin Province.
In 1961 there were only four units in Kirin Province working on
biological control; three of these on *Beauveria* and one on *Trichogramma*. In 1975 there were 200 units, with 120 on *Beauveria*, and
80 on *Trichogramma*. There are about 400 people working in these
two programs. *Beauveria bassiana* has been studied in Kirin Province
at least since the 1960's and was isolated from an <u>endemic</u> infection
of the European cornborer in Kirin in 1970. The utility of this
find was quickly investigated (Hsiu *et al.* 1973). The use of
*Beauveria* and *Trichogramma* in combination was expanded from 500
acres in 1971 to ca 66,000 acres in 1974 and to ca 148,000 acres
in 1975. Studies are underway also to see if *Beauveria*, or
*B.t.* and *Trichogramma* can be used against soybean insects. Mainly
cultural and chemical controls are now employed.

Releases of *Trichogramma* for corn borer control in Kirin
were reported to be at the level of 90,000 to 120,000 per acre,
with the distribution being made at 30 points per acre. Releases
are initiated before moth egg laying begins and are continued
through the egg laying period. (See (a) Seed and Plant Protection
Section, Tung Hua Region, Kirin Province (1975), and (b) Agricultural
Experimental Station, Sanke-yushu People's Commune, Tung Hua County,
Kirin Province (1975))· In parts of Kirin, *Trichogramma* are used
with *Bacillus thuringiensis* at different times during the season.
In single generation areas *Trichogramma* alone are usually sufficiently
effective but *Beauveria* may also be used. Cost of the *Trichogramma*
program is about $0.60 per acre.

The Commune center parasitizes the eggs used in the fields
and sends them to the brigades which make the releases at specified
points per acre (per mu). One 4-year evaluation showed that
corn borer populations per 100 plants were reduced, as compared
to the controls, by 69.6% to 83.4 %. The total acreage relying
on *Trichogramma* in three Kirin counties alone, increased from
183 acres in 1971 to cover 89,000 acres in 1974.

*Trichogramma* for Pine Caterpillar Control in Kirin. Since
1964 the Forest Protection Institute of Kirin's Bureau of Forestry has devoted much effort to the use of *Trichogramma* for control
of the spiny pine caterpillar, *Dendrolimus sibericus* (Tshetverikov),
a pest of all pines and of larch and spruce in the Province.
The former non-biological control methods were unsatisfactory.
Their present satisfactory program embraces in part chemical
control, but to an increasing extent, the use of *Trichogramma*.
The Chinese believe that the use of chemicals brings on various
other pest problems and resurgence of target pests due to interference
with natural enemies.

It was here that methods of mass producing *Trichogramma* on eggs
of the giant silkworm moths (above) were first developed. A single

giant silkworm egg produces an average of ca 60 *Trichogramma*.  (See further NAS, 1977).

Many hosts for *Trichogramma* were tried, but they settled on the giant silkworms.  They are readily available and cheap, bought from the silkworm culture industry in the Commune.  They are good hosts for mass production, and the eggs are readily available the year round.

The fecundity of *T. dendrolimi* is 105 as an average.  At the production facility, there is an automated belt equipped with flanges and flaps for holding a large supply of giant silkworm cocoons (Fig. 1) which are nearing time of emergence.  The belt can be rotated by electric motor to put any of the freshly emerged moths into position for collection.  Virgin gravid females are collected and passed through a good sized electric meat grinder (regular sausage grinder). This squeezes out eggs and all body parts into a container, with some eggs being broken but many coming through intact.  The eggs are cleaned of "gunk" and glued onto cards for parasitization (Fig. 2). As they are infertile, they can be readily stored and used over a period of time for production of *Trichogramma*.  They rear some *T. dendrolimi* in an outdoor screened insectary under natural variable temperatures and humidity, to expose them to more natural conditions for feeding back "hardier" individuals into the main stock culture.  A 10 percent per year input is

Figure 1.  Giant silkworms emergence box containing an automated belt equipped with flanges and flaps that bring freshly emerged moths into position for collection.

Figure 2. Infertile eggs of the giant silkworm glued to large
cards in groups capable of being cut and separated.

made in this way to the culture stock. I doubt, however, that
this would remedy the main problem that might arise from continuing
isolated culture of such an insect in the laboratory.

The time of releasing *Trichogramma* is most important. Emergence
in the field must coincide with the ovipositing period of the
host insect. For this program, they use ca 4 release points
per acre and ca 1000 *Trichogramma* per point. In experimental
releases they commonly get 95% parasitization, but for large-
scale forest management releases they get ca 80%. The cost reported
for production of *Trichogramma* for pine caterpillar control is
$0.28 per acre, which is $0.32 less than for corn borer; because
fewer are needed per release and fewer releases are required.
We were shown a stand of young larch where releases of *T. dendrolimi*
were initiated in 1965 when the caterpillars averaged 40 per
tree, with 60% of the trees infested. By 1975, the density had
declined to 2.6 per tree, with only 6.2% of the trees infested.
They attributed this change largely to the *Trichogramma* releases.

*Trichogramma* Use in Hunan, Chekiang and Kwangtung Provinces

In the Shanghai area, *Trichogramma* are used against certain

pests of rice, cotton and truck crops.  The Shanghai County People's
Commune use *Trichogramma* against the rice leaf roller when it
is necessary.  The "middle-school" children there produce the
*Trichogramma*.

In Hunan, a main rice area of China, *Trichogramma* is used
primarily for control of rice leaf roller on ca 16,500 acres
of rice. Both *T. australicum* (the most effective species) and
*T. dendrolimi* are used.  The methods of production developed
in Kirin (above) are used.  The rate of distribution at 30 points
per acre depends upon the density of leaf roller eggs.  The parasitism reported ranged from 70% to 85%.  If leaf roller egg densities
are less than ca 300,000 per acre, they commonly release ca 60,000
*Trichogramma* per acre; if egg densities are ca 300,000-600,000
per acre they release ca 120,000 *Trichogramma* per acre; if egg
densities are greater than 600,000 per acre they release 180,000-
240,000 *Trichogramma* per acre.

In Hunan, *Trichogramma* are also effectively used against
a sugar cane borer (*Argyroploce schistaceana* Snellen), European
corn borer and pine caterpillars. *Trichogramma* were reported
to be widely used also in Honan Province which we did not visit.
In 1974, they were used in 7,500 acres in Lai-kan County alone
against European corn borer in corn (NAS report, 1977).

In Kwangtung Province at Chung Shan University, *Trichogramma*
studies were conducted in the early 1950's, and by 1956 they
were being widely used against *A. schistaceana* (above).  No data
on percent kill were given.  The university later developed a
distinct biological control laboratory which in 1972 became the
Kwangtung Entomological Institute.  This institute does extensive
work with *Trichogramma* for use against rice pests and sugar cane
borers, and is initiating a range of other studies (see below).
It was here that the excellent Chinese movie on *Trichogramma*
was produced.  *Trichogramma* is used yearly on ca 1/5 of the rice
acreage in Kwangtung Province, and in 2/3 of the counties (196
stations).  It is the accepted method of combating the leaf roller.
Four species of *Trichogramma* are studied and their culture methods
are being developed for Commune use. These are *T. australicum*,
*T. ostrineae* and *T. dendrolimi*, reared on giant oak silkworm
moth eggs (above), and *T. japonicum* which is more host specific
and is produced on eggs of the rice grain moth, *Corcyra cephalonica*
(Stnt.).  *T. japonicum* is considered the most effective although
its use has not yet been widely adopted.

Use of *Trichogramma* in Kwangtung Province for rice pest
and sugar cane borer control is considered a major success.
Success is said to rest on several factors:  1) good production
and distribution methods; 2) adequate short-term prediction of
weather and target pest status, and 3) use of the best species,

and/or strain for the geographic area and crop. Temperature and humidity studies on four species of *Trichogramma* showed that against rice leaf rollers, *T. japonicum* is efficient mainly south of the Yangtze River, *T. australicum* south of the Yellow River, and *T. dendrolimi* throughout China where this pest occurs. In general, control has been reported "as good or better than by use of chemicals". In 1971-1973, use of *Trichogramma* in the Lang Tung Commune was extended from limited use to over 3,300 acres--over 70% of the total area in rice! Parasitism usually exceeded 80%. Yield loss in 1971 in the absence of *Trichogramma* releases was reported to be about 20%. Experiments showed that by *Trichogramma* releases damage could be reduced to ca 1%. Insecticide use has been cut by half, a fact reported to be much welcomed by the paddy workers as their personal hazards are thereby reduced.

In this Institute studies are also being conducted on the effects of *Trichogramma* releases on other rice insects. In one study area in 1974, adequate releases over the period May 5 to June 3, and in another area from May 18 to May 30, gave parasitization of the green rice caterpillar, *Naranga aenescens* Moore, on June 19 and June 22, respectively as 91.7% and 83.9%. In other areas where very few *Trichogramma* were released (there were no areas where none was released), parasitization was only 10 to 15%.

*Trichogramma* releases have recently formed a significant part of the experiments in the general integrated approach to control of the whole complex of rice insect pests in Kwangtung (see below). They have made releases against *Cnaphalocrosis medinalis* Guenee, *N. aenescens* (above) and yellow paddy borer, *Tryporyza incertulus* Walker. Some 30,000,000 *Trichogramma* were released in these experiments, involving 267 acres of rice. While the chemicals used in these experiments for control of certain other pests would have depressed the effects of *Trichogramma*, damage by the yellow paddy borer was reduced by the various measures from ca 4% to 1% and less. Parasitism by *Trichogramma* of both the rice leaf roller and *Naranga* was about 70% in 1974. In 1975 the heavy rains reduced the parasitism considerably.

Big Sand Commune in Kwangtung Province produces its own *Trichogramma*. *T. japonicum* is produced on the rice grain moth, *Corcyra cephalonica*. The rice moth is reared in a rice meal medium in shallow reed trays. This culture is housed in a darkened chamber at 26°-30°C and 75-80% relative humidity. Host eggs are collected from the oviposition units, cleaned of debris and sprinkled onto and stuck to glue coated paper sheets 10" x 10" in size. One sheet contains enough eggs to produce 60,000 *T. japonicum* (ca 2 to 5 per egg--vs an average of 50 or 60 for an egg of the oak silkworm moth using *T. dendrolimi*).

When parasitized, the sheet of parasitized eggs is cut into squares of 5,000 *Trichogramma* for field release.  From 120,000 to 300,000 per acre are used in rice, depending upon the species of *Trichogramma* and the density of the leaf roller.


## MISCELLANEOUS BIOLOGICAL CONTROL EFFORTS

While probably many other efforts to expand the use of biological control, both by augmentations and otherwise, are underway in China, our committee was exposed to only a few examples.  These included use of various entomopathogenic microbes, use of these agents and various parasites and predators in integrated control efforts and a few main augmentation efforts directed against special- ty crop pests, such as the lichee stink bug, *Tessaratoma papillosa* Drury, and the noctuid, *Eublemma amabilis* Moore, which is a serious pest of lac culture.  In their integrated control programs for rice, citrus, stone and pome fruits, and cotton the Chinese have included substantial elements of biological control, some augmenta- tive in nature, but these are dealt with elsewhere (NAS 1977).


### Biological Control of a Noctuid Pest of the Lac Industry

The purple lac insect, *Tachardia lacca* Kerr, is cultured on forest trees of *Dalbergia balansae* and other Papilionaceae in Kwangtung Province.  Infestations of appropriate densities are manually initiated on the branches of selected trees. It was ● said the noctuid, *E. amabilis*, eats the lac and probably feeds directly on the lac insect itself.  If a density of 100 *Eublemma* per m of "stick" occurs at an early nymphal stage of the lac insect, there is a complete loss of the crop, but if attack is delayed to a more mature stage, yield and quality are affected without loss of the entire crop.  Chemical control is too expensive and otherwise impracticable for use in this rough terrain.

Negi *et al.* (1946) had reported from India, possibilities of biological control of *Eublemma* by the parasite *B. greeni*. Chinese scientists of the People's Republic of China introduced this parasite from Hainan Island to Kwangtung Province in 1972. It became readily established.

*B. greeni* is produced in cultures on *Eublemma* itself or on the pink bollworm.  The rate of release of *B. greeni* is determined by the density of *E. amabilis*.  *B. greeni* has effectively controlled *E. amabilis* in a short period of time.  Parasitism in release areas has been ca 60% (maximum of 90%).  Lac yield in a release area was increased 20 to 40% and quality was twice as good, contrast- ed to "control" areas.  Due to the disruptive aspects of lac culture and harvesting, and the cold winters, it is necessary to artifici-

ally add the pest species itself into the lac area at times of
scarcity of *Eublemma*, in order to maintain the *B. greeni* population.
Pest-augmented areas had parasitism of 54% while non-augmented
areas had parasitism of only 7.4%.

Obtaining better pest control by artificially augmenting
a pest population at critical times, to improve biological control,
was shown experimentally to be feasible for control of cyclamen
mites on strawberries in California (Huffaker and Kennett 1956)
and for control of the cabbage butterfly in Missouri (Parker 1971),
but in neither case were those methods widely adopted.  The Chinese
expect soon to use this system for all of the lac pest problems.
This would be the first widescale commercial application of the
concept of adding the pest itself as a means of augmenting the
numbers and effectiveness of natural enemies.


Biological Control of the Lichee Stink Bug, *Tessaratoma papillosa*

In Kwangtung Province, the lichee stink bug is a serious pest
of lichee.  Scientists of Chung Shan University and lichee production
personnel in the communes have developed a program of biological
control of this pest.  They produce the eupelmid egg parasite,
*Anastatus* sp., on eggs of the giant silkworm as described for
*Trichogramma* (above).  Parasitism of the eggs of the giant silkworm
by *Anastatus* is obtained in the laboratory in the Fall and they
are placed in outdoor conditions to overwinter.  In the Spring
they are brought back into an outdoor insectary for emergence
and parasitization of fresh batches of giant silkworm eggs.  The
parasitized egg sheets are then placed by hand in the lichee trees.
In 1966 and 1967, they conducted successful large scale field
demonstrations in two counties in Kwangtung.  In 1970 they cultured
and released wasps in 12 communes.  Parasitism reached 85.5% to
98.7%, in contrast to only 10 to 14% in non-release areas.  In
1973, they had over 50 *Anastatus* rearing stations in areas of
Kwangtung where lichee is grown (Huang *et al.* 1974).


LITERATURE CITED

Agricultural Experimental Station, Sanke-yushu People's Commune,
      Tung Hua County, Kirin Province.  1975.  The control of European
      corn borer by using trichogrammatid egg parasites.  Acta
      Entomol. Sinica 18: 10-6.
Collaborative Research Group of Biological Control of Rice Pests,
      Kwangtung Province.  1974.  The control of rice leaf roller,
      *Cnaphalocrocis medinalis* Guenee by trichogrammatid egg
      parasites.  Acta Entomol. Sinica 17: 269-80.  [with English
      summary]
Hsiu Cheng-fung, Chang Yung, Kwei Cheng-ming, Han Yu-mei and Wang

Hwei-hsien. 1973. Field application with *Beauveria bassiana* (Bals.) Vuill. for European corn borer control. Acta Entomol. Sinica 16: 203-6.

Huang Ming-dau, Mai Siu-hui, Wu Wei-nan and Poo Chih-lung. 1974. The bionomics of *Anastatus* sp. and its utilization for the control of lichee stink bug *Tessaratoma papillosa* Drury. Acta Entomol. Sinica 17: 362-75. [with English summary]

Huffaker, C. B. and C. E. Kennett. 1956. Experimental studies on predation: (1) predation and cyclamen mite populations on strawberries in California. Hilgardia 26: 191-222.

Klassen, W., ed. 1973. Biological methods of protecting fruit and vegetable crops from pests, diseases and weeds as bases for integrated systems. [Summaries of Reports, All-Union Scientific Research Institute of Biological Methods of Plant Protection, Kishinev, USSR, Oct. 1971]. ARS, USDA. 173 pp. Mimeo.

National Academy of Sciences. 1977. Insect Control in the People's Republic of China. CSCPRC (Committee on Scholarly Communication with the People's Republic of China) Report No. 2. National Academy of Sciences, Washington, D.C. 217 pp.

Negi, P. S., S. N. Gupta, M. P. Misra, T. V. Venkataram and R. K. De. 1946. Biological control of *Eublemma amabilis* Moore by one of its indigenous parasites, *Microbracon greeni* Ashmead. Indian J. Entomol. 7: 37-40.

Pang Xion-fei and Chen Tai-lu. 1974. *Trichogramma* of China (Hymenoptera: Trichogrammatidae). Acta Entomol. Sinica 17: 441-54 [with English summary].

Parker, F. D. 1971. Management of pest populations by manipulating densities of both hosts and parasites through periodic releases, pp. 365-76. In "Biological Control", C. B. Huffaker, ed. Plenum Press, N.Y.

Seed and Plant Protection Station, Tung Hua County, Kirin Province. 1975. The experience and realization of large scale control of European corn borer by using trichogrammatid egg parasites. Acta Entomol. Sinica 18: 7-9.

CHAPTER 11

AUGMENTATION OF NATURAL ENEMIES IN WESTERN EUROPE

Emile Biliotti

Inspecteur Général
Institut National de la Recherche Agronomique
75341 Paris Cedex 07, France

The idea of using entomophagous species to contain pests is an old one and suggestions in this direction exist, for example, in the well known "Mémoires pour servir à l'histoire des insectes" of Reaumur but the first practical attempt towards conservation and augmentation of natural enemies in Western Europe was probably done by Decaux (1899). This author planned a complete program of pest management in orchards, in the frame of which apple buds attacked by *Anthonomus* were collected and put into containers divised in such a way that they allowed for the emergence of Ichneumonid parasites and prevented the release of adult pests. Other examples of the use of native species are listed and analysed by Marchal (1907).

After this early work, biological control in western Europe was mainly devoted to the introduction of parasites and predators. The introductory effort had its beginning with the classical *Rodolia cardinalis* Muls. against *Icerya purchasi* Mask., first in Portugal in 1897, then in France (Marchal 1913) and other countries. The most recent success is the use of *Cales noaki* How. to control the wooly whitefly, *Aleurothrixus flocusus* Mask. on citrus in Southern France (Onillon 1973), in Spain and also in Morocco and the Canaries Islands. Franz (1961) and Greathead (1976) have reviewed many of the biological control successes of Western and Southern Europe.

A great deal of experimental work has been done on the possibility of using other techniques of biological control, but few of them have led to large scale practical utilization. Mass breeding followed by periodical release was used in attempts to control the olive fly, *Dacus oleae* Gmel. in southern France using the braconid parasite, *Opius concola* Szepl. imported from Morocco

341

(Biliotti and Delanoue 1959) with partial success (Delanoue 1960, 1970). Good results were also obtained in Sicily, using the local strain of *O. concolo siculus* Mon. (Monastero and Delanoue 1967, Monastero 1968, Liotta and Mineo 1968). Olive fall was reduced from 100% to 30% after three successive years of releases. These operations which, in 1967, covered 2.800 ha and used 32 million parasites, were made possible by the development of special mass breeding techniques using *Ceratitis capitata* Wed. reared on artificial media, as a laboratory host (Feron *et al.* 1958, Monastero and Genduso 1962). These efforts were realised in the framework of an international cooperative effort between Italy and France.

Methods of mass breeding the only two important natural enemies of the olive moth, *Prays oleae* Bern have been developed in France (Arambourg 1970) on laboratory hosts and experimental releases of *Chelonus elaeaphilus* Silv. and *Ageniaspis fuscicollis praysincola* Silv. demonstrated that limitation of the pest by their use was feasible.

The first establishment of *Prospaltella perniciosi* Tow. against the San Jose scale, *Quadraspidiotus perniciosus* Comst. was obtained in Germany, Switzerland and France after several years of releases from mass breeding on infested melons in specialised insectaries (Benassy *et al.* 1968). Similar programs are going on in Spain and Turkey.

As in many other parts of the world, different species of egg parasites of the genus *Trichogramma* have been tried as possible control agents in agriculture and forestry. Much of the work has been carried out in Germany using chiefly indigenous species as *T. evanescens* Westw., *T. embryphagum* Htg., *T. cacaeciae* March, but also the nearctic species *T. minutum* Riley. The most interesting trials were conducted against the orchard pest *Cydia pomonella* L. in apple orchards. Stein (1960) showed that it was possible to obtain a significant reduction of damage and that the best results were achieved when all the parasites were released at the beginning of the flight period of the moth. Interesting methods of distribution of the parasite were developed by Schutte and Franz (1961) using specially designed spray guns which allowed a more even distribution of the *Trichogramma* in tree crowns. In spite these progress, practical application did not follow, because the damage by codling moths remained in excess of the 5% injury level tolerated in commercial orchards.

Generally speaking, the failures in the use of *Trichogramma* were mainly due to a lack of knowledge of the specific identity of insects used and of the ecological potentialities of the different available strains, despite the early recognition by Marchal (1936) of the diversity of biotypes in this genus. More recently,

the use of a selected strain of *Trichogramma* against the European
corn borer, *Ostrinia nubilalis* Hubn. in North-eastern France gave
good economic control of the pest as compared with the use of chemical pesticides or *Bacillus thuringiensis* (Voegelé *et al.* 1975).
Encouraging results were also obtained against *Prays citri* Mill
in Sicily (Mineo *et al.* 1975).

Releases of field-collected parasites have been used in several
cases concerning mainly forest and orchards pests.  The most massive operation was probably the attempt made to control *Diprion
pini* L. in Eastern Spain in 1948-49 by Ceballos and Zarco (1952).
In this case 2 tons of cocoons were collected during October and
November 1948 and exposed in February 1949; an estimated 3 million
parasites emerged from these cocoons and, together with some
700,000 more *Dahlbominus fuliginosus* obtained by parasitisation
of field-collected healthy cocoons, realised a 64% parasitism in
the field in 1949, which decreased markedly the infestation.

Augmentation of populations of general predators has been
developed using ants of the *Formica rufa* group (Klimetzek 1970).
Basic studies with ants were undertaken in Germany by Gosswald
(see for example Gosswald 1951) and Wellenstein (1952, 1967) and
large scale experiments on the possibility of "transplantation"
were undertaken in Italy (Pavan and Ronchetti 1965).  It was shown
for example that ants collected in the Alpes area were able to
establish viable colonies in Sardinia and in the south of Italy.
The exact value of the control exerted by ants on forest pests
has been discussed by Adlung (1966).

Among non-arthropod natural enemies many species have been
an object of interest (Buctiner 1966), but the role of birds was
the most frequently studied, in England, Netherlands, Germany,
etc.  A very large scale experimentation on augmentation of populations of tits was initiated in Spain some years ago (Ceballos
1972) and is still going on.

Manipulation of pest number has been shown to provide means
to avoid a pronounced reduction of natural enemies.  In Yougoslavia,
Maksimovic *et al.* (1970) have shown that by distributing egg masses
of *Porthetria dispar* L. in a forest at the period of lowest density
of the pest, it was possible to maintain adequate populations of
parasites and predators avoiding a pest outbreak which was observed
later in the control area.  Further progress in the use of indigenous
natural enemies depends on a better knowledge of the ecological
requirements of species involved and a better understanding of
the factors governing host-parasite competition.

The bionomics of some widely distributed predators and parasites
are often poorly known.  Such factors as spatial distribution,
seasonal migration and hibernation of coccinellids, were clearly un-

derstood at least for some of the most important species only
a few years ago (Hodek 1973). This offers possibilities to their
rational use in agroecosystems (Iperti 1965). The techniques
to be used include protection of the species on the hibernation
sites (Iperti 1966), periodical colonation of fields and orchards,
distribution of supplementary food, etc. It has been recently
shown that it is possible to collect diapausing univoltine coccinellids in summer, reactivate them in the laboratory and release them
in the field to obtain a supplementary Fall generation (Iperti
1976, Laudeho and Katsoyannos 1976).

During the past twenty years, most of the activity in the
field of biological control in Western Europe has been coordinated
through working groups of the O.I.L.B. (Organisation Internationale
de Lutte Biologique), previously C.I.L.B. (Commission Internationale
de Lutte Biologique), now Western Palearctic Regional section
of the International Organization for Biological Control (WPRS-
IOBC). The potentialities of many control agents have been investigated (see C.I.L.B. 1957, 1958, 1962) and a certain number of
projects more or less connected with augmentation of natural enemies
is under way. The most advanced of these deal with biological
pest control in greenhouses.

Many European workers have taken part in the biological control
programs in Africa and other parts of the world, but few of them
were oriented towards augmentation of natural enemies, we shall
indicate only one example: the use of *Trichogramma* in Madagascar.

After 3 years of study on the possible use of the indigenous
*T. australicum* Gir. to control the sugarcane borer, *Proceras sacchariphagus* Bos., a release of over 2 million parasites on a field
of 10 ha was undertaken in 1970, but failed to give a satisfactory
control. Breniere (1965a, b) concludes that, under the condition
of sugarcane cultivation in Madagascar, there is little hope of
success by augmentation of local *Trichogramma* populations. On
the other hand, Bournier and Peyrelongue (1973) obtained good
results against lepidopterous pests (*Heliothis*) of cotton by periodical releases of mass bred *T. brasiliensis* Ashm. from South America
in the area of Tulear in the South of Madagascar. Interesting
and promissing trials were also undertaken by the same author
in Senegal during the past two years.


LITERATURE CITED

Adlung, K. G. 1966. A critical evaluation of the European Research
     on use of Red wood Ants (*Formica rufa* Group) for the protection
     of the forests against harmful insects. Z. ang. Ent. 57, 167-
     189.
Arambourg, Y. 1970. Techniques d'élevage et essais expérimentauxde

lâchers de *Chelonus eleaphilus* Silv. parasite de *Prays oleae*
Bern. (Teigne de l'olivier). *In*: Colloque franco-soviétique
sur l'utilisation des entomophages, Antibes, 13–18 Mai 1968.
Ann. Zool. Ecol. anim., n° hors ser.:57–6

Bénassy, C., G. Mathys, H. Neuffer, H. Bianchi and E. Guignard.
1968. L'utilisation pratique de *Prospaltella perniciosi* Tow.,
parasite du Pou de San José *Quadraspidiotus perniciosus* Comst.
Entomophaga, Mém. hors ser n° 4: 28 pp.

Biliotti, E. and P. Delanoue. 1959. Contribution à l'étude
biologique d'*Opius concolor* Szepl. (Hym. Braconidae) en
élevage de laboratoire. Entomophaga 4:7–14.

Bournier, J. P. and J. Y. Peyrelongue. 1973. Introduction, élevage
et lâchers de *Trichogramma brasiliensis* Ashm. (Hym. Chalcididae) en vue de lutter contre *Heliothis armigera* Hbn. (Lep.
Noctuidae) à Madagascar. Coton et Fibres trop. 28:231–237.

Breniére, J. 1965a. Les trichogrammes parasites de *Proceras sacchariphagus australicum* Gir., parasite autochtone. Effets du ren-
forcement de la population parasite. Entomophaga 10:83–96.

Breniére, J. 1965b. Les trichogrammes parasites de *Proceras sacchariphagus* Boj., borer de la canne à Madagascar. 2. Etude biologique de *Trichogramma australicum* Gir. 3. Réalisation de
l'élevage massal du parasite. Entomophaga 20:99–117; 119–131.

Buctiner, C. H. 1966. The role of vertebrate predators in the
biological control of the forest insects. Ann. Rev. Entomol.
II, 449–470.

Ceballos, P. 1972. Proteccion de las aves insectivoras alimenta-
cion natural de *Parus major* y *P. caeruleus*. Mem. Real Acad.
Ciencas Madrid XXV. I. 61 pp.

Ceballos, G. and E. Zarco. 1952. Ensayo de lucha contra una plaga
de *Diprion pini* L. en masas de *Pinus silvestris* de la Sierra
de Albarracin. Inst. esp. Entomol., Madrid:39 pp.

CILB. 1957. Vier Jahre Massenzucht von *Prospaltella perniciosi*.
CILB 3éme Réunion du groupe de travail: Lutte biologique
contre le Pou de San José, Stuttgart 1957:12 pp.

CILB. 1958. Pou de San José. CILB Assemblée générale, Paris,
Fev. 1958:29.

CILB. 1962. Compte rendu des séances de l'Assemblée Générale de la
Commission Internationale de Lutte biologique contre les
ennemis des cultures. Tunis, 26 mars–4 avril 1962.

Decaux, F. 1899. Destruction rationnelle des ins ectes qui atta-
quent les arbres fruitiers par l'emploi simultané des insecticides, des insectes auxiliaires, et par la protection et l'élevage de leurs ennemis naturel les parasites. J. Soc. hort. Fr.
22:158–184.

Delanoue, P. 1960. Lâchers expérimentaux d'*Opius concolor* Szepl.
en vue de la lutte contre *Dacus oleae* Gmel. dans les Alpes
Maritimes. C. R. Ac. Agric. Fr. 46:712–718.

Delanoue, P. 1970. Utilisation d'*Opius concolor* Szepl. en vue de
la lutte contre *Dacus oleae* Gmel. (Mouche de l'olive). Colo-
que franco-soviétique sur l'utilisation des entomophages.

Antibes 13-18 mai 1968. Ann. Zool. Ecol. anim., n° hors ser.
:6"-69.

Féron, M., P. Delanoue and F. Soria. 1958. L'élevage massif arti-
ficiel de *Ceratitis capitata* Wied. Entomophaga 3:45-53.

Franz, J. M. 1961. Biological control of pest insects in Europe.
Ann. Rev. Entomol. 6:183-200.

Gosswald, K. 1951. Die rote Waldameise im Dienst der Waldhygiene
Forstwirtschaflichte Bedeutung Nutzung Lebensweise Zucht
Wermehrung und Schutz. Kinau Verlag Lüneburg, 160 p.

Greathead, D. J. 1976. A review of biological control in western
and southern Europe. Techn. Commun. n°7, Commonw. Inst. Biol.
Control Farnham Royal, Slough, 182 pp.

Hodek, I. 1973. Biology of Coccinellidae. W. Junk, The Hague,
Academia Praha:260 pp.

Iperti, G. 1965. Perspective d'utilisation rationnelle des
Coccinelles aphidiphages dans la protection des cultures.
90éme Congr. Soc. Savantes, Nice 1965:543-555.

Iperti, G. 1966. La protection de *Semiadalia un decimnotata*
contre *Beauver bassiana In* Ecology of aphidiphagous insectes.
Proc. Symp. Liblice near Prague 1965., Academie, Prague,
360 pp.

Iperti, G. 1976. La diapause chez les Coccinelles. Colloque
francotchèque, Prague 1974. Ann. Zool. Ecol. anim. 8:381-8.

Klimetzek, D. 1970. Zur Bedeutung des Kleinstandorts für die
Verbreitung hügelbauender Waldameisen der *Formica rufa* group
(Hymenoptera Formicidae). Z. ang. Ent. 66, 84-95.

Laudeho, Y. and P. Katsoyannos. 1976. Lâcher d'*Exochomus quadri-
pustulatus* L. pou lutter en automne contre la Cochenille noire
de l'olivier (*S. olea*) en grèce. (sous presse).

Liotta, G. and G. Mineo. 1968. Lotta biologica artificiale contro
la mos delle olive a mezzo dell'*Opius concolor siculus* Mon.
in Sicilia nel 1968. Boll. Ist. Entomol. agric. Palermo
7:183-196.

Maksimovic, M., P. Bjegovic and L. Vasiljevic. 1970. Maintening
the density of the Gypsy moth enemies as a method of biological control. Zast. Bilja 21:1-15.

Marchal, P. 1907. Utilisation des insectes auxiliaires entomophages dans la lutte contre les insectes nuisibles à
l'agriculture. Ann. Inst. Nat. agron. (2éme ser.) 6:281-354.

Marchal, P. 1913. L'*Icerya purchasi* en France et l'acclimata-
tion de s  ennemi d'origine australienne, le *Novius cardinalis*. Ann. Epiphyti 1:13-26.

Marchal, P. 1936. Recherches sur les *Trichogrammes*. Ann.
Epiphyties et Cytogen 2:448-550.

Mineo, G., R. Pralavorio, G. Maniglia, J. Voegelé and Y. Arambourg.
1975. Prove di controlle biologico del *Prays citri* Mill.
(Lep. Hyponomeudae) con *Ageniaspis fuscicollis* Dalm. (var.
Praysincola) Silv. (Hym. Encyrtidae) e *Trichogramma evan-
escens* Westw. (Hym. Trichogrammatid) sul limone in Sicilia.
Boll. Ist. Entomol. agric. Palermo 9:143-160.

Monastero, S.  1968.  Nouvelle expérimentation à grande é chelle
    de la lutte biologique contre la Mouche de l'olive (*Dacus
    oleae* Gmal.) au moyen d'*Opius concolor* Szepl. *siculus* Mon.
    dans les Iles Eolien (Sicile) en 1965.  Entomophaga 11:
    411-432.
Monastero, S. and P. Delanoue.  1967.  Premiére expérimen tation à
    grande échelle de lutte biologique contre la mouche de
    l'olive (*Dacus oleae* Gmel.) au moyen d'*Opius concolor* Szepl.
    *siculus* Mon. en Sicile.  Entomophaga 12:381-398.
Monastero, S. and P. Genduso.  1962.  La lotta biologica contro la
    mosca delle olive.  Possibilita di allevamento e diffusione
    degli *Opius* trovati in Sicilia.  Boll. Ist. Entomol. agr.
    Palermo 4:31-51.
OILB.  1968.  Comptes-rendus de la 4éme Assemblée générale de
    l'OILB, Pa 26-29 mars 1968:37 pp.
OILB.  1970.  Groupe de travail: Cochenilles des agrumes.  Maroc.
    26-31 octobre 1970.  Al Awamia 37:109 pp.
OILB.  1975.  Progrès en lutte biologique et intégrée.  Bull.
    SROP/WPRS 1975, n°1:152 pp.
OILB.  1975.  Lutte biologique contre les cochenilles et aleu-
    rodes des agrumes.  Comptes-rendus de la 3ème réunion du
    groupe de travail tel à Palerme du 24 au 27 septembre 1974.
    Bull. SROP/WPRS 1975, n°5:165-283.
Onillon, J. C.  1973.  Possibilités de régulation des populations
    d'*Aleurothrixus floccosus* Mask; (Homopt. Aleurodidae) sur
    agrumes par *Cales noacki* How. (Hymenopt. Aphelinidae).
    OEPP.EPPO Bull. 3:17-26.
Pavan, M. and G. Ronchetti Eds.  1965.  Studi ed esperienze practice
    di protezione biologica delle foreste. Min. Agric. Foreste,
    Roma, Collana Verde 16.
Schutte, F. and J. M. Franz.  1961.  Untersuchungen zur apfelwicklers
    (*Carpocapsa pomonella* L.) mit Hilfe *Trichogramma embryophagum*
    Hartig. Entomophaga 4:237-247.
Stein, W.  1960.  Versuche zur biologischen Bekampfung des Apfel-
    wicklen (*Carpocapsa pomonella* L.) durch Eiparasiten der
    Gattung *Trichogramma*.  Entomophaga 5:237-247.
Voegelé, J., M. Stengel, G. Schubert, J. Daumal and J. Pizzol.  1975.
    Les *Trichogrammes*. V(a). Premiers résultats sur l'introduct-
    tion en Alsace sous forme de lâchers saisonniers de l'écotype
    moldave *Trichogramma evanescens* Westw. contre la pyrale du
    mais *Ostrinia nubilalis* Hbn.  Ann. Zool. Ecol. anim. :535-551.
Wellenstein, G.  1952.  Ergebnisse 25 jähriger Grundlagenforschung
    zur Bedeutung der Roten Waldameise (*Formica rufa* L.). Mitt.
    Biol. Zentralanst. Berlin-Dahlem, 75:125-133.
Wellenstein, G.  1967.  Zur Frage der Standortansprüche hügelbau-
    ender Waldameise (*F. rufa* Group).  Z. ang. Zool., 54:139-166.

# CHAPTER 12

# THE INTRODUCTION OF NATURAL ENEMIES FOR PEST CONTROL

# IN GLASSHOUSES: ECOLOGICAL CONSIDERATIONS

N. W. Hussey and N. E. A. Scopes

Entomology Department
Glasshouse Crops Research Institute
Sussex, England

## THE PEST PROBLEM

Despite many widely held opinions to the contrary most pest problems are peculiar to individual glasshouses because infestations stem from contamination by survivors from earlier crops. This carryover occurs either because weeds remain as sources of infestation or because the pests survive as pupae in the soil or overwinter in cracks or crevices of the structure.

Such contamination is, of course, unnecessary and was less likely to occur in former years when growers paid more attention to effective cleansing of glasshouse structures between crops. It is accepted, nowadays, that labour input must be minimized to reduce costs and growers must recognise that the pest problem in one year is often governed by the control achieved in the previous year.

The fact that glasshouse pest populations are ecologically isolated has important implications for chemical control. The environment provided to grow the crop affords ideal conditions for pest reproduction. Temperature is generally high, varying between only narrow limits, and there is no wind, which outdoors may interfere with complex mating flights. Further, there are no rainstorms which drastically reduce some pest populations in sudden downpours. These equable conditions encourage rapid breeding so that many species complete several generations in the life of a single crop.

The grower normally counters these rapidly increasing populations by applying pesticides. Most of the important glasshouse pests -

mites, aphids or whiteflies are largely protected from pesticides
in their habitat; on the undersides or folds of leaves. It is,
therefore, technically impossible to kill all pests with a single
pesticide application. It follows that commercial control is
possible only when a series of applications are made. Growers
tend to continue using the most effective materials and so unwit-
tingly apply a severe and continuous selection pressure on the
pest population. This situation leads to a unique demonstration
of Darwin's principle of natural selection. The few individuals
in the pest population which have the ability to detoxify or
avoid the pesticide, survive. As successive applications are
made the proportion of these 'resistant' individuals increases —
very slowly at first and then more rapidly. Control failure
occurs quite suddenly in that the period between the first signs
of a reduction in the efficiency of a particular pesticide and
complete failure may occur within a few months. For instance,
resistance to organophosphate aphicides in *Myzus persicae* Sulzer
appeared to 'spread' to most nurseries growing year-round chrysanthemums within six months.

As glasshouse pests must be repeatedly controlled it is
difficult to avoid selecting resistance to different pesticides
in rapid succession especially if the alternative materials are
related. Most pests tend, initially, to be susceptible to two
or three 'families' of pesticides — e.g. aphids to organophosphates,
carbamates and pyrethrins. Tolerance to one material may confer
cross-resistance to all other members of the family so reducing
the potential aphicidal range by one third. A very serious,
if not impossible, situation may, therefore, occur. The whitefly,
*Trialeurodes vaporariorum* (Westwood), for instance, is now immune
to most commercially available pesticides on many nurseries in
the United Kingdom.

Another aspect of chemical usage restricting choice is the
problem of residual deposits. Lack of rain encourages the persistence of residues on the harvested products which, in the case
of food crops, have to be picked every few days over several
months. Many candidate pesticides are therefore ruled out.
Chemicals of short persistence are therefore obligatory and this
puts a severe limitation on control efficiency.

References has already been made to the management requirement
of minimal labour input in glasshouse culture. This restriction
has encouraged the rapid exploitation of ultra-low volume spraying
techniques. Unfortunately, although it is now possible to apply
pesticides very economically as smokes, aerosols and fogs, their
biological efficiency is poor. Although the droplet spectrum
achieved by a variety of machines, permits the pesticide to penetrate
dense leaf canopies, lack of aerial turbulance prevents their
impaction on the lower leaf-surfaces. Many pests are therefore

virtually unaffected.  This unsatisfactory situation is exacerbated
by the protective niches which some pests occupy.  Spider mites,
for instance, spin a fine silken web between the leaf-veins below
which they move freely.  All too often fine pesticidal droplets
are trapped on these threads and fail to reach the target.

Further, many fogging machines are capable of treating large
glasshouses from a single application point.  The temptation
to complete treatments as rapidly as possible overlooks the limitations on leaf-coverage referred to above.  Unless the machines
are moved through the crop, coverage is even less because efficiency
drops markedly a few feet from the machine.

The limitations of low-volume applications encourage successive
treatments so that food-crops such as tomatoes or cucumbers may
be treated more than 25 times in a season.  As well as encouraging
resistance this excessive pesticide programme has severe effects
on the plant itself.  For too long phytotoxicity has been recognised
only as visible damage to the foliage - largely in the form of
spots or marginal scorch.  However, when biological control permitted
direct comparisons between chemical and non-chemical treatments,
yield reductions of 10-20% were demonstrated (Addington 1966).
These losses apparently occur through interference with fertilisation
so that a proportion of the fruits fail to set.  Losses of this
order, sometimes amounting to more than 4,000/acre, put the apparent
labour-saving advantages of low-volume techniques in a far less
favourable light.

Traditional chemical control therefore has severe limitations
in the protected environment.  Some of the problems could be
overcome with adequate choice of alternatives, but this choice
becomes yearly more limited.  The development costs of new pesticides
have increased alarmingly and can only be recouped by large markets.
The glasshouse industry which uses only minute quantities of
pesticides can only adopt materials which have been developed
for quite different purposes on major outdoor crops.

This restriction inevitably calls for a new approach to
pest control under glass and the commercial implementation of
biological procedures is an important step in this direction.


BIOLOGY OF THE MAJOR GLASSHOUSE PESTS AND THEIR NATURAL ENEMIES

Perhaps the most serious pest problem is posed by the glasshouse
whitefly, *T. vaporariorum*.  The adults which are 1 mm long and
covered with waxy scales, feed almost entirely on the undersides
of the apical leaves.  Females lay yellowish-green eggs (0.1
mm long) erectly on the abaxial surface which after 2-4 days
turn grey and then black.  There are three nymphal stages, the

first being motile, though individuals seldom move more than
1-2 cm from their egg shells. After the first moult the crawler
flattens to the leaf becoming almost transparent. In the second
and third instars the legs and antennae become vestigial. As
the pupa develops the scale thickens-its perpendicular sides
formed of waxen rods. The adult subsequently emerges from a
T-shaped opening on the upper surface. There are normally twice
as many females as males.

The life-cycle varies with temperature, being completed
in 63 days at 15.6°C, but in only 23 days at 24°C. Between 4-
7 eggs are laid daily during the summer but during the winter
fecundity falls to only 2 per day. The adults are strongly attracted
to yellow-green light (540-550 mµ) so ensuring their establishment
on the young leaves most suitable for oviposition.

Outdoors, adults and eggs survive long periods of cold weather
and have been found alive after more than 30 days below freezing.
This ability to survive outside glasshouses in average winters
is thought to encourage infestations on early planted crops.

*T. vaporariorum* affects yield in two ways, first by reducing
crop production and second by contaminating the fruits with sooty
moulds which must be wiped off before marketing. On tomatoes,
crop loss occurs when the scale density exceeds 2,500 per leaf.
Both larvae and adult produce copious quantities of honeydew,
on which the sooty moulds develop. These moulds reduce leaf
transpiration and so, indirectly, the yield. As few as 18 adults
and 60 larvae per leaf produce sufficient honeydew to induce
sooty mould growth if the RH reaches at least 90% for eight hours
daily over two weeks. Under more humid conditions moulds may
develop within 3 days. On cucumbers, populations of more than
40 adults/leaf cause sooty moulds.

Biological control is achieved by the parasite, *Encarsia
formosa* Gahan, a chalcid wasp first discovered in Britain in
1926. The adult is 0.6 mm long with a dark-greenish head and
thorax and yellow abdomen. Although parthenogentic, males are
occasionally found, being easily recognised by a black abdomen.

The adult is capable of flying hundreds of feet apparently
attracted by volatiles produced by the developmental stages of
the host and the honeydew which they produce. The complexities
of searching are not fully understood but it appears that two
distinct processes are involved. Directional flight to whitefly
(Ledieu 1977a) infestations and random searching on the leaves
of the infested plant (van Lenteren *et al*. 1976). The former
process, which does not readily occur on dull winter days unless
the temperature is high, appears to be stimulated by sunshine
(Ledieu 1977a).

Glasshouse experience suggests that the wasp lives for 10–
14 days laying about 50 eggs though, in the laboratory, individuals
have lived for more than 5 weeks laying up to 350 eggs.

Unlike its host the parasite is killed by temperatures below
13°C (Scopes *et al.* 1973), at which temperature the life-cycle
is completed in 90 days compared with 30 days at 18.3°C and only
17 days at 24°C.  At 18.3°C, whitefly fecundity is several times
greater than that of the parasite though their rates of development
are equal but at 26°C while their fecundity is equal and parasite
develops in half the time required by its host.

The parasite is most efficient at low host densities, for
on excessive honeydew adults spend much time cleaning their feet
so reducing their parasitic efficiency (Table 1); dense populations
of whiteflies are therefore rarely controlled by parasites.

Plant morphology also influences parasite efficiency – *E.
formosa* is unable to establish satisfactorily on *Nicotiana glutinosa*
L. – though if the leaf-hairs are removed parasitism is improved.
It might be assumed that, on account of the high growing temperatures,
parasite control of whitefly on cucumbers would be simple, but
the converse is true.  In Holland, for instance, *Encarsia* is
not used commercially on this crop.  In fact, whiteflies are
more fecund on this host (O'Reilly 1975), while the general hairiness
of the leaves hinders movement of the parasite and hence its
parasitic efficiency.

*Encarsia* must, therefore, be established early in the growth
of the crop when numbers of *T. vaporariorum* are low so that control
is obtained after two, or at most three, generations before whitefly
populations have reached critical levels.

The other major glasshouse pest is *Tetranychus urticae* Koch,
commonly known in Britain as the red spider mite on account of
the bright orange colour of the overwintering form.  In America,
it is perhaps more usually called the two-spotted mite, a name

Table 1.   Effect of different scale populations on parasitism.

| Mean scale population | % parasitism |
|-----------------------|--------------|
| 25                    | 70.6         |
| 160                   | 60.9         |
| 360                   | 52.1         |

coined from the appearance of the summer feeding form in which
dark green food is visible through the body wall as two lateral
spots.  Its facility to diapause is ideally suited to the glasshouse
environment for, under monoculture, there is often an interval
during the winter in which no crops are grown.

Winter diapause in *T. urticae* is induced by daylength, temperature and nutrition which act on the developmental stages in
complex inter-relationship.  The critical daylength, below which
diapause is induced, depends on latitude and in Britain is approximately 12.5 hours.  In practice, if any two of the factors are
tending to induce diapause, most mites will enter hibernation
as they mature.  The role of nutrition is difficult to define
for, even in mid-winter, some mites may continue to develop so
long as the host plants are vigorous and healthy (Parr & Hussey
1966).  Emergence from diapause begins soon after the glasshouses
are re-heated in late winter - so long as the mites have experienced
a certain degree of chilling below their developmental threshold
temperature.  The pattern of emergence depends on the manner
in which the mites were 'triggered' into diapause the previous
year (Geyspits 1960).

Female mites lay between 6 and 8 eggs daily so producing
about 100 within their normal life-span of two weeks.  At 20°C,
the mean generation time is 24 days with a multiplication rate
of 31 times per generation.  Egg laying is depressed by high
humidities, 4.9 per female per day, compared with 7.3 eggs in
a dry atmosphere (35% RH).  This fact is the basis of using water
mists to control spider mites (Tulisalo 1975).

Mites feed on the cell chloroplasts so producing minute,
yellowish feeding marks on the upper surface of infested leaves.
This fine speckling coalesces to form bronzed areas as the population
increases.  Finally, leaves shrivel and die.  As mite numbers
become excessive individuals become negatively geotropic and
migrate to the apical leaves.  Here they congregate on the highest
leaf-tip which bends under the weight and from which the mites
descend on silken 'ropes'.

On account of their small size and rapid rate of multiplication,
mite numbers are best estimated by relating them to damage symptoms.

The (LDI) leaf damage index for an individual plant is the
mean of the values for individual leaves (Table 2).

The economic threshold of damage on established cucumber
plants, above which crop loss occurs, is 1.9.  An index of 2.5
will cause a 40% crop loss after five weeks (the period of development
for cucumber fruits) while 3.0 will lead to an 80% crop loss.
On cucumbers, the maximum increase in leaf damage is 1.0 in 12

Table 2.  Definition of intensities of mite damage to cucumber leaves.

| Damage index | Definition | Mite population (adults plus nymphs) per sq. in. |
|---|---|---|
| 0 | No damage | 0 |
| 1 | Incipient damage, one or two 1/2" feeding patches | 3 |
| 2 | Feeding patches tending to coalesce, only 2/5 of leaf affected | 12 |
| 3 | 2/3 of leaf with feeding marks as chlorotic patches | 107 |
| 4 | Dense feeding marks over entire leaf, but appearance still green | 228 |
| 5 | As 4 but leaf blanched and starting to shrivel | 592 |

days ($\equiv$ 0.09/day) so that the timing of pesticide treatments or manipulations of spider mite populations by predators can be predicted.

On tomatoes a similar threshold of 2.0 (equivalent to approximately 30% of the phytosynthetic area damaged) causes crop loss, while the maximum rate of increase recorded is 2.7 in 16 days ($\equiv$ 0.1/day). Though tomatoes are grown at 16°C rather than 21°C for cucumbers the mites have a similar rate of damage increase, suggesting that tomatoes are more susceptible to damage.

The implementation of biological control on these crops depends on the correct interpretation of leaf damage.

*Phytoseiulus persimilis* Athias-Henriot was first recognised in Europe in 1959 on plant material imported into Germany from Chile. Its potential was quickly appreciated and by 1967 it was being commercially exploited on a limited scale. It is a shiny, orange/red, pear-shaped mite which, unlike its prey, does not hibernate. Feeding only on spider mites the predator must be reintroduced into glasshouses every season.

At 21°C, each female lays 50-60 eggs at a rate of 3-4 per day: they hatch in 2-3 days. The complete life-cycle takes about 7 days – twice as fast as its prey. At 18°C, eggs are laid at a rate of 2-3 per day and the life-cycle is completed in about 10 days. *P. persimilis* has a high searching capacity, often rapidly transversing many uninfested plants – ten tomato plants in ten days. Each female consumes as many as 5 adults or 20 young *T. urticae* daily. However many spider mites are present on a plant, the predator will eliminate them in due time. Commercial control takes advantage of this certainty by introducing predators on low spider mite populations whose damage level is then maintained below the economic threshold.

Temperature governs the speed of control (Fig. 1) as it has a marked effect on population growth of predators, for example, at 20°C the rate of increase is 300 times every 30 days, while at 25°C the increase rises to 200,000 times. However, *P. persimilis* cannot survive above 30°C at which temperature *T. urticae* will continue to develop unchecked.

*P. persimilis* can tolerate low temperatures surviving for 2-3 months at 7°C so long as abundant food was provided. When

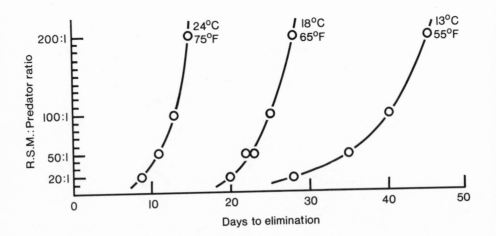

Figure 1. The effect of temperature on the elimination of red spider mites by *Phytoseiulus*.

placed on a mite infested leaf the predator will not, normally, migrate until all the prey have been consumed; only when ill-nourished do they disperse in search of other pest infestations. Eggs left at the original site hatch and the nymphs survive on the scattered eggs of the surviving *T. urticae*.

Once *P. persimilis* has eliminated its prey it normally survives for about 3 weeks to re-colonise fresh mite infestations. Spider mites establishing after this period must be controlled by further releases of predators.

## PRODUCTION AND USE OF NATURAL ENEMIES

The commercial use of natural enemies relies on suitable methods of mass production, and two species are now widely used (Table 3), while a third, *Aphidius matricariae* Haliday, is currently used on a limited scale. The primary centres of production are in Holland and Britain.

## PRODUCTION OF *E. FORMOSA*

The rearing methods in current use in Britain (Anon. 1975), were originally developed to cater for the "pest in first" method, where the crop is evenly infested with small numbers of the pest before, after a suitable interval, fixed numbers of the natural enemy are released. This leads to a precise and predictable interaction. To achieve these interactions, parasite rearing

Table 3.  Glasshouse pests and their natural enemies.

| Natural Enemy | Host Insect | Crops |
| --- | --- | --- |
| *Encarsia formosa* | *Trialeurodes vaporariorum* | Tomatoes<br>Cucumbers<br>Sweet Peppers |
| *Phytoseiulus persimilis* | *Tetranychus urticae* | Tomatoes<br>Cucumbers<br>Sweet Peppers<br>Chrysanthemums |
| *Aphidius matricariae* | *Myzus persicae* | Sweet Peppers<br>Chrysanthemums |

systems were designed to provide large numbers of pupae of similar age, to emerge as adults shortly after introduction (Scopes and Biggerstaff 1971). In practice, "pest in first" is unacceptable to many growers especially for controlling *T. vaporariorum*, so that parasites are normally sold as pupae on detached tobacco leaves and evenly distributed within the glasshouses soon after planting the crop, irrespective of the presence of pests. These releases continue regularly until the parasite establishes on the crop.

*E. formosa* is produced intensively on *Nicotiana tabacum* L. (cv. White Burley) and a relatively small unit (about 300m$^2$) can produce 1/2-1 million parasites, enough to make regular releases on 50-100 acres each week.

The production system (Figure 2) takes 34 days to produce adult parasites. In addition, 8-11 weeks are needed to produce clean tobacco plants. Thus, from sowing to parasite harvest takes 13-16 weeks. Half the plants are used to produce parasites (●) while half the remainder provide food for adult whiteflies (■) and half maintain whitefly stocks (▲). Precise temperature control is essential as parasite emergence must be timed to coincide with the development of susceptible 3rd instar whitefly larvae.

In contrast, *E. formosa* is produced by an extensive "ranching" system in Holland. Cucumber plants are infested with both *T. vaporariorum* and parasites, the latter being harvested progressively up the plant as they pupate within "black scales". This system requires considerably more space – 0.4 ha producing about 2.5 million parasites a week over 2 1/2 – 3 months. The harvested leaves carry parasites at all stages of development, a definite advantage when they are released onto commercial crops where hosts may be few and conditions for dispersal unsuitable. However, the production of parasites on cucumber plants precludes their use on this crop because of the dangers of transmitting other phytopathogenic diseases.

## PRODUCTION OF *P. PERSIMILIS*

The production of the red spider mite predator, *P. persimilis*, involves three distinct processes (Figure 3) which must be physically separated.

Though each process must be rigidly isolated, each is dependent on the previous stage. One week after sowing, seedling Broad beans, *Vicia faba* L., are infested with a few hundred *T. urticae* directly they appear above the soil. Three weeks later, at 23°C, massive populations will have developed and the plants begun to shrivel. These infested plants are laid on pots of mature

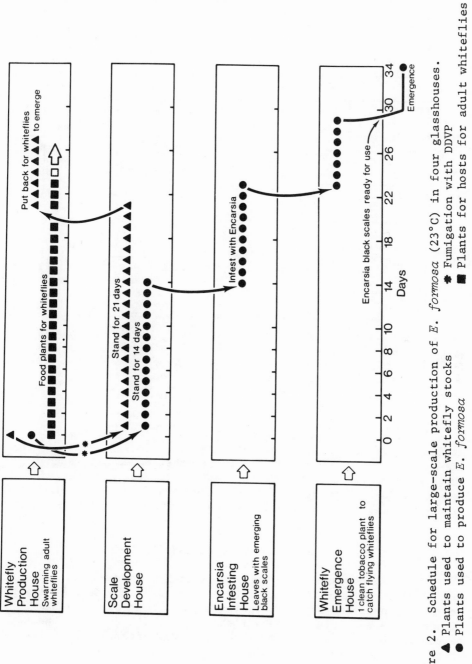

Figure 2.   Schedule for large-scale production of *E. formosa* (23°C) in four glasshouses.
▲ Plants used to maintain whitefly stocks   ✱ Fumigation with DDVP
● Plants used to produce *E. formosa*        ■ Plants used for hosts for adult whiteflies

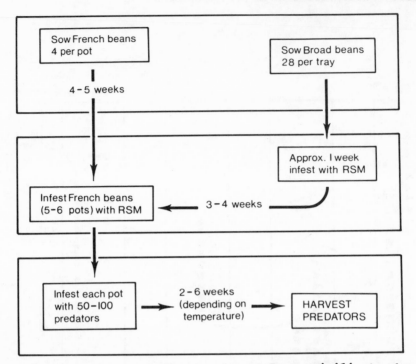

Figure 3. Schedule for the production of *P. persimilis* in three glasshouses. 1) Production of clean (unifested) plant material. 2) Production of the host (pest) *T. urticae* on these clean plants. 3) Production of *P. persimilis* and *T. urticae*.

French beans, *Phaesolus vulgaris* L., which the *T. urticae* will attack within 36 hours. These pots are transferred to the third glasshouse and 50-100 *P. persimilis* are introduced onto each. Three weeks later the spider mites will have been consumed and between 2500-5000 predators may be harvested. This rate of predator production may easily be changed by modifying temperature.

## RELEASE OF NATURAL ENEMIES

*Encarsia formosa*. The life-cycle of *T. vaporariorum* ranges from 3 wk at 24°C to 10 wk at 15°C and so the process of biological control is necessarily slow, taking up to 3 months. For this reason it is essential to establish the parasite early in the season while *T. vaporariorum* is sparsely distributed. Parasites established at this time readily search for their prey throughout the glasshouse. Further, an even (and sparse) distribution of both pest and natural enemy is essential if local patches of severe whitefly infestation are to be avoided. As mentioned

earlier these are not amenable to parasitic control by virtue
of the dense honeydew deposits associated with them.

Uniform populations of *T. vaporariorum* and *E. formosa* may
be obtained by the "pest in first" technique on both cucumbers
(Anon. 1976a) and tomatoes (Anon. 1976b). On cucumbers, 10
whitefly pupae are introduced onto every fifth plant followed,
14 days later, by 100 black parasitized scales on the same plants.
On tomatoes, 10 whitefly pupae are introduced onto one plant
in every 100, followed by three introductions of 150, 150 and
75 three, five and nine weeks after introducing the pest (Gould
*et al.* 1975). These parasite introductions are timed to coincide
with the presence of 3rd instar whitefly scales, the occurrence
of which at normal growing temperatures can be predicted.

To overcome grower resistance to the "pest in first" method
of controlling *T. vaporariorum* on tomatoes an alternative method
of releasing parasites has been developed (Stacey 1977) similar
to that used by Stary (1976) for aphid parasites. Tomato plants
(75 per hectare) bearing a thriving population of *E. formosa*
parasitizing about 90% of the whiteflies are planted among the
crop. These each produce approximately 10,000 *E. formosa* over
8-10 weeks. The parasites distribute themselves effectively
because the adults emerge into a 'natural' situation with food
in the shape of honeydew (carbohydrates) and young whitefly
scales (protein) readily available. By contrast, parasites
emerging on dried tobacco leaves (as received from commercial
producers) without scales and honeydew close at hand may be
unable to take flight and search so readily.

*Phytoseiulus persimilis*. The success of this predator
over the last ten years has given growers confidence in its
ability to control *T. urticae*, and so the "pest in first" method
is widely used in Britain especially on cucumbers. Alternated
plants are infested with approximately 10 *T. urticae* followed
by 2 *P. persimilis* on the same plants about 10 days later when
the mean leaf damage index has reached 0.4 (Anon 1976b). Large
numbers of predators will have developed and distributed themselves
by the time most spider mites emerge from diapause in the glasshouse
structure. As well as controlling *T. urticae* early in the season,
*P. persimilis* must eliminate the pest by late summer (August/September in Britain) to prevent red spider mites re-entering diapause
as the daylength shortens.

Success with this predator, as with parasites, depends
on obtaining a sparse and even distribution of both pest and
predator for only then will efficient searching occur. Predators
introduced to heavily infested plants will not disperse until
they have eliminated their prey several weeks later.

On tomatoes, the predator is usually established by infesting one-fifth of the plants in the propagating house, when the plants are close together. Thirty spider mites are placed on each and about 10 days later 4 predators are added to each and the plants subsequently distributed evenly through the crop (1 in 5). This 'inoculation' controls invasions of ex-diapausing mites over the ensuing month.

Where this practice is not followed mites will establish erratically leading to 'patches' of damage on groups of plants in particularly warm sites or below favourable hibernating niches in the structure. Infested leaves from these early infested plants are distributed through the rest of the crop (1 in every 10 plants) and when the damage index on these has reached 0.4, five predators are released on each.

It is important to recognise that the tactics of predator usage should be governed by an estimate of the size of the overwintering population. The "pest in first" approach is rarely necessary in the year following a season of effective biological control.

On chrysanthemums uniform distribution of *P. persimilis* is achieved by infesting boxes of rooted cuttings with both *T. urticae* and *P. persimilis*, prior to planting (Scopes and Biggerstaff 1973). This ensures control until the flowers are harvested.

*Aphidius matricariae*. This parasite may also be effectively distributed throughout a crop of chrysanthemums by 'box-seeding'. In this case, however, aphids that have been previously exposed to parasitism by *A. matricariae* are introduced into the boxes of cuttings at a rate of 1 per 10 plants. The aphids disperse and are eliminated by the parasite within 2-3 weeks - though, more usually, a sparse interaction is maintained by emigrant aphids for at least a month or two.

In Holland, natural enemies are used on about 1000 ha of glasshouses, while, in Britain, the treated area is currently about 300 ha. Other rearing units supplying commercial growers are operating in Denmark, Sweden, Finland and Ireland.


CHEMICAL USAGE WITHIN AN INTEGRATED CONTROL PROGRAMME

If biological agents are present they must survive despite applications of pesticides for the control of other pests and phytophatogenic fungi. It is widely known that many hymenopterous parasites are susceptible to insecticidal residues and so the successful development of integrated control programmes demands that all pesticides intended for use therein are carefully screened.

Such evaluation is, in the first instance, conducted in the laboratory.

Pesticides may have a direct contact effect which is tested by spraying natural enemies by means of a Potter Tower.  Except in the case of completely selective materials, such as dimenthirimol (Plant Protection 675) and diflubenzuron, a substantial proportion of adult parasites and predators are killed by most pesticides. Predator eggs and pupal parasites are often less susceptible though  newly emerged nymphs and adults may be affected by residual deposits on the leaf surface.

It is equally simple to assay for these residual effects by exposing adult parasites and the developmental stages of predators to deposits from applications at commercial rates. These are allowed to "decay" for different periods under the same environmental conditions as are required for the culture of the crop for which the programme is being prepared (Ledieu 1977b).  Deposits may also affect the behaviour of searching predators and parasites.  Some natural enemies are capable of avoiding discontinuous deposits of pesticides.  While such avoidance may permit their survival, despite the presence of lethal deposits, their efficiency may be sharply reduced through the extra time taken for 'avoidance' reactions.  Such effects may not always be deleterious.  Irving & Wyatt (1973) used an interesting technique to study the reaction of *E. formosa* to such behavioural effects. Whitefly scales (Instar III), the stage most susceptible to parasitism, were removed from their host plants and placed on a cellophane film lying on dampened green blotting-paper within a plastic box.  The box was upturned so that the host-scales were in their natural 'abaxial' orientation.  This technique allowed successive experiments to be conducted with the scales arranged to a consistent design.  A variety of tests can evaluate the effects of different patterns of pesticide deposits and even investigate whether the natural enemy is affected by pesticidal contamination of the host itself.

With this technique Jackson (1973) discovered that contamination of red spider mite eggs with the relatively innocuous arcaricide, malathion protected them from attack because the predator was repelled once its palpi had detected this irritant pesticide.

Some pesticides are sufficiently volatile to kill natural enemies by a fumigant action without coming into direct contact with the toxicant.  Babikir (1977) has devised a neat technique for assessing such effects which permits the simultaneous comparison of exposure to clean and contaminated air.  The deleterious effect of dichlorvos on the active stages of the predator is a significant example.

Undoubtedly the most intriguing side effect yet to be established is known as 'food-chain' toxicity. This phenomenon was first brought to light by high-volume sprays of the systemic fungicide benomyl which caused a rapid dislocation in the pattern of predatory control of red spider mites on cucumbers. When benomyl superficially contaminates the prey the predator imbibes the pesticide as it sucks 5-6 adult mites daily. Within a few days the egg production of the female predators drops sharply. When the fungicide is applied to the roots of the host plant, however, insufficient benomyl is absorbed in the mite tissues to affect the predator. This method of fungicide usage can, therefore, be harmonized with biological control.

Other examples of this 'food-chain' mechanism occur when systemic pesticides are applied to plant roots.

By supporting shoots of broad bean in small conical flasks with a collar of foam plastic Binns (1971) was able to isolate the test predators from fumes associated with volatile pesticides. When these shoots supported dense colonies of spider mites five predators were placed on each. More than 90% of the predators were killed within 48 h by dementon-S-methyl, methomyl and thionazin, whereas Menazon, Isolan and pirimicarb were harmless. In a further test, where only spider mite eggs were available on test leaves treated with thionazin and demeton-S-methyl, predators died in 24 h. The predator was, in these cases, killed by the pesticide diffusing or vaporizing from the internal tissues to the leaf-surface.

Spider mites collected after feeding on similarly treated leaflets for 24 h were offered to predators in clean glass vials closed with a cap of nylon gauze. After 48 h all the predators feeding on prey reared on thionazin treated shoots died but only 54% of those exposed in this way to demeton-S-methyl died.

These demonstrations of food-chain effects are especially relevant to the current problems posed by aldicarb resistance on chrysanthemums. Susceptible spider mites are rapidly killed on treated test plants but if predators are fed dying mites they are unaffected. This suggests that the predator might survive when feeding on aldicarb resistant mites. During 1976, resistance to this compound occurred on year-round chrysanthemums in the U.K. Slide-dip tests by Stone *et al.* (1977) suggested that the spider mites could tolerate 10 times the normal commercial dosage.

If predators are to be used on surviving resistant mites it is necessary to establish whether the aldicarb is completely detoxified or whether the tolerant individuals contain a non-lethal but perceptible concentration of the toxicant. When

test plants were placed in solutions containing 100 ppm aldicarb,
there was an obvious reduction in the numbers of eggs laid by
10 predators over 24 h, ranging from 56 in the controls to only
4 at 50 ppm.  Despite this clear effect on predator fecundity
all the eggs and larvae survived.  It is, therefore, believed
that the predator could be used when spider mites appear on
chrysanthemum beds treated with aldicarb because the concentration
of toxicant in the leaf tissues continuously declines after
the single post-planting treatment.

From laboratory assays of the types described the investigator
is faced with data which he must evaluate collectively to classify
the material as safe, or harmful, for use in integrated programmes.
At the Glasshouse Crops Research Institute it is felt that judge-
ments based solely on laboratory assays, however, carefully
conducted, tend to be unduly severe.  As pesticide coverage
in the glasshouse is inevitably poor many doubtfully lethal
compounds can be harmonized with natural enemies.  Each compound
is therefore tested by comparing the pattern of host/natural
enemy interactions on plants receiving successive sprays of
the candidate pesticide with those receiving only water.

In the case of *Phytoseiulus* its rate of population increase
over the 2 wk following pesticide treatments is compared with
untreated controls (Table 4), whereas control of whitefly by
*Encarsia* is similarly measured after seven weekly pesticide
treatments (Table 5).

Tests of this kind provide essential assessments of the
hazards associated with the use of those pesticides which must
be used for the control of fungi or minor pests, such as leaf-
hoppers or thrips.  Where such pesticides are harmless, as in

Table 4.  Effects of pesticides on glasshouse populations of *Phytoseiulus*.

| Pesticide | Reduction in numbers 2 wk after treatment controls |
|---|---|
| Water | Nil |
| Drazoxolon (Mil-Col, Plant Protection 781) | Nil |
| Chlorothalonil (Daconil) | -5% |
| Dodemorph (BAS 238 F) | -99% |
| Benomyl | -99% |
| Carbendazim (Bavistan, BAS 3460 F) | -93% |
| Triforine (Saprol) | -5% |

Table 5.  Glasshouse tests on effect of pesticides on *Encarsia*.
Two whiteflies and five parasites/plant.  Spraying began 3 wk later
and continued weekly for 7 wk.

Comparison with control on plants treated with
water sprays 16 wk after infestation

| No effect | Marked reduction in parasitism |
|---|---|
| Dipel (*B. thuringiensis*) | Delnav (dioxathion) |
| Dimilin (diflubenzuron) | Morestan (quinomethionate) |
| Plictran (cyhexatin) | Maneb |
| Childion (tetradifron) | |
| Rovral (Rome Poulenc RP 26019) | |
| Daconil (chlorothalonil) | |
| Ronilan (vinclozalin) | |

the case of pirimicarb, their place in a fully integrated programme
is simplified.  However, even if the chemical has undesirable
effects it may be used 'ecologically' by separating its site
of action from the natural enemy.  *Thrips tabaci* Lindeman is
an apt example for this species descends from the plant to the
soil to pupate.  Treatment of the soil with drenches of HCH
(lindane) or diazinon, which are lethal to both predators and
parasites, can be safety applied since these pesticides are
not systemic.  In large commercial glasshouses these materials,
though volatile, have no effect on natural enemies on the foliage
above because the degree of air movement prevents the build
up of a toxic aerial concentration.

    Agromyzid leaf-miners present particularly difficult problems
in integrated control, but careful consideration of their life-
habits indicates techniques which may permit successful harmonization
of lethal chemicals.  Conventionally, leaf-miners are controlled
by pesticides which kill young larvae within the leaf mines.
The most efficient material is dimenthoate but this remains on
the leaf-surface for many weeks and interferes with both parasites
and predators.  A less effective, but otherwise selective, alternative would be pirimicarb.  In the case of *Phytomyza syngenesiae*
(Hardy) on chrysanthemums a preferable technique takes advantage
of the fact that all the eggs laid on the upper surfaces of
the apical leaves.  Newly hatched larvae are vulnerable to many
pesticides but the most potent is dioxathion which, while killing
a high proportion of the larvae in the mines, also deters further

oviposition by surviving adults. Dioxathion is lethal to *Phytoseiulus* and aphid parasites but, if applied by misting above dense chrysanthemum beds, it is predominately settled on the upper surface of the terminal growth. Meanwhile the natural enemies, predators and aphid parasites, are safely sheltered on the abaxial surfaces of the lower growth.

A similar species, *Liriomyza solani* (Her.) which attacks tomatoes poses a more difficult problem. In the Channel Islands, where this species is endemic, control depends on preventing oviposition by the first generation emerging from pupae which have survived in the soil from the previous season. This is achieved by dusting HCH below the plastic sheeting on which the polythene growing bags are placed. As the glasshouse temperatures rise the HCH vapourises and a lethal concentration builds up below the sheet. This control is supplemented by low volume applications of dimethoate which, as mentioned earlier, are undesirably persistent.

An alternative method of preventing early establishment of this pest, which lays eggs even in the cotyledons, is to use dimethoate as a systemic following incorporation in the seedling compost at only 2ppm. Predators can survive 5 weeks after this treatment (Hussey, *et al*. 1975). Even within a few days of treatment whitefly oviposition is unaffected though only 69% of the eggs matured compared with 96% on untreated plants. However, *Encarsia* is unaffected by the time the third-instar scales are available. These relatively minor delays enable the "pest in first" technique to be safely exploited despite this leaf-miner treatment.

Caterpillars constitute another difficult problem for they are notoriously difficult to kill except when very young. DDT, carbaryl and methomyl are most widely used for control but these broad-spectrum compounds are highly toxic to natural enemies.

A selective alternative is urgently required. No doubt the insect growth regulator diflubenzuron will prove invaluable. This chemical has to be ingested by the insect and, once in the alimentary canal, it interferes with the moulting process of all larval instars. Most affected individuals are unable to moult and so die prematurely. As the material must be eaten it affects only insects with biting mouthparts so that parasites and predatory mites are not threatened by its application.

An interesting alternative is provided by *Bacillus thuringiensis* Berliner. Commercial preparations of this pathogen are normally more expensive than chemical pesticides and so are rarely regarded by growers as economic alternatives. This comparison is irrelevant in the case of integrated control programmes where the whole

package results in a considerable saving in pest control costs.
On one U.K. nursery implementation of a full integrated programme
on cucumbers reduced a pesticide bill of  260 per acre to less
than  1 and yet increased the crop yield by  1,500 as a result
of lessened phytotoxicity.  In these highly profitable circumstances
increased costs in one component of the programme are of no consequence.

However, even this unfavourable comparison is removed by
the most recent development-application of *Bacillus* by thermal
fogging machines.  Despite the high barrel temperature of most
of these foggers the material is only heated for about a second
and so almost all the spores survive.  The spores deposited are
almost entirely on the upper leaf surface where it would normally
be ineffective but, fortunately, the young larvae eat minute
holes in the leaves and so consume spores.  As fogs can be applied
within a few minutes, with only limited labour costs compared
with those involved in lengthy high-volume applications, successive
bacterial treatments therefore become economic.

Another important technique in integrated control is the
use of lethal materials when parasites are protected within the
mummified bodes of their hosts.  Once *Encarsia* larvae have eaten
out the contents of the whitefly host, i.e. the scale has become
black, the parasite is impervious to most pesticides.  When *Encarsia*
is released according to the programmes described earlier the
appearance of these black scales is predictable so that pesticide
applications can be safely timed.  This technique can be exploited
in two ways.  If parasites are to be released when excessive
numbers of whiteflies or aphids are present then a non-persistent
pesticide such as pyrethrin or pirimiphos-methyl can be applied
to alter the age structure of the population before the parasite
is released.  Alternatively, parasites may be released, and when
established, a large proportion of the unparasitized population
may be eliminated by non-persistent pesticides.

Parr & Stacey (1977) controlled an infestation of whiteflies
on tomatoes as high as 80 adults/leaf.  Two fogs of Pynosect
(a mixture of pyrethrin and resmethrin) were applied within four
days before releasing *Encarsia* at the rate of 5/plant.  When
their progeny had matured to black scales another fog was applied
to remove the bulk of another generation of whitefly adults and
the subsequent biological control was successful for the rest
of the season.

The strategy of pesticide usage within integrated programmes
is complex and harmonization with biotic agents obviously depends
on a detailed knowledge of their many side-effects (Table 6).

Table 6.  Recommendations on pesticide usage in integrated control (Glasshouse Crops Research Institute).

| Effect on predators | Pesticide | Effect on parasitised whitefly scales |
|---|---|---|
| Harmful | Actellifog (pirimiphos-methyl) | Harmful |
| Harmful | Benlate (benomyl) spray | None |
| Minimal | Benlate (benomyl) drench | None |
| Harmful | gamma-BHC (HCH) spray | Harmful |
| None | gamma-BHC (HCH) drench | None |
| Harmful | Bladafume (sulfotepp) | - |
| Minimal (some effect on oviposition) | Daconil (chlorothalonil) | None |
| Harmful | DDT sprays | Harmful |
| - | DDT/BHC smokes | Harmful |
| Harmful | Derris (rotenone) | - |
| Harmful | Diazinon spray | Harmful |
| Minimal | Diazinon drench | Minimal |
| None | Elvaron (dichlofluanid) | None |
| Minimal | Ethrel E (ethephon) | None |
| Harmful | Kelthane (dicofol) | None |
| Harmful | Malathion | Harmful |
| None | Maneb | - |
| Harmful | Morestan (quinomethionate) | - |
| Minimal | Nicotine shreds | None |
| None | Captan (orthocide) | - |
| Harmful) for at | Parathion spray | Harmful) for at |
| Harmful) least | Parathion smoke | Harmful) least |
| Harmful) 14 days | Parathion drench | Harmful) 14 days |
| Minimal | Petroleum oil emulsions | Minimal |
| None | Pirimor (pirimicarb) spray | None |
| None | Pirimor (pirimicarb) drench | None |
| Minimal | Plictran (cyhexatin) | None |
| Harmful | Pynosect 30 miscible spray | Minimal at half rate |
| Harmful | Pynosect 30 ULV fog | None |
| Harmful | Resmethrin (turbair ULV) | Minimal |
| Harmful | Roger E (dimethoate) | Harmful |
| Minimal as localised spray | Tedion (tetradifon) | None |
| Harmful | Undene (propoxur) | None |
| Harmful | Vapona strips (dichlorvos) | - |

THE IMPACT OF CROP HUSBANDRY ON INTEGRATED CONTROL

It is a mistake to believe that effective pest control will automatically follow the release of the correct numbers of natural enemies and the use pesticides chosen to harmonize safely with them. Growers and advisers have, for many years, believed that pest and disease control must be achieved without putting any restrictions on crop culture. This widespread belief is fostered by a continuance of the attitudes which developed in the immediate post-war era when the new synthetic organochlorine and organophosphorous pesticides provided certain and rapid control. This false security encouraged the adoption of short-cut methods which are still used though, more recently, in the name of labour saving. Indeed, most pest problems are believed to be caused by invasions from outdoor sources and so not within the grower's control. This is largely a myth - almost every pest problem occurs through a failure to appreciate some basic biological fact.

This unsatisfactory situation is highlighted by the red spider problem. Despite a concerted effort by the official research and advisory services over the past twenty years only recently have growers widely appreciated that the scale of the problem depends upon how many mites were allowed to hibernate the previous season.

The size of this overwintering population is not wholly determined by the effectiveness of biological or chemical control techniques. Although spider mites are 'switched' to the bright red diapausing form by the length of the photoperiod experienced by the proto- and deutonymphal instars the proportion of the population so affected is governed by their nutrition. Where plants remain in good condition in late summer most mites remain as active summer-feeding forms even in October. This fact can be used to obtain a most effective control. If the crop is cut out while still in good health and removed rapidly from the glasshouse any dislodged mites will die so long as no living plants remain. Complete mortality is assured because summer-feeding forms die if deprived of food for 24 h. Such adults cannot enter diapause, however strong the inducing pressures, which affect only the developmental stages. In earlier days, growers wisely fumigated their glasshouses with sulphur to kill mites before they reached their overwintering sites. Reference has already been made to other pests which survive from one year to the next, namely, thrips, leaf-miners and caterpillars. Even when soil sterilization was standard practice the effectiveness of steam or methyl-bromide had to be carefully monitored for, where telltale weeds appeared along the outer-walls or beneath heating pipes, then some, or all, of these pests appear in considerable numbers. Now that soil sterilization is being phased out these problems will doubtless increase.

Another cultural change bringing unexpected problems is the trend to peat/sand rather than loam composts. Two difficulties have followed from this development. Highly organic soils are especially attractive to sciarids whose pest status has increased sharply in recent years. So, while growers have opted for these composts on the understanding that they were providing an almost sterile growing medium, a rather intractable pest problem has been inherited instead. This pest, which damaged roots and so encourages root pathogens which also thrive in the sterile medium, can be controlled only by incorporating chlorfenvinphos or difluben- zuron in the composts before use. Other pesticides, such as diazinon, usually advised for sciarid control are rendered virtually ineffective by the highly organic nature of the compost. This absorption creates other problems, especially in respect of systemic pesticides which are released only erratically.

Mention has been made elsewhere of the limitations of parasite control in low winter light. This may be conveniently avoided by the use of oxamyl or aldicarb which should keep crops free of spider-mites, aphids and whiteflies for several weeks. On organic media, however, this control is erratic and so pests may re-establish at any time from two to ten weeks after treatment. These circumstances create planning problems for biological control. Similar restrictions apply to the systemic fungicides - benomyl and dimethirimol.

Later, when natural enemies have been successfully established, the pattern of control may be unexpectedly disturbed by the inadver- tant removal of unemerged parasites on leaf-trimmings. These leaves are removed to hasten ripening of the fruit and growers tend to adopt rather rigid systems which remove large quantities of leaf-material at one time. It is essential that sufficient foliage is left on the plants to support mature parasite scales. There are a variety of ways of achieving this objective. In Holland, newly trimmed leaves are left on the ground to desiccate on the heating pipes and so release parasites before they are removed from the glasshouse. Alternatively, where the crops are grown in double rows it is possible to leave sufficient foliage between the plants and so compromise between the requirements of hastened fruit ripening and parasite survival.

Other modifications of cultural practice are necessary to safeguard the first introductions of natural enemies. Such introduc- tions are made on leaf-material supplied by the rearing companies and particular care must be taken to avoid dislodging these leaves from the crop plants with excessive damping-down or overhead watering. Growers find it difficult to accept such restrictions to their normal practice but such is the price of integrated control.

Finally, it is important to recognize that within integrated control programmes the method of application, as well as the choice of pesticide, is an essential element. The whole pest and disease control programme can be decided before the crop is planted, including the chemical element which must be highly predictive. Such predictability is achieved only by a realistic understanding of the limitations of certain pesticide techniques. It is, therefore, essential to be properly equipped with spraying and drenching equipment to achieve the desired control of target pests. Included in these considerations is an understanding of the relative performance of different ultra-low volume machines. Research at the Glasshouse Crops Research Institute has shown quite clearly that the crop must be 'walked', even with fogging machines, if there is to be any prospect of a reasonable coverage with the pesticide. With almost all ULV methods the underleaf cover is poor and so, for practical purposes, one must assume that only flying adults will be killed even with successive applications.

Integrated control demands that all decisions affecting culture and protection are, themselves, integrated.

## ENVIRONMENTAL CONSIDERATIONS

The most obvious environmental factor to be considered is temperature. In Britain, most glasshouse crops are grown to a 'blue-print' which lays down the conditions which should prevail at each stage of crop growth. These stipulated temperatures, devised after much developmental work by the Agricultural Development and Advisory Service of the Ministry of Agriculture, are the basis of the predictable techniques of biological control referred to earlier. It must, however, be appreciated that the recommended temperatures are only minimum values sustained by the heating systems against lower outside ambient conditions. In sunshine, the glasshouse air temperatures rise above these minimum levels and they can be only partially reduced by opening the ventilators.

The high-leaf temperatures which occur in hot, sunny weather may create unexpected effects which must be anticipated in pest management techniques. The predator, *Phytoseiulus*, is one example, for this species avoids excessive heat such as occurs at the tops of cucumber plants, especially in older houses with a low roof. In these conditions predators leave the apical foliage and shelter on the lowest leaves or even on ground weeds. They must either be replaced in the evenings or the red spider mites controlled with sprays of cyhexatin applied to the uppermost (2 ft.) of growth. On the other hand it is apparent that these excessively hot conditions tend to occur in late June and July. If spider mites have been eliminated by this time, by effective

manipulation of the predator in late May and early June, the
problem will not occur.  This manipulation is usually achieved
by introducing spider mites to sustain adequate numbers of predators
following the initial control of ex-diapause populations which
should be completed by April.

Another interesting aspect of the environment is the influence
of dull weather in later winter and early spring when *Encarsia*
is first released to control whitefly on early tomato crops.
It is not uncommon to make six or seven successive releases of
parasites before they become established on the crop.  The exact
mechanism is still not fully understood but several important
leads have been explored.

The newly emerged parasites require food to lay viable eggs.
Introductions of parasites are made in the form of parasitized
scales on desiccated tobacco leaves which lack suitable food.
The parasite must, therefore, fly and find what is hopefully
a very low whitefly population in less than one per hundred host
plants.  Glasshouse observations have shown that this dispersal
requires a period of bright sunshine - either to provide solar
heat or higher light intensities.  At this time of year, in the
climate of Western Europe, the sun may be obscured by cloud for
several days.  These conditions naturally reduce the predictability
of the methods of parasite release described herein and it is
for this reason that the 'banker' method of introduction was
developed.  These plants, infested by both whiteflies and parasites,
ensure the availability of food so that the adult parasite can
survive for several days until sunshine encourages flight.

These limitations on parasite dispersal are most obvious
on tomatoes where the normal ambient temperature is only of the
order of 62°F but on cucumbers, where the minimum temperatures
are at least 5° higher, these limitations do not apply and the
parasite can be successfully established even in January.

Another environmental factor which is consciously manipulated
by the grower is the concentration of carbon dioxide.  This does
not directly affect natural enemies though the burners sometimes
emit sulphurous fumes which are lethal to predators.  On the
other hand, the maintenance of high carbon dioxide concentrations
induces very soft growth in young plants during periods of low
winter light.  Such plants are easily damaged by pesticides and
so the choice of chemicals for use as a prelude to biological
control is severely limited.  The tyroglyphid mite, *Tyrophagus
putrescentiae* (Schrank), or French Fly, which swarms from the
straw bales on which cucumbers are grown must be controlled by
parathion, dicofol or cyhexatin if the growing cucumber plant
is to be protected.  These alternatives will normally be reduced
only to parathion in the presence of optimal carbon dioxide

concentrations.

It is widely recognized by growers that the first spider
mites to emerge from hibernation do so in local hot-spots, along
heating pipes, near colorifiers or in the heat plumes from burners.
This fore-knowledge enables the grower to judge where to put
additional predators and also to use such vulnerable plants as
sources of spider mites to 'even-up' their distribution before
the predator is introduced.

Possibly the most neglected factor in the environment are
the plants themselves. Innumerable pest problems are initiated
by weeds which have been allowed to remain overwinter to provide
a 'carry-over' host to contaminate the newly propagated crop.
Thorough hygiene is a vital prelude to any integrated programme.

The crop, itself, may consist of one or several cultivars
whose susceptibility to pests varies. In the case of cucumbers
susceptibility to whitefly depends on the density of hairs on
the abaxial leaf-surface (Kowalewski and Robinson 1977). Regrettably,
the parasite is also adversely affected and so biological control
on this crop has to be very precise and the 'pest-in-first' approach
is strongly recommended. In the case of chrysanthemums, however,
where there is a wide range in the susceptibility of the commonly
grown cultivars the rates of population growth of M. *persicae*
differ widely and so the speed and effectiveness of the parasite
A. *matricariae* can be greatly enhanced.

In general, it can be said that the plant breeders tend
to seek high levels of resistance which would provide control
unaided. However, if biological control is to be superimposed
on this cultivar tolerance then even partial resistance can be
exploited effectively.

## COMMERCIAL PROBLEMS IN THE PRACTICAL UTILISATION

## OF BIOLOGICAL CONTROL

Natural enemies are perishable so production should, ideally,
equal demand, a situation rarely achieved at present. Maximum
demand occurs in the early spring (February-April) often with
a peak of only 2-4 weeks. Producers, therefore, cool-store *E.*
*formosa* at 55°F (Scopes *et al*. 1973) to cater for unexpected
variations in demand or production. Early in the year, production
problems are aggravated by the fact that plants are not of top
quality having been raised during the winter when the natural
light intensity is low.

The development of pest management programmes for glasshouse

use demands a range of expertise in plant husbandry, biology,
entomology and crop protection especially in the integration
of chemical controls for minor pests and diseases. As with
rearing, there is a peak period of demand for skilled man-power
early in the year, with virutally no input required from June
to December.

Attempts are, however, being made to diversify and prolong
the season of demand for these natural enemies by controlling
*T. urticae* on chrysanthemums and strawberries as well as *M. persicae*
on chrysanthemums and peppers. The pests of ornamental plants
in amenity and local authority glasshouses could provide a new
outlet for production companies. In Britain, a country of small
gardeners with many thousands of greenhouses and conservatories,
there is also an attractive amateur market awaiting exploitation
for there is a reluctance to use pesticides in these situations.

In many instances biological pest management has yet to
be '*sold*' to growers, a task which is inversely related to the
scale of the pest control problems experienced during previous
seasons. The failure of pesticides to control whiteflies in
recent years has encouraged many converts but it is difficult
for new initiates to accept that control takes two months rather
than two days. An interesting and new approach is, therefore,
being made by some rearing firms in that they offer 'supervised'
control or 'control by contract' rather than the mere postal
sale of natural enemies. This service is provided by experienced
operators or professional entomologists and if commercially viable
should ensure optimal performance of the complete integrated
programmes. There are, of course, problems in providing such
services in that the geographical distribution of nurseries willing
to pay for the control may entail excessive travelling costs
and restrict the time available to monitor the populations as
a basis of advice.

The continuing implementation of biological control demands
a permanent research input into the side-effects of new crop
production techniques and new pesticides. A total involvement
in the growing scene by the research team is of paramount importance.

One of the least recognized limitations of biological pest
control methods follows from the dramatic reduction in the number
of pesticides sprays so permitting hitherto minor or undetected
pests to become troublesome.

The decision to produce and release natural enemies is only
the beginning of a long and ever changing road to the implementation
of rational pest control.

## LITERATURE CITED

Addington, J.  1966.  Satisfactory control of red spider mite on
    cucumbers.  Grower 66:726-7.
Anon.  1975.  Biological pest control.  Rearing parasites and
    predators.  Growers Bull. 2:12 pp.  Glasshouse Crops Research
    Institute.
Anon.  1976a.  The biological control of cucumber pests.  Growers
    Bull. 1:19 pp.  Glasshouse Crops Research Institute.
Anon.  1976b.  The biological control of tomato pests.  Growers
    Bull. 3:23 pp.  Glasshouse Crops Research Institute.
Babiker, T.  1977.  The effects of fungicides on the predatory
    efficiency of *Phytoseiulus persimilis*.  Ph.D. Thesis, Univ.
    of Bradford.
Binns, E. S.  1971.  The toxicity of some soil applied systemic
    insecticides to *Aphis gossypii* and *Phytoseiulus persimilis*
    on cucumbers.  Ann. appl. Biol. 67:211-22.
Geyspits, K. F.  1960.  The effect of the conditions in which pre-
    ceding generations were reared on the photoperiodic reaction
    of geographical forms of the Cotton Spider Mite.  Trans.
    Peterhof. Biol. Inst. L.S.U. 18:169-77.
Gould, H. J., W. J. Parr, H. C. Woodville and S. P. Simmonds.
    1975.  Biological control of glasshouse whitefly on cucumbers.
    Entomophaga 20:285-92.
Hussey, N. W., D. L. Stacey and W. J. Parr.  1975.  Control of the
    leaf-miner, *Liriomyza bryoniae*, within an integrated programme
    for the pests and diseases of tomato.  Proc. 8th Brit. Insect.
    and Fungic. Conf. 109-16.
Irving, S. N. and I. J. Wyatt.  1973.  Effects of sub-lethal doses
    of pesticides on the oviposition behaviour of *Encarsia formosa*.
    Ann. Appl. Biol. 75:57-62.
Jackson, G. J.  1973.  The feeding behaviour of *Phytoseiulus per-
    similis* particularly as affected by certain pesticides.  Ann.
    Appl. Biol. 75:165-71.
Kowalewski, E. and R. W. Robinson.  1977.  Whitefly resistance in
    *Cucumis*.  Bull. 3. I.O.B.C. West Polarcti Regional Section.
    149-53.
Ledieu, M.  1977a.  Dispersal of the whitefly parasite, *Encarsia
    formosa*, within glasshouse crops.  Rep. Glasshouse Crops Res.
    Inst. 1976.
Ledieu, M.  1977b.  Integration of pesticides with the whitefly
    parasite, *Encarsia formosa*.  Rep. Glasshouse Crops Res. Inst.
    1976.
O'Reilly, C. J.  1975.  Investigation on the biology and biological
    control of the glasshouse whitefly.  Ph.D. Thesis, Nat. Univ.
    Ireland.
Parr, W. J. and N. W. Hussey.  1966.  Diapause in the Glasshouse
    red spider mite (*Tetranychus urticae*): a synthesis of present
    knowledge.  Hort. Res. 6:1-21.
Parr, W. J. and D. L. Stacey.  1977.  Integrated control of whitefly.

Rep. Glasshouse Crops Res. Inst. 1976.

Scopes, N. E. A. and S. M. Biggerstaff. 1971. The production, handling and distribution of the whitefly, *Trialeurodes vaporariorum*, and its parasite *Encarsia formosa*, for use in biological control programmes in glasshouses. Pl. Path. 20:111-16.

Scopes, N. E. A. and S. M. Biggerstaff. 1973. Progress towards integrated pest control on year-round chrysanthemums. Proc. 7th Brit. Insect. and Fungic. Conf. 227-35.

Scopes, N. E. A., S. M. Biggerstaff and D. E. Goodall. 1973. Cool storage of some parasites used for pest control in glasshouses. Pl. Path. 22:189-

Stacey, D. L. 1977. "Banker" plant production of *Encarsia formosa* and its use in the control of glasshouse whitefly on tomatoes. Pl. Path. (In press).

Stary, P. 1976. Biology of aphid parasites with respect to integrated control. W. Junk, The Hague, 641 pp.

Stone, L. E. W., A. Lane, R. Hammon, H. J. Gould and R. F. Potter. 1977. Observations on the control of the glasshouse red spider mite, *T. urticae* on year-round chrysanthemums by aldicarb in 1976. Pl. Path. (In press).

Tulisalo, U. 1975. Control of two-spotted mite by high air humidity and direct contact with water. Ann. Ent. Fenn. 40:158-62.

Van Lenteren, J. C. H. W. Nell, L. A. Lelie Van Der Sevenster and J. Wofts. 1976. The host parasite relationship between *Encarsia formosa* and *Trialeorodes vaporariorum*.
I.   Ent. Exp. and Appl. 20:123-30.
II.  Z. Angew Ent. 81:372-6.
III. Z. Angew Ent. 81:377-80.

CHAPTER 13

# AUGMENTATION OF NATURAL ENEMIES FOR CONTROL

# OF PLANT PESTS IN THE WESTERN HEMISPHERE

R. L. Ridgway, E. G. King and J. L. Carrillo

Agricultural Research Service, U.S. Department of
Agriculture, Beltsville, Maryland 20705 and
Stoneville, Mississippi 38776, U.S.A., and Institute
Nacionale de Investigaciones Agricolas, Chapingo, Mexico

Augmentation is defined as periodically increasing either the number of parasites or predators or the supply of their food resources to assure that adequate numbers of parasites and predators are present to provide the desired level of pest control. Similar definitions have been used previously (Ridgway 1972, Rabb et al. 1976). The release of natural enemies has also been classified as being either inoculative or inundative (DeBach and Hagen 1964). Inoculative releases appropriately describe releases of natural enemies for the purpose of colonizing populations to regulate pest populations through in-field reproduction of the released species. However, the use of inundative releases to describe all other augmentative releases may be misleading since by definition inundative means "to flood or to overwhelm by great numbers."

In our opinion, the goal of most releases of natural enemies should be to augment existing populations of natural enemies only to the extent necessary to maintain the pest population at an acceptable level. Therefore, augmentation appropriately describes a number of activities including (1) the release of natural enemies (inoculative releases, seasonal colonization, supplemental releases, strategic releases, programmed releases, or inundative releases), (2) the use of supplemental foods, or (3) the use of kairomones or other behavioral chemicals.

The importance of having an ecological basis for developing augmentation programs has been discussed previously (DeBach and Hagen 1964, Ridgway 1969, Huffaker 1971, Huffaker and Messenger 1976); and the biological requirements for effective releases have been summarized (Ridgway et al. 1974) as follows: (1) an understanding of the major ecological parameters governing the principal

interactions between the parasite or predator to be released and
the pest to be controlled; (2) an ability to rear predictable quantities of insects of known quality; and (3) an ability to store, transport, and release the parasites and predators in such a manner that
they will become a competitive part of the life system in which they
are expected to operate. Many aspects of these requirements as
well as different approaches to augmentation have been discussed
in previous chapters; therefore, emphasis in this chapter will be
placed on reviewing examples of successful releases of natural
enemies for control of insects and mites attacking plants in the
Western Hemisphere. A discussion of mechanical methods for release
of parasites and predators and a description of some model systems
for utilizing augmentation of natural enemies are also included.

## EXPERIMENTAL AND PRACTICAL APPLICATIONS

Successful demonstrations or applications of augmentations of
natural enemies will be discussed for the following categories
of commodities: glasshouse crops, fruit crops, vegetable crops,
field crops, and forests.

### Glasshouse Crops

Although most of the research and utilization of natural
enemies in glasshouses has been done in Europe or the Soviet
Union, significant work also has occurred in the United States
and Canada. Perhaps the first effort to use augmentations to control pests in greenhouses in the Americas was that in which the
parasite *Encarsia formosa* Gahan was used to control the greenhouse
whitefly (McLeod 1938, 1939). Later McClanahan (1970, 1972)
developed an integrated control program utilizing parasites to control whiteflies on greenhouse cucumbers and tomatoes in Canada.

Doutt (1951) was successful in reducing a very heavy infestation of *Planococcus* (=*Pseudococcus*) *citri* (Risso) on commercial
greenhouse gardenias to a low level and maintained control by introducing several species of parasites and predators. The most
effective species were the parasites *Exochomus flavipes* Thunberg
and *Anagyrus kivuensis* Compere and the predator *Chrysopa carnea*
Stephens. Less than 1 percent of the terminals were infected in
the greenhouse in which natural enemies were released compared with
about 50 percent infected terminals in the "control" greenhouse.
More recently, Helgesen and Tauber (1974, Table 1) developed a
reliable and economical biological control system for greenhouse
poinsettias by defining the time of parasite introduction, the
ratio of parasites to hosts, and the greenhouse temperature.

TABLE 1.--Results of releasing natural enemies in glasshouses
for control of insects.

*Encarsia formosa* (1:30 parasite-host) against the
greenhouse whitefly on poinsettas (Helgesen and
Tauber 1974)

| Time in no. of days | No. whiteflies/leaf | | % reduction |
| | Release | Control | |
| --- | --- | --- | --- |
| 50 | 110 | 110 | 0 |
| 60 | 110 | 130 | 15 |
| 70 | 35 | 120 | 71 |
| 80 | 20 | 120 | 83 |
| 90 | 27 | 110 | 75 |

Release of *Chrysopa carnea* against the green peach aphid
on snapdragons (Harbaugh and Mattson 1973)

| No. *Chrysopa* larvae/plant | Mean no. of aphids | Mean % control |
| --- | --- | --- |
| 0 | 152 | – |
| 2 | 61 | 60 |
| 6 | 40 | 74 |

Harbaugh and Mattson (1973) utilized releases of larvae of
*Chrysopa carnea* on greenhouse snapdragons for control of the green
peach aphid, *Myzus persicae* (Sulzer).  High levels of control were
obtained with releases of 2 or more larvae per plant (Table 1).

Current practical applications of releases of natural enemies
in glasshouses in the Western Hemisphere are apparently limited
to the releases of *Encarsia* to control whiteflies in Canada.

### Fruit Crops

Citrus.  One of the first attempts to use augmentation on
citrus was made with an exotic coccinellid *Cryptolaemus montrouzieri*
Mulsant  that had been introduced into the United States.  Armitage
(1919) reported that an insectary had been established for producing
this beetle for distribution in citrus groves to control mealybugs.  Preliminary data indicated that the beetle was effective,
and Armitage (1929) and Smith and Armitage (1931) reported that
releases of 10 beetles per citrus tree controlled mealybugs attacking citrus.  Since control of the mealybug was accomplished by

progeny from the released insects, 60 to 90 days were required before
control was complete. A complete documentation of the historical
development of the use of *C. montrouzieri* is given by DeBach and
Hagen (1964). Although rearing and release of *C. montrouzieri* for
control of mealybugs continues, this use has been reduced,
apparently because of the effective control provided by several
introduced parasites. Another use of augmentations on citrus
involved the release of 2 species of *Aphytis* in California. This
use of *Aphytis chrysomphali* (Mercet) and *Aphytis lingnanensis*
Compere for control of California red scale, *Aonidiella aurantii*
(Maskell) was reviewed by DeBach and Hagen (1964). Although DeBach
et al. (1949) showed that releases of 400,000 *Aphytis* per acre
would control *A. aurantii*, the use of this parasite was apparently
phased out by 1971 after the establishment of another parasite,
*Aphytis melinus* DeBach (Lorbeer 1971). At the present time, *A.
melinus* is released on a continuing basis in the Fillmore Citrus
Protective District in Ventura County, California (Pennington
1975).

An encyrid parasite, *Microterys flavus* (Howard), was tested in
periodic releases against brown soft scale, *Coccus hesperidum* L.,
on citrus (Hart 1972). Apparently, this parasite and other natural
enemies are largely eliminated by drift from insecticide treatments
in nearby cotton fields. When chemical treatment was stopped and
*M. flavus* was reintroduced, parasitization occurred, and progeny
from the release generation continued to provide control of the
soft brown scale.

Pears. Probably the most definitive studies involving augmentations were the studies by Doutt and Hagen (1949, 1950) in which
they reported control of *Pseudococcus* sp. on pears was achieved by
distributing eggs of *Chrysopa californicus* Conquillett *(=Chrysopa
carnea)* in the forks of trees.

Avocado. McMurtry et al. (1969) reported that multiple
releases of *Stethorus picipes* Casey against the avocado brown mite.
*Oligonychus punicae* (Hirst), at a rate totaling 400 to 500 adult
beetles per tree resulted in an earlier buildup of *S. picipes* in
release areas, a lower peak population of *O. punicae*, and a lower
percentage of heavily bronzed leaves in the release plots compared
with check plots receiving no predator releases (Table 2). Factors
restricting commercial practice of this approach included the
critical need for precise timing of beetle releases and expense of
mass producing it.

Apples. Croft and McMurtry (1972) showed that a release of
128 predaceous mites, *Typholodromus occidentalis* Nesbitt, per apple
tree gave season-long economical control of *Tetranychus mcdanieli*
McGregor, and they suggested that this may be an economically
feasible approach for control of this pest on apple (Table 2).

TABLE 2.--Results of releasing natural enemies on fruit crops for control of insects and mites.

*Chrysopa californicus* (=*carnea*) against mealybugs on pears (Doutt and Hagen 1950)

| Total eggs/tree | No. of colonizations | % fruit infested | |
|---|---|---|---|
| | | Release | Control (avg.) |
| 250 | 1 | 32 | 64 |
| 1000 | 1 | 40 | 64 |
| 750 | 3 | 12 | 64 |

*Stethorus picipes* (400–500/tree) against avocado brown mites (McMurtry et al. 1969)

| Sample dates | No. ♀/10 leaves | | % reduction |
|---|---|---|---|
| | Release | Control | |
| Aug. 09 | 115 | 257 | 55 |
| Aug. 23 | 475 | 1178 | 60 |
| Sep. 07 | 329 | 807 | 59 |

*Typhlodromus occidentalis* (128/tree) to control *Tetranychus mcdanieli* on apple (Croft and McMurtry 1972)

| Sample dates | No. prey/100 leaves | | % reduction |
|---|---|---|---|
| | Release | Control | |
| Jul. 12 | 0.06 | 3.82 | 98 |
| Aug. 21 | 3.40 | 34.76 | 90 |
| Sep. 17 | 0.41 | 104.10 | 99 |

*Neoaplectana dutkyi* against the codling moth on apples (Dutky 1974)

| Sample dates | No. codling moths/tree | | % reduction |
|---|---|---|---|
| | Total | Parasitized | |
| Sep. 22, 1960 | 3.35 | 2.70 | 81 |
| Nov. 02, 1960 | 6.85 | 5.35 | 78 |
| May 10-11, 1961 | 0.62 | 0.19 | 31 |
| May 21, 1961 | 0.68 | 0.50 | 74 |

Croft and Hoying (1975) and Meyer (1975) have demonstrated that phytoseiids that have developed resistance to certain pesticides can be released into apple trees and will readily establish. Further, the resistance can be retained, thus enabling the predator mite to survive under pesticide pressure in an integrated control program. Thus, these mites can be introduced into other areas and used in augmentation programs where pesticides are also used (Croft 1976).

A nematode, *Neoaplectana dutkyi* Jackson provided over 70 percent reduction of overwintering codling moths, *Laspeyresia pomonella* (L.), in experimental tests in West Virginia and Indiana when the limbs and trunks of apple trees were sprayed with *N. dutkyi* Dutky 1974, Table 2). Nickle (1974) also reviewed several experiments in which this nematode has been used in tests against a number of other insect pests.

Peaches. Extensive research has been conducted on using the braconid *Macrocentrus ancylivorus* Rohwer in periodic releases to control the oriental fruit moth, *Grapholitha molesta* (Busck). Most of this work has been summarized by DeBach and Hagen (1964). It was generally conceded that the parasite releases reduced fruit injury, but the advent of organic phosphorus insecticides precluded the use of parasites in many areas.

Strawberries. Huffaker and Kennett (1956) demonstrated experimentally that stocking strawberry plantings with the predatory mites, *Typhlodromus cucumeris* Oudemans, or *T. reticulatus* Oudemans, or both, in the first year or early in the second year after planting gave consistent control of the cyclamen mite, *Steneotarsonemus (=Tarsonemus) pallidus* Banks. Control was achieved by early establishment of predator-prey equilibrium before the season of population surge of the prey species. Once equilibrium was achieved, it was relatively permanent in absence of detrimental chemical treatments. They suggested deliberate introduction of clippings bearing predators and cyclamen mites from old strawberry fields into first year plantings to achieve rapid equilibrium.

Oatman et al. (1968) reported that mass releases of *Phytoseiulus persimilis* Athias-Henriot at the equivalent rate of 320,000 total per acre at weekly intervals for 8 weeks effectively suppressed *Tetranychus urticae* Koch populations on strawberry, and fruit yield was 7.4 percent higher in release plots, but this was not statistically significant.

Vegetable Crops

Cabbage. Parker (1971) summarized research conducted by him-

self and others (Parker et al. 1971, Parker and Pinnell 1972) in
which both the host *Pieris rapae* (L.), and parasites *Apanteles
rubecula* Marshall and *Trichogramma evanescens* Westwood were released
in cabbage.  When all 3 insects were released, the host population
was controlled sufficiently so that larval damage remained below the
level causing economic injury; the check plots (no parasite or host
releases) had *P. rapae* populations that exceeded this level (Table
3).  They concluded that there were 2 critical requirements for
suppression of host populations by parasites: (1) a consistent supply
of hosts that was adequate to maintain the parasite population, and

TABLE 3.--Results of releasing natural enemies for control of
insects on vegetable crops.

*Trichogramma evanescens* and *Apanteles rubecula* against
imported cabbageworms in the field (Parker 1971)

| No. parasites per acre per season | | No. cabbageworms per 100 plants | | |
|---|---|---|---|---|
| Trichogramma | Apanteles | Release[a] | Control[b] | % reduction |
| 2,727,000 | 11,300 | 41 | 1481 | 97 |
| 2,180,000 | 8100 | 46[c] | 1481 | 97 |
| 282,000 | 7800 | 263[c] | 1481 | 82 |

a/ Avg. of 1 or 2 one to 1.5 acre plots.
b/ Avg. of 4 one-half acre plots.
c/ Fertile hosts released.

*Trichogramma pretiosum* against *Heliothis zea* on tomato
in the field (Oatman and Platner 1971)

| | % parasitism[a] | | | | | |
|---|---|---|---|---|---|---|
| | Tomato fruitworm | | Cabbage looper | | Tomato hornworm | |
| Survey date | Release | Control | Release | Control | Release | Control |
| Jun 3, 6 | 26 | 0 | 16 | 6 | 38 | 0 |
| 17,20 | 60 | 0 | 30 | 4 | 33 | 0 |
| Jul 1, 3 | 69 | 19 | 67 | 7 | 33 | 0 |
| 15,18 | 94 | 71 | 73 | 75 | 89 | 90 |
| Aug 5, 8 | 70 | 61 | – | 67 | 100 | 91 |

a/ 465,000 parasites per acre; 40,200 per acre per week
from May 20 to August 8.

TABLE 3.--(Continued)

========================================================================

*Trichogramma pretiosum* against lepidopteran pests on
tomatoes, bell peppers, cabbage and collards in a field
cage (Martin et al. 1976)

| Crop | % parasitism | |
|------|---------------|-------------|
|      | Release[a/]   | Pre-release |

|                                  *Plusinae*             |||
| Tomatoes     | 68 | 2 |
| Bell peppers | 56 | 2 |
| Cabbage      | 58 | 2 |
| Collards     | 51 | 2 |

|                                  *Heliothis* spp.       |||
| Tomatoes     | 68 | 5 |
| Bell peppers | 50 | 5 |

a/ Three releases of 378,000 parasites per acre per
release.

========================================================================

Results of releasing *Voria ruralis* against the cabbage
looper on collards in field cages (SooHoo et al. 1974)

| No. mated ♀ *Voria*/cage | No. larvae recovered/cage | % parasitized |
|--------------------------|---------------------------|---------------|
| 1  | 64 | 9  |
| 5  | 73 | 42 |
| 10 | 66 | 62 |
| 50 | 64 | 85 |

========================================================================

(2) release of effective parasites when existing ones are inadequate.
Martin et al. (1976) also obtained relatively high rates of parasitism of the cabbage looper, *Trichoplusia ni* (Hubner) after release
of *Trichogramma* (Table 3).

    Tomatoes, peppers, and collards. Release of *Trichogramma*
*pretiosum* Riley on tomatoes increased parasitism of eggs of the
tomato fruitworm, *Heliothis zea* (Boddie), cabbage looper, and tomato
hornworm, *Manduca quinquemaculata* (Haworth) (Oatman and Platner 1971,
Table 3). Similar results were obtained by Martin et al. (1976) with
releases on tomatoes, peppers, and collards (Table 3). In other
studies on collards in field cages, SooHoo et al. (1974) obtained
high rates of parasitism of the cabbage looper with releases of a
tachinid parasite, *Voria ruralis* (Fallen) (Table 3).

Potatoes. Shands et al. (1972) briefly reviewed a series of papers on research that had been conducted to develop augmentation procedures for using predators for controlling aphids on potatoes. Generally small plots were inadequate, and finally the tests were transferred to large cages to provide better control of the experiments. Coccinella septempunctata L. and C. transversoguttata Feldermann were introduced onto caged potato plants infested with aphids at various times and rates. There was a trend toward decreasing aphid abundance with increasing predator numbers, and this was correlated with a trend toward increased tuber production. However, none of the treatments gave adequate commercial control of the aphids.

Peas. Halfhill and Featherston (1973) reported on a study in which the parasite Aphidius smithi Sharma & Subba Rao was reared in field cages of the pea aphid, Acyrthosiphon pisum (Harris) and allowed to escape into surrounding alfalfa fields (rate unknown) through temperature-controlled vents in the cage roofs. The objective of this study was to prevent migration of aphids from their overwintering site in alfalfa to peas grown for processing. When parasite, natural, and chemical control systems were compared in 1967, averages of 13, 56, and 82 aphids/100 sweeps, respectively, were obtained from fields receiving different treatments. Thus, where parasites were released, the aphid population remained low; where insecticides were used aphid numbers varied greatly; and with natural control the initial aphid population was high though it later dropped to an acceptable level. Unfortunately, the parasite control system was not considered commercially feasible because insecticides were needed for other insect pests; and the parasite production and distribution requirements appeared to be prohibitive.

## Field Crops

Cotton. Natural enemies play an extremely important role in regulating populations of the bollworm, Heliothis zea Boddie, and the tobacco budworm, H. virescens F. (Whitcomb and Bell 1964, Ridgway 1969, Ridgway and Lingren 1972). Consequently, there has been considerable interest in augmentation for control of Heliothis on cotton.

Probably more effort has been devoted to use of egg parasites, Trichogramma spp., than to use of any other group of parasites. Trichogramma has been available from commercial insectaries in the United States for several decades. However, most of the definitive data on the possible impact of releases of Trichogramma on Heliothis populations has been obtained within the past 10 years. Lingren and Kim (1970) clearly demonstrated that releases of Trichogramma could be used to effectively control Heliothis populations on cotton. Releases of about 200,000 Trichogramma per acre at 2- to 3-day intervals resulted in parasitism of about 60 percent of Heliothis eggs (Table 4). However, they concluded that season-long control probably could not be obtained in most areas with a total release of less than 1 million parasites per acre.

TABLE 4.--Results of releasing *Trichogramma* on cotton for
control of the bollworm and tobacco budworm.

| No. of *Trichogramma* acre (in 1000's)[a] | Percent parasitism of eggs | |
|---|---|---|
| | Release | Control |
| Lingren and Kim (1970) in field | | |
| 200 | 58 | 11 |
| Ashley et al. (1974) in field cages | | |
| 82 | 50 | - |
| 164 | 80 | - |
| Stinner et al. (1974) in field | | |
| 19 | 33 | 5 |
| 77 | 55 | 5 |
| 155 | 59 | 5 |
| 387 | 81 | 5 |
| Ridgway, Kinzer, and Jones (original data) in field[b] | | |
| 50-100 | 73 | 0 |
| 50-100 | 64 | 5 |
| 50-100 | 24 | 5 |
| 50-100 | 49 | 5 |
| 50-100 | 49 | 7 |

a/ Numbers are per release; several releases made.

b/ Control values are for pre-release.

Studies in 1971 by Stinner et al. (1974) in replicated plots
in which 19,000 to 387,500 *Trichogramma* were released resulted in
parasitism of *Heliothis* eggs of from 33 to 81 percent (Table 4).
However, these studies were somewhat complicated by insecticidal
drift that caused high mortalities of adult parasites that were in
the field when aerial applications of methyl parathion were made to
nearby fields. Further studies with releases of 50,000 to 100,000
*Trichogramma* per acre on 5 cotton farms resulted in an average rate

of parasitism of 51 percent (Table 4).  On one farm where insecti-
cides were not used for control of boll weevils, *Anthonomus grandis*
Boheman, throughout the season, the population of *Heliothis* larvae
was maintained well below the economic threshold even though large
*Heliothis* egg populations were present for approximately 3 weeks
(Ridgway et al. 1973).

Pacheco et al. (1971) and Pacheco (1972) made releases of
*Trichogramma* for control of the bollworm on cotton in Mexico.
Results were somewhat variable and frequently complicated by
insecticide applications; however, releases of about 3 million
wasps per hectare increased parasitism to as high as 84 percent.
Although not entirely supported by research data, an extensive
effort is underway in Mexico to utilize *Trichogramma* to control
lepidopteran pests on cotton as well as on other crops.  Since
the initiation of this program by the Administration for Plant
Health in 1964, about 28 billion *Trichogramma* have been produced
for use on cotton and other crops (Jimenez 1975).  Similar, but
less extensive, programs for the release of *Trichogramma* are also
underway in El Salvador, Columbia, and Nicaragua (Gonzalez 1976,
Vaughn 1975).

A number of important studies have been conducted to investigate the possible use of releases of larval or egg-larval parasites
for control of *Heliothis* and the pink bollworm, *Pectinophora
gossypiella* (Saunders) on cotton.

Braconids have been suggested for use in periodic releases
for control of *Heliothis* species.  Lingren (1969) reported that
his field tests demonstrated that *Apanteles marginiventris* (Cresson)
had considerable potential for use in an inundative release-type
program.  Jackson et al. (1970) reported 57.5 percent parasitization
of third stage *H. virescens* larvae in cages where 1200 (equivalent)
*Microplitis croceipes* (Cresson) female wasps were released per acre
(Table 5).  Lewis et al. (1972) reported that 80 percent parasitism
of *H. virescens* larvae feeding on cotton could be attained with 400
to 600 *Cardiochiles nigriceps* Viereck females per acre.  Their calculations were based on a 2-year study during which visual estimates
of total actively searching females of *C. nigriceps* per acre were
made and correlated with percentage of *H. virescens* larvae parasitized by *C. nigriceps*.

As a result of cage and laboratory studies, it was proposed
that the ichneumonid, *Campoletis sonorensis* (Cameron), had considerable potential in periodic releases for control of *Heliothis*
larvae in cotton (Noble and Graham 1966, Lingren et al. 1970, Lingren
and Nobel 1972, Lingren 1977, Table 5).  When this parasite was
released at the rate of 680/day for 10 consecutive days in a one-
half acre cage infested with tobacco budworm larvae, about 85

TABLE 5.--Results of releasing larval parasites for control of the
bollworm or tobacco budworm on cotton in field cages.

| Number ♀ parasites/acre | Number larvae/acre | % parasitism |
|---|---|---|
| *Campoletis sonorensis* (Noble and Graham 1966) | | |
| 807 | 4500 | 52 |
| 1613 | 8600 | 82 |
| *Campoletis sonorensis* (Lingren 1977) | | |
| 835 | - | 85-95 |
| *Micropletis croceipes* (Jackson et al. 1970) | | |
| 1200 | - | 58 |
| *Eucelatoria* sp. (Jackson et al. 1970) | | |
| 2500 | 1850 | 57 |
| *Palexorista laxa* (Jackson et al. 1970) | | |
| 2500 | 1313 | 51 |

percent of the host larvae were parasitized for 9 consecutive weeks.
Host studies indicated that *Heliothis* larvae were preferred
over several other lepidopteran larvae, and early-stage larvae
were prefferred over late-stage larvae.

Bryan et al. (1971) reported that cage studies indicated
that *Bracon kirkpatricki* (Wilkinson) would provide early season
control of pink bollworms if they were released at the rate of
10,000/acre at a time coinciding with appearance of the first host
larvae and at weekly intervals thereafter throughout the first
generation. Larger scale studies were conducted in 1971 (Bryan
et al. 1973a) and in 1972 (Bryan et al. 1973b) utilizing the
parasites *B. kirkpatricki* and *Chelonus blackburni* Cameron in
periodic releases.

During both years, *B. kirkpatricki* generally reduced the
early (bloom) infestation, but *C. blackburni* was not as effective

in reducing the late (boll) infestation, possibly because of the low numbers released. Once the pink bollworm larvae enter the bolls, they are no longer accessible to *B. kirkpatricki*. Thus, a highly efficient egg parasite or a parasite that can seek out host larvae in bolls will probably be required before augmentation procedures with parasites for control of the pink bollworm will be practical.

Jackson et al. (1970) reported from their studies with tachinid flies that if *Eucelatoria* sp. and *Palexorista laxa* were released at the rate of 2500 female flies per acre on cotton containing 5000 *Heliothis* larvae per acre, then about 50 percent parasitization should occur in 2 days (Table 5). D. E. Bryan (pers. comm.) considers *Eucelatoria* sp. an excellent candidate for use in the augmentation approach for control of *Heliothis* larvae because this parasite will parasitize late-stage larvae, is nearly host specific, and gives a relatively high progeny return per host larvae.

A number of significant studies involving augmentation of predators for control of insect pests on cotton, especially *Heliothis*, has been conducted. Lingren et al. (1968a) conducted studies in field-cages in which releases of *Chrysopa carnea* (Table 6) substantially reduced populations of *Heliothis*. Later, van den Bosch (1969) obtained similar results in predator-free cages infested with first-instar larvae of *H. zea*. Lopez et al. (1976) also demonstrated the effectiveness of releases of *C. carnea* in field cages.

Because of the high levels of reduction of *Heliothis* obtained in field cages with releases of larvae of *C. carnea* and the availability of methods of rearing this insect, further studies were conducted to explore its possible use on cotton.

Ridgway and Jones (1968, 1969) reduced tobacco budworms and bollworms 74 to 99 percent by releasing *C. carnea* in field cages and in the field (Table 6). In one experiment in field cages, very high levels of alternate prey reduced the efficiency of *C. carnea* against tobacco budworm larvae. The releases of eggs were generally successful in the field cages, but releases of eggs in the field have been less successful (Ridgway, unpublished data). Likewise, releases of adults in the field have had limited success (Kinzer, unpublished data). However, releases of 2- to 3-day-old larvae have consistently produced significant reductions of *Heliothis* on cotton. For example, some reduction in *Heliothis* was obtained by releasing 10,000 *C. carnea* larvae per acre, and high levels of reduction were obtained by releasing 100,000 per acre (Ridgway et al. 1973, 1974, Ridgway and Kinzer 1974, Kinzer 1976).

TABLE 6.--Results of releasing larvae of the predator *Chrysopa
        carnea* on cotton for control of the bollworm and tobacco
        budworm .

| No. of predators per acre (in 1000's) | No. of bollworm or budworm larvae per acre (in 1000's) | | % reduction |
|---|---|---|---|
| | Release | Control | |
| *Lingren et al. (1968a) in cages* | | | |
| 420 | 0.6 | 17.0 | 96 |
| *Ridgway and Jones (1968) in cages* | | | |
| 400 | 0.3 | 58.6 | 99 |
| 400 | 14.4 | 58.6 | 75 |
| 25 | 1.6 | 6.1 | 74 |
| 100 | 1.6 | 6.1 | 74 |
| 300 | 0.6 | 6.1 | 90 |
| *Ridgway and Jones (1969) in the field* | | | |
| 146 | 0.7 | 18.0 | 96 |
| *Ridgway and Jones (original data) in the field* | | | |
| 200 | 1.7 | 9.5 | 82 |
| 92 | 0.8 | 7.5 | 89 |
| *Kinzer (1976) in the field* | | | |
| 10 | 4.2 | 6.3 | 33 |
| 30 | 2.9 | 6.3 | 54 |
| 100 | 1.1 | 6.3 | 83 |

After a comprehensive study of the feeding and searching
behavior of *C. carnea* with special reference to its predatory
effects on *Heliothis* spp. on cotton, Boyd (1970) reported that the
kind of prey available was not particularly important, but a minimum
quantity of prey was essential for survival of larvae; also, the
availability of prey was particularly important to the survival of
small larvae. However, plant nectar would aid survival when quantities of prey were low. In related laboratory studies, the preference
of *C. carnea* for prey was found to be, in descending order, 1st-
instar *Heliothis* larvae; cotton aphid, *Aphis gossypii* Glover;
*Heliothis* eggs; and carmine spider mites, *Tetranychus cinnabarinus*
(Boisduval). All instars of *C. carnea* larvae were capable of

destroying *Heliothis* eggs and 1- to 3-day-old *Heliothis* larvae.
The average percentage efficiency of kill of 1- to 5-day-old
*Heliothis* larvae was 52 for 1st-, 54 for 2nd-, and 90 for 3rd-
instar *C. carnea* larvae. Thus, it is important to have adequate
numbers of *C. carnea* larvae present when *Heliothis* larvae are
small if one is to obtain maximum efficiency of predation.

Boyd (1970) also demonstrated that *C. carnea* is most active
at about 27°C when light intensity is low and prey is limited.
Also, larvae of *C. carnea* were found inside the bracts of cotton
squares where they apparently were sheltered from high temperatures
and light intensity. Since *Heliothis* larvae are most often found
feeding on squares, they are thus a primary target for predation
by *Chrysopa* larvae.

Field-cage studies were conducted by Lingren et al. (1968a),
van den Bosch (1969), and Lopez et al. (1976) in which *Geocoris
punctipes* (Say), *Nabis americoferus* Carayon and *Podisus
maculiventris* (Say) were released on cotton. The results demonstrated the ability of these hemipteran predators to suppress populations of *Heliothis* spp. (Table 7). The unique ability of nymphs
and adults of *P. maculiventris* to prey on large larvae was demonstrated in the study by Lopez et al. (1976).

Tobacco. Lawson et al. (1961) reported that predation of
tobacco hornworms could be substantially increased by erecting
nesting shelters for vespids, *Polistes* spp. near tobacco fields.
He also showed that local predators could be supplemented by transporting shelters occupied by *Polistes* spp. to tobacco fields. Fye
(1972) conducted studies on overwintering and holding of the wasp
*Polistes exclamans arizonensis* Snelling and suggested the possibility of caging them near cotton fields for control of lepidopteran
larvae. Further, since these wasps return to their nest at night,
they could be confined when toxic chemicals were being applied or
hazardous agronomic practices were employed.

Elsey (1971, 1975, Table 7) increased the seasonal density of
*Jalysus spinosus* (Say) in tobacco by making early-season releases in
North Carolina. Although numbers of tobacco budworms were not reduced
by the releases, some suppression of tobacco hornworms was noted.

Soybeans. Periodic releases of an eulophid parasite *Pediobius
foveolatus* (Crawford) imported from India have been effective in the
control of the Mexican bean beetle, *Epilachna varivestis* Mulsant.
Stevens et al. (1975) reported on preliminary field releases of
the parasite in 1972 and 1973 and a large scale suppression attempt
of the Mexican bean beetle in 1974. All of the tests were conducted
in Maryland. High rates of parasitization of Mexican bean beetle
larvae by *P. foveolatus* were recorded after releases in 1972 and 1973.

TABLE 7.--Results of releasing hemipteran predators for control
of lepidopterous pests on cotton and tobacco.

*Geocoris punctipes* in field cages against the tobacco
budworm on cotton (Lingren et al. 1968a)

|  | No. budworm larvae per acre in 1000's | | |
| No. days after release | Release | Control | % reduction |
| --- | --- | --- | --- |
| 4 | 1.5 | 52 | 71 |
| 6 | 22.0 | 33 | 33 |
| 8 | 18.0 | 47 | 62 |

*Nabis americoferus* against the bollworm on cotton in
field cages (van den Bosch 1969)

| No. adult predators per cage | No. bollworms per cage | % reduction |
| --- | --- | --- |
| 0 | 235 | - |
| 100 | 52 | 77.9 |
| 200 | 28 | 88.1 |

*Podisus maculiventris* (100,000 per acre) against tobacco
budworms on cotton in field cages (Lopez et al. 1976)

|  | No. larvae in 1000's | | |
| No. days after release | Release | Control | % reduction |
| --- | --- | --- | --- |
| 5 | 54.2 | 49.1 | 0 |
| 7 | 22.0 | 27.4 | 20 |
| 9 | 1.4 | 18.4 | 92 |
| 14 | 0 | 3.5 | 100 |

*Jalysus spinosus* against hornworms on tobacco in the
field (Elsey  1975)

|  | Percent predation on eggs | |
| Sampling date | Release | Control |
| --- | --- | --- |
| July 14-20 | 22.6 | 1.1 |
| 21-24 | 34.4 | 9.3 |
| 28-31 | 42.3 | 6.1 |

In 1974, 3 to 5 small snap-bean plots adjacent to soybean fields were established in each of 11 counties.  Since Mexican bean beetles prefer the snap beans as an oviposition site, large populations developed which in turn could support large numbers of *P. foveolatus*.  Thus, parasites released in the bean plot provided an early inoculum in each county, and, in addition, parasites were released into other soybean fields infested with Mexican bean beetles.  About 488,900 parasites were released over 232,600 acres of soybeans and average percentage parasitization by *P. foveolatus* for the 11 counties ranged from 22 percent (Aug. 26 - Sept. 9) to 84 percent (Sept. 25 - Oct. 4).  Two years prior to release of the parasite about 166,000 acres of the soybeans in the 11 counties were treated with insecticides for Mexican bean beetle control, but in 1974 only about 30,000 acres were treated.  Since *P. foveolatus* does not overwinter in many areas where the Mexican bean beetle is a pest, including Maryland, yearly releases will be necessary.  This parasite has also been distributed to other States such as Florida, South Carolina, Virginia, and Delaware where separate studies have begun.

Sugarcane.  Periodic releases of tachinids have been used most extensively in the Carribean Islands, Latin America, and South America for control of lepidopteran larvae attacking sugarcane with emphasis on *Diatraea* spp.  Reports prior to 1969 have been reviewed by Bennett (1969).  Augmentation procedures have been conducted in Cuba, Haiti, Peru, Antigua, Guadeloupe, Venezuela, and Bolivia by using the tachinids *Lixophaga diatraea* (Townsend), *Metagonistylum minense* Townsend, and *Paratheresia claripalpis* (Wulp).  Releases have traditionally been inoculative with only a few parasites being released per acre or the rate was unknown.  However, Bennett questioned the validity of reported successes because of lack of a check area or the fact that the number of parasites released was so low that the size of the natural population probably exceeded the number released.

The use of periodic releases of parasites in Columbia, Brazil, Venezuela, and Cuba for control of *Diatraea* spp. is apparently continuing though documentation of the efficacy of these releases is lacking.  T. E. Summers (pers. comm.) reports that insectaries are operating in Columbia for production of *L. diatraea* for use in periodic releases for control of the sugarcane borer, *Diatraea saccharalis* (F.).

F. Ferrer W. (pers. comm. 1976) stated that the tachinid *Metagonistylum minense* was introduced into Venezuela and later used in inoculative release augmentation programs.  In some areas this procedure has been followed for about 20 years with a reported reduction of stalk internodes bored from 14 to 3 percent.

Procedures used by one commercial company "Servico de Control
Integrade" include observing infestations and percentage parasitization and then making releases of *M. minense* when the crop is
about 5 months old.

Teran (1976) reported that in Brazil tachinids are reared and
released in areas of sugarcane fields with high *D. saccharalis*
populations. This practice of inoculative releases was carried out
in several thousand acres of sugarcane in 1974 and 1975 and resulted
in a 50 percent reduction of damage to sugarcane by *D. saccharalis*.

Montes (1970) reported that the number of *L. diatraea* released
annually among 6 Cuban provinces had increased from 824,208 (1961)
to 4,456,650 (1966) but reported no corresponding reduction in
damage by the sugarcane borer. He concluded that the releases
were effective in inoculating sugarcane borer populations, but
once a "balance" between the host/parasite population had been
achieved, then additional parasites did not increase the parasitization rate.

Knipling (1972) proposed in a theoretical study that releases
of *L. diatraea* at the rate of 1000/acre during the second generation of *D. saccharalis* in Louisiana would result in 69 percent
parasitization of the larvae. With recycling, sufficient parasites
would be produced to cause 85 percent parasitization of the third
generation and 80 percent parasitization of the fourth generation.
His theoretical study was based on earlier reports of the seasonal
cycle of the parasite-host buildup. Further, there was good reason
to be optimistic about this parasite since it had been introduced
into a number of countries, and low level releases had resulted in
incredible reductions in damage by sugarcane borer larvae to sugarcane (Bennett 1969). For example, Scaramuzza (1951) reported that
by releasing only 2 to 4 *L. diatraea* females per hectare for 4 consecutive years in Cuba, the percentage bored joints in a 3250
hectare area was reduced from 15.4 to 1.8 percent. This report was
confirmed somewhat by a report (though not from the Western Hemisphere) by Boedijono (1974) on *Diatraeophaga striatalis* Townsend.
This parasite is routinely released at the rate of 15 females per
hectare twice each season for control of *Proceras (=Chilo) sacchariphagus* Bojer and *Chilo auricilia* Dudgeon. By this method joint
infestation at harvest time is suppressed from 16.3 to 4.3 percent.

Releases of *L. diatraea* conducted during summer months in
Louisiana resulted in low parasitization (less than or equal to 10
percent) during 1973 and 1974 and failure of the flies to recycle
(McPherson and Hensley 1976, R. D. Jackson et al. unpubl. data).
Bonfils and Galichet (1974) reported increases in parasitization
by *L. diatraea* immediately after periodic releases in Guadeloupe,
French West Indies, but they found that the number of parasites in

the release area dropped to the pretreatment level in the following
generation.  McPherson and Hensley attributed low parasitization
in field cage tests in part, to fly mortality due to high temperatures during the test period.  Montes (1970) reviewed the literature on *L. diatraea* seasonal cycle and previous releases in Cuba.
He generally concluded that high parasitization rates by *L.
diatraea* could occur during the spring and fall but that parasitization was reduced during the summer due to higher temperatures.
R. D. Jackson (unpubl. data) conducted studies in Louisiana where
flies were caged on individual plants thereby elminating the
"hotcap effect" common in cage studies; fly survival approached
that for flies held in the laboratory.

R. D. Jackson and E. G. King (unpubl. data) recorded higher
rates of parasitization in Louisiana when *L. diatraea* was released
in May on first generation sugarcane borer larvae.  After releases
of 50 and 350 mated female flies/acre at 3 locations during 1976,
about 24 percent of the *D. saccharalis* larvae were parasitized
by *L. diatraea*.  Recycling occurred, but about the time insecticide
applications began for control of *D. saccharalis* in some of the fly
release areas and in adjacent fields, parasitization by *L. diatraea*
was reduced.

Insecticides are a major component in the Louisiana sugarcane
borer management program, and growers may use 3 applications of
azinphosmethyl per year (Hensley 1971).  Thus, the failure of
summer releases of *L. diatraea* may also be attributed, in part,
to the use of insecticides in adjacent sugarcane fields.  Insecticides are not applied for control of the sugarcane borer first
generation.  It is likely as stated by Bryan et al. (1973a) that
until research of this nature can be conducted in the absence of
insecticidal control for any pest in adjacent areas, such results
will be difficult to assess, and the assessments will be regarded
with reservations.

Summers et al. (1976) reported on releases of *L. diatraea* in
Florida at 3 locations for control of the sugarcane borer during
1973 and 1974.  In each test high rates of parasitization by
*L. diatraea* occurred after releases, and the flies recycled.  Parasitization rates of 20 percent or greater were therefore maintained.
In 1974 after release of 600 flies per acre, the parasitization
rate was high (78 percent), and the flies continued to be active
in the release and in surrounding areas for the remainder of the
season.  The increased rate of parasitization in the treated area
and in adjacent areas was accompanied by decreased numbers of
bored joints.  The number tended to increase with increasing distance from the release area.

Releases of *L. diatraea* were made in Florida during 1976 by
hand and airplane (J. W. Smith, T. E. Summers, E. G. King,
D. F. Martin, unpubl. data). Test plots were about 720 acres in
size and separated by 15 to 30 miles. Flies were released 4 weeks
earlier in the hand release area than in the aerial release area.
Once releases were initiated, about 80 flies per acre per week
were released in both areas, releases were terminated after the
week of August 23, 1976.

Before flies were released, the sugarcane borer larval population was highest in the aerial release plot and lowest in the
check area (Table 8). *L. diatraea* were detected in the test areas
prior to release. Parasitized sugarcane borer larvae were detected
within one week after releases were initiated, but large numbers
were not detected until sufficient time (ca. 30 days) had elapsed
for recycling to occur. Suppression of the sugarcane borer population by *L. diatraea* was evident in the hand release area after
about 9 weeks, and apparently a reduction occurred in the aerial
release area during the latter part of the season. A corresponding
decrease in the sugarcane borer population in the check area did
not occur. Less than one-half of the sugarcane borer larvae in
either release area were usually parasitized during any week,
though the number parasitized per week was greater in the aerial
release area. The number of parasitized sugarcane borer larvae
per acre would probably have been greater if egression from the
release area, as reported by Summers et al. (1976), could have
been controlled. Additionally, food consumption by sugarcane borer
larvae parasitized by *L. diatraea*, regardless of host larval stage
parasitized, is less than consumption by nonparasitized larvae, an
indication of benefit other than host population reduction (E. D.
Brewer, E. G. King, unpubl. data).

TABLE 8.--Results of releasing *Lixophaga diatraea* on sugarcane
for control of the sugarcane borer in Florida; 80 flies
released per acre on 720-acre test plots.

| Date | Percent parasitism[a] | | Control |
| | Aerial release | Hand release | |
|---|---|---|---|
| Jun. 14 | 0.0 | 0.0 | 0.0 |
| Jul. 12 | 0.0 | 0.7 | 0.0 |
| Aug. 09 | 0.7 | 38.0 | 0.0 |
| Sep. 13 | 50.6 | 40.0 | 0.7 |
| Oct. 11 | 48.7 | 52.8 | 0.9 |

a/ Parasite releases initiated on June 10 by air and on
July 5 by ground; all releases terminated on August 23.

Grain sorghum. Starks et al (1975) confirmed early conclusions by DeBach and Hagen (1964) when they released *Hippodamia
convergens* Guerin-Meneville in grain sorghum for greenbug,
*Schizaphis graminum* (Rondani) control. Cooke (1963) made 4 massive
releases (40,000 to 150,000 beetles/release) of *H. convergens*
against the pea aphid in alfalfa fields and obtained some reduction
in the aphid population. However, the beetles quickly left the
fields whether or not he had fed them diet (honey and protein
hydrolysates) prior to release.

Starks et al. (1976) reported on a 2-year study in which the
feasibility of using the parasite *Lysiphlebus testaceipes* (Cresson)
in periodic releases for control of the greenbug was evaluated on
grain sorghum. In 1972 the parasites were released at rates of
7,000, 14,000, and 28,000/acre. Release areas were compared with
a nonrelease area and with an area treated with insecticide.
Rates of parasitization were greater at the 2 higher rates of
release (14,000 and 28,000), and harvest results showed that these
areas and the insecticide plots produced significantly higher
yields when compared with the untreated control. A comparison of
control of greenbugs by insecticide, parasite release, and plant
resistance indicated that the latter was most desirable. However,
the greenbug has developed resistance to insecticides and if the
plant resistance breaks down, parasite releases may be the only
viable alternative.

## Forestry

Weseloh and Anderson (1975) reported consistently greater
percent parasitism (average = 6.3 to 26.0 percent) of gypsy moth
larvae, *Lymantria dispar* (L.), by *Apanteles melanoscelus* (Ratzeburg) in areas where the parasite was released than in areas where
it was not released. Recycling occurred where there was a high
host population, but in areas with a low host population, percentage parasitization returned to the rate in untreated areas
after 3 weeks. Results reported by Grimble (1975) and later by
Ticehurst and Fusco (1976) disagreed with those of Weseloh and
Anderson. Grimble reported no difference in parasitism of gypsy
moth larvae following releases of *Apanteles melanoscelus* and Ticehurst and Fusco reported that releases of *Apanteles melanoscelus,
Apanteles porthetriae* Muesebeck, and *Apanteles liparidis* Bouche
had no immediate effect on host parasitism, host reduction, or
foliage protection. However, Hoy (1975) was successful in
increasing parasitism with releases of a triple hybrid of *A.
melanoscelus*.

### MECHANIZED METHODS FOR RELEASE

Most augmentations of parasites and predators in the past have
been accomplished by manual releases.  In those situations where
agriculture is labor-intensive, manual releases may be practical.
However, under conditions of highly mechanized agriculture, more
rapid methods that utilize less labor are needed.  Some significant progress has been made in mechanizing the releases of *Chrysopa
carnea*, *Trichogramma* spp., and *Lixophaga*.

Following successful experiments with manual releases of
*Chrysopa* eggs in field cages and of *Chrysopa* larvae in field plots
(Ridgway and Jones 1968, 1969), a number of approaches to more
efficient methods of distribution were explored.  Several efforts
to distribute *Chrysopa* eggs mechanically on cotton were tested,
but none were satisfactory (Jones and Ridgway 1976).  Related
studies were conducted to mechanize the release of *Chrysopa* larvae.
A procedure was developed in which *Chrysopa* were reared by mixing
*Chrysopa* eggs and eggs of the Angoumois moth, *Sitotroga cerealella*
(Oliver), in kiln-dried sawdust.  By using carefully controlled
ratios of eggs and sawdust and proper temperatures, satisfactory
yields of 2- to 3-day *Chrysopa* larvae were produced (Kinzer 1976).
A back-pack unit was then designed for distributing the sawdust mixture on cotton plants (Reeves 1975).  The same principles were then
used in the development of a 4-row motorized unit mounted on a high
clearance sprayer chassis (Reeves 1975, Fig. 1).  Both types of
equipment were used successfully to distribute *Chrysopa* larvae on
cotton for control of the bollworm and tobacco budworm.  Preliminary
studies also indicated that the same sawdust mixture could be distributed by aircraft; however, wind conditions had considerable
effect on distribution patterns.

Mechanization developed for food packaging was used in the
packing of *Trichogramma* for manual release (Stinner et al. 1974).
Later, a paper triangular-shaped package developed for food packaging was used for packaging *Trichogramma* since it was better suited
to mechanical release.  A machine was designed and constructed to
mechanically release packaged *Trichogramma* from an airplane
(Reeves 1975, Fig. 2).

Packaging of insects for aerial release has the potential for
providing efficient releases.  However, bulk handling and release is
more desirable because of the cost of packages and labor and the
space requirements associated with packaging.  The transition from
packaging to bulk handling of aerial releases was first accomplished
with the adult screwworm flies, *Cochliomyia hominivorax* (Coquerel),
and later used to release pink bollworms (Higgins 1970).  Bulk handling has now been adapted to a number of other insects including
*Lixophaga diatraea*.  The machine used in such aerial releases consisted of a refrigerated box that houses a mechanism to store and

meter out the parasites at a predetermined rate (Fig. 3).  The
insects are stored inside the box on a stack of trays positioned
above a metering device.  The trays can be released independently,
and the insects are dropped into the metering unit.  A variable
speed endless belt in the metering unit carries the insects out of
the refrigerated box into a tube that carries them out of the plane.
The tube is positioned horizontal to the line of flight and rearward
at about 45 degrees.  The outboard end is positioned to prevent
released insects from contacting any part of the aircraft.

Preliminary studies with *L. diatraea* indicated that 6-day-old
flies held at 4.4°C in both the laboratory and the plane could be
free-dropped from a plane traveling 120 m.p.h.  Flies recovered
after chilling and dropping from a plane lived as long as untreated
flies, and there was no difference in egg production and parasitization rate.  (King unpubl. data).

Fig. 1.  Motorized equipment for field application of *Chrysopa
carnea* larvae and sawdust mixture on cotton.

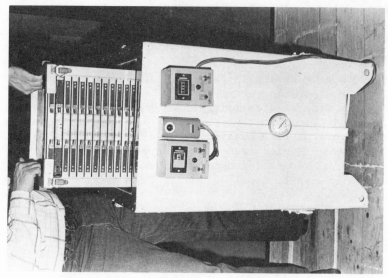

Fig. 3. Equipment used in aerial releases of *Lixophaga diatraea* (A. controls for tray counter; B. thermostat; C. remote control; D. thermometer; E. removable tray assembly).

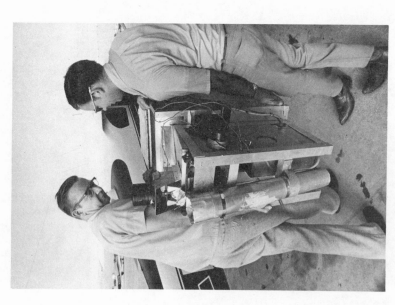

Fig. 2. Equipment for aerial release of packaged *Trichogramma* from an airplane.

The aerial release method was used during 1975 and 1976 in Florida over about 2,000 acres of sugarcane. There was no indication that hand-released flies performed better than aerial-released flies in 1976; 30 to 50 percent of the host larvae were parasitized in both plots.

Recently, progress has been made in the development of a bulk aerial release system for *Trichogramma*. Eggs of the Angoumois grain moth containing pharate adults of *Trichogramma pretiosum* attached to wheat bran flakes with a mixture of mucilage and water and mechanically broadcast on cotton produced satisfactory rates of parasitism of the eggs of *Heliothis* (Jones et al. In press). Preliminary efforts to adapt this method to aerial release also appear promising.

Although a bulk handling and release method has been adapted for use with several insects, different approaches will be required for the many different specialized parasites and predators that may be used in augmentation.


## INTEGRATION OF AUGMENTATIONS INTO MANAGEMENT SYSTEMS

The results of releases of natural enemies for control of plant pests in the Western Hemisphere have been summarized in a rather simplistic fashion in order to emphasize the potential impact that such releases can have on pest populations. However, the practical use of augmentations will, in most cases, require the careful selection of the natural enemy or enemies to be augmented and the development of a total pest management or crop management system that will complement their use.

The importance of integrating augmentations of natural enemies into a total management system and the illustration of how this might be done is particularly well documented by Hussey and Scopes (Chapter 12). Obviously, glasshouses offer the specific advantages of a confined ecosystem where the variables are more easily controlled, but increased information about unconfined ecosystems should make possible the practical use of augmentation under more varied conditions.

Systems for the practical use of augmentation against field crop pests in the U.S.S.R., particularly against pests in cooler climates where only about one generation is involved, have apparently been developed and placed into large-scale operation (Beglyarov and Smetnik, Chapter 9). Extensive practical uses are also reported from the People's Republic of China (Huffaker, Chapter 10). However, the systems developed in these countries may not be easy to adapt to conditions in other countries.

## Model Systems for U.S. Agriculture

A brief review of some existing or potential management systems with different characteristics should be useful in viewing the potential of augmentations in the U.S.  The following will be considered:  (1) collective management of a perennial crop, (2) intensive management of several annual crops in small units, (3) extensive management of an annual crop in large units, (4) extensive management of 2 or more annual crops in large units.

Collective Management of a Perennial Crop.  An effective pest management system including augmentation is in operation on about 9,000 acres of citrus in Ventura County, California.  The system has evolved over a period of about 50 years.  The principal pests are California red scale and black scale; the minor pests include mealybugs, citrus thrips, and aphids.  The control program consists of the use of established exotic parasites and supervised control that includes augmentation of populations of *Aphytis melinus*, a parasite of the California red scale; *Metaphycus helvolus* (Compere), a parasite of black scale, *Saissetia oleae* Bernard; and *Cryptolaemus montrouzieri*, a predator of mealybugs.  Spray oils are used as needed on nearly one-half of the acreage, and selective insecticides are used on 5 to 15% of the acreage for control of outbreaks of scales or thrips.  The principal natural enemy used for augmentations is *A. melinus*; about 16,000 per acre were released in 1975.  The pest management program is operated by a cooperative with an annual operating budget of about $90,000.  About 99% of the over 350 growers in the area participate voluntarily in the cooperative (Pennington 1975 and pers. comm.).

Much of the technical guidance that resulted in the development of this successful program came from scientists at the University of California at Riverside and has been summarized by DeBach (1964, 1974).  Although experimental evidence is not available to quantify precisely the impact of each management tactic on the pest populations, experience has shown this system to be highly desirable, and it can serve as a model for developing other systems.  In fact, the system has several characteristics that favor maximum use of natural enemies.  First, citrus is a perennial crop that produces less drastic changes in the ecosystem than do annual crops.  Also, there is a year-round outlet for reared natural enemies, and the seasonal labor requirements for both production and pest management are relatively stable.  In addition, the high level of participation in the cooperative by producers in a somewhat isolated area greatly reduces the possibility of such interferences as drift that might result from unnecessary use of insecticides on adjoining crops (Hart 1972, Stinner et al. 1974)

Intensive Management of Several Annual Crops in Small Units.
Although the production of a high value perennial crop on sizable

acreage in an isolated area such as citrus in Ventura County provides excellent conditions for maximum utilization of natural
enemies, other production systems can be identified wherein
conditions suitable for augmentation may exist.

For example, there are some 35,000,000 home gardeners in
the U.S. (Detwiler et al. 1976) and perhaps 500,000 small diversified farms (5 to 100 acres) (Goldstein 1977) where a wide range
of crops are grown. These production units can take advantage of
the benefits of diversity in establishing a more stable ecosystem.
In addition, they are often characterized by being labor intensive
in contrast to the machinery intensive operations of larger farms.
Thus, the special (personal) attention often required if augmentation of natural enemies is to be used effectively may be uniquely
available in these small units. To date, research and development
efforts directed toward augmentations for pest control in these
units has been very limited, but perhaps adequate information is
available to explore the potential.

Some of the annual crops commonly grown in small units
include tomatoes, beans, peppers, lettuce, and cabbage. Through
the proper selection of cool and warm season crops, some plants
can be maintained throughout most of the year so as to provide a
reservoir for beneficial insects (Olkowski and Olkowski 1976).
By selecting resistant varieties, when available, the threat of
pests can be further reduced (Yepson 1976). A number of cultural
and mechanical controls also can be helpful (Lewis and Turney 1977).
In addition, microbial agents such as the delta-endotoxin produced
by *Bacillus thuringiensis* Berliner (Falcon 1971) can be used for
controlling many lepidopteran larvae without harming naturally
occurring beneficial insects. When additional natural enemies are
needed, they may be purchased from a number of commercial firms
(see Appendix). The natural enemies primarily available to the
gardener and small farmer in the U.S. are *Trichogramma* egg parasites for control of lepidopteran insects; green lacewings, *Chrysopa*
spp., for control of most soft-bodied insects such as aphids and
lepidopteran eggs and larvae; and predacous mites for control of
phytophageous mites. A general bulletin explaining the use of
these natural enemies has been prepared (Rincon-Vitova Insectaries,
Inc. undated).

Although the home gardeners and small diversified farms
currently provide only a very small percentage of U.S. agricultural
production, the number of these units, particularly home gardens,
is increasing. The future impact of this type of agriculture is
uncertain; however, because of the keen interest of the operators
of these small units in biological controls they are potentially
important users of augmentation of natural enemies.

Extensive Management of an Annual Crop in Large Units.
Perennial crops and intensively managed annual crops provide some
unique opportunities for use of augmentation of natural enemies,
but the vast majority of U.S. agricultural production occurs in
large units that depend on mechanization and management of capital
to produce adequate food and fiber with a small work force.  If
augmentations of natural enemies is to have major impact on pest
control practices in such systems, they will have to be adaptable
to large-scale mechanized agriculture

A total management system for control of insect pests on
cotton that would include augmentation offers considerable potential
benefit because some 47% of the insecticides used in agriculture
in the U.S. is applied to cotton (Andrelinas 1974).  Also, a great
deal of technology is becoming available that can be used in such
a practical pest management system.

The principal insect pests in most U.S. cotton growing regions
are the boll weevil, the bollworm, the tobacco budworm, plant bugs,
*Lygus* spp., and the cotton fleahopper, *Pseudatomascelis seriatus*
(Reuter).  As noted previously, natural enemies play an extremely
important role in regulating populations of the bollworm and tobacco
budworm on cotton, but the application of insecticides for control
of other pests such as plant bugs or boll weevils often destroys
these parasites and predators.  However, a number of new technolo-
gies are available or are on the horizon that may make it possible
to utilize these natural enemies more effectively.  By careful
selection of plant varieties that will produce an acceptable yield
in a shorter season and/or by the use of a resistant variety, problems associated with most major insect pests can be reduced (Lacewell et al. 1976, Meredith 1976, Anderson et al. 1976).  In addition, when the cotton fleahopper or other plant bugs are present,
foliar applications of selective insecticides should only be used
if the cotton plants are not fruiting properly.  Then insecticides
should be selected and applied that will have minimum effect on the
naturally occurring beneficial insects (Lingren et al. 1968b).

In the past, control of boll weevils has been accomplished
primarily with in-season use of broad-spectrum insecticides.  However, the recently developed insect growth regulator, diflurbenzuron
(Dimlin), which inhibits synthesis of chitin in many insects,
though it has not yet been approved for use on cotton, seems to
have great potential controlling boll weevils since it has only
limited effects on beneficial insects (Taft and Hopkins 1975,
Lloyd et al. 1977, Ables et al. In press).

The management program for optimum use of natural enemies on
cotton then consists of: (1) the production of an early crop
through selection of varieties and other management practices,

(2) the use of resistant or tolerant varieties where available, and (3) a method of controlling injurious plant bugs and boll weevils that will have the minimum effects on natural enemies. The next critical step involves precise decision making concerning the control measures that should be used for suppression of *Heliothis*. Models can be used to predict peak periods of oviposition of *Heliothis* eggs, and field inspections can be made more efficient (Hartstack et al. 1977). Also predator-prey ratios can be used to assess the expected impact of naturally occurring parasites and predators (Hartstack et al. 1975), and plant models (Jones et al. 1975) can be used to estimate potential effects on yield as a basis for determining whether augmentations are needed. With such tools, decisions can be made about the number of *Trichogramma*, *Chrysopa*, or other natural enemies that may be needed to increase populations of natural enemies to the level that will assure adequate control of *Heliothis*. However, some organized area-wide programs probably will be required if this system is to work effectively because of the adverse effects insecticide drift can have on naturally occurring and released parasites and predators. Also, current methods of producing the numbers of natural enemies necessary for effective control of *Heliothis* on cotton are not cost effective.

Extensive Management of Two or More Annual Crops in Large Units. Although large-scale agriculture often emphasizes a single crop (monoculture), there are production systems that involve two or more annual crops such as cotton and grain sorghum; cotton, soybeans, and/or corn; cotton and alfalfa; corn and small grains, etc. In such cases, the use of highly selective parasites over large areas for control of such pests as the *Heliothis* complex is likely to be the most efficient system of augmentation. Moreover, augmentation can probably be used more easily to suppress populations over a large area early in the season, so it may be compatible with systems that utilize some insecticides later in the year. Since the *Heliothis* complex is probably the most destructive group of insect pests in the U.S. and feeds on cotton, corn, and soybeans, and a wide range of other field crops and vegetables, perhaps release of selective parasites over a large area could provide a particularly economical approach to augmentation in the U.S. (Knipling, Chapter 3).

## FUTURE PROSPECTS

The future prospects for augmentations of natural enemies, particularly in the U.S. have been reviewed recently (Rabb et al. 1976, Stinner 1977) as have the research support needs (Stinner 1977). Factors influencing expanded uses of any pest control strategy can be grouped into 3 categories: (1) biology,

(2) economics, and (3) sociology. However, when one is considering the future of augmentations, one should perhaps make the
focal point the establishment of the minimum numbers of natural
enemies required to obtain acceptable levels of pest control.
(Of course, the factors associated with the determination of this
minimum include biological, economical, and social considerations.)

Most of the factors that will be needed in determining the
numbers of natural enemies needed for augmentations are discussed
throughout the preceding chapters, but they are highlighted in
certain ones as follows: selection of species and design of integrated pest management systems (Huffaker et al., Chapter 1),
nutrition including supplemental food resources (House, Chapter 5),
mass production and storage (Morrison and King, Chapter 6), rearing
quality insects (Boller and Chambers, Chapter 7), and retention and
stimulation of released natural enemies (Vinson, Chapter 8).

The numerical relationships that have been presented, particularly in Tables 1 through 8, tend to reflect a need to concentrate
on the development of more efficient systems for rearing and
utilizing natural enemies. The theoretical presentation by Knipling
(Chapter 3) and other evidence (particularly that by Beglyarov and
Smetnik, Chapter 9) would indicate that there is need for additional
research and development to improve the efficiency of mass-reared
natural enemies. At the same time, those systems, for example, the
release of *Pediobius foveolatus* for control of Mexican bean beetle
(Stevens et al. 1975), that depend primarily on inoculative
releases may already be cost-effective.

Certainly, the resources available for research to improve
technology for mass rearing, storage, and distribution of quality
insects and for the development of improved management systems
are key elements in the rate of expansion of the use of augmentations. Likewise, there must be willingness to pay the increased
costs that may initially be associated with biological controls
and to encourage the development of management styles and institutional arrangements favoring biological controls (Starler and
Ridgway, Chapter 15).

## DISCLAIMER

Mention of a commercial (or proprietary) product in this paper
does not constitute an endorsement of this product by the USDA.

## REFERENCES CITED

Ables, J. R., S. L. Jones, and D. L. Bull. 1977. Effect of
diflubenzuron on beneficial arthropods associated with
cotton. Southwest Entomol. In press.

Anderson, J. M., R. R. Bridge, A. M. Heagler, and G. R. Tupper.
1976. The economic impact of recently developed early-
season cotton strains on firm and regional cropping systems
and income. p. 90-100 In Proc. Beltwide Cotton Prod. Res.
Conf.

Andrilenas, P. A. 1974. Farmers Use of Pesticides in 1971--
Quantities. U.S. Dept. of Agr. Agr. Econ. Rep. No. 252.

Armitage, H. 1919. Controlling mealybugs by the use of their
natural enemies. Calif. Stat. Hort. Comm. Monthly Bulls.
8: 257-60.

Armitage, H. M. 1929. Timing field liberations of *Cryptolaemus*
in the control of the citrophilus mealybug in the infested
citrus orchards of southern California. J. Econ. Entomol.
22: 910-5.

Ashley, T. R., J. C. Allen, and D. Gonzalez. 1974. Successful
parasitization of *Heliothis zea* and *Trichoplusia ni* eggs
by *Trichogramma*. Environ. Entomol. 3: 319-322.

Bennett, F. D. 1969. Tachinid flies as biological control agents
for sugarcane moth borers, pp. 117-18. In Pests of Sugar
Cane, J. R. Williams, J. R. Metcalfe, R. W. Mungomery,
and R. Mathes (eds.). New York: Elsevier Publ. Co. 568 pp.

Boedijono, W. A. 1974. An attempt to control sugarcane stemborers
with the dipterous parasite *Diatraeophaga striatalis* (Towns.).
Int. Soc. Sug. Technol.

Bonfils, J. and P. F. Galichet. 1974. Effects d'introductions
renouvelies de *Lixophaga diatraeae* [Diptera, Tachinidae] sur
des populations naturelles du parasite et de son hate,
*Diatraeae saccharalis*. Entomophaga 19(1): 67-73.

Boyd, J. P. 1970. Feeding and searching behavior of *Chrysopa
carnea* Stephens. Ph.D. Dissertation. Texas A & M University.
108 pp.

Bryan, D. E., C. G. Jackson, R. Patana, and E. G. Neeman. 1971.
Field cage on laboratory studies with *Bracon kirkpatricki,*
a parasite of the pink bollworm. J. Econ. Entomol.
64(5): 1236-41.

Bryan, D. E., R. E. Fye, C. G. Jackson, and R. Patana. 1973a.
Releases of parasites for suppression of pink bollworms in
Arizona. U.S. Dept. Prod. Agr. Res. Rpt. 8 pp.

Bryan, D. E., R. E. Fye, C. G. Jackson, and R. Patana. 1973b.
Releases of *Bracon kirkpatricki* (Wilkinson) and *Chelonus
blackburni* Cameron for pink bollworm control in Arizona.
U.S. Dept. Agr. Prod. Res. Rpt. 24 pp.

Cooke, W. C.  1963.  Ecology of the pea aphid in the Blue Mountain
    area of eastern Washington and Oregon.  Tech. Bull. U.S.
    Dept. Agric. 1287, 48 pp.

Croft, B. A.  1976.  Establishing insecticide-resistant phytoseid
    mite predators in deciduous tree fruit orchards.
    Entomophaga 21: 383-399.

Croft, B. A. and S. A. Hoying.  1975.  Carbaryl resistance in
    native and released populations of *Amblyseius fallacis*.
    Environ. Entomol. 4: 895-8.

Croft, B. A. and J. A. McMurtry.  1972.  Minimum releases of
    *Typhlodromus occidentalis* to control *Tetranychus mcdanieli*
    on apple.  J. Econ. Entomol. 65: 188-91.

DeBach, P.  1974.  Biological control by natural enemies.  Cambridge
    Univ. Press.  323 pp.

DeBach, P., and K. S. Hagen.  1964.  Manipulation of entomophagous
    species, *In*:  P. DeBach, (ed.) Biological Control of Insect
    Pests and Weeds.  pp. 429-58.  London: Chapman and Hall
    Ltd. 844 p.

DeBach, P., E. J. Dietrick, C. A. Fleschner, and T. W. Fisher.
    1950.  Periodic colonization of *Aphytis* for control of the
    California red scale.  Preliminary tests, 1949.  J. Econ.
    Entomol. 43: 783-802.

Detwiler, D., T. Thompson, and J. Loomis.  1976.  News about
    community gardens.  Gardens for All, Inc., Shelburne, VT.
    7 pp.

Doutt, R. L.  1951.  Biological control of mealybugs infesting
    commercial greenhouse gardenias.  J. Econ. Entomol. 44: 37-40.

Doutt, R. L., and K. S. Hagen.  1949.  Periodic colonizations of
    *Chrysopa california* as possible control of mealybugs.  J.
    Econ. Entomol. 42: 560.

Doutt, R. L., and K. S. Hagen.  1950.  Biological control
    measures applied against *Pseudococcus maritimus* on pears.
    J. Econ. Entomol. 43: 94-6.

Dutky, S. R.  1974.  Nematode parasites, pp. 576-90.  In Pro-
    ceedings of the Summer Institute on Biological Control of
    Plant Insects and Diseases, ed. F. G. Maxwell and F. A.
    Harris.  Univ. Press, Jackson, MS.  647 p.

Elsey, K. D.  1971.  Stilt bugs predation on artificial infestation
    of tobacco hornworm eggs.  J. Econ. Entomol.  64(3): 772-773.

Elsey, K. D.  1975.  *Jalysus spinosus*:  Increased numbers produced
    on tobacco by early-season releases.  Tobacco Sci. 19: 13-15.

Falcon, L. A.  1971.  Use of bacteria for control of insects.  p.
    67-95.  In H. D. Burges and N. W. Hussey, eds. Microbial
    Control of Insects.  Academic Press, N.Y.

Fye, R. E.  1972.  Manipulation of *Polistes exclamans arizonensis*.
    Environ. Entomol. 1(1): 55-7.

Goldstein, J.  1977.  Up the organic ladder with a spade and hoe.
    Organic Gardening and Farming.  July.  pp. 30-6.

Gonzalez, R. H. 1976. Crop protection in Latin America, with special reference to integrated pest control. FAO Plant Protection Bull. 24(3): 65-75.

Grimble, D. 1975. Releases of *Apanteles melanoscelus* for gypsy moth suppression. Appl. For. Res. Note 15: 1-6.

Halfhill, J. E. and P. E. Featherston. 1973. Inundative releases of *Aphidius smithi* against *Acyrthosiphon pisum*. Environ. Entomol. 2: 469-72.

Harbaugh, B. K. and R. H. Mattson. 1973. Lacewing larvae control aphids on greenhouse snapdragons. J. Amer. Soc. Hort. Sc. 98: 306-9.

Hart, W. G. 1972. Compensatory releases of *Microterys flavus* as a biological control agent against brown soft scale. Environ. Entomol. 1: 414-419.

Hartstack, A. W., J. A. Witz, and R. L. Ridgway. 1975. Suggested applications of a dynamic *Heliothis* model (MOTHZV-1) in pest management decision making. Proc. of Beltwide Cotton Prod. Res. Conf. 118-22 pp.

Hartstack, A. W., J. L. Jenson, I. A. Witz, J. A. Jackman, J. P. Hollingsworth, and R. E. Frisbie. 1977. The Texas program for forecasting *Heliothis* spp. infestations on cotton. In Proc. Beltwide Cotton Prod. Res. Conf. In press.

Helgesen, R. G., and M. J. Tauber. 1974. Biological control of greenhouse whitefly, *Trialeurodes vaporariorum* (Aleyrodidae: Homoptera), on short-term crops by manipulating biotic and abiotic factors. Can. Ent. 106: 1175-1188.

Hensley, S. D. 1971. Management of sugarcane borer populations in Louisiana, a decade of change. Entomophaga 16:133-46.

Higgins, A. H. 1970. A machine for free aerial release of sterile pink bollworm moths. USDA. ARS 81-40.

Hoy, M. A. 1975. Forest and laboratory equations of hybridized *Apanteles melanoscelus* (Hym.: Braconidae), a parasitoid of *Porthetria dispar* (Lep.: Lymantriidae). Entomophaga 20: 261-8.

Huffaker, C. B. 1971. Biological Control. Plenum Press, N.Y. and London. 511 pp.

Huffaker, C. B. and C. E. Kennett. 1956. Experimental studies on predation: Predation and cyclamen-mite populations on strawberries in California. Hilgardia 26: 191-222.

Huffaker, C. B. and P. S. Messenger, eds. 1976. Theory and Practice of Biological Control. Academic Press. N.Y. 788 pp.

Jackson, C. G., D. E. Bryan, E. G. Neeman, and A. L. Wardecker. 1970. Biological control: Results of field cage tests with parasites of *Heliothis* spp. Third Quarterly Report. Cotton Insects Biological Control Investigations. Tucson, Arizona.

Jimenez, J., Eleazar. 1975. The action program for biological control in Mexico. Presented at the Work Conf. of the U.S. and Mexico on "Biological control of pests and weeds," Brownsville, TX. February 20-21. 3 pp.

Jones, J. W., J. D. Hesketh, R. F. Colwick, H. C. Lane, J. M. McKinion, and A. C. Thompson. 1975. Predicting square, flower, and boll production of cotton at different stages of organogenesis. In Proc. Beltwide Cotton Prod. Conf.

Jones, S. L. and R. L. Ridgway. 1976. Development of methods for field distribution of eggs of the insect predator *Chrysopa carnea* Stephens. USDA, ARS-S-124. 5 pp.

Jones, S. L., R. K. Morrison, and J. R. Ables. A new and improved technique for the field release of *Trichogramma pretiosum*. Southwestern Entomol. In press.

Kinzer, R. E. 1976. Development and evaluation of techniques for using *Chrysopa carnea* Stephens to control *Heliothis* spp. in cotton. Ph.D. Dissertation. Texas A&M University. 60 pp.

Knipling, E. F. 1972. Simulated population models to appraise the potential for suppressing sugarcane borer populations by strategic releases of the parasite *Lixophaga diatraeae*. Environ. Entomol. 1: 1-6.

Lacewell, R. D., J. M. Sprott, G. A. Niles, J. K. Walker, and J. R. Gannaway. 1976. Cotton growing in an integrated production system. Trans. Am. Soc. Agric. Engr. 19: 815-18.

Lawson, F. R., R. L. Rabb, F. E. Guthrie, and L. G. Bowery. 1961. Studies of an integrated control system for hornworms on tobacco. J. Econ. Entomol. 540: 93-7.

Lewis, K. R., and H. A. Turney. 1977. Insect controls for organic gardeners. Texas Agric. Ext. Sta. MP-1284. 10 pp.

Lewis, W. J., A. N. Sparks, R. L. Jones, and D. J. Barras. 1972. Efficiency of *Cardiochiles nigriceps* as a parasite of *Heliothis virescens* on cotton. Environ. Entomol. 1: 468-71.

Lingren, P. D. 1969. *Apanteles marginiventris* as a parasite of the cabbage looper and *Heliothis* spp. attacking cotton. Meeting Entomol. Soc. Amer., Chicago, IL.

Lingren, P. D. 1977. *Campoletis sonorensis*: Maintenance of a population on tobacco budworms in a field cage. Environ. Entomol. 6: 72-76.

Lingren, P. D. and J. G. Kim. 1970. Inundative releases of *Trichogramma* sp. for control of bollworm and tobacco budworm attacking cotton. Presented at ESA mtg., Miami, Florida. November 30-December 3.

Lingren, P. D. and L. W. Noble. 1972. Preference of *Campoletis perdistinctus* for certain noctuid larvae. J. Econ. Entomol. 65: 104-7.

Lingren, P. D., R. L. Ridgway, and S. L. Jones. 1968a. Consumption by several common arthropod predators of eggs and larvae of two *Heliothis* species that attack cotton. Ann. Entomol. Soc. Amer. 61: 613-8.

Lingren, P. D., R. L. Ridgway, C. B. Cowan, J. W. Davis, and W. C. Watkins. 1968b. Biological control of the bollworm and tobacco budworm by arthropod predators affected by insecticides. J. Econ. Entomol. 61: 1521-25.

Lloyd, E. P., R. H. Wood, and E. B. Mitchell. 1977. Boll weevil
    suppression with TH-6040 applied in cottonseed oil as a
    foliar spray. J. Econ. Entomol. In Press.
Lopez, J. D., R. L. Ridgway, and R. E. Pinnell. 1976. Comparative
    efficacy of four insect predators of the bollworm and tobacco
    budworm. Environ. Entomol. 5: 1160-1164.
Lorbeer, H. 1971. Integrated biological control in Fillmore
    citrus groves. Calif. Citrograph. April.
McClanahan, R. J. 1970. Integrated control of the greenhouse
    whitefly on cucumbers. J. Econ. Entomol. 63: 599-601.
McClanahan, R. J. 1972. Integrated control of the greenhouse
    whitefly. Can. Dep. Agric. Publ. 1469. 7 pp.
McLeod, J. H. 1938. The control of the greenhouse whitefly in
    Canada by the parasite *Encarsia formosa* Gahan. Scient.
    Agric. 18: 529-35.
McLeod, J. H. 1939. Biological control of greenhouse pests.
    Rept. Entomol. Soc. Ont. 70: 62-68.
McMurtry, J. A., H. G. Johnson and G. T. Scriven. 1969.
    Experiments to determine effects of mass release of
    *Stethorus picipes* on the level of infestation of the avocado
    brown mite. J. Econ. Entomol. 6: 1216-1221.
McPherson, R. M., and S. D. Hensley. 1976. Potential of *Lixophaga
    diatraeae* for control of *Diatraea saccharalis* in Louisianna.
    J. Econ. Entomol. 69: 215-8.
Martin, P. B., P. D. Lingren, G. L. Greene, and R. L. Ridgway.
    1976. Parasitization of two species of Plusiinae and *Heliothis*
    spp. after releases of *Trichogramma pretiosum* in seven crops.
    Environ. Entomol. 5: 991-5.
Meredith, W. R. 1976. Nectariless cottons. p. 34-37 In Proc.
    Beltwide Cotton Prod.-Mech. Conf.
Meyer, R. H. 1975. Release of carbaryl resistant predatory mites
    in apple orchards. Environ. Entomol. 4: 49-51.
Montes, M. 1970. Estudios sobre bionomia V la biometria de
    *Lixophaga diatraeae* (Towns.). (Diptera, Tachinidae).
    Ciencias Biologicas 4(7), 60 pp.
Nickle, W. R. 1974. Nematode infections. In: Insect Diseases,
    Vol. II, pp. 327-376, ed. G. E. Cantwell. Marcel Delker,
    Inc., New York.
Noble, L. W., and H. M. Graham. 1966. Behavior of *Campoletis
    perdistinctus* (Vierek) as a parasite of the tobacco budworm.
    J. Econ. Entomol. 59: 1118-20.
Oatman, E. R., and G. R. Platner. 1971. Biological control of
    the tomato fruitworm, cabbage looper, and hornworms on pro-
    cessing tomatoes in southern California, using mass releases
    of *Trichogramma pretiosum*. J. Econ. Entomol. 64: 501-6.
Oatman, E. R., J. A. McMurtry, and V. Voth. 1968. Suppression
    of the two-spotted spider mite on strawberry with mass
    releases of *Phytoseiules persimilis*. J. Econ. Entomol.
    61: 1517-1521.

Olkowski, H. and W. Olkowski. 1976. How to control garden pests without killing almost everything else. Horticulture Magazine. June. 5 pp.

Pacheco, M., Francisco. 1971. Evaluation of the parasitism of the wasp *Trichogramma* spp. on eggs of the bollworm in Sonora - 1971. Prog. Rpt. No. 3. 14 pp.

Pacheco M., Francisco, J. L. Carrillo S., J. Monge C., and R. Covarrubias G. 1971. Adelantos sobre la evaluaction del parasitismo de la avispita *Trichogramma* spp. sobre huevecillos de gusano bellotero en sonora durante 1969. Agricultura Technia en Mexico, Organo del INIA, SAG, Vol. III, No. 2. pp. 53-57.

Parker, F. D. 1971. Manipulation of pest populations by manipulating densities of both hosts and parasites through periodic releases. pp. 365-76. In Biological Control. C. B. Huffaker, ed. Plenum Press. New York. 511 p.

Parker, F. D., and R. E. Pinnell. 1972a. Effectiveness of *Trichogramma* spp. in parasitizing eggs of *Pieris rapae* and *Trichoplusia ni*. 1. Field studies. Environ. Entomol. 1: 785-789.

Parker, F. D., and R. E. Pinnell. 1972b. Further studies of the biological control of *Pieris rapae* using supplemental host and parasite releases. Environ. Entomol. 1: 150-157.

Parker, F. D., F. R. Lawson, and R. E. Pinnell. 1971. Suppression of *Pieris rapae* using a new control system: mass releases of both the pest and its parasite. J. Econ. Entomol. 64: 721-735.

Pennington, N. 1975. Managers report, Fillmore Citrus Protective District. Jan. 1 to Dec. 31. 3 pp.

Rabb, R. L., R. E. Stinner, and R. van den Bosch. 1976. Conservation and augmentation of natural enemies. p. 233-54. In C. B. Huffaker and P. S. Messenger, eds. Theory and Practice of Biological Control. Academic Press, N.Y.

Reeves, B. G. 1975. Design and evaluation of facilities and equipment for mass production of facilities and equipment for mass production and field release of an insect parasite and an insect predator. Ph.D. Dissertation. Texas A&M University. 180 pp.

Ridgway, R. L. 1969. Control of the bollworm and tobacco budworm through conservation and augmentation of predaceous insects. Proc. Tall Timbers Conf. on ecol. animal control by habitat mgmt. 1: 127-144.

Ridgway, R. L. 1972. Use of parasites, predators, and microbial agents in management of insect pests of crops. Proc. National Extension Insect-Pest Management Workshop, Purdue U. pp. 51-62.

Ridgway, R. L. and S. L. Jones. 1968. Field-cage releases of *Chrysopa carnea* for suppression of populations of the bollworm and the tobacco budworm on cotton. J. Econ. Entomol. 61: 892-898.

Ridgway, R. L. and S. L. Jones. 1969. Inundative releases of
*Chrysopa carnea* for control of *Heliothis* on cotton. J.
Econ. Entomol. 62: 177-180.

Ridgway, R. L. and R. E. Kinzer. 1974. Chrysopids as predators
of crop pests. Entomophaga, Mem. H. S., 7. pp. 45-51.

Ridgway, R. L. and P. D. Lingren. 1972. Predaceous and parasitic
arthropods as regulators of *Heliothis* populations. Southern
Coop. Series Bull. No. 169. pp. 48-56.

Ridgway, R. L., R. K. Morrison, R. E. Kinzer, R. E. Stinner, and
B. G. Reeves. 1973. Programmed releases of parasites and
predators for control of *Heliothis* spp. on cotton. Proc.
1973 Beltwide Cotton Prod. Res. Conf. pp. 92-94.

Ridgway, R. L., R. E. Kinzer, and R. K. Morrison. 1974. Production
and supplemental releases of parasites and predators for con-
trol of insect and spider mite pests of crops, pp. 110-6. In
F. G. Maxwell and F. A. Harris (eds.). Proceedings of the
Summer Institute on Biological Control of Plant Insects and
Diseases. Univ. Press. Mississippi, Jackson. 647 pp.

Rincon-Vitova Insectaries, Inc. Undated. Rincon-Vitova
Insectaries offers integrated pest management programming.
Oakview, CA., USA. 8 pp.

Scaramuzza, L. C. 1951. El control biologico y sus resultados
en el lucha contra el barrenador a perforador de la cana
*Diatraea saccharalis* (Fabr.) en Cuba por medio de la mosca,
*Lixophaga diatraeae* (Towns.) Asambl. latino amer.
Fitoparasitologia, 1: 282-92.

Shands, W. A., G. W. Simpson, and C. C. Gordon. 1972. Insect
predators for controlling aphids on potatoes. 5. Numbers
of eggs and schedules for introducing them in large field
cages. J. Econ. Entomol. 65: 810-17.

SooHoo, C. F., R. S. Seay, and P. V. Vail. 1974. *Voria ruralis*:
Field cage evaluation of four densities of the larval
parasite against *Trichoplusia ni*. Environ. Entomol. 3: 439-40.

Smith, H. S. and H. M. Armitage. 1931. The biological control
of mealybugs attacking citrus. Univ. Calif. Bull. 509, 74 pp.

SooHoo, C. F., R. S. Seay, and P. V. Vail. 1974. *Voria ruralis*:
Field cage evaluation of four densities of the larval parasite
against *Trichoplusia ni*. Environ. Entomol. 3: 439-40.

Starks, K. J., E. A. Wood, Jr., R. L. Burton, and H. W. Somsen.
Release of parasitoids to control greenbugs on sorghum.
USDA-ARS-S-91.

Stevens, L. M., A. L. Steinhauer, and J. R. Coulson. 1975.
Suppression of Mexican bean beetle on soybeans with annual
inoculative releases of *Pediobius foveolatus*. Environ.
Entomol. 4: 947-52.

Stinner, R. E. 1977. Efficacy of inundative releases. Ann.
Rev. Entomol. 22: 515-31.

Stinner, R. E., R. L. Ridgway, J. R. Coppedge, R. K. Morrison, and W. A. Dickerson, Jr. 1974. Parasitism of *Heliothis* eggs after field releases of *Trichogramma pretiosum* in cotton. Environ. Entomol. 3: 497-500.

Summers, T. E., E. G. King, D. F. Martin, and R. D. Jackson 1976. Biological control of *Diatraea saccharalis* in Florida by periodic releases of *Lixophaga diatraeae*. Entomophaga 4: 359-66.

Taft, H. M. and A. R. Hopkins. 1975. Boll weevils: field populations controlled by sterilizing emerging overwintering females with TH-6040 sprayable bait. J. Econ. Entomol. 68:183-5.

Teran, F. O. 1976. Perspectivas do controle biologico da broca da cana-de-Acucar. Boletim Tecnico Copersucar 2: 5-9.

Ticehurst, M. and R. A. Fusco. 1976. Release of three *Apanteles* spp. against the gypsy moth in Pennsylvania. J. Econ. Entomol. 69: 307-8.

van den Bosch, R., T. E. Leigh, D. Gonzalez, and R. E. Stinner. 1969. Cage studies on predators of the bollworm in cotton. J. Econ. Entomol. 62: 1486-9.

Vaughn, Mario R. 1975. El parasito tricogramma: revision monografica. Banco Nacional de Nicaragua, Departmento Tecnico. 23 pp.

Weseloh, R. M. and J. F. Anderson. 1975. Inundative release of *Apanteles melanoscelus* against the gypsy moth. Environ. Entomol. 4: 33-6.

Whitcomb, W. H. and K. Bell. 1964. Predaceous insects, spiders and mites of Arkansas cotton fields. Arkansas Exp. Sta. Bull. 698.

Yepsen, R. B. 1976. Organic plant protection. Rodale Press, Inc., Emmaus, PA. 688 pp.

CHAPTER 14

AUGMENTATION OF NATURAL ENEMIES FOR CONTROL OF INSECT PESTS OF MAN

AND ANIMALS IN THE UNITED STATES

D. E. Weidhaas and P. B. Morgan

Insects Affecting Man Research Laboratory, Agricultural
Research Service, U.S. Department of Agriculture
Gainesville, Florida 32604 U.S.A.

Although biological control in its broadest sense encompasses
the use of methods or organisms to reduce the density of insects,
we are here concerned only with the biological control produced by
the use of predators and parasites. Also, the present paper does
not presume to review biological control of all the many pest
species of insects and arthropods that are of concern in the field
of medical and veterinary entomology. Our purpose is simply to
note general review articles that will lead the reader to the general literature on biological control of insects and arthropods
affecting man and animals and then to review in more detail the
background concerning the parasites and predators that are being
considered for use in augmentation systems.

A large variety of predators is associated with natural regulation of populations of insects and arthropods such as mosquitoes,
various biting and non-biting flies and gnats, ticks, chiggers,
fleas, bed bugs, and cockroaches. Jenkins (1964) published an
annotated list and bibliography of pathogens, parasites, and predators of medically important insects. Chapman (1974) published a
review of biological control of mosquito larvae in which he listed
188 references relating to the subject. Chapman et al. (1971, 1972)
and Laird (1970, 1971) reviewed the status of biological control of
mosquitoes and medical insects. Gerberich and Laird (1968) and Bay
(1967) reviewed fish as mosquito predators. Also, recently, Bay
et al. (1976) gave an up-to-date summary of the status of biological
control.

Generally, the species of insects and arthropods affecting man
and animals are characterized by high biotic potentials, i.e., a
single female is capable of laying hundreds of thousands of eggs.

417

There are exceptions such as the tsetse fly.  However, in natural populations, these high biotic potentials simply insure the survival of the species because populations vary in density over time and growth rates are normally far below the biogic potential.  For example, if a female deposited an average of 100 eggs in a lifetime, 98 or 80% of the resulting insects would have to die during some stage of development to restrict population growth to 1X or 10X, respectively.  The great proportion of the mortality that occurs is caused by existing, natural biological and physical control factors.  Thus one appraoch to the use of biological control agents for control of pest insects is a system of augmentation:  parasites, predators, or pathogens are reared and released in an attempt to increase population reduction over the normal rate.  Excellent progress has recently been made in the reduction of pest insect population by augmentation with microhymenopteran parasites of flies and nematode parasites of mosquitoes.  Attempts are now beginning to relate the control provided by the use of these agents to the life cycle of the hosts and in some cases to disease transmission.

Some representatives of microhymenopteran wasps that offer potential for fly control are *Spalangia endius* Walker, *S. cameroni* Perkins, *S. nigra* Latreille, *S. nigroaenea* Curtis, *S. muscidarum* Richardson (=*S. nigroaenea*), *Muscidifurax raptor* Girault and Sanders, *M. zaraptor* Kogan and Legner, *Nasonia vitripennis* (Walker), *Mormoniella vitripennis* (Walker) (=*N. vitripennis* (Walker)), *Pachycrepoideus vindemiae* (Rondani) (=*P. dubius* Ashmead and *Tachinaephagus zealandicus* Ashmead.  The wasps that parasitize *M. domestica* were studied by Beard (1964), Legner (1967a, 1972), Legner et al. (1965, 1967), McCoy (1965), Legner and McCoy (1966), Legner and Gerling (1967), Legner and Olton (1968), Legner and Greathead (1969), Ables and Shephard (1974a,b, 1976) and Olton and Legner (1974), Morgan (1977), Morgan et al. (1975a,b, 1976a,b,c), Morgan and Patterson (1975a,b).  The potential of these parasites to control field populations of house flies was discussed by Legner and Brydon (1966) and Legner and Detrick (1972) and Keiding (1974).  Those that parasitize *S. calcitrans* were studied by Pinkus (1913) and Monty (1972).  Those that parasitize *Haematobia irritans* by Lindquist (1936), Combs and Hoelscher (1969), Depner (1968), and Peck (1974).  Those that parasitize *M. autumnalis* were studied by Turner et al. (1968), Burton and Turner (1968), Hayes and Turner (1971), Hair and Turner (1965).

In general a female parasite wasp is ready to mate and oviposit immediately upon emergence from the host puparia; moreover, she obtains her nourishment by ingesting the haemolymph of the host as it exudes from the oviposition wound (Gerling and Legner 1968).  Thus, the typical sequence has 4 phases - finding the "host area"; locating host pupae; drumming and drilling; and feeding.  For example once the female *M. vitripennis* has found a pupa, she systematically examines the surface while drumming with the tips of the antennae.  Then she begins tapping with the tip of the abdomen on the surface of the puparium. This activity apparently places the tip of the ovipositor in position for drilling, a proceudre that requires from 10 minutes to 1 hour. When the wall of the puparium is pierced, the entire length of the ovipositor is inserted, and an egg is deposited on the developing pupa (Edwards 1955).

More information is needed about the egg laying habits of parasites in fly pupae, the number of eggs laid per pupa, and the number developing per pupa, but many observations have been made. Wylie (1971a) observed that as many as 25 larvae of *M. vitripennis* can mature on a house fly pupa; Whiting (1967) reported as many as 200 developing from 1 *Sarcophaga* sp. pupa. Holmes (1972) found that genetically marked female *M. vitripennis* which were the 2nd or 3rd to oviposit on *Sarcophaga bullata* Parker produced fewer progeny than marked females which were the first to oviposit on the host. Olton and Legner (1974) observed that 18 larvae of *T. zealandicus* can mature on 1 host. In contrast, Morgan and Patterson (1975b) rarely observed as many as 2 *M. raptor, S. endius* or *S. cameroni* maturing on a host. Wylie (1971b) observed that *M. zaraptor* chose to oviposit on unparasitized *M. domestica* pupae rather than on pupae already attacked by *M. vitripennis* or *S. cameroni* and noted the discrimination greater against parasitized *M. domestica* by *M. zaraptor* and least against those attacked by *S. cameroni*. However, Wylie (1965) determined that *M. vitripennis* laid fewer eggs on parasitized than on unparasitized pupae of *M. domestica*.

Once the parasite egg has been deposited, temperature has an effect on the developmental time of the parasite. For example, Gerling and Legner (1968) found the developmental time from initial oviposition by *S. cameroni* to eclosion of the adult parasite was 24-27 days at 26°C and 90-100% RH. Morgan et al. (1975a,b) documented a similar period for *S. endius*, and they found that by increasing the temperature to 27.8°C, the life cycle of *S. endius* was reduced to 18-20 days; however, they noted that 33-35 days are required for the life cycle of *S. endius* in the field. Also, it was observed by Pinkus (1913) that *S. muscidarium* required a developmental time of 84 days at 13.9°C and on 2 occasions required 106 and 109 days. Legner and Gerling (1967) found that it took 17-22 days at 26°C and 90-100% RH. Increasing the temperature above 27.8°C did not further reduce the developmental time of either *S. endius* or *M. raptor* (Morgan and Patterson 1975b). Nstvik (1954) concluded that the developmental time for *P. vindemiae* was 19 days at 25°C and 65% RH; Crandell (1939) determined that an average of 18 days (19-23) was needed. Nagel and Pimental (1963) reported 14 days at 26.7°C for *M. vitripennis*; Whiting (1967) reported a generation time of 10 days at 28°C.

The average total progeny per female *M. vitripennis* was 139 (Nagel and Pimental 1963). Nstvik (1954) found the average total progeny per female *P. vindemiae* to be 298. Ables and Shephard

(1974b) reported the average progenies per female *S. endius* and *M. raptor* were 15-40 and 5-30, respectively.  Morgan et al. (1975a, 1976b) observed that one *S. endius* female produced an average of 9.46-9.6 progeny from 1- and 2-day-old house fly pupae and 16.1 progeny from 0- to 4-day-old stable fly pupae.  They also reported 2.6 progeny per wasp per day and a female-to-male sex ratio of 2:1; unmated females produced males only.  This correlated with an average lifespan of 3.88 days for each female *S. endius*, when the wasps obtained their protein from the haemolymph of host pupae, and with a 33.15% daily loss rate, which would allow an average of 36.7 ovipositions and 10 $F_1$ progeny per female *S. endius*.  These data agree closely with those Lindquist (1936) obtained with *S. muscidarium var stomoxysiae* Girault (=*S. endius*):  66% of emerged *S. muscidarium* were females, unfertilized females produced male progeny; and females obtaining protein from the haemolymph of host *Scarcophaga* or *Cryptolucilia* pupae had an average longevity of 8.4 days.  Girault and Sanders (1910) reported sex ratios in *N. vitripennis* ranging from 1 male to 3 females and 1 male to 9 females; Nagel and Pimental (1963) observed that the longevity of *N. vitripennis* males averaged 1.62 days and that of the females averaged 6.96 days; the average progeny per female parasite was 139.8.  Both Wylie (1967) and Legner (1969) concluded that total oviposition was correlated with high host densities and host size.

When reviewing the general biology of the wasps, *S. endius* appears to have advantages as a biological control agent against *M. domestica*.  *M. raptor* confines its searching primarily to the upper levels of manure; *S. endius* is active at all levels.  Crandell (1939) found *P. vindemiae* only a weak competitor of *M. raptor*.  Also, *N. vitripennis* is apparently less competitive than *P. vindamiae*, and Whiting (1967) observed that the males were apterus and the females were inefficient fliers.  Moreover, Hair and Turner (1965) were unable to culture *N. vitripennis* on *M. autumnalis*, and Burton and Turner (1968), though they could propagate *M. raptor* on face flies, found that only 5% of the adult wasps were able to emerge unaided.

In evaluating the activity of parasites against *M. domestica, S. calcitrans, Fannia canicularis* (L.), and *F. femoralis* (Stein) at sites in the western hemisphere, Legner (1967b) found that 92% of the house fly pupae collected in Uruguay were parasitized by 6 species of parasites (*M. raptor, S. cameroni, S. endius, S. nigroaenea, Tachinaephagus giraulti* Johnson and Tiegs and *Trichopria* n. sp.); in New Brunswick, Canada, 90.7% of the house fly pupae were parasitized by *M. raptor* and *S. nigroaenea*.  He also found that a host such as *S. calcitrans* with the habit of pupating nearer the surface in the breeding site was generally more heavily parasitized and that the parasites did not disperse rapidly from a release site.  In fact, dispersal was usually slow beyond host-free barrier zones, as from 1 poultry ranch to another.  Legner (1972) also concluded that it was practical to utilize exotic strains for inundative releases by crossing strains from climatically similar but geographically isolated areas.

In other tests made over a period of 18 months, Legner and Brydon (1966) evaluated parasitism at 2 poultry ranches in southern California. Although 6 species of parasites were active, *M. raptor* and *S. endius* accounted for 95% of observed parasitism in *F. femoralis* and *Ophyra leucostoma* (Wiedemann). Similarly, from March through June 1970, Legner and Detrick (1972) released thousands of *S. endius*, *M. raptor*, and *T. zealandicus* at 6 poultry ranches to control *F. canicularis* and *F. femoralis*. Samples taken in June 1970 revealed a 6.5 times lower density (13.8 to 2.1) of these flies and the percentage parasitism had almost doubled (12.9 to 22.5%). Additional inoculative releases of *S. endius*, *M. raptor*, *M. zaraptor* Kogan and Legner, and *T. zealandicus* were made by Legner and Detrick (1974) over a 20-month period on poultry ranches located in southern California. They observed significant reduction in average densities of Diptera: *M. domestica*, *Muscina stabulans* (Fallen), *F. canicularis*, *F. Femoralis*, *Ophyra leucostoma*, *S. calcitrans*, and *Phaenicia*. Also, releases made during the spring had greater effect on fly populations than did similar releases in the summer. Olton and Legner (1975) made inoculative releases of *T. zealandicus*, *S. endius*, and *M. raptor* from December through April in an enclosed poultry house in southern California. The result was 46% parasitism of *M. domestica* but only 16% of *F. femoralis*.

In addition, McCoy (1965) released *M. raptor* but found that parasitism of fly pupae never exceeded 25%. Mourier (1972) also released *S. cameroni*, and *M. raptor* on 6 farms (10,000 parasites/farm) in northern Denmark. Although the parasite population built up faster than normal, it was still insufficient to reduce host populations to an acceptable level.

Monty (1972) released *S. endius*, *S. nigra*, *M. raptor*, *P. vindemiae*, and *Sphegigaster* sp. against populations of muscoid flies on Mauritius. *Spalangia* sp. was recovered in greater numbers than any other species released: 68% of the house flies were parasitized and 44.4% of *S. calcitrans*. However, no parasites were recovered from *Stomoxys nigra*. The inundative releases decreased the populations of *S. calcitrans* and *M. domestica*, but the percentage of parasitism dropped as soon as the releases were stopped. He therefore concluded that the parasites, even when they are well established, cannot maintain themselves at densities high enough to effect control.

These early experiments with releases of parasites against pest flies therefore demonstrated that some degree of control could be obtained. However, the results were not consistent. Then in all probability, consistent rapid, and reproducible high levels of control would depend on sustained releases of one or more parasites in much higher numbers over some interval of time.

Recent studies by P. B. Morgan and colleagues in Florida have demonstrated that high, rapid, and effective levels of control of house flies can be obtained by the sustained release of large numbers of a single parasite, *S. endius*, over one to two months. It was therefore necessary to develop: (1) efficient rearing procedures for the parasites; (2) effective methods of releasing parasites so that a continuing number of healthy, vigorous parasites would be available over time; (3) methods of analyzing the effectiveness of the parasites; and (4) methods of determining the degree of fly control obtained.

Morgan et al. (1977) have described their procedures for mass-producing *S. endius*. The effort depends primarily on the ability to mass-produce house flies. Briefly, house fly pupae are transferred to parasitization cages, specialized escape-proof, portable, clear plexiglass cages. The house fly pupae constitute the only source of food for the parasites. The cages are held in rooms maintained at 24°C and 60% RH. One single cage is stocked initially with 160,000 2-day-old house fly pupae and 60,000 *S. endius* adults (40,000 females and 20,000 males). Then, since the daily mortality rate of *S. endius* is about 33%, ca. 20,000 new wasps are added each day. However, each group of house fly pupae is exposed to the parasites for only one day so each day a new group is placed in the cage. Suitable handling procedures for the parasitized pupae have been developed as described by Morgan et al. (1977).

The relatively long development time of the immature stage of the parasite *S. endius* in house fly pupae (about 2 times the time for development of the immature stages of house flies is probably one of the reasons the parasite does not provide satisfactory control of house flies naturally. However, this long development time makes sustained releases of the parasite practical. For example, newly parasitized house fly pupae can be held for 5 days to ensure that any fertile house flies (from unparasitized pupae) are removed before release. Then, it will still be 13 to 15 days before adult parasites emerge from the pupae. This period provides ample time for packaging and transport of parasites of a variety of ages. Therefore, one does not have to provide for daily releases of parasites, but can release packages of parasites that provide for continuous release of parasites for several days to a week or more. The technique readily, easily and inexpensively allows the production of about 1,000,000 parasites per colony cage per 7-day week. Much larger numbers can be produced at relatively low cost by increasing the number of colony cages and the production of house flies.

Recently, Morgan and colleagues have conducted several augmentation releases with *S. endius* in which they demonstrated total sup-

pression and complete parasitization of fly pupae.  Their first experiment (Morgan et al. 1975a) was preliminary and was conducted in
an enclosed building containing 20 hens.  Manure was allowed to
accumulate, and a population of house flies was started.  Introduction of *S. endius* caused 100% parasitism and eliminated the population.  Actual field trials were then conducted at a dairy installation and a poultry installation.  At the commercial dairy in Alachua
County, Florida, they (Morgan et al. 1976c) released *S. endius* 3
times a week for 5 weeks, a total of 90,000 parasites.  By the end
of the release, all house fly pupae sampled were parasitized, and
the fly population was reduced by 93%.  The cost per week calculated
from the rearing costs of the parasite ($500/million) (Morgan et al.
1975a) was $6 per week exclusive of overhead, facilities, and hand-
ling.  In the test at the poultry installation (three open-sided
poultry houses containing 6700 caged layers), parasites (*S. endius*)
were released for 10 weeks.  After 4 weeks, 100% parasitism was
observed and house flies were completely suppressed within 35 days
(Morgan et al. 1975b).

Thus, we have found that releases of *S. endius* can give complete suppression and essentially complete parasitism of house
flies when sufficient numbers are released at relatively short
intervals over an extended period.  However, house flies do not
always breed in the absence of other muscoid or filth-breeding
flies, and there are other species of parasites than *S. endius* that
may be useful in augmentation systems.  It will be important to
obtain further data about pupal parasites other than *S. endius* to
study the integration of these augmentation procedures with other
methods of control, and to determine the requirements for control
when several species of muscoid flies breed in proximity.  Methods
of estimating accurately the absolute density of pest flies are
also needed so the numbers of parasites to be released can be quan-
titated for specific locations.  In fact, Weidhaas et al. (1977)
have attempted to relate the biology and life history of house flies
and *S. endius* to the control obtained by Morgan and associates
through simulation models.  Such models agree in general with field
studies, tend to synthesize existing biological data, and point up
the need for additional biological, ecological and control studies.

Chapman (1974) has given an excellent summary of the status of
nematodes as biological control agents for mosquitoes.  He indicated
that there were then nematodes reported from 63 mosquito species
worldwide.  Only a few of these have been studies in detail, and
only a few have undergone the necessary research to determine that
they may be useful in augmentation systems.

Nematode eggs hatch into preparasitic stages (juvenile preparasites) that attack mosquito larvae, penetrate the cuticle, and
enter the hemocoele.  Growth of the juvenile preparasites is rapid.
In fact, some nematodes that are effective against mosquitoes (and

thereby kill) the fourth-stage mosquito larvae. Thus, the nematode acts as a larval killing system. Other nematodes may remain in the mosquito into the adult stage and then prevent or reduce egg production of the female. In both cases, the emerging parasites molt and mate, and the female may lay several thousand eggs. The basic killing system can thus be increased by recycling of the nematode in mosquito breeding habitats.

One species of mosquito nematode, *Romanomermis culicivorax* Ross and Smith has shown considerable potential for mosquito control and has been developed for augmentation systems. Methods of mass-producing this nematode (Petersen and Willis 1972) on *Culex p. quinquefasciatus* Say in the laboratory were therefore developed. As a result, large quantities of the nematode can be made available for field studies. Briefly, the procedure is as follows: First, pre-parasitic forms are allowed to infest *C. p. quinquefasciatus* larvae. Then, the nematode eggs are collected in sand and sprayed with conventional hand sprayers in breeding areas. Such field treatment (releases) of nematodes (*R. culicivorax*) have been made in ditches, ponds, and pools in Louisiana and in rice fields in California and Louisiana. Infection levels in mosquito larvae have been high and have ranged from 40 to 80-85%. The nematodes have also recycled in some treated areas and in some areas with natural infections for several years. Thus, with a nematode such as *R. culicivorax* we have a potentially useful augmentation system for some species of mosquitoes. However, *R. culicivorax* has not been effective in research trials in certain polluted or highly saline waters, and other methods or species will be needed in these areas. According to Chapman (1974) limited trials of *R. culicivorax* in Taiwan indicated that three species of *Culex* there were also somewhat resistant to it. In addition, *C. p. fatigans* in highly polluted ditches and drains in Bangkok, Thailand, were not highly infected when nematodes were sprayed at very high dosages.

For filth-breeding flies and selected species of mosquito in habitat favorable to nematode parasites, research has shown that augmentation systems are potentially useful in biological control schemes. Further developmental research on numbers and patterns of release should make these agents more practical for use and more efficient at lower cost.

## REFERENCES CITED

Ables, J. R. and M. Shephard. 1974a. Hymenopterous parasitoids associated with poultry manure. Environ. Entomol. 3: 884-6.
Ables, J. R. and M. Shephard. 1974b. Responses and competition of the parasitoids *Spalangia endius* and *Muscidifurax raptor* (Hymenoptera: Pteromalidae) at different densities of house fly pupae. Can. Entomol. 106: 825-30.

Ables, J. R. and M. Shephard. 1976. Seasonal abundance and activity of indigenous Hymenopterous parasitoids attacking the house fly (Diptera: Muscidae). Can. Entomol. 108: 841-4.

Bay, E. C. 1967. Mosquito control by fish: a present day appraisal. WHO Chron. 21: 415-23.

Bay, E. C., C. O. Berg, H. C. Chapman, and E. F. Legner. 1976. Biological control of medical and veterinary pests. p. 457-474 In C. B. Huffaker and P. S. Messenger, ed. *Theory and Practice of Biological Control*. Academic Press, New York.

Beard, R. L. 1964. Parasites of muscoid flies. Bull. WHO 31: 491-3.

Burton, R. P. and E. C. Turner, Jr. 1968. Laboratory propagation of *Muscidifurax raptor* on face fly pupae. J. Econ. Entomol. 61: 1380-3.

Chapman, H. C. 1974. Biological control of mosquito larvae. Annu. Rev. Entomol. 19: 33-59.

Chapman, H. C., J. J. Petersen, and T. Fukuda. 1972. Predators and pathogens for mosquito control. Am. J. Trop. Med. Hyg. 21: 777-81.

Chapman, H. C., J. J. Petersen, D. B. Woodard and T. Fukuda. 1971. Current status of biological control of mosquitoes. Proc. Gulf Coast Conf. Mosq. Supp. Wildl. Manage. 2: 2-3.

Combs, R. L., Jr., and C. E. Hoelscher. 1969. Hymenopterous pupal parasitoids found associated with the horn fly in northeast Mississippi. J. Econ. Entomol. 62: 1234-5.

Crandall, H. A. 1939. The biology of *Pachycrepoideus dubius* Ashmead (Hymenoptera), a pteromalid parasite of *Piophila casei* Linne. Ann. Entomol. Soc. Am. 32: 632-54.

Depner, K. R. 1968. Hymenopterous parasites of the horn fly, *Haematobia irritans* (Diptera; Muscidae) in Alberta. Can. Entomol. 100: 1057-60.

Edwards, R. L. 1955. The host-finding and oviposition behavior of *Mormoniella vitripennis* (Walker) (Hymenoptera: Pteromalidae), a parasite of muscoid flies. Behavior 7: 88-112.

Gerberich, J. B. and M. Laird. 1968. Bibliography of papers related to the control of mosquitoes by the use of fish (an annotated bibliography for the years (1901-1966). FAO Fisheries Tech. Paper No. 75. FAO United Nations, Rome. 1-70.

Gerling, D., and E. F. Legner. 1968. Developmental history and reproduction of *Spalangia cameroni*, parasite of synanthropic flies. Ann. Entomol. Soc. Am. 61: 1436-43.

Girault, A. A. and G. E. Sanders. 1910. The chalcidoid parasites of the common house or typhoid fly (*Musca domestica* L.) and its allies. Psyche 17: 9-28.

Hair, J. A., and E. C. Turner, Jr. 1965. Attempted propagation of *Nasonia vitripennis* on the face fly. J. Econ. Entomol. 58: 159-60.

Hayes, C. G. and E. C. Turner, Jr. 1971. Field and laboratory evaluation of parasitism of the face fly in Virginia. J. Econ. Entomol. 64: 443-8.

Holmes, H. B.   1972.   Genetic evidence for fewer progeny and a
    higher percent males when *Nasonia vitripennis* oviposits in
    previously parasitized hosts.   Entomophaga 17: 79-88.
Jenkins, D. W.   1964.   Pathogens, parasites, and predators of med-
    ically important arthropods.   Bull WHO 30.   150 p.
Keiding, J.   1974.   House flies (*Musca domestica*).   In R. Pal and
    R. H. Wharton, ed. *Control of Arthropods of Medical and
    Veterinary Importance*.   Plenum Press, New York and London.
    138 p.
Laird, M.   1970.   Integrated control of mosquitoes.   Am. Zool. 10:
    573-8.
Laird, M.   1971.   Microbial control of arthropods of medical impor-
    tance.   p. 347-406.   In H. D. Burges, ed. *Microbial Control of
    Insects and Mites*.   Academic Press, New York.
Legner, E. F.   1967a.   The status of *Nasonia vitripennis* as a
    natural parasite of the house fly, *Musca domestica*.   Can.
    Entomol. 99: 308-9.
Legner, E. F.   1967b.   Behavior changes the reproduction of *Spalangia
    cameroni, S. Endius, Muscidifurax raptor*, and *Nasonia vitri-
    pennis* (Hymenoptera: Pteromalidae) at increasing fly host den-
    sities.   Ann. Entomol. Soc. Am. 60: 819-26.
Legner, E. F.   1969.   Adult emergence interval and reproduction in
    parasitic Hymenoptera influenced by host size and density.
    Ann. Entomol. Soc. Am. 62: 220-6.
Legner, E. F.   1972.   Observations on hybridization and heterosis
    in parasitoids of synanthropic flies.   Ann. Entomol. Soc. Am.
    65: 254-63.
Legner, E. F., and H. W. Brydon.   1966.   Suppression of dung-inhab-
    iting fly populations of pupal parasites.   Ann. Entomol. Soc.
    Am. 59: 638-51.
Legner, E. F., and E. I. Detrick.   1972.   Inundation with parasitic
    insects to control filth breeding flies in California.   Proc.
    40th Annu. Conf. Calif. Mosq. Contr. Assoc., Inc., January 31-
    February 2.   p. 129-30.
Legner, E. F., and E. I. Detrick.   1974.   Effectiveness of super-
    vised control practices in lowering population densities of
    synanthropic flies on poultry ranches.   Entomophaga 19: 467-78.
Legner, E. F., and D. Gerling.   1967.   Host feeding and oviposition
    on *Musca domestica* by *Spalangia cameroni, Nasonia vitripennis,*
    and *Muscidifurax raptor* (Hymenoptera: Pteromalidae) influences
    their longevity and fecundity. Ann. Entomol. Soc. Am. 60: 678-91.
Legner, E. F., and D. J. Greathead.   1969.   Parasitism of pupae in
    east African populations of *Musca domestica* and *Stomoxys cal-
    citrans*.   Ann. Entomol. Soc. Am. 62: 128-33.
Legner, E. F., and C. W. McCoy.   1966.   The housefly, *Musca domes-
    tica* Linneaus, as an exotic species on the western hemisphere
    incites biological control studies.   Can. Entomol. 98: 243-8.

Legner, E. F., and G. S. Olton. 1968. Activity of parasites from Diptera: *Musca domestica, Stomoxys calcitrans*, and species of *Fannia, Muscina* and *Ophyra*. 11. At sites in the eastern hemisphere and pacific area. Ann. Entomol. Soc. Am. 61: 1306-41.

Legner, E. F., E. D. Bay, and C. W. McCoy. 1965. Parasitic natural regulatory agents attacking *Musca domestica* L. in Puerto Rico. J. Agric. Univ. PR 49: 368-76.

Legner, E. F., E. C. Bay, and E. B. White. 1967. Activity of parasites from Diptera: *Musca domestica, Stomoxys calcitrans, Fannia canicularis*, and *F. femoralis*, at sites in the western hemisphere. Ann. Entomol. Soc. Am. 60: 462-8.

Lindquist, A. W. 1936. Parasites of horn fly and other flies breeding in dung. J. Econ. Entomol. 29: 1154-8.

McCoy, C. W. 1965. Biological control studies of *Musca domestica* and *Fannia* sp. on southern California poultry ranches. Proc. 33rd Ann. Conf. Calif. Mosq. Contr. Assoc., Inc. January. p. 40-2.

Monty, J. 1972. A review of the stable fly problem in Mauritius. Rev. Agric. Sucriere Maurice 51: 13-29.

Morgan, P. B. 1977. The parasitic wasp - A research update. Proc. 36th Ann. Fla. Poultry Institute, May. p. 10.

Morgan, P. B., and R. S. Patterson. 1975a. Field parasitization of house flies by natural populations of *Pachycrepoideus vindemiae* (Rondani), *Muscidifurax raptor* Girault and Sanders, and *Spalangia nigroaenea* Curtis. Fla. Entomol. 58: 202.

Morgan, P. B., and R. S. Patterson. 1975b. Possibilities of controlling stable flies and other muscoid flies with parasitic wasps. Proc. 46th Ann. Mtg. Fla. Anti-Mosq. Assoc., April 13 through 16. p. 29-35.

Morgan, P. B., and R. S. Patterson. 1977. Facilities for culturing microhymenopteran pupal parasitoids of muscoid flies. In USDA Tech. Bull. (In Press).

Morgan, P. B., A. Benton and R. S. Patterson. 1976a. The potential use of parasites to control flies in the Caribbean area. Virgin Islands Agriculture and Food Fair. p. 43.

Morgan, P. B., R. S. Patterson, and G. C. LaBrecque. 1976b. Host-parasitoid relationship of the house fly, *Musca domestica* L., and the protelean parasitoid *Spalangia endius* Walker. J. Kans. Entomol. Soc. 49: 483-88.

Morgan, P. B., R. S. Patterson, and G. C. LaBrecque. 1976c. Controlling house flies at a dairy installation by releasing a protelean parasitoid *Spalangia endius* Walker. J. GA Entomol. Soc. 11: 39-43.

Morgan, P. B., R. S. Patterson, G. C. LaBrecque, D. E. Weidhaas, A. Benton, and T. Whitfield. 1975a. Rearing and release of the house fly pupal parasite *Spalangia endius* Walker. Environ. Entomol. 4: 609-11.

Morgan, P. B., R. S. Patterson, G. C. LaBrecque, D. E. Weidhaas, and A. Benton. 1975b. Suppression of a field population of house flies with *Spalangia endius*. Science (USA) 189: 388-9.

Morgan, P. B., G. C. LaBrecque, and R. S. Patterson. 1977. Mass culturing the microhymenopteran parasite, *Spalangia endius* Walker. J. Med. Entomol. (In Press).

Mourier, H. 1972. Release of native pupal parasitoids of house-flies on Danish farms. Vidensk. Medd. Dan. Naturhist. Foren. 135: 129-37.

Nagel, W. P. and D. Pimental. 1963. Some ecological attributes of a Pteromalid parasite and its housefly host. Can. Entomol. 95: 208-13.

Nstvik, E. 1954. Biological studies of *Pachycrepoideus dubius* Ashmead (Chalcidoidea: Pteromalidae), a pupal parasite of various Diptera. Oikos 5: 196-204.

Olton, G. S. and E. F. Legner. 1974. Biology of *Tachinaephagus zealandicus* (Hymenoptera: Encytridae) parasitoid of synan-trophic Diptera. Can. Entomol. 106: 785-800.

Olton, G. S. and E. F. Legner. 1975. Winter inoculative releases of parasitoids to reduce houseflies in poultry manure. J. Econ. Entomol. 68: 35-8.

Peck, O. 1974. Chalcidoid (Hymenoptera) parasites of the horn fly, *Haematobia irritans* (Diptera: Muscidae), in Alberta and else-where in Canada. Can. Entomol. 106: 473-7.

Petersen, J. J. and O. R. Willis. 1972. Procedures for the mass rearing of a mermithid parasite of mosquitoes. Mosq. News 32: 226-30.

Pinkus, H. 1913. The life history and habits of *Spalangia musci-darum* Richardson, a parasite of the stable fly. Psyche 20: 148-58.

Turner, E. C., Jr., R. P. Burton, and R. R. Gerhardt. 1968. Natural parasitism of dung-breeding Diptera: a comparison between native hosts and an introduced host, the face fly. J. Econ. Entomol. 61: 1012-5.

Weidhaas, D. E., D. G. Haile, P. B. Morgan, and G. C. LaBrecque. 1977. A model to simualte control of house flies with a pupal parasite, *Spalangia endius*. Environ. Entomol. (In Press).

Whiting, A. R. 1967. The biology of the parasitic wasp *Mormoniella vitripennis (=Nasonia brevicornis)* (Walker). Q. Rev. Biol. 42: 333-406.

Wylie, H. G. 1965. Discrimination between parasitized and un-parasitized house fly pupae by females of *Nasonia vitripennis* (Walker) (Hymenoptera: Pteromalidae). Can. Entomol. 97: 279-86.

Wylie, H. G. 1967. Some effects of host size on *Nasonia vitripen-nis* and *Muscidifurax raptor* (Hymenoptera: Pteromalidae). Can. Entomol. 99: 742-48.

Wylie, H. G. 1971a. Observations on intraspecific larval competi-tion in three hymenopterous parasites of fly puparia. Can. Entomol. 103: 137-42.

Wylie, H. G. 1971b. Oviposition restraint of *Muscidifurax zaraptor* (Hymenoptera: Pteromalidae) on parasitized housefly pupae. Can. Entomol. 103: 1537-44.

Section IV

ANALYSIS OF CURRENT USES

AND PROSPECTS FOR EXPANSION

CHAPTER 15


ECONOMIC AND SOCIAL CONSIDERATIONS FOR THE UTILIZATION OF

AUGMENTATION OF NATURAL ENEMIES

N. H. Starler and R. L. Ridgway

Economic Research Service and Agricultural Research
Service, U. S. Department of Agriculture, Washington,
D. C. 20250 and Beltsville, Maryland 20705  U.S.A.

Augmentation of natural enemies is a tool which can be part of
integrated pest management programs that include classical biological techniques (the importation of natural enemies), cultural
techniques to create favorable environments for natural enemies,
genetic modification of pests, pest-resistant strains of plants,
pheromones, microbial agents, growth regulators, and the use of
chemical pesticides to control outbreaks.

In a recent analysis of the control of insect pests of plants
in the United States, the use of insecticides was confirmed to be
the predominant method of insect control with an estimated expenditure of about $500 million on 50 to 60 million acres (Ridgway et
al. In press).  However, non-insecticidal methods also are used
extensively with insect resistant varieties being planted on perhaps 40 million acres (Shalk and Ratcliffe 1975).  Natural enemies,
including parasites, predators, and microbial agents also are used
on a significant amount of acreage.

Although chemical insecticides are currently the most widely
used insect control agents, several factors are causing many
producers to assess the benefits and costs of integrated approaches
to pest control.  Three of the most important factors are: (1) the
development of pest resistance to chemicals, (2) the increased
costs associated with the development, production and application
of pesticides, and (3) the adoption of stringent regulations due to
public concern over human health and environmental implications of
using chemical pesticides.

If integrated pest management is to be used widely, we can
expect expanded utilization of a number of different pest control

431

strategies including the use of biological control agents, in
general, and augmentation of natural enemies in particular, can
be expected.  This chapter is designed to explore some of the
economic and social issues associated with augmentations of natural
enemies.  The following topics are considered: benefits and costs
of biological methods, extent of production of natural enemies,
costs of rearing natural enemies, extent of use of augmentations,
cost of augmentations, and institutional options for encouraging
the use of augmentations.

## BENEFITS AND COSTS OF BIOLOGICAL METHODS

Biological control methods are basically preventative rather
than corrective.  Therefore, such methods as augmentation of natural
enemies can normally be justified economically only when there is
a high probability that a problem will develop without their use.
One, therefore, is faced with the difficulty of measuring the
benefits derived from control of a pest situation that has not yet
occurred (USDA In press).

Since there have not been definitive benefit-cost analyses of
biological control systems, it is not known how biological control
systems would fare in a carefully conducted long-term analysis
that accounted for all of the private and public benefits and
costs.  Currently, a purchaser of insecticides generally gives only
limited attention to the fact that his use of insecticides may
contribute in the long run to problems of pest resistance and
increased environmental and human health risks.  Also, a neighbor's
use of insecticides can obviate his efforts to implement a program
of biological control (Hart 1972, Stinner et al. 1974).  In many
instances, the cited factors as well as the difficulty of
partitioning benefits of the use of biological agents from those
due to good production practices provide a disincentive for choosing
biological agents over chemicals.

Although the individual consumer of agricultural commodities
is seldom directly involved in the choice of pest control agents,
he can, by his willingness to pay higher dollar costs for commodities produced with biological agents, influence the farmers' pest
control decisions.  His purchase of these commodities which may be
of equal nutritive value, but in some cases of less desirable
appearance, can contribute to more ecologically efficient pest
control since the use of biological agents reduces the risks to
non-target organisms (Pimental et al. 1977).  However, up to the
present time this type of action by consumers has been limited.
Moreover, the additional insect damage to commodities that are
produced with biological agents may result in increased spoilage
during storage and transportation, which would be a significant
obstacle to the marketing of these commodities.

## EXTENT OF PRODUCTION OF NATURAL ENEMIES

Numerous examples of experimental and practical applications of augmentations have been reviewed (Section III, Chapters 9-14, this treatise). However, quantifying the extent of production of natural enemies in the United States and in the world is difficult because of the availability of only limited data. Some information has been assembled on production of parasites and predators in the U.S. by making contacts with known commercial suppliers and other knowledgeable individuals. From these we have been able to obtain estimates of numbers of parasites and predators that are mass-reared for use in practical applications.

The production of natural enemies in the U.S. for practical use (Table 1) is confined to a limited number of private firms, cooperatives, and government agencies. The egg parasites, *Trichogramma* spp., are reared in the largest quantities with an annual production of over 3 billion insects each year. These parasites are released for control of lepidopterous insects on a wide range of crops. Other natural enemies reared in substantial numbers for which estimates could be obtained include the *Aphytis melinus* DeBach (203 million), a parasite of the California red scale; *Cryptolaemus montrouzieri* Mulsant (26 million), a predator of mealybugs; and *Chrysopa carnea* Stephens (18 million), a general predator of most soft bodied insects. Other parasites and predators known to be reared for use in practical releases in the U.S. are also listed (Table 1).

In addition to the natural enemies that are mass-reared, adults of the convergent lady beetles, *Hippodamia convergens* Guerin-Meneville, and egg cases of the praying mantid, are field collected in relatively large numbers and sold. However, there are some serious questions about the effectiveness of releases of these particular natural enemies (DeBach 1974).

Information about numbers of mass-reared insects produced in countries other than the U.S. is available from various reports, but it cannot be considered complete. For example, reports of annual production of 50 billion egg parasites, primarily *Trichogramma*, in the U.S.S.R. (Beglyarov and Smetnik, Chapter 9), and of production of 28 billion *Trichogramma* in Mexico (Jiminez 1975) over the past decade are indications of significant practical use of these natural enemies. Also, the production of 170 million *Encarsia* annually in Britain and the Netherlands (Hussey 1977) is reported as is the production of 32 million *Opius* in Sicily (Billoti, Chapter 11), and over 4 million *Lixophaga* in Cuba (Montes 1970).

TABLE 1.--Some natural enemies available for use in a practical
          augmentation in the United States.[a] Names and
          addresses of suppliers are listed in the Appendix.

| Natural enemy | Target pest | Estimated annual no. in millions |
|---|---|---|
| *Insectary reared* | | |
| **Parasites** | | |
| *Trichogramma* spp. | Lepidopterous eggs | 3,264 |
| *Aphytis melinus* DeBach | California red scale | 203 |
| *Pauridia* sp. | Black scale | 8 |
| *Metaphycus helvolus* (Compere) | Mealybugs | 6 |
| *Pediobius foveolatus* (Crawford) | Mexican bean beetle | - |
| *Apanteles scutellaris* Musebeck | Tomato bean beetle | - |
| *Chelonus blackburni* Cameron | Pink bollworm | - |
| *Chelonus texanus* Cresson | Armyworm | - |
| *Comperiella bifasciata* Howard | California red scale | - |
| *Tachinaephagus zealandicus* Ashmead | Flies | - |
| *Muscidifurax raptor* Girault | Flies | - |
| *Apalangia endius* Walker | Flies | - |
| *Pachycrepoidens vindemmiae* (Rond.) | Flies | 1 |
| *Reesimermis nielseni* Tsai & Grund | Mosquitoes | - |
| **Predators** | | |
| *Cryptolaemus montrouzieri* Mulsant | Mealybugs | 26 |
| *Chrysopa carnea* Stephens | General predator | 18 |
| *Toxorhychites* sp. | Mosquitoes | - |
| *Phytoseiulus persimilis* Athias-Henriot | Mites | - |
| *Typhlodromus hibisci* (Chant) | Mites | - |
| *Typhlodromus occidentalis* Nesbitt | Mites | - |
| *Field collected* | | |
| *Hippodamia convergens* Guerin-Meneville | General predator | 7 |
| Praying mantid | General predator | 1 |

[a] Source: Ridgway et al.  In press, May and Foulke 1977,
Gerberg 1977, Fairfax Biological Laboratory, Inc. 1977, Dively
1976, Ridgway et al. 1974, DeBach 1974.

## COST OF NATURAL ENEMIES

Likewise, when specific costs of rearing natural enemies are considered, again only partial information is available. Detailed technical analyses including costs of research and development, buildings and equipment, labor and materials, overhead, return on investment, etc. have either not been conducted or have not been published. However, some insight into the costs of parasites and predators in the United States can be obtained by analyzing published price lists and estimates that have been made by biological scientists (Table 2). As might be expected, there are considerable differences between retail prices and scientists' estimates. These differences indicate that improved mass rearing technology and expanded markets would probably reduce retail prices. Conversely, some of the hidden costs such as research and development, overhead, and return on investment may not be included in the scientists' estimates. Also, the cost of the natural enemies is just a part of the cost of augmentations. Costs of storage, distribution, and release as well as possible losses due to mortality or deterioration and possible additional pest management service costs must also be included to determine the total cost of augmentations.

TABLE 2.--Costs of natural enemies reared in the United States.[a]

| Natural enemies | Cost per 1,000 in U.S. dollars |
|---|---|
| **Price list** | |
| *Trichogramma* spp. | $ 0.125 |
| *Aphytis melinus* | 1.25 |
| *Spalangia endius* | 1.25 |
| *Reesinermis nielseni* | 0.02 |
| *Chrysopa carnea* (eggs) | 1.25 |
| *Phytoseiulus persimilis* | 50.00 |
| | |
| **Estimated potential costs** | |
| *Trichogramma* spp. | 0.03 |
| *Spalangia endius* | 0.50 |
| *Brachymeria intermedia* (Nees) | 3.00 |

[a] Source: Rincon-Vitova Insectaries, Inc. (1976), Fairfax Biological Laboratory, Inc. (1977), Weidhaas and Morgan (Chapter 14), Knipling (Chapter 3).

## EXTENT OF USE OF AUGMENTATIONS

A review of available information indicates that more quantitative information is available concerning the extent of use of
augmentations than on the number of natural enemies reared or their
cost.  In fact, a summary of the extent of the U.S. and worldwide
use of augmentations indicates that in a number of specialized
situations augmentations are used on a substantial portion of the
crops treated for pests (Table 3).

The use of augmentations in the controlled environment of
glasshouses is having a significant impact on pest control practices in Western Europe and in the U.S.S.R.  For instance,
augmentations with *Encarsia* and *Phytoseiulus* are used in the
growing of over 800 acres of tomatoes, cucumbers, chrysanthemums,
and sweet peppers in Britain; similar practices are used on over
2,500 acres in the Netherlands (Hussey 1977) and on 3,200 acres of
glasshouse crops in the U.S.S.R. (Beglyarov and Smetnik, Chapter 9).

Apparently, the most extensive use of augmentations of natural
enemies is in the U.S.S.R. where *Trichogramma* are released on 18.5
million acres of crops.  Field crops including wheat, corn, sugar-
beets and hemp receive the vast majority of these releases which
are used to combat cutworms and corn borers as well as other insect
pests living close to the ground.  *Trichogramma* also are released
for control of insects on vegetables such as cabbage and potatoes
(Kravchenko and Dolan 1975, Beglyarov and Smetnik, Chapter 9).

Substantial use of *Trichogramma* also occurs in the Peoples
Republic of China where releases are made on 1.7 million acres of
cotton to control infestations of lepidopterous pests.  *Trichogramma*
are employed in fields of rice and corn.  The extent of these
releases is unknown, but in one province, there are 148,000 acres
of corn involved (Huffaker, Chapter 10).

Among the more definitive uses of augmentations in the U.S.
are those involving the releases of *Aphytis melinus* and other
natural enemies for control of citrus pests in the Fillmore Plant
Protection District (Ridgway et al., Chapter 13).  Some 9,000 acres
are involved in this district.  This acreage together with other
acreages of citrus known to receive augmentations (Table 3) represents about 3 percent of the citrus grown in the United States.

Releases of *Trichogramma* in apple orchards were attempted in
Germany but were unsuccessful because codling moth damage could not
be kept below the 5 percent injury level.  However, in 1967 some
32 million brachonid parasites released on 6,900 acres achieved
partial success against the olive fly in Sicily.  These parasites

TABLE 3.—Some worldwide examples of practical uses of augmentation of natural enemies by commodity.

| Commodity | Pest | Natural Enemies | Country | No. of Acres | Source |
|---|---|---|---|---|---|
| **Glasshouse crops** | | | | | |
| Cucumbers | Spidermites | *Phytoseiulus* sp. | U.S.S.R. | 3200 | Beglyarov & Smetnik (Chpt. 9) |
| | | | Britain | 200 | Hussey (1977) |
| | | | France | 30 | Kravchenko & Dolan (1975) |
| | | | Netherlands | 900 | Hussey (1977) |
| | | | Sweden | 12 | Kravchenko & Dolan (1975) |
| | | | Norway | 12 | Kravchenko & Dolan (1975) |
| | | | Finland | – | Kravchenko & Dolan (1975) |
| | Whiteflies | *Encarsia* sp. | Britain | 200 | Hussey (1977) |
| | | | Netherlands | 20 | Hussey (1977) |
| | | | Canada | 5 | Kravchenko & Dolan (1975) |
| | Spidermites, whiteflies, and thrips | *Aphidoletes* sp. | U.S.S.R. | 60 | Beglyarov & Smetnik (Chpt. 9) |
| Tomatoes | Spidermites | *Phytoseiulus* sp. | Britain | 150 | Hussey (1977) |
| | | | Netherlands | 100 | Hussey (1977) |
| | | | Romania | 5 | Kravchenko & Dolan (1975) |
| | | | Poland | 5 | Kravchenko & Dolan (1975) |
| | | *Encarsia* sp. | Netherlands | 1500 | Hussey (1977) |
| | | | Britain | 600 | Hussey (1977) |
| Peppers | Spidermites | *Phytoseiulus* sp. | Britain | 5 | Hussey (1977) |
| | | | Netherlands | 100 | Hussey (1977) |
| Lettuce, celery and dill | Aphids | *Chrysopa* sp. | U.S.S.R. | 5 | Kravchenko & Dolan (1975) |

TABLE 3.—Continued

| Commodity | Pest | Natural Enemies | Country | No. of Acres | Source |
|---|---|---|---|---|---|
| Chrysanthemum | Aphids | *Aphidus* sp. | Britain | – | Hussey & Scopes (Chpt. 12) |
|  | Spidermites | *Phytoseiulus* sp. | Britain | 10 | Hussey (1977) |

**Fruit crops**

| Commodity | Pest | Natural Enemies | Country | No. of Acres | Source |
|---|---|---|---|---|---|
| Citrus | Red scale | *Aphytis* sp. | U.S.A. | 11,000 | May & Foulke (1977), (1975), Pennington (1975) |
|  | Mealybugs | *Chrysopa* sp. | U.S.A. | 7,000 | May & Foulke (1977), Lindsay (1975) |
|  | Black scale | *Cryptolaemus* sp. | U.S.A. | 14,000 | May & Foulke (1977), Lindsay (1975), Pennington (1975) |
| Melons | Scale, aphids | *Prospaltella* sp. | Germany | – | Biliotti (Chpt. 11) |
|  |  |  | Switzerland | – | Biliotti (Chpt. 11) |
|  |  |  | France | – | Biliotti (Chpt. 11) |
| Olives | Olive fly | *Opius* sp. | Sicily | 6,900 | Biliotti (Chpt. 11) |
| Apples | Codling moth | *Trichogramma* sp. | U.S.S.R. | – | Beglyarov & Smetnik (Chpt. 9) |

**Vegetable crops**

| Commodity | Pest | Natural Enemies | Country | No. of Acres | Source |
|---|---|---|---|---|---|
| Cabbage | Cutworms | *Trichogramma* sp. | U.S.S.R. | – | Kravchenko & Dolan (1975) |
| Potatoes | – | *Trichogramma* sp. | U.S.S.R. | – | Beglyarov & Smetnik (Chpt. 9) |
| Misc. vegetables | – | *Trichogramma* sp. | U.S.A. | – | May & Foulke (1977) |

**Field crops**

| Commodity | Pest | Natural Enemies | Country | No. of Acres | Source |
|---|---|---|---|---|---|
| Wheat, corn, hemp & others | Cutworms, gamma moth, corn borer | *Trichogramma* spp. | U.S.S.R. | 14,200,000[a] | Beglyarov & Smetnik (Chpt. 9) |

TABLE 3.--Continued

| Commodity | Pest | Natural Enemies | Country | No. of Acres | Source |
|---|---|---|---|---|---|
| Sugar beets & others | Cabbage moth, gamma moth | *Trichogramma* sp. | U.S.S.R. | 3,700,000[a] | Beglyarov & Smetnik (Chpt. 9) |
| Rice | Rice leaf roller | *Trichogramma* sp. | China | – | Huffaker (Chpt. 9) |
| Corn | Corn borer | *Trichogramma* sp. | China | 148,000[b] | Huffaker (Chpt. 9) |
| Cotton | Lepidopterous pests | *Trichogramma* sp. | China | 1,700,000 | Huffaker (Chpt. 9) |
| | Lepidopterous pests | *Trichogramma* sp. | Mexico | – | Jimenez (1975) |
| | Lepidopterous pests | *Trichogramma* sp. | U.S.A. | – | May & Foulke (1977) |
| Wheat | Cutworms | *Trichogramma* sp. | Turkey | – | Kravchenko & Dolan (1975) |

a/ Information not available to divide acreage treated in the U.S.S.R. (18,500,000 acres) by commodity; these values are general estimates.
b/ Includes acreage from Kirin Province only.

were released on an experimental basis in southern France, but
the size of the area treated was not reported and current status
of this program is unclear (Biliotti Chapter 11).

Although vegetables are second to orchards in numbers of acres
treated with augmentations in the U.S., they constitute a very
small part of commercial enterprises. However, experience in the
U.S.S.R. and Europe with greenhouse use suggests that vegetables
and ornamentals grown in U.S. greenhouses might provide a logical
commercial expansion of augmentation. On the other hand, use of
augmentations in home vegetable gardens and orchards may accelerate
given the public's interest in this productive passtime (Kaitz and
Weimer 1976).

Presently, the employment of natural enemies of pests of man
and animals is an emerging use of augmentation. Parasites of
flies are being used in a small number of dairies, livestock
operations and poultry farms in the U.S., and parasitic nematodes
are being used to a limited extent for mosquito control (Nickle
1976). Numbers of domestic animals and acres of land receiving
these augmentations are unknown at this time.

## COSTS OF AUGMENTATIONS

To the user of pest controls the principal inputs for his
benefit-cost calculations are the amounts he must expend to treat
a unit of the commodity he is producing. To our knowledge,
comparative data to equate the cost effectiveness of augmentations
relative to other pest control methods within the various countries
where augmentations are used are unavailable. The data that follow
provide some insight into cost of augmentations but probably are
not meaningful for making comparisons among countries.

Releases of *Chrysopa* on glasshouse cucumbers and celery in the
U.S.S.R. reportedly achieved controls at significantly lower costs
than would have been expended for chemical alternatives. In the
Netherlands, it is reported that the use of *Phytoseiulus* and
*Encarsia* in glasshouses against spider mites and white flies costs
30 percent less than would chemical treatment (Hussey 1977). The
benefit-cost comparison of augmentations relative to chemicals
appears to be similarly favorable in Finland since 250 out of 500
greenhouses augment with *Phytoseiulus* and *Encarsia* at a cost of
5.4 marks per meter$^2$ ($54 U.S. per acre). Canada reports a cost
of $12 per 50 meter$^2$ ($971 per acre) for glasshouses treated with
*Encarsia* (Kravchenko and Dolan 1975).

For field release of *Trichogramma* the reports from U.S.S.R.
indicate a cost of 0.3 to 0.6 roubles per hectare ($0.16 to $0.32

per acre) (Kravchenko and Dolan 1975), and reports from China
indicate a cost of $0.50 to $0.70 per acre to control the corn
borer (Huffaker Chapter 9).  The extensive use of *Trichogramma*
in both countries implies that the benefit-cost ratio for augmentations relative to other controls is favorable.  Although reports
from Turkey indicate a cost of 0.1 to 0.25 Turkish lira ($.006 U.S.
to $.014 U.S.) per *Trichogramma* for field releases on wheat
(Kravchenko and Dolan 1975), the extent of utilization is unknown.

For releases of the scale parasites, *Metaphycus* and *Aphytis*
along with the mealy bug predator, *Cryptolaemus*, in citrus orchards,
the Fillmore Citrus Protective District, expends less than $10.00
per acre (Pennington 1975).  The Associates Insectary expended
about $12 per acre ($0.10 per tree) in 1974-75 to release
*Cryptolaemus* (Lindsay 1975).

Release of fly parasites from 1969 to 1974 cost an average of
2-1/2 cents per chicken per season and 65 cents per cow per season
for weekly or biweekly release of fly parasites (DeBach 1974).


## INSTITUTIONAL OPTIONS FOR DEVELOPMENT AND USE OF AUGMENTATION

The extensive use of *Trichogramma* on field crops in the U.S.S.R.
and the People's Republic of China, the commercial use of *Encarsia*
and *Phytoseiulus* in glasshouses in Europe and the U.S.S.R., and the
release of *Aphytis* on citrus in the United States over an extended
period indicate that in some cases there are implicitly favorable
benefit-cost ratios for augmentations.  Apparently, these favorable
ratios occur when sufficient quantities of natural enemies of
adequate quality can be reared and distributed to sites where the
user has a measure of control over the ecological system surrounding the area where the natural enemies are used.  When the user
exercises this control, he is in a position to realize the potential
benefits from augmentation (surmounting and/or avoiding resistance
problems, achieving effective pest control, and reducing environmental and human health risks) and to overcome the costs associated
with the consequences of incompatible pest control actions of
adjoining crops.

There is considerable evidence that augmentation of natural
enemies is favored by certain institutional arrangements.  Some
aspects of institutional options applicable to pest management in
the United States were considered by Hepp (1976) in his study of
alternative delivery systems available to farmers wishing to obtain
integrated pest management services.  Although some of these
delivery systems are not particularly suited to implementation of
augmentation, Hepp's approach to delineation of alternative
systems should be helpful in considering the various options for

making augmentations a more viable part of pest management systems
in the U.S.  Therefore, some institutional options will be reviewed
within the context of who rears the predators and parasites, who
distributes these natural enemies, who coordinates the release
based on knowledge of commodity and/or pest management systems,
and who bears the venture risks (uncertainties associated with
rearing and distribution) and the user risks (possible user losses
if the natural enemies do not provide adequate pest control).

In the U.S., the following institutional arrangements have
been observed: (1) private insectary selling directly to users who
make their own release decisions, (2) private insectary producing
the natural enemies and providing the commodity/pest management
service to the user, (3) private consultant purchasing the natural
enemies and using them in a pest management service, (4) grower
cooperative (voluntary participation) providing the natural enemies
either by purchasing or by rearing them and distributing the
natural enemies, (5) pest control districts (special taxing
authority or involuntary participation) receiving or purchasing
the natural enemies, distributing them in a delineated area and
assessing the residents, and (6) State and Federal agencies
producing and releasing biological agents.

### Private Insectary Selling Directly to User

The private insectary (production company or manufacturer)
selling directly (or through a distributor who provides minimum
technical assistance) to the user is much like the company that
sells chemical insecticides.  Stent (1975) conducted a commercial
feasibility study of new generation pesticides from the standpoint
of a private company and concluded that the profitability of such
products as viruses, bacteria, and pheromones was limited because
of (1) small market size, (2) small profit margin which is related
in part to patentability, (3) temporal efficacy (how quickly the
product acts) and (4) cost effectiveness.  Earlier, Stanford
Research Institute (1974) identified the various elements of a
venture decision on augmentation of natural enemies.  A schematic
developed in this effort (Figure 1) should be useful in analyzing
the venture decision faced by a private company that is considering
rearing and distributing natural enemies.

Venture risks may be divided into revenue and cost risks.  On
the attached diagram, revenue risks are associated with the market
considerations and cost risks with the production considerations.

To start with, the size of the market for biological agents is
smaller than for broad spectrum chemical pesticides.  The potential
market size is based on crop acres and level of infestation.  It

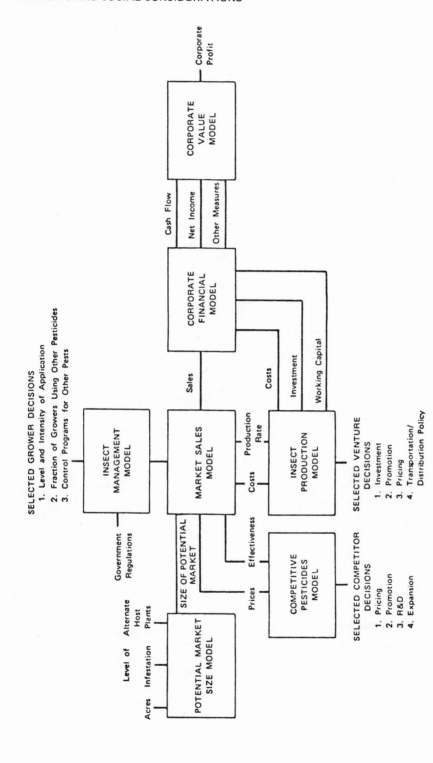

Fig. 1. Schematic of the venture decision for commercialization of augmentation of natural enemies (Stanford Research Institute 1974).

becomes limited very quickly by the specific nature of the biological agent and by the existence of competitive controls. The
inability to acquire a patent for a given use allows new entrants
into the biological control industry to reduce the projected market
size. The use of chemical pesticides on adjoining sites reduces
the size of the market even further by inhibiting the effectiveness
of the natural enemies.

On the cost side, there are the risks associated with attempting to identify the scale economies associated with insect production
technology to ensure sufficient quantity and quality of natural
enemies. The problems of seasonal peak load demands for the insect
and the relatively short shelf life of natural enemies compound the
risk. Fortunately, a nematode of mosquitoes can be stored rather
easily (Fairfax 1976). However, storage of many other natural
enemies requires rather sophisticated manipulation (Morrison and
King, Chapter 6). The risk can be reduced to the extent that the
seasonal demands can be smoothed out. Diversification is one
method that is being tried by *Encarsia* producers. They are
encouraging *Encarsia* release on glasshouse ornamentals as well as
on glasshouse vegetables (Hussey Chapter 12). After considering
the risks associated with potential revenue and costs, the private
producer must consider the availability and cost of capital as well
as the impact of government regulations on use, rearing and
transport.

In addition to the risks to the producer associated with
rearing and distribution, there is the risk of loss that is
perceived by the user which affects the production company's
markets. At the present time, some home gardeners and organic
gardeners and farmers apparently are more willing to assume these
risks than are many large-scale farmers. Therefore, most direct
sales in the U.S. are made to these users. Apparently, some
additional incentives will be necessary before major expansion of
private insectaries selling directly to the user can be expected.

## Private Insectary Providing Management Services

Company representatives of the largest supplier of parasites
and predators in the U.S. stress that the effective application of
biological agents usually requires more sophistication on the part
of their users (growers of fruits, vegetables, field crops, and
ranchers and poultry farmers). The distributor, therefore, must
be in a position to enhance the knowledge of his clients. In
the case of this particular company the knowledge base includes
crop/livestock production practices as well as techniques of pest
control. Knowledge is a difficult service to sell at a profit
because once it is delivered to the purchaser it is freely
available to nonpurchasers. Nevertheless, this company feels that

such knowledge is vital both to the success of augmentations and to
its own economic viability.  For these reasons, the rearing of
natural enemies has been combined with a pest control, crop/livestock
production consulting service since 1960 (Dietrick 1972, DeBach
1974).  However, pest management services involving natural enemies
may be able to operate effectively without being integrally associated with the rearing operation.

## Pest Management Consultants

For the most part, private pest management consultants
currently functioning in the U.S. rely primarily on scouting and
the use of chemical pesticides.  However, some consultants provide
a total crop management program that includes advice on fertilizer
and seed variety.  These services usually are rendered on a per
acre basis.  Among these consultants there are a few that employ
augmentation techniques.  As augmentation technology develops it
can be expected that there will be an increased number of situations
in which an economic incentive exists for private consultants to
include augmentations of natural enemies along with the other pest
control techniques.  However, it is often difficult to effectively
utilize alternative pest control techniques without formal coordination among producers.  Voluntary grower organizations are one
way to achieve this necessary coordination.

## Grower Cooperatives

Grower cooperatives are emerging in some places as a viable
arrangement for applying current methods of pest control (Vogelsang
1977).  The Fillmore Citrus Protective District, Ventura County,
California was organized in 1922 to chemically eradicate the red
scale.  It now depends mainly on the augmentation of natural enemies
to control most of the insect problems.  Currently, it releases
scale parasites *Aphytis* and *Metaphycus* and a mealybug predator,
*Cryptolaemus*, on over 9,000 acres (Pennington 1975).  It is one of
the unique examples of long term use of augmentations.  Associates
Insectary, a cooperative organized in a similar manner, operating
since 1928 in California, employs chemicals and biological agents
on a fifty-fifty basis (Lindsay 1975).

Growers in other parts of the U.S. are forming cooperatives to
provide integrated pest management systems.  Biological control can
be a part of such systems.  This possibility is illustrated by the
Safford Valley Cotton Growers Cooperative's reliance on native
beneficial insects and cultural practices for the control of pink
bollworm on cotton.  It would seem that augmentation could be
implemented under the auspices of this type of cooperative as well

as under the other types of cooperatives identified recently (Good et al. 1977, Vogelsang 1977). However, for each of these types, the option of subcontracting the pest and/or crop management functions to private consultants is available and in many instances, may be the most efficient way to proceed. Even the issue of bearing risks of user losses has been faced by some of these cooperatives. Their responses have ranged from disclaimers of responsibility to liability insurance to indemnify officers for mistakes made in pest management decisions. In the opinion of the authors, it is important to provide such protection for these key people in the cooperative, but the issue of grower risks must be solved directly if farmers are to be given the incentive to use augmentation as well as other new techniques.

## Pest Control Districts

With the possible exception of glasshouses, dairy barns, livestock and poultry operations and other situations in which the user has control over the immediate environment, the use of natural enemies for pest control usually is enhanced by cooperation among all the producers of agricultural commodities in an area. Informal, voluntary cooperation among individuals in a region may not be adequate. Thus, it may be difficult for producers to participate individually in effective augmentation programs or to form effective organizations. In these situations, the formation of pest control districts may be desirable since the group can pursue pest control policies consistent with augmentations of natural enemies or other control strategies and can utilize the legal power of the State in enforcement of necessary procedures.

The use of the concept of pest control districts in the United States is primarily restricted to mosquito control. There are some 260 organized mosquito control districts that expend some $45 million annually (NAS 1975, Vol. V). The mosquito control districts utilize habitat management and insecticides as their primary control methods. However, a number of districts also stock mosquito breeding areas with fish that prey on mosquitoes.

Enabling legislation that provides for the organization of pest control districts to control crop pests exists or has been proposed in a number of States (NAS 1975, Vol. III). However, districts requiring mandatory participation in the use of augmentation of natural enemies for the control of crop pests do not currently exist in the United States.

## State and Federal

Besides providing a regulatory and enforcement framework within which institutions of the type described may operate, State and Federal governments in the U.S. have taken active roles in operating certain pest control programs. Perhaps the most significant example has to do with the use of sterile screwworms, *Cochleomyia hominivorax* (Coquerel), for the eradication of the screwworm from the Southeastern U.S. and for the control of the screwworm in the Southwestern U.S. This program, which is funded jointly by livestock producers and Federal and State agencies, is operated by the U.S. Department of Agriculture (Bushland 1971). The program in the Southwestern United States has involved the rearing and release of 50 to 200 million sterile flies each week for over 10 years and has provided effective suppression of a major pest (Waterhouse et al. 1976).

There is now increasing interest and activity among Federal and State agencies dealing specifically with mass rearing and release of natural enemies. Until recently, research agencies have been primarily involved in distributing imported natural enemies. Now, however, parasites of such insects as the cereal leaf beetle and the gypsy moth are being reared and distributed by Federal and State action groups. In addition, a research and development program involving augmentation by inoculative releases of *Pediobius foveolatus* (Crawford), a parasite of the Mexican bean beetle, *Epilachna varivestis* Mulsant, (Stevens et al. 1975, Dively 1976) has been transferred into an annual action program by the State Department of Agriculture in Maryland. Expansion of this program into other States has been proposed. Similar programs are likely to be undertaken by Federal and State action agencies in the future.

## CONCLUSIONS

An informed decision to employ biological controls, in general, and augmentations of natural enemies in particular, should involve a calculation of the benefit-cost ratio of biological agents relative to other methods. The current state of pest control technology and existing institutional arrangements is such that chemical agents dominate the pest control mix. Moreover, the lack of incentive to use biological controls is based at least partially on user perception of the favorable benefit-cost ratio of pesticides. Development of some new methodolgy as well as in-depth case studies are needed in order to determine more meaningful benefit-cost ratios for biological methods.

The current use of augmentations represents only a small portion of worldwide pest control practices. However, augmentations of natural enemies do have a major impact on pest control practices in specialized situations. Their use for control of pests of glasshouse crops in Europe and the U.S.S.R. and for control of field crop pests in the U.S.S.R. and the People's Republic of China appear to be particularly noteworthy.

The expansion of the use of augmentations of natural enemies will require continued research and development activities to reduce costs. Also, additional incentives for involvement of the private sector may be needed. In addition, development of new institutional arrangements more suited to implementation of augmentations should encourage expansion.

## REFERENCES CITED

Bushland, R. C. 1971. Sterility principle for insect control: historical development and recent innovations. p. 3-14 In Sterility Principle for Insect Control or Eradication. Int. Atomic Energy Agency, Vienna. 542 pp.

DeBach, P. 1974. Biological control by natural enemies. Cambridge University Press, London. 323 pp.

Dietrick, E. M. 1972. Private enterprise pest management based on biological controls. Proc. Tall Timbers Conf. on Ecological Animal Control by Habitat Management. p. 7-21.

Dively, G. P. 1975. Pilot soybean scouting and parasite release program on the Delmarva Penninsula. Mimeograph. 6 pp.

Fairfax Biological Laboratory, Inc. 1977. Label for product "Doom" and related technical information.

Gerberg, E. J. 1977. Personal communication. Insect Control and Research, Inc., Baltimore, Maryland.

Good, J.M., R. E. Hepp, P. O. Mohn, and D. L. Vogelsang. 1977. Establishing and operating grower owned organizations for integrated pest management. Ext. Serv., USDA. PA 1180.

Hart, W. G. 1972. Compensatory releases of *Microterys flavus* as a biological control agent against brown soft scale. Environ. Entomol. 1: 414-19.

Hepp, R. E. 1976. Alternate delivery systems for farmers to obtain integrated pest management services. Agr. Econ. Rpt. No. 298. East Lansing, Michigan. 24 pp.

Hussey, N. W. 1977. Personal communication. Acreage under biological control. Glasshouse Crops Res. Inst. May 10.

Jimenez, J., Eleazar. 1975. The action program for biological control in Mexico. Presented at the work conf. of the U.S. and Mexico on biological control of pests and weeds. Brownsville, Texas. Feb. 20-21. 3 pp.

Kaitz, E. F., and J. P. Weimer. 1976. Home grown fruits and vegetables and their use. USDA. TVS-201. August. 4 pp.

Kravchenko, I., and V. Dolan. 1975. Biological measure of plant protection. Committee on Agr. Prob. Econ. Comn. for Europe. U.S. Econ. and Social Council. May 28. 17 pp.

Lindsay, J. B. 1975. Annual report of Associates Insectary. San Paula, California. 8 pp.

May, C., and K. Foulke. 1977. Personal communication. The rearing of natural enemies in Ventura and Riverside Counties, California.

Montes, M. 1970. Estudios sobre bionomia V la biometria de *Lixophaga diatraeae* (Towns.). (Diptera, Tachinidae). Ciencias Biologicas 4(7): 60 pp.

National Academy of Sciences. 1975. Contemporary pest control practices and prospects. Wash., D. C. Vol. III, 138 pp. and Vol. V, 282 pp.

Nickle, W. R.   1976.   Toward the commercialization of a mosquito
    mermithid.   Proc. First Internatl. Colloquium Invert. Path.
    p. 241-4.

Pennington, N.   1975.   Managers report.   Fillmore Citrus Protect.
    Dist.   Jan. 1 to Dec. 31.   3 pp.

Pimentel, D. et al.   1977.   Pesticides, insects in foods, and
    cosmetic standards.   BioScience 27(3): 178-85.

Rincon-Vitova Insectaries, Inc.   1976.   Price List.   P.O. Box 95,
    Oak View, California.   1 p.

Ridgway, R. L., R. E. Kinzer, and R. K. Morrison.   1974.   Produc-
    tion and supplemental releases of parasites and predators for
    control of insect and spider mite pests of crops.   p. 110-116
    In F. G. Maxwell and F. A. Harris, eds.   Proc. of the Summer
    Ist. on Biol. Control of Plant Insects and Diseases.
    University Press Mississippi, Jackson.   647 pp.

Ridgway, R. L., N. H. Starler, and P. A. Andrilenas.   In press.
    Extent of use, costs, and trends in the control of plant
    pests: insects.   In W. Ennis, ed.   Crop Protection.
    Amer. Soc. Agron.

Schalk, J. M., and R. Ratcliffe.   1975.   An evaluation of ARS
    program on alternative methods of insect control: use of crop
    varieties with resistance to insects.   Bul. Entomol. Soc.
    Amer. 22: 7-10.

Stanford Research Institute.   1974.   Proposal for a preliminary
    assessment of the supplemental release of natural predators
    and parasites for the control of the bollworm and tobacco
    budworm on cotton.   Menlo Park, California.   10 pp.

Stent, P. D.   1975.   Commercial feasibility of new generation
    pesticides.   Proc. Conf. on Substitute Chem. Program.
    Environmental Protection Agency.   p. 73-86.

Stevens, L. M., A. L. Steinhauer, and J. R. Coulson.   1975.
    Suppression of Mexican bean beetle on soybeans with annual
    inoculative releases of *Pediobius foveolatus*.   Environ.
    Entomol. 4: 947-52.

Stinner, R. E., R. L. Ridgway, J. R. Coppedge, R. K. Morrison, and
    W. A. Dickerson, Jr.   1974.   Parasitism of *Heliothis* eggs
    after field releases of *Trichogramma pretiosum* in cotton.
    Environ. Entomol. 3: 497-500.

U.S. Department of Agriculture.   1977.   In press.   Biological
    agents for pest control: current status and future prospects.

Vogelsang, D.   1977.   Local cooperatives in integrated pest manage-
    ment.   USDA.   Farmers Coop. Serv.   Res. Rpt. No. 37.

Waterhouse, D. F., L. E. LaChance, and M. J. Whitten.   1976.   Use
    of autocidal methods.   p. 637-59 In C. B. Huffaker and
    P. S. Messenger, eds.   Theory and practice of biological
    control.   Academic Press, N.Y. 788 pp.

APPENDIX


COMMERCIAL SOURCES OF NATURAL ENEMIES IN THE U.S. AND CANADA


(Known Producers Shown in Bold Type)

ASSOCIATES INSECTARY
1400 Santa Paula Street
Santa Paula, California 93040

BENEFICIAL INSECT COMPANY
383 Waverly Street
Menlo Park, California 94025

Beneficial Insect Company
P.O. Box 1578
Oroville, California 95965

Bio-Control Company (Cal. Bug Co.)
Route 2, Box 2397
Auburn, California 95603

Bio Insect Control
P.O. Box 915
Plainview, Texas 79072

BIOLOGICAL FIELD SERVICE COMPANY
1308 South Pinkham Road
Visalia, California 93277

Biosystems Fly Control
1525 63rd Street
Emeryville, California 94608

Biotactics, Inc.
22412 Pico Street
Colton, California 92324

Bo-Bio Control
P.O. Box 154
Banta, California 95302

Bobitrol
54 South Bear Creek Drive
Merced, California 95350

Butterfly Breeding Farm
389 Rock Beach Road
Rochester, New York 14617

Bowen, Carry
54 South Bear Circle Drive
Merced, California 95340

Burpee Seed Company
Warminster, Pennsylvania 18974
Clinton, Iowa 52732
Riverside, California 92502

Cal-Ag Services
480 South 4th Street
Kerman, California 93630

California Green Lacewings
P.O. Box 2495
Merced, California 95340

Comeaux, Howard P.
Route 2, P.O. Box 259
Lafayette, Louisiana 70501

Ecological Insects Service
15075 W. California Avenue
Kerman, California 93630

Entomological Services
1333 Rose Avenue
Modesto, California 95350

Ferndale Gardens
Fairibault, Minnesota 55021

FILLMORE CITRUS PROTECTIVE DIST.
1003 West Sespe Avenue
Fillmore, California 93015

Fountain's Sierra Bug Company
P.O. Box 114
Rough & Ready, California 95975

Garden Guards Company
P.O. Box 16
Bootstown, Ohio 44272

Gothard, Inc.
P.O. Box 332
Canutillo, Texas 79835

GOTHARD, JACK, JR.
P.O. Box 7
Valentine, Nebraska 69201

Eastern Biological Control Company
Route 5, P.O. Box 379
Jackson, New Jersey 08527

FAIRFAX BIOLOGICAL LABORATORY, INC.
Electronic Road
Clinton Corners, New York 12514

GREENHOUSE BIOCONTROLS
P.O. Box 1722
Kingsville, Ontario, Canada NOR140

Harris, Paul
P.O. Box 1495
Marysville, California 95901

INSECT CONTROL & RESEARCH CO., INC.
1330 Dillon Heights Avenue
Baltimore, Maryland 21228

Kings Labs
P.O. Box 69
Limerick, Pennsylvania 19468

Lakeland Nurseries Sales
340 Poplar Street
Hanover, Pennsylvania 17331

Ladybug Sales
Route 1, P.O. Box 93A
Biggs, California 95917

Linwood Gardens
Department 743
Linwood, New Jersey 08221

Natural Systems Technology
P.O. Box 350
Caldwell, New Jersey 07006

ORCON
5132 Venice Boulevard
Los Angeles, California 90019

PEST MANAGEMENT SYSTEMS
Route 1, P.O. Box 49
Archer, Florida 32618

Pyramid Nursery
P.O. Box 5274
Reno, Nevada 89102

Pyramid Nursery and Flower Shop
4640 Attawa Avenue
Sacramento, California 95822

Mantis Unlimited
625 Richards Road
Wayne, Pennsylvania 19087

Mincemoyer Nursery
West County Line Road
Jackson, New Jersey 08527

Montgomery Ward and Co., Inc.
Catalog Sales

Red Leopard Ladybugs
1000 41st Street
P.O. Box 19008
Sacramento, California 95819

RINCON-VITOVA INSECTARIES, INC.
P.O. Box 95
Oak View, California 93022

Robbins, Robert
424 N. Courtland
E. Stroudsburg, Pennsylvania 18301

Schnoor's Sierra Bug Company
P.O. Box 114
Rough & Ready, California 95975

Sears
Catalog Sales

Supervised Control Services
405 Oak Street
Shafter, California 93263

Unique Nursery
4640 Attawa Avenue
Sacramento, California 95822

Western Biological Control Laboratories
P.O. Box 1045
Tacoma, Washington 98401

West Coast Ladybug Sales
Route 1, P.O. Box 93A
Biggs, California 95917

CONTRIBUTORS

G. A. Beglyarov
All-Union Scientific Research
   Institute of Phytopathology
Moscow, U.S.S.R.

E. Biliotti, Director
National Agricultural
   Research Council
Route de St Cyr
78000 Versailles, France

E. F. Boller
Plant Protection Section for
   Arboriculture, Viriculture
   and Horticulture
Wadenswil, Switzerland

J. L. Carrillo
Department of Entomology
National Institute of
   Agricultural Investigations
Chapingo, Mexico

D. L. Chambers
Agricultural Research Service
U.S. Department of Agriculture
Gainesville, Florida

G. Gordh
Agricultural Research Service
U.S. Department of Agriculture
Washington, D.C.

H. L. House
Agricultural Canada Research Br.
Smithfield Experimental Farm
Trenton, Ontario, Canada

C. B. Huffaker
Division of Biological Control
University of California
Berkeley, California

N. W. Hussey
Entomology Department
Glasshouse Crops Res. Institute
Sussex, England

E. G. King
Agricultural Research Service
U.S. Department of Agriculture
Stoneville, Mississippi

E. F. Knipling
Agricultural Research Service
U.S. Department of Agriculture
Beltsville, Maryland

J. A. Logan
Department of Entomology
North Carolina State University
Raleigh, North Carolina

P. B. Morgan
Agricultural Research Service
U.S. Department of Agriculture
Gainesville, Florida

R. K. Morrison
Agricultural Research Service
U.S. Department of Agriculture
College Station, Texas

R. L. Rabb
Department of Entomology
North Carolina State University
Raleigh, North Carolina

R. L. Ridgway
Agricultural Research Service
U.S. Department of Agriculture
Beltsville, Maryland

N. E. A. Scopes
Entomology Department
Glasshouse Crops Res. Institute
Sussex, England

E. M. Shumakov
All-Union Research Institute
   for Plant Protection
Leningrad, U.S.S.R.

A. I. Smetnik
Central Research Laboratory
  For Plant Quarantine
Moscow, U.S.S.R.

N. H. Starler
Economic Research Service
U.S. Department of Agriculture
Washington, D.C.

S. B. Vinson
Department of Entomology
Texas A & M University
College Station, Texas

D. E. Weidhaas
Agricultural Research Service
U.S. Department of Agriculture
Gainesville, Florida

Abdullaeva, E., 60
Ables, J. R., 406, 418, 419
Adashkevich, B. P., 306, 307, 310, 313
Addington, J., 351
Adlung, K. G., 343
Afanes'eva, O. V., 284
Alam, M. M., 128
Alcock, J., 249
Alekseeva, Ya. A., 284
Alexandrov, N., 293
Allen, H. W., 139
Allen, J. C., 266
Alloway, T. M., 262
Altman, P. L., 158, 160
Anderson, J. F., 399
Anderson, J. M., 406
Andreeva, L. A., 305
Andrelinas, P. A., 406
Andrewartha, H. G., 7
Anikina, N. E., 312
Antonenko, O. P., 44, 46
Apple, J. M., 208
Arambourg, Y., 342, 343
Areshnikov, B. A., 46, 51
Armitage, H. M., 193, 381
Arthur, A. P., 245a, 247, 248, 251, 255, 256, 262, 267, 268
Ashley, T. R., 139, 266
Askew, R. R., 129, 130, 247
Asyakin, B. I., 310
Atallah, Y. A., 164
Atwal, A. S., 166

Babikir, T., 363
Badgley, M. E., 47, 191
Bagchee, S. N., 255
Bagglioni, M., 62, 64
Baier, M., 248
Bailey, V. A., 7, 80, 83
Baker, J. L., 133
Bakker, K., 255
Barfield, C. S., 248, 251, 252, 253, 254, 255, 262, 265, 267

Barlow, J. S., 161, 166, 170, 171, 173
Barras, D. J., 389
Bartlett, B. R., 226, 246a, 249, 252, 268
Bashadze, 305
Batsch, W. W., 156, 158
Bay, E. C., 417, 418
Beard, R. L., 418
Beck, S. D., 151
Bedard, W. D., 249
Beglyarov, G. A., 305, 307, 309, 310, 311, 312, 403, 408, 433, 436
Beirne, B. P., 140, 155, 184
Bell, J. U., 203
Bell, K., 387
Bell, M. G., 203
Belov, V. K., 309, 310
Benassy, C., 342
Bennett, F. D., 128, 203, 395, 396
Benton, A., 418, 419, 420, 423
Benzer, S., 262
Berg, C. O., 417
Beroza, M., 93, 248, 252, 253, 254, 255
Bianchi, H., 342
Bichina, T. A., 62
Bierl, B. A., 248, 252, 253, 254, 255
Biggerstaff, S. M., 196, 353, 358, 374
Biliotti, E., 342, 433, 440
Binns, E. S., 364
Birch, L. C., 7
Bishop, G. W., 263
Bjegovik, P., 343
Blum, M. S., 239
Bodenheimer, F. S., 7
Boedjonó, W. A., 396
Boldt, P. E., 186
Boldyreva, E. P., 294
Boller, E., 139, 169, 184, 220, 221, 225, 229, 230, 408

Bondarenko, N. V., 305, 307, 309, 310, 312
Bonfils, J., 396
Boucek, Z., 129
Bournier, J. P., 344
Bowery, L. G., 393
Bowman, M. C., 248, 252, 255
Boyd, J. P., 392, 393
Box, H. E., 139
Bracken, G. K., 164, 168, 172, 173
Bragg, D., 246a
Brambell, M. R., 154
Bravenboer, L., 206
Brazzel, J. R., 208
Breniere, J., 128, 344
Brewer, F. D., 205
Bridge, R. R., 406
Briggs, G. M., 158
Briolini, G., 62
Brothers, D. J., 250
Brown, L., 200, 203
Brown, L. E., 249
Brown, L. G., 19
Brown, W. L., 243
Brues, C. T., 152, 154, 163
Brun, P., 184, 187
Bryan, D. E., 89, 184, 185, 199, 389, 390, 391, 397
Brydon, H. W., 418, 421
Buctiner, C. H., 343
Buleza, V. V., 55
Bull, D. L., 406
Bullini, L., 136
Burekova, V. I., 46, 50, 51, 52
Burgess, A. F., 93
Burks, M. L., 251, 252, 255, 268
Burnett, T., 80
Burns, J. M., 136
Burov, V. N., 44, 45, 46
Burton, R. L., 184, 187, 188, 199, 208, 399
Burton, R. P., 418, 420
Bush, G. L., 225, 230, 232, 233
Bushchik, T. I., 312
Bushland, R. C., 447
Butt, B. A., 208, 230

Bynum, E. K., 89

Cade, E., 239
Calvert, D., 268
Cameron, E. A., 93
Camors, F. B., 245a, 247
Cantwell, G. E., 186
Carlson, G. A., 4
Carrillo, J. L., 389
Carter, W., 140
Carton, Y., 246a, 248
Case, J. J., 250
Castillo Chacon, I. A. R., 184
Ceballos, G., 343
Celli, G., 62
Chalkov, A. A., 312
Chambers, D. L., 220, 221, 222, 225, 230, 408
Chang, C. H., 264, 332
Chant, D. A., 4, 24
Chapman, H. C., 417, 423, 424
Chabora, P. C., 249
Chernova, O. A., 43
Chesnut, T. L., 262
Chochia, 298
Christensen, J. B., 20
Christenson, L. D., 228
Chumakova, B. M., 301, 302
Ciampolini, M., 62, 64
Clark, L. R., 5, 7
Clausen, C. P., 79, 126, 129, 238, 263
Cody, M. L., 8
Combs, R. L., 418
Coluzzi, M., 130, 136, 225
Colwick, R. F., 407
Compere, H., 129
Conway, G. R., 6, 8
Cooke, W. C., 399
Coppedge, J. R., 388, 400, 404, 432
Corbet, P. S., 246a, 249, 253
Cothran, W. R., 20
Coulson, J. R., 118, 408, 447
Covarrubias, R., 389
Cowan, C. B., 406
Crandell, H. A., 419, 420
Croft, B. A., 382, 384

Cross, W. W., 262
Crossman, S. S., 93
Croze, H., 249
Crozier, R. H., 137
Cushman, R. A., 134, 238

Dadd, R. H., 159, 160
Daumel, J., 184, 187, 343
Davidson, G., 137
Davis, G. R. F., 161
Davis, J. W., 406
De, R. K., 337
DeBach, P., 25, 29, 40, 79,
    125, 127, 128, 129, 130,
    132, 134, 135, 137, 138,
    139, 172, 184, 185, 186,
    192, 193, 200, 222, 225,
    226, 255, 263, 264, 265,
    379, 382, 384, 399, 404,
    433, 434, 441, 445
Decaux, F., 341
Deevey, E. S., 230
Delanoue, P., 342
DeLoach, C. J., 246a, 254
Delucchi, V. L., 126, 129
DeMichele, D. W., 19
Depner, K. R., 418
Dethier, V. G., 152
Detrick, E. I., 418, 421
Detwilder, D., 405
Diamond, J. M., 8
Dickerson, W. A., 388, 400,
    407, 432
Dietrick, E. M., 445
Dirsh, V. M., 284
Dittmer, D. S., 158, 160
Dively, G. P., 434
Doane, C. C., 93, 95
Dobzhansky, T., 225
Dodd, A. P., 264
Dohanion, S. M., 265
Dolan, V., 436, 440, 441
Doutt, R. L., 29, 125, 126,
    127, 134, 137, 222, 225,
    239, 256, 380, 382
Dowden, P. B., 93
Doymm, 55
Dupuis, C., 44
Dutky, S. R., 186, 384
Dyadechko, 289

Dysart, F. J., 200
Dzyuba, Z. A., 294

Eckman, D. P., 208
Economopoulos, A. P., 229,
    230
Edwards, R. L., 249, 250,
    255, 419
Eisner, T., 243
Elfimov, V. I., 53, 54
Elsey, K. D., 393
Etienne, J., 186
Evans, A. C., 152
Evans, H. E., 134, 135

Falcon, L. A., 20, 405
Farkas, S. R., 239
Farley, R. D., 251
Fazaluddin, M., 131
Featherston, P. E., 198, 199,
    387
Fedchenko, M. A., 46
Feder, G., 20
Fedorinchik, N. S., 286
Fedotov, D. M., 43
Fedotova, K. M., 52
Feeney, P. P., 245a
Feron, M., 342
Ferriere, C., 133
Feshchin, D. M., 51
Finney, G. L., 131, 183, 185,
    188, 189, 190, 191, 198,
    199, 263
Fisher, R. C., 260, 263, 264
Fisher, T. W., 127, 131, 183,
    185, 188, 193
Fiske, W. F., 7, 40, 80
Flanders, S. E., 47, 131,
    133, 134, 185, 193, 198,
    199, 200, 240, 284
Flaherty, D. L., 30
Fleschner, C. A., 238
Foerster, A., 133
Force, D. C., 139, 140
Foulke, K., 434
Francisco, J. L., 389
Frankel, G., 153
Fransz, H. G., 8, 11, 12
Franz, J. M., 58, 128, 341,
    342

Frazer, B. D., 19
Frazer, H. G., 19
Fried, M., 228
Frisbie, R. E., 407
Fukuda, T., 417
Fusco, R. A., 399
Fye, R. E., 184, 185, 199,
    390, 393

Galichet, P. F., 396
Gannaway, J. R., 406
Gantt, C. W., 204, 205
Gaprindashvili, N. K., 59,
    298, 305
Garber, M. J., 139, 173
Gardner, T. R., 263
Gast, R. T., 208
Gause, G. F., 17
Geier, P. W., 5, 7
Genduso, P., 342
Gerberg, E. J., 434
Gerberich, J. B., 417
Gerhardt, R. R., 418
Gerling, D., 134, 249, 418,
    419
Geyspits, K. F., 354
Gilbert, N., 19
Gill, R. J., 248, 252, 253,
    255
Gilmore, J. E., 83, 87, 115
Ginney, G. L., 30
Gokhelashvili, R. D., 62
Goldstein, J., 405
Goncharenko, E. G., 59
Gonzalez, D., 131, 139, 226,
    255
Gonzalez, R. H., 389
Good, J. M., 446
Goodall, D. E., 353, 374
Goodpasture, C., 136, 137
Gordh, G., 130, 133, 134,
    135, 250
Gordon, C. C., 387
Gordon, H. T., 165
Goryunova, Z. S., 294, 301,
    302
Gossard, T. W., 19
Gosswald, K., 343
Gould, H. J., 361, 364
Gradwell, G. R., 80

Graham, H. M., 389
Granett, J., 266
Greany, P. D., 131, 245a, 247,
    248, 251, 252, 253, 255,
    263, 264
Greathead, D. J., 341, 418
Green, G. L., 386
Griffiths, K. J., 264
Grimble, D. G., 96, 399
Grissell, E. E., 137
Gross, Jr., H. R., 96, 186,
    200, 230, 241, 249, 256,
    257, 258, 260, 261, 263,
    266, 267
Guerra, M., 203
Guignard, E., 342
Guillot, F. S., 261, 264
Gupta, S. N., 337
Gusev, G. B., 291, 293, 294,
    304
Gutierrez, A. P., 19, 20, 251
Guthrie, F. E., 393

Hafez, M., 127, 134, 137
Hagen, K. S., 25, 138, 164,
    165, 168, 189, 190, 192,
    200, 248, 251, 255, 263,
    268, 379, 382, 384, 399
Hagov, E. M., 304
Haile, D. G., 423
Hair, J. A., 418, 420
Hale, R., 168
Halfhill, J. E., 198, 199,
    387
Hall, J. C., 127, 128, 132
Hammon, R., 364
Han, Y., 332
Harris, W. A., 262
Hart, W. G., 404, 432
Hartstack, A. W., 407
Hassan, S. A., 189
Hassell, M. P., 5, 6, 8, 12,
    19, 245a, 247
Hathaway, D. O., 230
Hawke, S. D., 251
Haydak, M. H., 203
Hayes, C. G., 418
Hays, D. B., 248
Heagler, A. M., 406
Heath, R. R., 253

Hegdekar, B. M., 245a, 248, 251, 255, 268

Helgesen, R. G., 380

Hendry, L. B., 245a, 248, 252, 253, 255, 262

Hensley, S. D., 396, 397

Henson, R. D., 245a, 248, 251, 252, 253, 254, 255, 262, 265, 267

Hepp, R. E., 441, 446

Herman, S. G., 17

Herrebout, W. M., 247

Herzog, G. A., 230

Hesketh, J. D., 407

Higgins, A. H., 400

Hirsch, J., 229

Hodek, I., 344

Hoefer, I., 262

Hoelscher, C. E., 418

Holdaway, F. G., 264

Holldobler, B., 250

Holling, C. S., 8, 17, 19, 90

Hollingsworth, J. P., 407

Holmes, H. B., 419

Hood, C. S., 93

Hopkins, A. R., 406

House, H. L., 160, 161, 163, 164, 165, 166, 167, 168, 170, 171, 173, 268, 408

House, V. S., 189, 190

Howard, L. O., 7, 40, 80

Hoy, M. A., 138, 173, 399

Hoying, S. A., 384

Hsiao, T., 264

Hsiu, C., 332

Huang, M., 338

Hubsch, 246a

Huettel, M. D., 220, 221, 222, 224, 228, 230, 232, 233

Huffaker, C. B., 4, 5, 7, 8, 9, 12, 14, 17, 19, 21, 28, 29, 30, 40, 80, 125, 200, 338, 378, 384, 403, 441, 436

Hughes, R. D., 5, 7

Hussey, N. W., 206, 354, 367, 403, 433, 436, 440, 444

Iperti, G., 344

Iving, S. N., 363

Ismailov, Ya. I., 290

Ivanov, S., 62, 64

Iwata, K., 134

Jackman, J. A., 407

Jackson, C. G., 89, 184, 185, 199, 389, 390, 391

Jackson, G. J., 363

Jackson, R. D., 397, 398

Jacobson, M., 253

Janzen, D. H., 22

Jaynes, H. A., 89, 263

Jenkins, D. W., 417

Jennings, J., 137

Jenson, J. L., 407

Jimenez, E., 389

Johnson, H. G., 382

Johnson, F. M., 136

Joiner, R. L., 264

Jones, F. G. W., 185, 249, 268, 269

Jones, J. W., 407

Jones, R. E., 19

Jones, R. L., 116, 238, 241, 245a, 248, 252, 253, 254, 255, 256, 257, 258, 259, 260, 261, 262, 263, 266, 267, 268, 389

Jones, S. L., 263, 391, 400, 406

Kaitz, E. F., 440

Kajita, H., 169

Kalashnikova, G. I., 290

Kamenkova, K. V., 51, 53, 57

Karl, K. P., 128

Kartavtsev, N. I., 286, 289, 294

Katsoyannos, P., 344

Keiding, J., 418

Kennedy, J. S., 22

Kennett, C. E., 21, 30, 338, 384

Kerrich, G. J., 129

Khasimuddin, S., 132, 134, 137, 139

Khloptseva, R. I., 312

Khloshcheva, R. I., 312
Kholchenkov, V. A., 58, 62, 63
Killebrew, R., 164
Kim, J. G., 387
King, E. G., 184, 203, 204, 205, 267, 397, 398, 408, 444
Kinzer, R. E., 183, 184, 187, 389, 391, 400, 434
Kiseleva, O. M., 286
Kitto, G. B., 225, 232
Klassen, W., 330
Klechkovskii, E. R., 46, 50, 51, 52
Klimetzek, D., 343
Knipling, E. F., 32, 80, 83, 84, 85, 87, 89, 93, 96, 102, 115, 185, 186, 200, 230, 267, 396, 407, 408, 435
Kochetova, N. I., 50, 56
Kolmakova, V. D., 286
Kolotova, 298
Koroleva, N. I., 285
Kowalewski, E., 374
Kozlova, T. A., 305
Kraus, M., 244
Krause, G., 268
Kravchenko, I., 436, 440, 441
Krebs, C. J., 16
Kogan, M., 133
Kovaleva, M. F., 286
Kovtun, I. V., 51
Kozlov, M. A., 43, 55, 292, 293
Kozlova, E. N., 58
Kozlova, T. A., 310
Kupershtein, M. L., 44, 45
Kurbanov, G. G., 305
Kuzina, N. P., 306, 307
Kwei, C., 332

Labeyrie, V., 248
LaBrecque, G. C., 418, 419, 420, 422, 423
LaChance, L. E., 447
Lacewell, R. D., 406
Lacey, L. A., 134

Laing, J. E., 29, 30
Laird, M., 417
Lane, A., 364
Lane, H. C., 407
Lapina, V. F., 284
Laudeho, Y., 344
Lawson, F. R., 393
Ledieu, M., 352, 363
Leeling, N. C., 266
Leetes, G. L., 184, 187, 188, 199
Legay, J. M., 171
Legner, E. F., 133, 138, 417, 418, 419, 420, 421
Lehman, W., 62
Leigh, T. F., 139
Leipzig, P. A., 20
Leius, K., 155, 164, 168
Leonard, D. E., 93, 95, 246a, 252
Leppla, N. C., 230
Levinson, Z. H., 175
Lewis, K. R., 405
Lewis, W. J., 96, 116, 186, 200, 230, 238, 241, 243, 248, 249, 252, 254, 255, 256, 257, 258, 259, 260, 261, 262, 263, 266, 267, 268, 389
Limon, S., 134
Lindauer, M., 262
Lindsay, J. B., 441, 445
Lindquist, A. W., 418, 420
Lingren, P. D., 268, 386, 387, 389, 391, 393, 406
Liotta, G., 342
Listkova, R. A., 312
Livshits, 58
Lloyd, D. C., 264
Lloyd, E. P., 406
Loew, W. B., 20
Logan, J. A., 33, 408
Longfield, C., 249
Loomis, J., 405
Loosli, J. K., 152
Lopez, J. D., 391, 393
Lorbeer, H., 199, 382
Lotka, A. J., 7
Lucas, A. M., 139
Lucas, H. L., 19

Luck, R. F., 8, 19, 29
Lukin, V. A., 60
Luppova, E. P., 304, 305
Lydin, L. V., 230

MacArthur, R. H., 8, 9, 22
Mackauer, M., 127, 135, 139,
    169, 183, 184, 220, 221,
    222
MacPhee, A. W., 29
Madden, J., 246a, 247
Mai, S., 338
Maksimovic, M., 343
Man, R. F. L., 239, 353
Mangum, C. C., 208
Maniglia, G., 343
Mansingh, A., 168
Marchal, P., 341, 342
Marston, N., 186
Martin, C. H., 248
Martin, D. F., 184, 203,
    204, 205, 397, 398
Martin, P. B., 386
Masner, L., 54
Mason, G., 137
Mathai, S., 166
Mathys, C., 342
Matkovskii, S. G., 293
Matovskaya, 292
Matthews, R. W., 135, 238
May, C., 434
May, R. M., 5, 6, 8, 19, 33
Maynard, L. A., 152
Mayr, E., 126, 127
McClanahan, R. J., 380
McCoy, C. W., 418, 421
McGuire, J. U., 32, 82, 83,
    84, 89, 102
McKinion, J. M., 407
McLeod, J. H., 380
McMurtry, J. A., 206, 207,
    382, 384
McPherson, R. M., 396
Meelis, E., 255
Meier, N. F., 48, 285, 288,
    292, 299, 301
Mendenhall, B., 249
Meredith, W. R., 406
Merritt, C. M., 20

Messenger, P. S., 5, 7, 8,
    9, 14, 19, 28, 29, 30,
    40, 126, 222, 379
Metcalf, J. R., 128
Metchell, S. N. T., 208
Meyer, R. H., 384
Micks, D. W., 137
Miles, L. R., 184, 203, 204,
    205
Miller, M. C., 251
Milne, A., 7
Mineo, G., 342, 343
Misra, M. P., 337
Mitchell, E. B., 246a, 406
Mitchell, W. C., 239, 253
Mohn, P. O., 446
Moiseev, E. G., 305, 307,
    309
Monastero, S., 342
Monge, J., 389
Monteith, L. G., 247
Montes, M., 203, 396, 397,
    433
Monty, J., 418, 421
Moore, B. P., 239
Moore, N. W., 249
Moran, V. C., 250
Morgan, P. B., 418, 419,
    420, 422, 423, 435
Mormyleva, V. F., 62, 63
Morris, R. F., 5, 7
Morrison, F. B., 152, 153
Morrison, R. K., 183, 184,
    185, 187, 189, 190, 191,
    192, 200, 203, 230, 267,
    388, 389, 400, 408, 432,
    434, 444
Morton, L. J., 230
Mourier, H., 421
Mudd, A., 246a, 253
Murdoch, W. W., 17, 33
Murphey, R. K., 249
Musebeck, C. F. W., 265

Nadler, G., 208
Nagaraja, H., 131, 132
Nagarkatti, S., 131, 132, 139
Nagel, W. P., 419, 420
Neck, R. W., 225, 232

Negi, P. S., 337
Neeman, E. G., 389, 390, 391
Nesser, S., 134
Nettles, W. C., 245a, 251,
    252, 255, 268
Neuffer, H., 342
Newsom, L. D., 39
Nicholson, A. J., 7, 16, 40,
    80, 83
Nickle, W. R., 384, 440
Nieble, B. W., 208
Niles, G. A., 406
Niva, 301
Noble, L. W., 268, 389
Nordlund, D. A., 96, 116,
    186, 200, 230, 238, 241,
    243, 256, 257, 258, 259,
    260, 261, 263, 266
Norton, W. N., 251
Novitskaya, G. N., 59
Nstvik, E., 419
Nyaes, A., 205, 307

Oaten, A., 17, 33
Oatman, E. R., 131, 255, 264,
    384, 386
Olhowski, H., 405
Olkowski, W., 405
Olton, G. S., 418, 419, 421
Oman, P. W., 129
Onillon, J. C., 187, 341
O'Reilly, C. J., 353
Orphanides, G. M., 255

Pacheco, R., 389
Parker, F. D., 228, 338, 385
Parr, W. J., 354, 361, 367,
    368
Parsons, P. A., 229
Patana, R., 89, 184, 185,
    199, 390
Patterson, R. S., 418, 419,
    420, 422, 423
Pavan, M., 343
Payne, T., 245a, 247, 255
Peck, O., 418
Pennington, N., 382, 404,
    441, 445
Perkins, W. D., 96, 186,
    200, 230

Perekrest, O. N., 313
Peterson, J. J., 417, 424
Petrova, V. K., 62, 64
Petrushova, 58
Pettey, F. W., 264
Peyrelongue, J. Y., 344
Phillips, A. M., 200, 203,
    230
Picard, F., 248
Pimental, D., 419, 420, 432
Pinkus, H., 418, 419
Pinnell, R. E., 228, 385,
    391, 393
Pizzol, J., 343
Platner, G. R., 131, 386
Pleshanov, A. S., 305
Plotnikov, V. F., 312
Poo, C., 338
Popov, G. A., 293
Popova, A. I., 299, 302
Potter, R. F., 364
Povel, G. D. E., 136
Pralavorio, R., 343
Pratt, J. J., 168
Price, P. W., 4, 16, 22,
    255, 260
Pritchard, G., 249
Pukinskaya, G. A., 294

Quednau, W., 131, 133, 246a,
    248, 254
Quinn, W. G., 262

Rabb, R. L., 4, 24, 25, 32,
    183, 379, 393, 407, 408,
Radetskii, A. F., 283, 284
Rajendram, G. F., 248, 251,
    255, 268
Rao, S. V., 127, 132
Ratcliffe, R., 431
Ratzeburg, J. T. C., 133
Raulston, J. R., 230
Ravelli, V., 62, 64
Reed, D. P., 245a
Reeves, B. G., 208, 389, 400
Regev, U., 20
Rehmet, A., 137
Remaudiere, G., 293
Remington, C. L., 139
Remund, U., 229, 230

Rice, R. E., 239, 253
Richerson, J. V., 246a, 254
Ridgway, R. L., 83, 184,
    185, 187, 189, 190, 191,
    192, 200, 203, 222, 230,
    248, 252, 253, 255, 379,
    386, 387, 388, 389, 391,
    393, 400, 404, 406, 408,
    431, 432, 434, 436
Ridgway, W. O., 208
Riordan, D. F., 173
Robacker, D. C., 255, 262
Robinson, R. W., 374
Rodendorf, B. B., 43
Rodin, J. O., 249
Rogachaya, L. B., 51
Rogers, D., 238
Rollins, C., 248, 255
Ronchetti, G., 343
Root, R. B., 245a
Rosen, D., 126, 128
Rosenberg, 61
Rosenzweig, M. L., 33
Rossler, Y., 137, 138, 230
Roth, J. P., 251, 252
Robama, T., 19
Rubtsov, I. A., 43, 47, 298,
    300
Rudinsky, J. A., 248
Ruesink, W. G., 20
Ryakhovskii, V. V., 46, 50,
    51, 52, 53, 54
Ryan, R. E., 248
Ryckman, R. E., 125

Saad, A. A. B., 263
Sabrosky, C. W., 126
Sagovan, 298
Sailer, R. I., 79
Salt, G., 185, 238, 240,
    244
Samson-Boshuizen, M., 255
Sang, J. H., 158
Sang, P., 159, 161, 164, 165,
    169
Sawall, E. F., 168, 263
Scaramuzza, L. C., 396
Schalk, J. M., 431
Schlinger, E. I., 125, 126,
    127, 128, 132, 134

Schmidt, G. T., 246a, 248,
    252, 254
Schneider, F., 263
Schoenheimer, R., 156
Schoenleger, L. G., 208
Schubert, G., 343
Schutte, F., 342
Schwartz, A., 249
Schwinch, I., 241
Scopes, N. E. A., 187, 196,
    206, 207, 353, 358, 374,
    403
Scriven, G. T., 206, 207, 382
Selander, R. K., 232
Sethi, S. L., 166
Shahjahan, M., 246a
Shands, W. A., 387
Shapiro, V. A., 51, 52, 54
Shchepetil'nikova, V. A., 43,
    46, 47, 48, 51, 53, 54,
    285, 286, 287, 288, 290,
    291, 292, 293
Shchichenkov, P. I., 290
Shea, K. P., 17
Shengelaya, E. S., 305
Shephard, M., 418, 419
Shewell, G. E., 239
Shoemaker, C. A., 20, 21
Shorey, H. H., 239, 241
Shteinberg, D. M., 58, 165
Shumakov, E. M., 43, 48
Shuvakhina, E. Ya., 306
Sidorovkina, E. P., 285, 305
Silverstein, R. M., 249, 255
Simmonds, F. J., 29, 128,
    139, 186
Simmonds, S. P., 361
Simpson, G. W., 387
Singh, P., 156, 158
Skeith, R., 19
Slavov, N., 62
Slifer, E. H., 251
Slobodchikoff, C. N., 248,
    255
Smetnik, A. I., 403, 408, 436
Smirnova, A. A., 58
Smith, B. C., 153, 168
Smith, C. N., 183
Smith, H. S., 7, 40, 80,
    185, 193, 198, 199, 381

Smith, J. M., 166
Snodgrass, R. E., 164
Snow, J. W., 268
Solomon, M. E., 5, 8, 9, 17, 40, 41, 80
Somsen, H. W., 399
Sonnet, P., 255
Soo Hoo, C. F., 153
Soper, R. S., 239
Soria, F., 342
Southwood, T. R. E., 4, 5, 6, 7, 8, 9, 11, 12, 21, 23, 24, 28, 40, 230, 242
Sparks, A. N., 116, 252, 253, 254, 261, 389
Spencer, H., 200, 203
Spradberry, J. P., 246a, 249, 269
Sprott, J. M., 406
Stacey, D. L., 361, 367, 368
Starks, K. J., 184, 187, 188, 199, 399
Starler, N. H., 408
Starostin, S. P., 45, 46
Stary, P., 187, 199, 244, 361
Stein, W., 184, 342
Steiner, J. E. R., 255
Steiner, L. F., 208
Steinhauer, A. L., 118, 393, 408, 447
Stengel, M., 343
Stent, P. D., 442
Stepanov, 298, 300
Sternlicht, M., 239, 253
Stevens, L. M., 118, 393, 408
Stinner, R. E., 4, 19, 24, 25, 32, 183, 185, 186, 187, 188, 200, 203, 228, 230, 379, 388, 389, 404, 407, 432
Stone, L. E. W., 364
Storozhkov, Yu. V., 309
Sumaroka, A. F., 294
Summers, C. G., 19, 20
Summers, T. E., 203, 397, 398
Sundby, R. A., 263
Sweetman, H. L., 264

Taft, H. M., 406
Talitskii, V. I., 63, 292, 293, 305
Tamaki, G., 230
Tanner, J. T., 33
Tassan, R. L., 164, 165, 168, 189, 190, 263
Tauber, C. A., 164, 168
Tauber, M. J., 164, 168, 380
Tawfik, M. F. S., 268
Taylor, J. S., 244
Telenga, I. A., 47, 286, 287, 288, 289, 294
Telford, A. D., 25, 263
Teran, F. O., 184, 396
Tetle, J. P., 239
Thompson, A. C., 407
Thompson, J. V., 186
Thompson, S. N., 156, 158, 160, 161, 163, 166
Thompson, T., 405
Thompson, W. R., 250
Thorpe, W. H., 185, 249, 268, 269
Ticehurst, M., 399
Tilden, P. E., 249
Tillyard, R. J., 264
Titova, E. V., 44
Tkachev, V. M., 59
Townes, H., 130
Traynier, R. M. M., 241
Trefrey, D., 93
Tron, 289
Tsintsadze, K. V., 307
Tsitsipis, J. A., 229, 230
Tulisalo, U., 354
Tumlinson, J. H., 253
Tupper, G. R., 406
Turnbull, A. L., 255
Turner, E. C., 418, 420
Turner, J. E., 230
Turner, W. K., 230
Turney, H. A., 405
Tvaradze, M. S., 305
Typpo, J. T., 158
Tyrrell, D., 239
Tyumeneva, V. A., 285

Ullyett, G. C., 238

Urquijo, L. P., 139
Ushchekov, A. T., 305, 307, 309, 310
Uspenskii, F. M., 305
Usinger, R. L., 125
Uvarov, B. P., 7, 151, 156

Van Abeelen, J. H. F., 229
van den Assem, J., 135, 136
van den Bosch, R., 20, 24, 25, 32, 126, 127, 128, 132, 183, 222, 263, 379, 391, 393, 407
van den Veer, J., 247
Vanderzant, E. S., 156, 157, 158, 160, 163, 175
van Lenteren, J. C., 255, 260, 263, 264, 352
van Swet, W. R., 255
Vardell, H., 208
Varley, G. C., 19, 40, 80
Vashadze, 305
Vasil'ev, G. A., 312
Vasil'ev, I. V., 283, 284, 292
Vasiljevic, L., 343
Vaughn, M. R., 389
Venkataram, T. V., 337
Vereshchagin, B. V., 64
Vereshchagina, V. V., 58
Viktorov, G. A., 40, 41, 42, 43, 45, 46, 47, 48, 49, 50, 53, 54, 55, 56, 57
Vinogradov, A. V., 58, 62, 63
Vinogradova, N. M., 48
Vinson, S. B., 221, 222, 238, 239, 240, 241, 242, 244, 245a, 248, 249, 251, 252, 253, 254, 255, 256, 260, 261, 262, 263, 264, 265, 266, 267, 268, 408
Vite, J. P., 250, 253
Vodinskaya, K. I., 313
Voegele, J., 96, 184, 186, 187, 200, 230, 343
Vogelsang, D. C., 445, 446
Volkov, V. F., 286, 287
Volterra, V., 7
Voronin, K. E., 294

Voth, V., 384

Wagner, R. P., 232
Waldbauer, G. P., 4, 22, 153
Wallis, R. L., 230
Walker, J. K., 406
Walker, M. F., 248
Wang, H., 332
Wang, Y., 19
Wardecker, A. L., 389, 391
Waterhouse, D. F., 447
Watkins, W. C., 406
Way, M. J., 4, 11, 24
Weaver, K. M., 255, 262
Weidhaas, D. E., 418, 419, 420, 423, 435
Weimer, J. P., 440
Wellenstein, G., 343
Wellington, W. G., 16, 23
Went, D. F., 268
Weseloh, R. M., 239, 246a, 248, 252, 253, 255, 399
Whitcomb, W. H., 387
White, E. B., 139, 173, 185, 186, 193, 418
White, L. D., 208
Whitfield, T., 418, 419, 420
Whiting, A. R., 419, 420
Whittaker, R. H., 243
Whitten, M. J., 139, 447
Wilbert, H., 249
Wilkes, A., 139, 170, 173
Williams, J. R., 246a, 250
Williamson, D. L., 250, 253
Willis, O. R., 424
Wilson, D. D., 246a, 248, 252, 253, 255
Wilson, E. O., 8, 9, 22
Wilson, F., 139
Witz, I. A., 407
Woets, J., 206, 207
Wood, D. L., 249
Wood, E. A., 184, 187, 188, 199, 399
Wood, H. R., 406
Woodard, D. B., 417
Woodville, H. C., 361
Woodworth, C. W., 7, 40
Wright, R. H., 237

Wu, W., 338
Wyatt, I. J., 363
Wygodzinsky, P., 125
Wylie, H. G., 264, 419,
    420

Yarynkina, T. V., 46
Yasnosh, V. A., 296
Yazgan, S., 156, 157, 161,
    163, 165, 166, 170
Yepson, R. B., 405
Yokoyama, T., 171, 173
Young, J. C., 255
Yuzbyashbyan, O. Sh., 305

Zaeva, I. P., 44, 46
Zangheri, S., 62, 64
Zatyamina, V. A., 52
Zatyamina, V. V., 46, 50, 51
Zayats, Uy. V., 304
Zech, E., 61
Zeleny, I., 305
Zhuravleva, A. M., 312
Zil'bermints, I. V., 312
Zinkevich, E. P., 55
Zlatanova, A. A., 60
Zommorodi, A., 293
Zorin, P. V., 309, 313
Zwolfer, H., 244

*Achroia grisella* (F.), 185

*Acyrthosiphon pisum* (Harris), 19, 153, 199, 306

*Aedes mariae* (Sergent and Sergent), 136

*Aedes zammitti* Theobold, 136

*Aelia acuminata* (L.), 293

*Aelia fieberi* Scott, 293

*Ageniaspis fuscicollis praysincola* Silv., 342

*Agria housei* Shewell (=*A. affinis* = *Pseudosarcophaga affinis*), 156, 157, 159, 160, 161, 162, 166, 167, 171, 175

*Agrotis* (=*Euxoa*) *segetum* Schiffermuller, 285, 286, 313

*Aleochara bilineata* Gyllenhal, 313

*Aleurothrixus flocusus* Mask., 341

*Allotropa burrelli* Muesebeck, 296

*Allotropa convexifrons* Muesebeck, 296

*Alophora* [=*Phasia*] *subcoleoptrata* (L.), 43, 44

*Anagasta* (*Ephestia*) *kuehniella* (Zeller), 185, 246a, 253

*Anagyrus kivuensis* Compere, 380

*Anarhopus sydneyensis* Timberlake, 265

*Anastatus* sp., 330, 331, 338

*Anastrepha suspensa* (Loew), 245a

*Anatis mali* Auct, 153

*Antherea perniyi* (Gnerin-Madneville), 331

*Anthonomus grandis* Boheman, 79, 186, 208, 245a, 341, 389

*Aonidiella aurantii* (Maskell), 128, 166, 185, 193, 246a

*Aonidiella citrina* (Coquillett), 134

*Apalangia endius* Walker, 434

*Apanteles chilonis* Munakata, 169

*Apanteles kazak* Telenga, 313

*Apanteles lautellus* Marshall, 64

*Apanteles liparidis* Bouche, 399

*Apanteles melanoscelus* (Ratzeburg), 138, 139, 239, 246a, 399

*Apanteles porthetriae* Muesebeck, 399

*Apanteles rubecula* Marshall, 385

*Apanteles scutellaris* Muesebeck, 434

*Apanteles subecula*, 385

*Apectnis rufata* (Gmelin), 244

*Aphaereta pallipes* (Say), 166

*Aphelinus* sp., 295

*Aphelinus* [*Aphelinus mali* (Haldeman)], 294

*Aphidius* sp., 135, 438

*Aphidius* (*Diaeretiella*) *rapae* (Curtis), 245a

*Aphidius matricariae* Haliday, 357, 362, 374

*Aphidius pulcher* Baker, 198

*Aphidius smithi* Sharma and Subba Rao, 187, 198, 387

*Aphidoletes* sp., 437

*Aphidoletes aphidimyza* (Rondani), 249, 309, 310

*Aphis craccivora* Koch (=*Aphis laburni* Kaltenbach), 306

*Aphis fabae*, 244

*Aphis gossypii* Glover, 310, 392

*Aphytis* sp., 194, 195, 438, 441, 445

*Aphytis chrysomphali* (Mercet), 128, 193, 382

*Aphytis coheni* DeBach, 246a

*Aphytis lingnanensis* Compere, 132, 135, 185, 186, 193, 382

*Aphytis maculicornis* Masi, 30, 132, 134, 139

*Aphytis melinus* DeBach, 193, 382, 404, 434, 435, 436

*Aphytis mytilaspidis* (LeBaron),
    128, 132, 138
*Aphytis proclia* (Walker), 302
*Archytas marmoratus* (Zeller),
    245a, 251, 252, 268
*Argyroploce schistaceana*
    Snellen, 335
*Arrhionomyia tragica* Mg., 61
*Arthrocnodax* sp., 309
*Aschersonia* sp., 312
*Ascogaster quadridentata* Wes-
    mael, 59, 60, 61
*Ascogaster rufipes* (Latreille),
    60
*Aspidiotus nerii* Bouche, 185,
    186, 194
*Aster pilosus*, 248
*Autographa* (=*Phytometra*) *gamma*
    L., 286

*Bacillus thuringiensis* Berlin-
    er, 330, 332, 343, 367, 368,
    405
*Bathyplectes curculionis*
    (Thomson), 20
*Beauveria bassiana* (Bals.) V.,
    330, 332
*Beta vulgaris*, 244
*Biosteres* (*Opius*) *longicadatus*
    (Ashmead), 245a, 247
*Bombyx mori* (L.), 296
*Brachymeria intermedia* (Nees),
    86, 87, 93, 94, 95, 102,
    109, 110, 111, 112, 113,
    114, 115, 118, 119, 120,
    246a, 435
*Bracon brevicornis* Wesmael,
    166
*Bracon greeni* Ashmead, 331,
    337, 338
*Bracon hebetor* Say, 313
*Bracon kirkpatricki* (Wilkin-
    son), 105, 198. 199, 390,
    391
*Bracon mellitor* (Say), 245a,
    251, 255, 262, 265, 266
*Braunsia* (=*Microdes*) *rufipes*
    (Nees), 59, 60, 61
*Bucculatrix crataegi* Zeller,
    63

*Cadra cautella* (Walker), 258
*Cales noaki* How., 341
*Calosoma sycophanta* (Linnaeus),
    154
*Campoletis sonorensis* (Camer-
    on), 246a, 252, 264, 389,
    390
*Carabidae* sp., 44
*Cardiochiles abdominalis*
    Guenee, 248
*Cardiochiles nigriceps* Viereck,
    96, 239, 240, 242, 244,
    245a, 249, 252, 256, 260,
    264, 268, 389
*Carpcoris* sp., 53, 293
*Carpocapsa pomonella* L., 155
*Casinaria ichnogaster* Thomson,
    60
*Ceratitis capitata* (Wiedemann),
    264, 342
*Cheiloneurus noxius* Compere,
    246a
*Chelonus blackburni* Cameron,
    390, 434
*Chelonus elaeaphilus* Silv.,
    342
*Chelonus texanus* Cresson, 252,
    266, 434
*Chilo auricilia* Dudgeon, 396
*Chilo suppressalis* Walker, 170
*Chilocorus bipustalatus* (L.),
    299
*Chilocorus reinpustulatus*
    (Scriba), 299
*Choristoneura murinana* Hubner,
    244
*Chrysopa* sp., 284, 400, 405,
    407, 437, 438, 440
*Chrysopa abbreviata* Curtis, 305
*Chrysopa albolineata* Killing-
    ton, 305
*Chrysopa californicus* Conquil-
    lett (=*Chrysopa carnea*),
    382, 383
*Chrysopa carnea* Stephens, 160,
    162, 164, 165, 187, 188,
    189, 190, 191, 208, 210,
    263, 304, 305, 306, 307,
    309, 310, 381, 391, 392,
    393, 401, 433, 434, 435

*Chrysopa formosa* Brauer, 305
*Chrysopa (Nineta) flava* (Scopoli), 305
*Chrysopa (Nineta) vittata* Wesmael, 305
*Chrysopa perla* (L.), 305
*Chrysopa phyllochroma* Wesmael, 305
*Chrysopa ventralis* Curtis (=*C. prasina* Burmeister), 305
*Cnaphalocrocis medinalis* Guenee, 331, 336
*Coccinella trifasciata* L., 19
*Cochliomyia hominivorax* (Coquerel), 208, 400, 447
*Coccinella septempunctata* L., 166, 305, 307, 310, 387
*Coccophagus gurneyi* Compere, 302, 303
*Coccophagoides utilis* Doutt, 30
*Coccygominus (=Pimpla) turionellae* (L.), 59, 157, 268
*Coccus hesperidum* L., 246a, 382
*Cochliomyia* sp., 207
*Coleomegilla maculata* (DeGeer), 164
*Coleomegilla maculata lengi* Timberlake, 153, 168
*Colias eurythme* Boisduval, 136
*Colias philodice* Godart, 136
*Colias transfersoguttata* Feldermann, 136
*Comperiella bifasciata* Howard, 134, 434
*Corcyra cephalonica* (Staunton), 166, 331, 335, 336
*Cryptolaemus* sp., 298, 438, 441, 445
*Cucurbita maxima* Dene, 194
*Culex p. fatigans*, 424
*Culex p. quinquefasciatus* Say, 424
*Cryptolaemus montrouzieri* Mulsant, 192, 193, 264, 297, 381, 382, 404, 433, 434
*Cryptolucilia* sp., 420
*Cydia pomonella* L., 342

*Cyrtorhinus mundulus* (Breddin), 256
*Cyzenis albicans* (Fall.), 245a, 247

*Dactylopius tomentosus* (Lamarck), 264
*Dacus cucurbitae* Coquillett, 208
*Dacus oleae* (Gmelin), 229, 341
*Dahlbominus fuliginosus*, 343
*Dalbergia balansae*, 337
*Dalbominus (=Microplectron) fuscipennis* (Zetterstedt), 170
*Dendroctonus frontalis* Zimmerman, 245a
*Dendrolimus sibericus* (Tshetverikov), 332
*Dialeurodes citri* (Ashmead), 313
*Diatraea grandiosella* (Dyar), 246a
*Diatraea saccharalis* (F.), 85, 120, 128, 185, 203, 397
*Diatreophaga striatalis* Townsend, 936
*Diprion pini* L., 343
*Dolycoris baccarum* (L.), 53, 54, 293
*Drino bohemica* Mesnill, 247
*Drosophila* sp., 138
*Drosophila melanogaster* Meigen, 155, 161, 169, 175

*Ectophasia* Townsend, 44
*Elasmostethus* sp., 54
*Elasmucha* sp., 54
*Eliozeta [=Clytiomyia] helluo* F., 43
*Encarsia* sp., 313, 365, 366, 367, 368, 373, 433, 436, 437, 440
*Encarsia formosa* Gahan, 184, 188, 196, 197, 198, 312, 353, 358, 360, 361, 363, 374, 380, 381
*Enonymus europaeus*, 244
*Ephestia kuehniella* (Zeller), 187

*Ephialtes extensor* Tachenberg,
   60
*Epilachna varivestis* Mulsant,
   118, 393, 447
*Epunctulatus carbonarius*
   (Christ), 60
*Eriosoma lanigerum* (Hausmann),
   294
*Eublemma*, 338
*Eublemma amabilis* Moore, 331,
   337
*Eucelatoria* sp., 86, 87, 89,
   91, 92, 97, 101, 102, 104,
   106, 118, 119, 390, 391
*Eulophidae* sp., 64
*Eupatorium glandulosum* Spren-
   gel, 264
*Euphasiopteryx ochracea*, 239
*Euplectrus plathypenaee*
   (Howard), 251
*Eurydema ventralis* (=ornatum),
   54
*Eurygaster austriaca*, 55
*Eurygaster integriceps* Puton,
   42, 43, 55, 65, 283, 291
*Eurygaster marus* L., 54, 55
*Exeristes comstockii* (Cresson),
   164
*Exeristes roborator* (Fabrici-
   us), 156, 158, 160, 161,
   162, 166
*Exochomus flavipes* Thunberg,
   380

*Fannia canicularis* (L.), 420,
   421
*Fannia femoralis* (Stein),
   420, 421
*Feltia subterranea* (F.), 249
*Formica polyctena* Forest, 250
*Formica rufa*, 343
*Frankliniella fusca* (Hinds),
   309

*Galleria mellonella* (L.), 120,
   185, 187, 188, 203, 205,
   245a, 268
*Geocoris punctipes* (Say), 393,
   394

*Grapholitha molesta* (Busck),
   185, 384
*Graphosoma* sp., 53
*Graphosoma italica* (Mueller),
   293
*Graphosoma italicus*, 54
*Graphosoma semipunctata* (F.),
   293
*Gryllus integer*, 239

*Haematobia irritans*, 418
*Heliothis* sp., 79, 83, 87, 89,
   96, 97, 98, 101, 102, 103,
   104, 105, 106, 119, 245a,
   256, 266, 287, 313, 344,
   388, 390, 391, 392, 393,
   407
*Heliothis armigera* (Hubner),
   244
*Heliothis virescens* (F.), 91,
   96, 118, 119, 239, 242,
   245a, 249, 252, 268, 387,
   389
*Heliothis zea* (Boddie), 91, 96,
   118, 119, 200, 241, 245a,
   246a, 258, 263, 266, 268,
   385, 386, 387, 391
*Helomyia antiqua* (Meigen), 313
*Helomyia lateralis* (Meigen),
   43
*Heydenia unica* Cook and Davis,
   245a, 247
*Hippodamia convergens* Guerin-
   Meneville, 192, 399, 433,
   434
*Hylemia brassicae* (Wiedemann),
   313
*Hyles* (=Celerio) *euphorbiae*
   (Linnaeus), 166

*Ibalia drewseni* Borries,
   246a
*Ibalia leucospoides* (Hochen-
   warth), 246a
*Icerya* sp., 299, 300
*Icerya purchasi* Maskell, 297
   341
*Ips confusus* (LeConte),
   249

*Itoplectis conquisitor* Say, 155, 156, 157, 158, 161, 162, 164, 165, 166, 170, 172, 245a, 247, 248, 256, 268

*Jalysus spinosus* (Say), 393, 394

*Kellymyia kellyi* (Aldrich), 157

*Laspeyresia nigricana* (Stephens), 286
*Laspeyresia pomonella* (L.), 42, 58, 65, 79, 384
*Leskia aurea* (Fallen), 60
*Leucoptera* (=*ceniostroma*) *scitella* (Zeller), 63
*Lindorus* sp., 299
*Lindorus lophantae* Blaisdell, 298
*Liotryphon punctulatus* (Ratzeburg), 61
*Liriomyza solani* (Her.), 367
*Lixophaga* sp., 115, 433
*Lixophaga diatraeae* (Townsend), 85, 86, 87, 96, 120, 185, 186, 187, 188, 203 , 204, 205, 206, 246a, 252, 395, 396, 397, 398, 400, 401, 402, 521
*Lygocerus* sp., 265
*Lymantria dispar* L., 87, 239, 246a, 399
*Lysiphlebus testaceipes* (Cresson), 187. 188, 198, 199, 399
*Lygus* sp., 406
*Lygus lineolaris* Palisot deBeauvois, 246a

*Macrocentrus ancylivorus* Rohwer, 185, 188, 198, 199, 384
*Macrocentrus marginator* (Nees), 60
*Macrocentrus nidulator* (Nees), 60
*Malacosoma americanum* (Fabricius), 155

*Mamestra* (=*Barathra*) *brassicae* (L.), 284, 286
*Manduca quinquemaculata* (Haworth), 386
*Melittobia chalybii* Ashmead, 135
*Metagonistylum minense* Townsend, 395, 396
*Metaphycus* sp., 441, 445
*Metaphycus helvolus* (Compere), 199, 404, 434
*Microbracon brevicornis* Wesman, 244
*Microplitis croceipes* (Cresson), 86, 96, 119, 241, 245a, 252, 259, 261, 264, 267, 389, 390
*Microterys flavus* (Howard), 251, 382
*Monodontomerus* Westwood, 136
*Monolinia fructicola*, 247
*Mormoniella* (*Nesonia*) *vitripennis* (Walker), 418, 419, 420
*Muscidifurax*, 133
*Muscidifurax raptor* Girault and Sanders, 418, 419, 420, 421, 434
*Muscidifurax zaraptor* Kogan and Legner, 418, 419, 421
*Musca autumnalis*, 418, 420
*Musca domestica* Linnaeus, 157, 419
*Muscina stabulans*, 421
*Myrmica* sp., 250
*Myzus persicae* (Sulzer), 87, 245a, 306, 309, 350, 357, 374, 375, 381

*Nabis americoferus* Carayon, 393, 394
*Naranga aenescens* Moore, 336
*Neoaplectana dutkyi* Jackson, 383, 384
*Nezara viridula* (L.), 246a
*Nicotiana glutinosa* L., 353
*Nicotiana tabacum* L., 358, 396

*Oligonychus punicae* (Hirst), 382

*Ooencyrtus telenomicida*, 50, 51, 52, 53, 54, 56

*Operoptera brumata* (L.), 245a

*Ophyra leucostoma* (Wiedemann), 421

*Opius* sp., 433, 438

*Opius concola* Szepl., 341

*Opius concolo siculus* Mon., 342

*Opius lectus*, 253

*Opius tryoni* (Cameron), 264

*Orgilus lepidus* Muesebeck, 245a

*Orgilus obscurator* Nees, 155

*Oryzaephilus surinamensis* (Linnaeus), 161

*Oscinella* sp., 313

*Ostrinia* (=*Pyrausta*) *nubilalis* Hubner, 79, 268, 343

*Pachycrepoideus vindemiae* (Rondani) (=*P. dubius* Ashmead), 418, 419, 420, 421, 434

*Palexoriata laxa*, 390, 391

*Palomena* sp., 53

*Palomena prasina* (L.), 54, 293

*Paratheresia claripalpis* (Wulp), 395

*Parlatoria oleae* (Colvee), 30, 134

*Pastinaca sativa* Linnaeus, 155

*Pauridia* sp., 434

*Pectinophora gossypiella* (Saunders), 79, 185, 389, 393

*Pediobius foveolatus* (Crawford), 118, 395, 408, 434, 447

*Pentatoma* sp., 54

*Perilampus tristis* Mayr, 61

*Perilitus coccinellae* (Schrank), 246a

*Perilloides bioculatus* (F.), 303, 304

*Perillus* sp., 303, 304

*Peristenus pseudopllipes* (Loan), 246a

*Phaeogenes cynarea* Bragg, 246a

*Phaesolus vulgaris* L., 360

*Phalangida* sp., 44

*Phasia crassipennis* (F.), 44

*Phasia* [=*Ectophasia*] *crassipennis* F., 43

*Phasia* Latreille, 44

*Phthorimaea operculella* (Zeller), 185, 245a

*Phyllonorycter corylifoliella* (Hubner), 63, 64

*Phyllonorycter* (=*Lithocolletis*) *pyrifoliella* Gram, 63, 64

*Phytomyza atricornis* Meigen, 239

*Phytomyza syngenesiae* (Hardy), 366

*Phytoseiulus* sp., 311, 312, 365, 367, 372, 436, 437, 438, 440, 441

*Phytoseiulus persimilis* Athias-Henroit, 184, 187, 206, 207, 310, 355, 356, 357, 358, 360, 361, 362, 384, 434, 435

*Pieris brassicae* (L.), 153, 246a, 385

*Pimpla* sp., 61

*Pimpla instigator* F., 246a

*Pimpla turionellae* L., 59, 157, 268

*Pinus resinosa* Vit., 247

*Pinus sylvestris* L., 247

*Pittosporum undulatum* Vent., 193

*Planococcus* (=*Pseudococcus*) *citri* (Risso), 193, 380

*Platyptilla carduidactyla* (Riley), 246a

*Podisus maculiventris* (Say), 393, 394

*Polistes* sp., 393

*Polistes exclamans arizonensis* Snelling, 393

*Porthetria dispar* L., 343

*Prays citri* Mill, 343

*Prays oleae* Bern, 342

*Pristomerus vulnerator* Pans, 59, 61

*Procecidochares utilis* (Stone), 264

*Proceras* (=*Chilo*) *sacchariphagus* Bojer, 344, 396

*Prospaltella* sp., 438

*Prospaltella fasciata* MaLenotti, 128
*Prospaltella perniciosi* Tower, 128, 134, 301, 302, 342
*Pseudaphycus* sp., 296
*Pseudaphycus malinus* Gahan, 295
*Pseudatomascelis seriatus* (Reuter), 406
*Pseudeucoila bochei* Weld, 255, 264
*Pseudoccus comstocki* (Kuwana), 295, 304
*Pterostichus crenuliger* Chaudoir, 44
*Pulvinaria aurantii*, 297

*Quadraspidiotus perniciosus* (Comstock), 128, 134, 301, 342

*Reesimermis nielseni* Tsai and Grund, 434, 435
*Rhagoletis cerasi* L., 230
*Rhagoletis pomonella* (Walsh), 253
*Rhaphigaster* sp., 54
*Rhopalosiphum maidis* (Fitch), 153
*Rodolia cardinalis* (Muslant), 299, 300, 341
*Romanomermis culcivorax* Ross and Smith, 424

*Saissetia oleae* Bernard, 404
*Samia* (=*Philosamia*) *cynthia* (Drury), 187, 331
*Saccaromyces fragilis*, 190
*Sarcophaga aldrichi* (Parker), 157
*Sarcophaga bullata* Parker, 419
*Scambus buoliance* (Hartig), 155
*Scarcophaga* sp., 420
*Schinia arcigera* Guenee, 248
*Schinus molle* L., 193
*Schizaphis graminum* (Rondani), 87, 399
*Sirex noctilio* F., 246a, 247

*Sitotroga cereallella* Oliver, 185, 188, 200, 201, 284, 305, 331, 400
*Solenotus begini* (Ashmead), 239
*Spalangia calcitrans*, 418, 420, 421
*Spalangia cameroni* Perkins, 418, 419, 421
*Spalangia drosophilae* Ashmead, 313, 423
*Spalangia endius* Walker, 188, 418, 420, 421, 422, 435
*Spalangia muscidarium* Richardson (=*S. nigroaenea*), 418, 419
*Spalangia muscidarium var stomoxysiae* Girault (=*S. endius*), 420
*Spalangia nigra* Latreille, 418, 421
*Spalangia nigroaenea* Curtis, 418, 420
*Sphegigaster* sp., 421
*Spodoptera exigua* (Hubner), 185
*Spodoptera frugiperda* (Smith), 249, 268, 419
*Steneotarsonemus* (=*Tarsonemus*) *pallidus* Banks, 384
*Stethorus picipes* Casey, 382, 383
*Stigmella malella* (Stainton), 63, 64, 65

*Tachardia lacca* Kerr, 337
*Tachinaephagus giraulti* Johnson and Tiegs, 420
*Tachinaephagus zealandicus* Ashmead, 418, 419, 421, 434
*Teleas reticulatus* Kieffer, 54
*Telenomus* sp., 54, 283, 291, 293, 294
*Telenomus chloropus* Thomson (=*sokolovi* Mayr), 47, 48, 50, 51, 5253, 54, 55, 292
*Tessaratoma papillosa* Drury, 337, 338

*Tetranychus cinnabarinus* (Boisduval), 392

*Tetranychus modanieli* McGregor, 382, 383

*Tetranychus urticae* Koch, 207, 309, 353, 354, 356, 357, 360, 361, 362, 375, 384

*Thanasimus dubius* (F.), 250

*Therioaphis trifolii* (Monell), 29

*Thrips tabaci* Lindeman, 366

*Toxorhychites* sp., 434

*Trialeurodes vaporariorum* (Westwood), 196, 312, 350, 351, 352, 353, 357, 358, 360, 361

*Trichomma enecator* Rossi, 59, 61, 62

*Trichogramma* sp., 28, 32, 83, 86, 87, 88, 89, 96, 102, 115, 118, 131, 132, 133, 184, 186, 200, 201, 203, 241, 260, 261, 283, 291, 313, 330, 332, 333, 334, 337, 343, 387, 388, 389, 400, 402, 403, 405, 407, 433, 434, 435, 436, 438, 439, 440, 441

*Trichogramma australicum* Gir., 331, 335, 336, 344

*Trichogramma brasiliensis* Ashm., 344

*Trichogramma cacaeciae* March, 342

*Trichogramma cacoeciae pallida* Meyer, 285, 286, 290

*Trichogramma californicum* Nafaraja and Nagarkatti, 268

*Trichogramma dendrolimi*, 331, 333, 335, 336

*Trichogramma embryophagum* Hartig, 286, 287, 290, 342

*Trichogramma evanescens* Westwood, 185, 245a, 252, 258, 284, 285, 286, 288, 289, 342, 385

*Trichogramma fasciatum* (Perkins), 128

*Trichogramma japonicum* Ashmead, 331, 335, 336

*Trichogramma minutum* Riley, 128, 287, 342

*Trichogramma ostrineae*, 331

*Trichogramma perkinsi* Girault, 139

*Trichogramma pretiosum* Riley, 140, 187, 202, 208, 209, 266, 385

*Trichoplusia ni* (Hubner), 79, 266, 386

*Trichopoda pennipes* (F.), 246a

*Trichopria* n. sp., 420

*Trissolcus grandis* (Thomson), 47, 48, 50, 51, 52, 53, 54, 55, 292

*Trissolcus pseudoturesis* Rjachovsky, 50, 52, 53, 55, 292

*Trissolcus scutellaris* (Thomson), 50, 54, 55, 292

*Trissolcus simoni* (Mayr), 51, 52, 54, 55

*Trissolcus vassilievi* (Mayr), 291, 292

*Trissolcus victorovi* Kozlov, 55

*Trissolcus volgensis* (Viktorov), 50, 52, 54, 55

*Trioxys pallidus* (Haliday), 132

*Trioxys utilis* Muesebeck, 132

*Troilus* sp., 54

*Typhlodromus cucumeris* Oudemans, 384

*Typhlodromus hibisci* (Chant), 434

*Typhlodromus occidentalis* Nesbitt, 382, 383, 434

*Typhlodromus reticulatus* Oudemans, 384

*Tyrophagus putrescentiae* (Schrank), 373

*Venturia (Nemeritis) canescens* (Gravenhorst), 185, 246a, 253

*Vicia faba* L., 358

*Voria ruralis* (Fallen), 386

SUBJECT INDEX

Alfalfa, 387, 399
Allomones, 243, 253
Animals, 417
Antigua, 395
Apple, 342, 382, 383, 436, 438
Artificial diets, 172, 298
Associates Insectary, 193, 441, 445
Autocidal control, 117
Avocado, 382, 383

Beans, 395, 405
Berries, 313
Biotypes, 242
Bolivia, 395
Brazil, 184, 396

Cabbage, 283, 288, 289, 307, 313, 384, 385, 386, 405, 436, 438
Canada, 303, 437
Cannibalism, 189
Caltalpa, 304
Celery, 437, 440
Chemical control, 283, 296, 299, 308-311, 332, 336, 351, 384, 447
China, 184, 200, 329-339, 441, 448, 449
Cryptic species, 126
Chrysanthemums, 357, 362, 364, 365, 374, 375, 436, 438
Citrus, 283, 298, 305, 307, 337, 381, 382, 404, 405, 436, 437, 441
Climatic adaption, 28, 287, 292, 295-297, 298-304, 421
Collards, 386
Columbia, 389
Commercial suppliers, 433, 451
Consultant, 442, 445, 446
Cooperatives, 442, 446
Corn, 288, 331, 335, 407, 436, 438, 439

Cost-benefit, 118, 432, 441
Cost, 118, 184, 186, 188, 196, 199, 207, 211, 296, 330-332, 334, 349, 423, 435, 440
Cotton, 305, 313, 337, 344, 387-394, 400, 403, 406, 407, 436, 439
Cuba, 395, 396, 397
Cucumbers, 196, 307, 309, 310, 351-354, 357, 358, 361, 372-374, 380, 436, 437, 440
Cultural control, 329, 332, 337, 371, 405
Cytogenetics, 139

Dairies, 423, 440
Density, 14, 40, 79, 81, 91, 97, 100, 103, 105, 107, 186, 259, 266, 290, 353
Deterioration, 184, 285
Development time, 352, 356, 419, 422
Diapause, 354, 370
Diets, 158, 189, 399
Dietetics, 156
Dill, 302, 437
Digestability, 153
Dispersal, 21, 22, 96, 97, 101, 116, 118, 225, 261, 295, 361, 377, 398, 420

Economics, 118, 184, 196, 297, 311, 444
Ecosystem, 3, 23, 79, 97, 112, 403, 405
Ecotypes, 301
Efficiency of natural enemies, 82, 83, 84, 87, 95, 96, 102, 115, 393, 423, 424
Eggplant, 306
Egypt, 297, 299
El Salvador, 389
England, 343, 349-377, 437, 438

Facelia, 302
Feasibility, 442

Fecundity, 81, 186, 288, 293
    295, 297, 302, 303, 333,
    353, 354, 365, 419, 420
Federal programs, 447
Fillmore Citrus Protective
    District, 436, 441, 445
Finland, 437
Food, 152, 155
Food consumption, 398
Food supplements, 168, 263, 379
Food toxicity, 364
Forests, 286, 399
France, 184, 293, 341, 342, 343
    437, 438, 440
Fruit, 283, 337

Gardenias, 380, 405, 440, 444
Genotype analysis, 139
Grain, 283, 313
Grapes, 298
Guadeloupe, 395

Habitat, 5, 21, 22, 30, 97,
    100, 105, 285, 286, 289,
    350
Haiti, 395
Hemp, 436, 438
Heterosis, 169
Host
    feeding, 164
    finding, 82, 95, 101, 102,
    247
    natural, 185
    plants, 91
    preference, 54, 133, 249,
    268, 392
    rejection, 264
    unnatural, 185
Host habitat location, 247
Host selection, 238-242, 245,
    246
Hybridization, 131, 138

India, 337
Innoculative releases, 32,
    379, 395, 396, 408, 421
Insect growth regulators,
    407
Insecticidal drift, 407

Insecticide resistance, 312, 350,
    364, 384, 431
Insecticides, 20, 101, 336, 388,
    389, 397, 398, 399, 404, 406,
    431, 444
Institutional options, 441
Insurance, 442, 446
Intraspecies differentiation,
    286
Inundative releases, 27, 32, 292,
    297, 307, 308, 313, 379, 389,
    420
Italy, 298, 342, 343

Kairomones, 55, 82, 116, 243,
    250, 252-254, 379
    behavior, 255, 268
    host marking, 56, 255, 260, 263
    learning, 255, 262
    rearing, 268
    retention, 259, 267
Karotypes, 137
Korea, 296

Lac, 331, 337
Larch, 332
Lettuce 405, 437
Lichee, 337, 338
Livestock, 440
Longevity, 186, 295, 297, 353,
    420

Madagascar, 344
Man, 417
Management, 404, 406, 444
Manipulations
    chemical, 261
    environment, 25
Mass production, 183, 292, 333,
    357, 408, 424, 433
Mass production techniques
    Aphelinidae, 193
    Braconidae, 198
    Chrysopidae, 189
    Coccinellidae, 192
    Encyrtidae, 199, 359
    Phytoseiidae, 206, 358
    Ptermalidae, 422
    Tachinidae, 203

Mass rearing, 101, 220, 283, 293, 297, 302, 305, 313, 336, 358, 400
Mechanical control, 405
Melons, 342, 438
Metachron, 237
Mexico, 389, 439
Microbial agents, 405
Millet, 331, 335
Models, 102
  population, 6, 19, 32, 85, 407
  parasite-host, 83, 87, 89, 97, 107, 112
  simulation, 423
Morocco, 341
Mulberry, 283, 304

Netherlands, 196, 206, 207, 343, 437, 440
Nicaragua, 389
Norway, 437
Nutrition, 151, 155, 157, 162, 165, 166
  balance, 165
  comparative, 159
  efficiency, 153
  host, 167
  parasitoid adult, 164
  value, 166

Olive, 342, 438
Orchards, 58, 63, 286, 289, 290, 299, 301, 302
Ornamentals, 440, 444
Oviposition, 248, 251

Packaging, 400
Pakistan, 302
Parasitism rates, 289
Parasite-host interactions, 80, 81
Parasite-host ratios, 196, 291
Patent, 444
Pears, 382, 383
Peas, 306, 387
Peppers, 306, 357, 375, 386, 405, 436, 437

Periodic releases, 25, 395, 399
Peru, 395
Pest control districts, 442, 446
Pesticide application, 350, 351, 372
Pesticides, 39, 45, 311, 329, 343, 349, 362, 364, 365, 366-371, 372, 375, 431, 442, 444
Pheromones, 117, 243, 253
Pine, 331, 332
Poinsettias, 380
Poland, 437
Portugal, 341
Potatoes, 288, 306, 387, 436, 438
Poultry, 423, 440, 441
Predator-prey ratios, 306
Programmed releases, 379
Population
  dynamics, 5, 7, 9, 11, 12, 18, 22, 40, 42, 49
  increase, 29, 300
  perspectives, 30
  regulation, 17, 40, 79
  response, 16
  self-perpetuation, 102, 424
  theory, 7

Quality control, 220
  adaptability, 222
  allozymes, 232
  behavior, 226
  components, 224
  competitiveness, 228
  genetic integration, 228
  genetic problems 224
  monitoring, 227, 228, 230
  performance testing, 229
  selection pressure, 231
  variation, 229
Quality traits, 226

Races, 127, 285, 286, 289, 301
Rates of release, 290, 291, 295, 391
Rearing, 101, 169, 173, 297, 424
Rearing mechanization, 309, 311, 313
Release techniques, 284, 291, 313, 342, 362, 398, 400-403

Reproductive isolation, 225
Resistant varieties, 329,
    374, 405-407
Rice, 330, 335-337, 436,
    439
Risk, 442, 444, 446
Romania, 437

Sanitation, 374
Scouting, 445
Searching behavior, 28, 29
    85, 249, 262, 287, 352,
    356, 393, 418, 420
Seasonal colonization, 285-
    287, 291, 297, 302, 304-
    306, 313, 379
Semispecies, 126
Sex ratio, 170, 420
Sibling species, 126
Sicily, 342, 436, 438
Snapdragons, 381
Sorghum, 331, 399
Soybeans, 332, 393, 407
Spain, 341
Specificity, 28, 80, 300
Spruce, 332
State programs, 447
Storage, 186, 187, 293, 296,
    374, 401, 408
Strains, 127
Strategies (r and k), 8, 10
Strawberries, 338, 375, 384
Sugarbeet, 283, 288, 289, 439
Sugarcane, 331, 395, 396, 397,
    403, 436
Superparasitism, 260
Survival, 22, 97, 103, 352,
    357, 370, 392
Sweden, 437
Switzerland, 342, 438
Synchronization, 28
Symbiosis, 156
Systematics, 43, 125, 127
    behavior, 134
    chromotography, 137
    courtship, 135
    isozymes, 136

Thresholds, 100, 102, 352, 354,
    386, 389
Timing of releases, 290, 293
Tobacco, 196, 358, 393, 394
Tomatoes, 306, 351, 355, 357,
    361, 362, 368, 373, 380,
    385, 386, 405, 436, 437
Turkey, 342, 439, 441

United States, 284, 295, 303,
    330, 379-416, 438-440, 442,
    444, 447
U.S.S.R., 39-78, 184, 201, 238-
    328, 330, 403, 437-439, 440,
    441, 448

Vegetables, 283, 306, 307, 308,
    407, 440, 444
Venezuela, 395
Vigor, 186, 188
Voracity, 303

West Germany, 184, 342, 343,
    436, 438
West Indies, 397
Wheat, 44, 436, 438, 439

Yugoslavia, 343